SCHAUM'S OUTLINE OF

THEORY AND PROBLEMS

OF

GRAPH THEORY

V. K. BALAKRISHNAN, Ph.D.

Professor of Mathematics
University of Maine
Orono, Maine

SCHAUM'S OUTLINE SERIES

McGRAW-HILL

New York St. Louis San Francisco Auckland Bogotá Caracas Lisbon London Madrid Mexico City Milan Montreal New Delhi San Juan Singapore Sydney Tokyo Toronto

V.K. BALAKRISHNAN is a professor of mathematics at the University of Maine, where he coordinates an interdepartmental program on operations research. He has an honors degree in mathematics from the University of Madras, a master's degree in pure mathematics from the University of Wisconsin at Madison, and a doctorate degree in applied mathematics from the State University of New York at Stony Brook. He is a Fellow of the Institute of Combinatorics and ts Applications and a member of the American Mathematical Society, Mathematical Association of America, and the Society for Industrial and Applied Mathematics. He is the author of *Introductory Discrete Mathematics* (1991), *Network Optimization* (1995), and *Schaum's Outline of Combinatorics* (1995).

Schaum's Outline of Theory and Problems of
GRAPH THEORY

6 7 8 9 10 11 12 13 14 15 16 17 18 19 20 CUS CUS 10

ISBN 0-07-005489-4

Sponsoring Editor: Barbara Gilson
Production Supervisor: Suzanne Rapcavage
Editing Supervisor: Maureen B. Walker

Library of Congress Cataloging-in-Publication Data

Balakrishnan, V.K., date
Schaum's outline of theory and problems of graph theory / V.K. Balakrishnan.
p. cm. – (Schaum's outline series)
Includes bibliographical references and index.
ISBN 0-07-005489-4 (pbk.)
1. Graph theory–Problems, exercises, etc. 2. Graph theory–Outlines. syllabi, etc. Title
QA166.B28 1997 97-90
511'.5'076–dc21 CIP

McGraw-Hill
A Division of The ***McGraw-Hill*** *Companies*

This book is dedicated to
W. T. Tutte

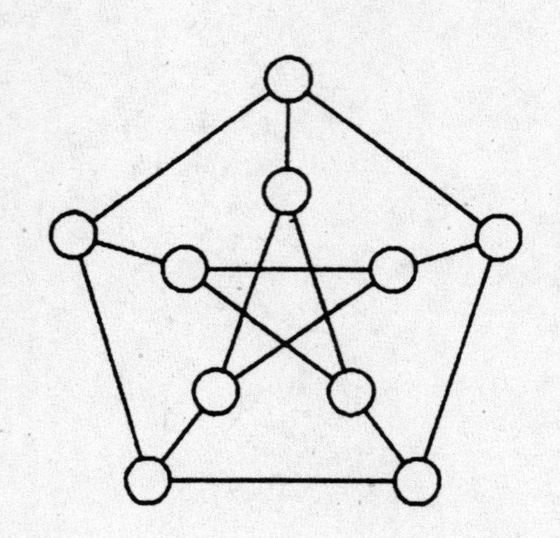

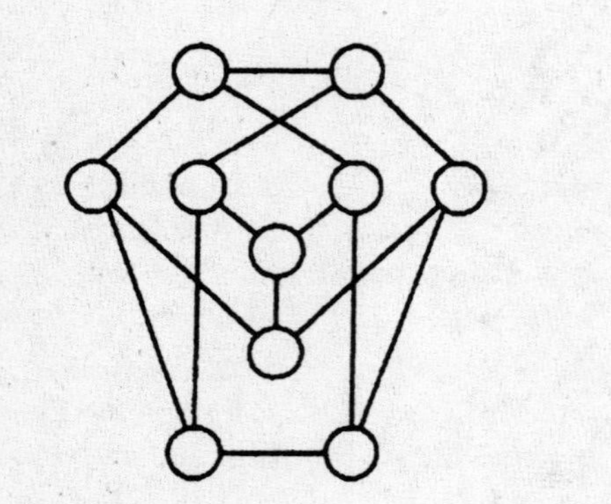

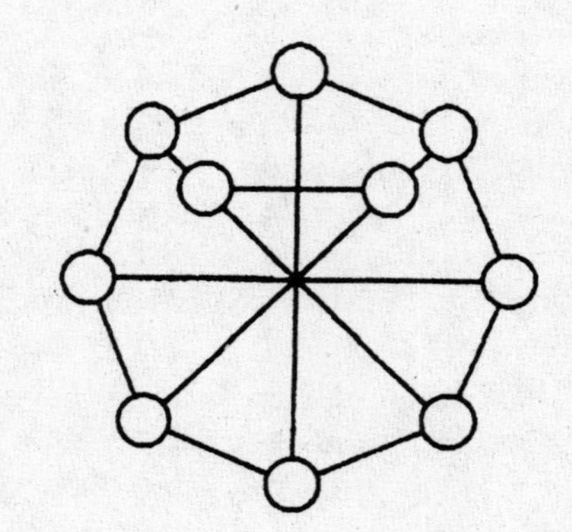

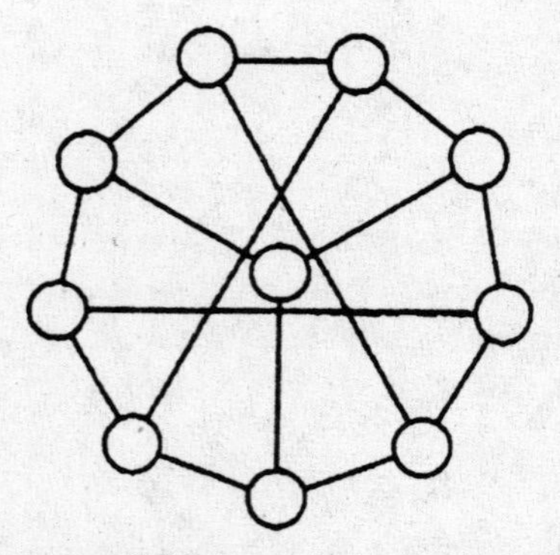

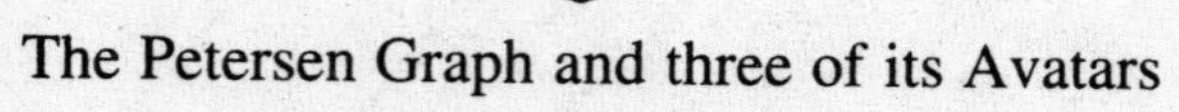

The Petersen Graph and three of its Avatars

Preface

The theory of graphs, with its diverse applications in natural and social sciences in general and in theoretical computer science in particular, is becoming an important component of the mathematics curriculum in colleges and universities all over the world. This book presents the basic concepts of contemporary graph theory in a sequence of nine chapters. It is primarily designed as a supplementary textbook for mathematics and computer science students with a wide range of maturity. At the same time it can also serve as a useful reference book for many academic and industrial professionals who are interested in graph theory.

Graph Theory can be considered a companion volume to *Combinatorics,* which was published as a Schaum Outline in 1995. The style of presentation of the material is the same in both outlines. In each chapter the basic concepts are developed in the first few pages by giving definitions and statements of the major theorems to familiarize the reader with topics that will be fully explored in the selection of solved problems that follow the text. The problems are grouped by topics and are presented in increasing order of maturity and sophistication. In some cases the results established as solutions of problems are some deep theorems and proofs of conjectures that have remained unsettled for several years.

In writing this book I have benefited enormously from the contributions of other mathematicians and scientists. My book brings together the main ideas of graph theory that I learned from the scholarly writings of others distinguished in the field, and no originality is claimed as far as the results presented in the outline are concerned. At the same time, if there are any errors, I accept complete responsibility for their occurrence, and they will be rectified in a subsequent printing of this outline once they are brought to my attention. Any feedback from the reader in this context will be gratefully acknowledged. My e-mail address is vkbal@gauss.umemat.maine.edu and may be used for this purpose.

Since this outline provides basic theory and solved problems, in many cases explicit references are not made to the source of the material. Many people deserve recognition for their specific contributions, and a partial list of books that helped me to prepare this outline is appended as a Select Bibliography for further study.

I am grateful to Dan Archdeacon and Lowell Beineke for the valuable suggestions they made during the course of reviewing parts of the manuscript. In this connection I would also like to thank Kenneth Appel and Douglas West for their helpful hints in the clarification of several results during the preparation of the manuscript. Paul Erdös is no longer with us to show the way. We all miss him dearly. I consider it a singular blessing that I could discuss with him some of the exciting results presented in this outline, and I am forever beholden to him for the kindness and warmth he bestowed on me as well as for his encouragement.

The credit for creating the artwork in this book goes to Dr. Arvind Sharma of the Los Alamos National Laboratory, and it is indeed a great pleasure to acknowledge my indebtedness to him in this regard.

In conclusion I would like to express my immense gratitude to the editorial and production staffs at McGraw-Hill and Progressive Publishing Alternatives for the unfailing cooperation and encouragement extended to me throughout the production of this outline.

V.K. Balakrishnan
University of Maine

Contents

Chapter 1

Graphs and Digraphs

1.1 INTRODUCTION

Many structures involving real-world situations can be conveniently represented on paper by means of a diagram consisting of a set of points (usually drawn as small circles or dots) together with lines (or curves) joining some or all pairs of these points. For example, the points in a diagram could represent different cities in a country, and a line joining two points that does not pass through a third point may indicate that there is direct air service between the two cities represented by those two points. In some instances, it may happen that there is direct air service from A to B but not from B to A. In such situations, an arrow (directed line or directed curve) is drawn from A to B so that the line joining A and B becomes oriented or directed. The possibility that there can be more than one line joining two points or that there is a line joining a point to itself cannot be ruled out in a more general setting. It is also possible that there can be an arrow from A to B and another arrow from B to A. The particular manner in which these lines and arrows are drawn on a piece of paper is not relevant for our investigations. What really matters is to know whether lines and arrows exist connecting the various points. In some situations, it may be pertinent to ask whether these lines can be drawn such that no two lines intersect except possibly at those points to which they are already joined. A mathematical abstraction of such structures involving points and lines leads us to the concept of graphs and digraphs.

A **graph** G consists of a set V of **vertices** and a collection E (not necessarily a set) of unordered pairs of vertices called **edges.** A graph is symbolically represented as $G = (V, E)$. In this book, unless otherwise specified, both V and E are finite. The **order** of a graph is the number of its vertices, and its **size** is the number of its edges. If u and v are two vertices of a graph and if the *unordered pair* $\{u, v\}$ is an edge denoted by e, we say that e **joins** u and v or that it is an edge between u and v. In this case, the vertices u and v are said to be **incident on** e and e is **incident to** both u and v. Two or more edges that join the same pair of distinct vertices are called **parallel** edges. An edge represented by an unordered pair in which the two elements are not distinct is known as a **loop.** A graph with no loops is a **multigraph.** A graph with at least one loop is a **pseudograph.** A **simple graph** is a graph with no parallel edges and loops. The term *graph* is used in lieu of *simple graph* in many places in this book. The **complete graph** K_n is a graph with n vertices in which there is exactly one edge joining every pair of vertices. The graph K_1 with one vertex and no edge is known as the **trivial graph.** A **bipartite graph** is a simple graph in which the set of vertices can be partitioned into two sets X and Y such that every edge is between a vertex in X and a vertex in Y; it is represented as $G = (X, Y, E)$. The **complete bipartite graph** $K_{m,n}$ is the graph (X, Y, E) with m vertices in X and n vertices in Y in which there is an edge between every vertex in X and every vertex in Y. The **union** of two graphs $G_1 = (V_1, E_1)$ and $G_2 = (V_2, E_2)$ is the graph $G = G_1 \cup G_2 = (V, E)$, where V is the union of V_1 and V_2 and E is the union of E_1 and E_2.

A **directed graph** or **digraph** consists of a finite set V of vertices and *a set A* of *ordered pairs* of distinct vertices called **arcs.** If the ordered pair $\{u, v\}$ is an arc a, we say that the arc a is **directed from** u to v. In this context, arc a is **adjacent from** vertex u and is **adjacent to** vertex v. In a **mixed graph,** there will be at least one edge and at least one arc. If each arc of a digraph is replaced by an edge, the resulting structure is a graph known as the **underlying graph** of the digraph. On the other hand, if each edge of a simple graph is replaced by an arc, the resulting structure is a digraph known as an **orientation** of the simple graph. Any orientation of a complete graph is known as a **tournament.**

Structures thus defined are called graphs because they can be represented graphically on paper. Such graphical representations of structures often enable us to understand and investigate many of their properties. Here are some examples of graphs and digraphs.

Example 1(a). In Fig. 1-1, we have a graph in which the vertex set is $V = \{1, 2, 3, 4, 5, 6, 7\}$. The order is 7 and the size is 8. This is a pseudograph with a loop at vertex 6 and three parallel edges between vertex 2 and vertex 4.

Example 1(b). Suppose each vertex of a graph represents either a recent college graduate or a firm that is hiring college graduates. Join a vertex representing a college graduate and a vertex representing a firm if and only if the firm is interested

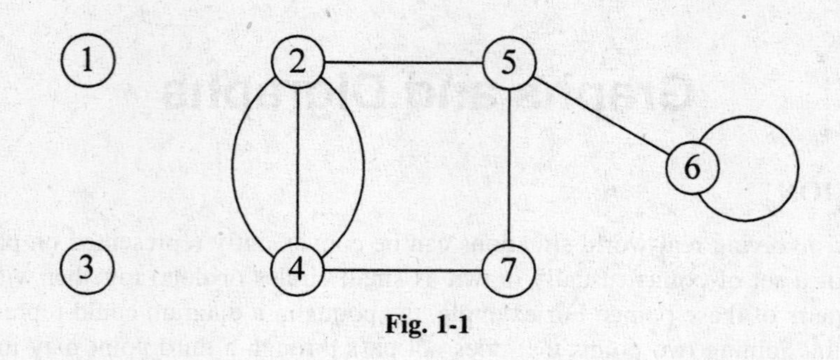

Fig. 1-1

in hiring that graduate. In this situation, we have a bipartite graph $G = (X, Y, E)$, where every vertex in X represents a graduate and every vertex in Y represents a firm.

Example 1(c). Suppose a certain commodity is manufactured at one location denoted by S (the source) and then is transported by trucks to another location denoted by T (the terminal or the sink) along different routes passing through several intermediate locations. This situation can be represented by a digraph in which the intermediate locations, the source, and the terminal are all vertices. We draw a directed edge from a vertex A to another vertex B if it is possible to transport the commodity from the location represented by A to the location represented by B in a truck that does not stop at another location en route.

Example 1(d). Here is an example of a graph in which the number of edges is not finite. Let $V = \{1, 2, 3, \ldots, n\}$ be the set of vertices. Corresponding to each real number in the open interval $(i, i + 1)$, where $i = 1, 2, \ldots, (n - 1)$, we draw an edge joining the vertex i and the vertex $(i + 1)$.

Example 1(e). Here is an example of a graph in which we have an infinite set of vertices. Let $V = \{1, 2, 3, \ldots\}$ be set of vertices. So each positive integer represents a vertex. Join the vertex representing a prime number p and the vertex representing the integer $p + 2$ if and only if $(p + 2)$ is also a prime number. It is not known whether the set of edges is finite or infinite.

Example 1(f). Let V be a finite set of open intervals of real numbers. Join the vertex i representing the open interval I and the vertex j representing another open interval J (which is not equal to I) if and only if the intersection of I and J is nonempty. The graph thus constructed is a simple graph known as an **interval graph.** For example, in the interval graph defined by the open intervals (3, 8), (7, 9), (3, 6), and (5, 10) represented by vertices A, B, C, and D, respectively, it is easy to see that there is an edge between every pair of vertices except between B and C.

Example 1(g). Fig 1-2(*a*) is the diagram of the complete graph with four vertices. An orientation of this graph describing a tournament with four vertices is shown in Fig. 1-2(*b*).

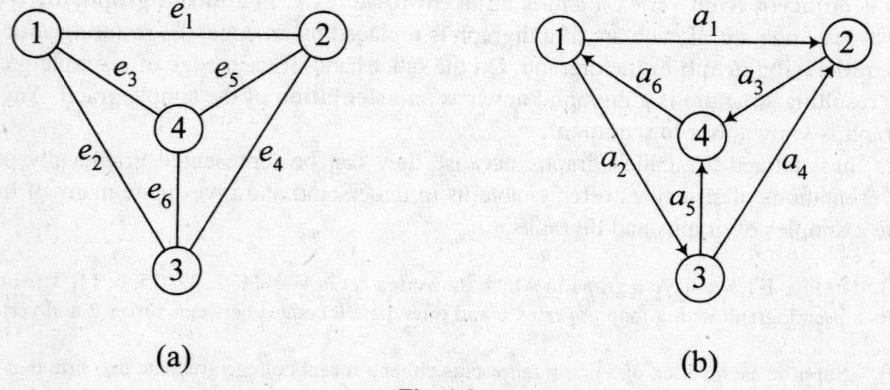

Fig. 1-2

1.2 GRAPH ISOMORPHISM

Two graphs $G = (V, E)$ and $G' = (V', E')$ are **identical** if $V = V'$ and $E = E'$. This rigid approach to the identification and classification of graphs is too restrictive. It is often possible that two graphs have the same structure even though they are not equivalent. In this case, we may consider them to be the same for all practical purposes. For example, any simple graph with vertex set $\{a, b, c\}$ and an edge set consisting of one edge is not structurally different from some other simple graph with vertex set $\{p, q, r\}$ and an edge set consisting of one edge. This structural equivalence between two nonequivalent graphs leads us to the concept of isomorphic graphs.

Two graphs $G = (V, E)$ and $G' = (V', E')$ are said to be **isomorphic** if there exists a one-to-one correspondence f, called an **isomorphism,** from V to V' such that there is an edge between $f(v)$ and $f(w)$ in G' if and only if there is an edge between v and w in G. For all practical purposes, two isomorphic graphs can be considered as one and the same graph. Obviously, two equivalent graphs are isomorphic, but the converse is not true, as pointed out in the previous paragraph. Thus any complete graph with n vertices is isomorphic to any other complete graph with n vertices.

In the **cyclic graph** $C_n = (V, E)$ (where $n > 2$), V is the set $\{1, 2, \ldots, n\}$ and E is $\{\{1, 2\}, \{2, 3\}, \ldots, \{n - 1, n\}, \{n, 1\}\}$. A **triangle** is a cyclic graph with three vertices.

Example 2. Consider the three graphs in Fig. 1-3. The graphs G_1 and G_2 are isomorphic because of the isomorphism f defined by

$$f(v_1) = w_1, \quad f(v_2) = w_4, \quad f(v_3) = w_3, \quad f(v_4) = w_6, \quad f(v_5) = w_5, \quad \text{and} \quad f(v_6) = w_2$$

G_3 is not isomorphic to either G_1 or G_2.

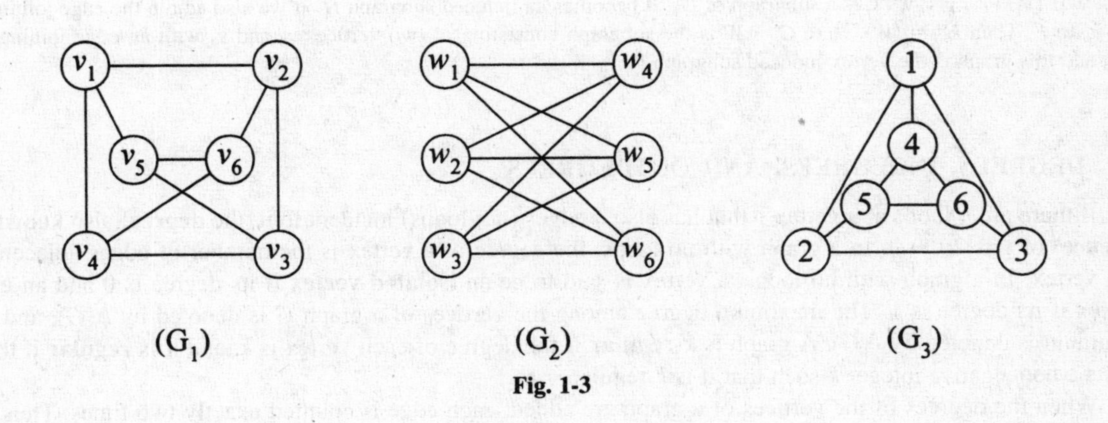

Fig. 1-3

To show that two graphs are isomorphic, one must point out an explicit isomorphism between them. The definition of isomorphism between two digraphs is analogous.

Since we consider two graphs to be the same if they are isomorphic, the trivial graph is the only graph of order one. A simple graph with two vertices may have no edge or one edge. Thus there are two nonisomorphic simple graphs of order two. Any simple graph with three vertices gives rise to four nonisomorphic cases: (1) with no edge at all, (2) with one edge joining a pair, (3) with two edges such that there is no edge joining a pair, and (4) with three edges. In other words, there are four nonisomorphic simple graphs of order 3.

Given two arbitrary simple graphs of the same order and the same size, the problem of determining whether an isomorphism exists between the two is known as the **isomorphism problem** in graph theory. In general, it is not all easy (in other words, there is no "efficient algorithm") to solve an arbitrary instance of the isomorphism problem.

1.3 SUBGRAPHS

The graph $H = (W, F)$ is a **subgraph** of the graph $G = (V, E)$ if W is a subset of V and F is a subset of E. (More generally, an arbitrary graph H can be considered as a subgraph of G if H is isomorphic to a subgraph of G.) If a subgraph H of the graph G is a cyclic graph, H is called a **cycle** in G. A complete subgraph of G is a **clique** in G. Any graph G' for which a given graph G is a subgraph is called a **supergraph** of G.

Any subgraph $H(V, F)$ of $G = (V, E)$ is a **spanning subgraph** of G. A **factor** of a graph is a spanning subgraph with at least one edge. If F is a set of edges in $G = (V, E)$, the spanning subgraph obtained by deleting the edges of F from G is denoted by $G - F$. If F consists of one edge f, we write $G - f$ instead of $G - \{f\}$. If W is a set of vertices in $G = (V, E)$, the graph obtained from G by deleting every vertex in W as well as any edge in E that is adjacent to a vertex in W is denoted by $G - W$. If W consists of a single vertex w, we write $G - w$ instead of $G - \{w\}$.

If $H = (W, F)$ is a subgraph of a graph $G = (V, E)$ such that an edge exists in F between two vertices in W if and only an edge exists in E between those two vertices, the subgraph H is said to be **induced by the set** W and is denoted by $\langle W \rangle$, which is the maximal subgraph of G with respect to the set W.

A subgraph G' of $G = (V, E)$ is a **vertex-induced subgraph** (or simply an **induced subgraph**) of G if there exists a set W of vertices in G such that $\langle W \rangle = G'$. Observe that if W is a subset of vertices of graph G, the subgraph $G - W$ is the induced subgraph $\langle V - W \rangle$, where the set $V - W$ is the relative complement of W in V.

We next consider the edge analog of induced subgraphs. Suppose F is a set of edges in $G = (V, E)$. The subgraph induced by F is the minimal subgraph $\langle F \rangle = (W, F)$, where v is in W if and only if v is adjacent to at least one edge in F. A subgraph G' of G is an **edge-induced subgraph** if there exists a set F of edges in G such that $G' = \langle F \rangle$.

Example 3. In graph G_1 shown in Fig. 1-3, the graph $H = (W, F)$, where $W = \{v_1, v_2, v_5, v_6\}$ and $F = \{\{v_1, v_2\}, \{v_1, v_5\}, \{v_2, v_6\}\}$, is a subgraph of G_1. It becomes an induced subgraph H' if we also adjoin the edge joining v_5 and v_6 to F. Then $H' = \langle W \rangle$. Here $G - W$ is the subgraph consisting of two vertices v_3 and v_4 with an edge joining the two, and this graph is the vertex-induced subgraph $\langle V - W \rangle$.

1.4 DEGREES, INDEGREES, AND OUTDEGREES

If there are p loops at a vertex v that has also q edges (not loops) incident to it, the **degree** (also known as **valence**) of v is $2p + q$. In a graph with no loops, the degree of a vertex is the number of edges adjacent to that vertex. In a graph with no loops, a vertex is said to be an **isolated vertex** if its degree is 0 and an **end-vertex** if its degree is 1. The maximum degree among the vertices of a graph G is denoted by $\Delta(G)$, and the minimum is denoted by $\delta(G)$. A graph is ***k*-regular** if the degree of each vertex is k, and it is **regular** if there exists a nonnegative integer k such that it is k-regular.

When the degrees of the vertices of a graph are added, each edge is counted exactly two times. Thus we have the following result, known as the first theorem of graph theory due to Euler.

Theorem 1.1. The sum of the degrees of a graph is twice the number of edges in it. (See Solved Problem 1.36.)

The result stated above (which implies that the sum of the degrees is even) is also known as the **handshaking lemma** because the number of hands shaken in a party is always even since each handshake involves exactly two hands of two different individuals. A vertex in a graph is an **odd vertex** if its degree is odd. Otherwise, it is an **even vertex.**

Theorem 1.2. Every graph has an even number of odd vertices. (See Solved Problem 1.38.)

Example 4. In graph G shown in Fig. 1-1, let d_i be the degree of vertex i, where $i \in V = \{1, 2, 3, 4, 5, 6, 7\}$. Then $d_1 = d_3 = 0$, $d_2 = 4$, $d_4 = 4$, $d_5 = 3$, $d_6 = 3$, and $d_7 = 2$. The sum of the degrees is 16, which is equal to twice the number of edges. The vertices v_5 and v_6 are odd.

In a digraph, the number of arcs adjacent to a vertex is the **indegree** of that vertex, and the number of arcs adjacent from a vertex is the **outdegree** of that vertex. When the outdegrees (or indegrees) of all the vertices are added, each arc is considered once in the counting process. We thus have the following result.

Theorem 1.3. In a digraph, the sum of the outdegrees of all the vertices is equal to the number of arcs, which is also equal to the sum of all the indegrees of all the vertices. (See Solved Problem 1.42.)

Example 5. The outdegrees of the four vertices 1, 2, 3, and 4 of the digraph in Fig. 1-2(b) are 2, 1, 2, and 1. The indegrees are 1, 2, 1, and 2. There are six arcs.

1.5 ADJACENCY MATRICES AND INCIDENCE MATRICES

Let $G = (V, E)$ be a graph where $V = \{1, 2, \ldots, n\}$. The **adjacency matrix of the graph** is the $n \times n$ matrix $A = [a_{ij}]$, where the nondiagonal entry a_{ij} is the number of edges joining vertex i and vertex j and the diagonal entry a_{ii} is twice the number of loops at vertex i. The adjacency matrix of a graph is obviously symmetric, that is, $a_{ij} = a_{ji}$ for every i and every j. The adjacency matrix of a simple graph is a binary matrix (0, 1 matrix) in which each diagonal entry is zero. Notice that in the adjacency matrix of the complete graph K_n, each nondiagonal entry is 1.

Since the n vertices of a graph can be labeled in $n!$ different ways and for each such labeling of vertices we have an adjacency matrix of the graph, by an abuse of notation any of these matrices is considered the adjacency matrix of the graph. At any rate, the adjacency matrix is uniquely determined apart from the ordering of its rows and columns. See Solved Problem 1.61 for more on this.

The **adjacency matrix of a digraph** with vertex set $\{1, 2, \ldots, n\}$ is the $n \times n$ binary matrix $A = [a_{ij}]$ in which $a_{ij} = 1$ if and only if there is an arc from vertex i to vertex j. Each diagonal entry in the adjacency matrix A of a digraph is zero, and A need not be symmetric.

Theorem 1.4. (i) In the adjacency matrix of a graph, the sum of the entries in a row (or a column) corresponding to a vertex is its degree, and the sum of all the entries of the matrix is twice the number of edges in the graph. (ii) In the adjacency matrix of a digraph, the sum of the entries in a row corresponding to a vertex is its outdegree, the sum of the entries in a column corresponding to a vertex is its indegree, and the sum of all the entries of the matrix is equal to the number of arcs in the digraph. (See Solved Problem 1.59.)

Example 6(a). The adjacency matrix of the graph of Fig. 1-1 is

$$\begin{bmatrix} 0 & 0 & 0 & 0 & 0 & 0 & 0 \\ 0 & 0 & 0 & 3 & 1 & 0 & 0 \\ 0 & 0 & 0 & 0 & 0 & 0 & 0 \\ 0 & 3 & 0 & 0 & 0 & 0 & 1 \\ 0 & 1 & 0 & 0 & 0 & 1 & 1 \\ 0 & 0 & 0 & 0 & 1 & 2 & 0 \\ 0 & 0 & 0 & 1 & 1 & 0 & 0 \end{bmatrix}$$

Example 6(b). The adjacency matrix of the digraph of Fig. 1-2(b) is

$$\begin{bmatrix} 0 & 1 & 1 & 0 \\ 0 & 0 & 0 & 1 \\ 0 & 1 & 0 & 1 \\ 1 & 0 & 0 & 0 \end{bmatrix}$$

Let $G = (V, E)$ be a simple graph where $V = \{1, 2, \ldots, n\}$ and $E = \{e_1, e_2, \ldots, e_m\}$. The **incidence matrix** $B = [b_{ij}]$ of G is defined as follows. Row i of B corresponds to vertex i for each i. Column k corresponds to edge e_k for each k. If e_k is the edge that joins vertex i and vertex j, the entries b_{ik} and b_{jk} are 1 and all the other entries in column k are zero. If G is a digraph and e_k is the arc from vertex i to vertex j, we define $b_{ik} = -1$ and $b_{jk} = 1$ in column k. Again, all the other entries in column k are zero. Thus the incidence matrix of a simple graph with n vertices and m edges is an $n \times m$ matrix in which each entry is 0 or 1, whereas the **incidence matrix of a digraph** with n vertices and m arcs is an $n \times m$ matrix in which each entry is 0 or 1 or -1. Each column of an incidence matrix has exactly two nonzero entries. We have the following analog of the previous theorem.

Theorem 1.5. (*i*) The sum of the entries in a row of the incidence matrix of a simple graph corresponding to a vertex is its degree, and the sum of all the entries in the matrix is twice the number of edges. (*ii*) The sum of the entries in a row of the incidence matrix of a digraph is its outdegree minus its indegree, and the sum of all the entries in the matrix is zero. (See Solved Problem 1.60.)

Example 7(a). The six edges in the simple graph of Fig. 1-2(*a*) are e_1 joining 1 and 2, e_2 joining 1 and 3, e_3 joining 1 and 4, e_4 joining 2 and 3, e_5 joining 2 and 4, and e_6 joining 3 and 4. The incidence matrix of this simple graph is

$$\begin{bmatrix} 1 & 1 & 1 & 0 & 0 & 0 \\ 1 & 0 & 0 & 1 & 1 & 0 \\ 0 & 1 & 0 & 1 & 0 & 1 \\ 0 & 0 & 1 & 0 & 1 & 1 \end{bmatrix}$$

Example 7(b). The six arcs in the digraph of Fig. 1-2(*b*) are a_1 from 1 to 2, a_2 from 1 to 3, a_3 from 2 to 4, a_4 from 3 to 2, a_5 from 3 to 4, and a_6 from 4 to 1. The incidence matrix of this digraph is

$$\begin{bmatrix} -1 & -1 & 0 & 0 & 0 & 1 \\ 1 & 0 & -1 & 1 & 0 & 0 \\ 0 & 1 & 0 & -1 & -1 & 0 \\ 0 & 0 & 1 & 0 & 1 & -1 \end{bmatrix}$$

1.6 DEGREE VECTORS OF SIMPLE GRAPHS

The **degree vector** $d(G)$ of a simple graph G is the sequence of degrees of its vertices arranged in nonincreasing order. If G and G' are isomorphic, $d(G) = d(G')$. But the converse is not true. The three graphs in Fig. 1-3 have all the same degree vector [3 3 3 3 3 3], but G_3 is not isomorphic to G_1 or G_2.

Every simple graph has a unique degree vector (which can be easily constructed), and the sum of the terms in the vector is even. Notice that each term in a degree vector with n components is nonnegative and is at most $(n - 1)$, where n is the order of the graph. On the other hand, a finite nonincreasing vector $v = [d_1 \ d_2 \ \cdots \ d_n]$ of nonnegative integers, where each $d_i \le (n - 1)$ and the sum of the terms is even, need not be the degree vector of a simple graph. Consider, for example, the vector $v = [3 \ 3 \ 3 \ 1]$. If there were a simple graph with v as a degree vector, the subgraph obtained by deleting the vertex of degree 1 (with three vertices and four edges) would not be simple.

A vector v is called a **graphical vector** if there exists a simple graph such that v is the degree vector of that graph. Thus [3 3 3 1] is not a graphical vector. The following theorem gives a necessary and sufficient condition for a vector to be a graphical vector.

Theorem 1.6 **(Hakimi–Havel).** Let $v = [d_1 \ d_2 \ d_3 \ \cdots \ d_k]$ be a nonincreasing vector of k (where k is at least 2) nonnegative integers such that no component d_i exceeds $(k - 1)$. Let v' be the vector obtained from v by deleting d_1 and subtracting 1 from each of the next d_1 components of v. Let v_1 be the nonincreasing vector obtained from v' by rearranging its components if necessary. Then v is a graphical vector if and only if v_1 is a graphical vector. (See Solved Problem 1.68.)

Example 8. Consider $v = [5 \ 4 \ 3 \ 3 \ 3 \ 3 \ 3 \ 2]$ with eight components in which no component exceeds 7. Delete the first component 5, and subtract 1 from the next five components of v. We obtain $v' = [3 \ 2 \ 2 \ 2 \ 2 \ 3 \ 2]$. By rearranging the components of v', we get the nonincreasing vector $v_1 = [3 \ 3 \ 2 \ 2 \ 2 \ 2 \ 2]$ with seven components.

According to Theorem 1.6, v is a graphical vector if and only if v_1 is a graphical vector. Proceeding further, we see in the next iteration that v_1 is a graphical vector if and only if $v_2 = [2\ \ 2\ \ 2\ \ 2\ \ 1\ \ 1]$ is a graphical vector. At the next iteration, we see that the vector v_2 is a graphical vector if and only if $v_3 = [2\ \ 1\ \ 1\ \ 1\ \ 1]$ is a graphical vector. Then v_3 is a graphical vector if and only if $v_4 = [1\ \ 1\ \ 0\ \ 0]$ is a graphical vector. At the next stage, we get $v_5 = [0\ \ 0\ \ 0]$. Now v_5, being the degree vector of a simple graph of order 3 with no edges, is a graphical vector. So the given vector v is also a graphical vector.

Using Theorem 1.6, it is possible to test whether an arbitrary vector with integer components is a graphical vector as outlined in the following algorithm.

Algorithm to Test Whether a Given Vector Is Graphical

The input is a nonincreasing vector v with integer components.

Step 0. (Initialization) The current vector is v.

Step 1. If the current vector with k components has a component that exceeds $(k - 1)$, go to step 5. Otherwise, go to step 2.

Step 2. If the current vector has a negative component, go to step 5. Otherwise, go to step 3.

Step 3. If the current vector is the zero vector, go to step 6. Otherwise, go to step 4.

Step 4. (Iteration) Rearrange the components of the current vector so that it becomes nonincreasing with d_1 as the first component. Delete d_1 from the rearranged vector, and subtract 1 from each of the next d_1 components of the rearranged vector. The vector thus constructed is the updated current vector. Go to step 1.

Step 5. The input vector is not graphical. Go to step 7.

Step 6. The input vector is graphical. Go to step 7.

Step 7. Stop.

(Note: The zero vector of step 3 with k components is graphical since it is the degree vector of a simple graph with k vertices and no edges.)

Example 9. Using this algorithm it can be verified that $v = [5\ \ 4\ \ 4\ \ 3\ \ 3\ \ 3\ \ 2]$ is a graphical vector.

Iteration 1:

$$v = [5\ \ 4\ \ 4\ \ 3\ \ 3\ \ 3\ \ 2] \quad \text{and} \quad v_1 = [3\ \ 3\ \ 2\ \ 2\ \ 2\ \ 2]$$

Iteration 2:

$$v = [3\ \ 3\ \ 2\ \ 2\ \ 2\ \ 2] \quad \text{and} \quad v_1 = [2\ \ 2\ \ 2\ \ 1\ \ 1]$$

Iteration 3:

$$v = [2\ \ 2\ \ 2\ \ 1\ \ 1] \quad \text{and} \quad v_1 = [1\ \ 1\ \ 1\ \ 1]$$

Iteration 4:

$$v = [1\ \ 1\ \ 1\ \ 1] \quad \text{and} \quad v_1 = [1\ \ 1\ \ 0]$$

Iteration 5:

$$v = [1\ \ 1\ \ 0] \quad \text{and} \quad v_1 = [0\ \ 0]$$

At the end of the fifth iteration we get the zero vector. So the given vector v is a graphical vector.

A simple graph for which the graphic vector v given above is the degree vector is shown in Fig. 1-4.

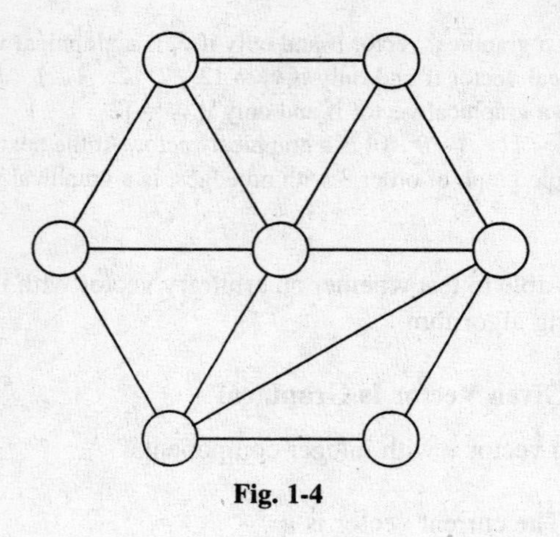

Fig. 1-4

Example 10. Using the algorithm, we now show that the vector $v = [3 \quad 3 \quad 3 \quad 1]$ is not a graphical vector.

At the end of the first iteration, we get $[2 \quad 2 \quad 0]$ as the current vector. At the end of the next iteration, we get $[1 \quad -1]$ as the current vector, which has a negative component. The conclusion is that v is not a graphical vector.

Even if a vector is graphical, it is not the case that it is the degree vector of a unique (up to isomorphism) simple graph. The vector $v = [3 \quad 3 \quad 3 \quad 3 \quad 3 \quad 3]$ is graphical, but both G_1 and G_3 in Fig. 1-3 have v as the degree vector.

Solved Problems

INTRODUCTION

1.1 Draw the diagram of each of the following graphs $G = (V, E)$:

(*a*) $V = \{1, 2, 3, 4, 5\}$ and $E = \{\{1, 2\}, \{1, 4\}, \{1, 5\}, \{2, 3\}, \{3, 4\}, \{4, 4\}\}$

(*b*) $V = \{1, 2, 3, 4, 5, 6\}$ and $E = \{\{1, 2\}, \{1, 4\}, \{1, 4\}, \{2, 3\}, \{2, 5\}, \{3, 5\}\}$

Solution. See Fig. 1-5.

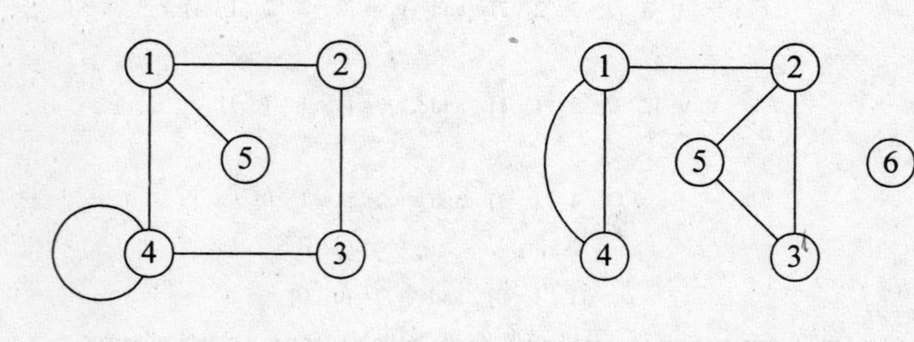

Fig. 1-5

1.2 Draw the diagram of each of the following graphs $G = (V, E)$:

(*a*) $V = \{1, 2, 3, 4, 5, 6\}$ and $E = \{\{1, 2\}, \{1, 3\}, \{1, 4\}, \{2, 5\}, \{2, 6\}, \{3, 5\}, \{3, 6\}, \{4, 5\}, \{4, 6\}\}$.

(*b*) $V = \{1, 2, 3, 4, 5\}$ and $E = \{\{1, 2\}, \{1, 4\}, \{2, 3\}, \{2, 4\}, \{2, 5\}, \{3, 4\}, \{3, 5\}\}$

Solution. See Fig. 1-6.

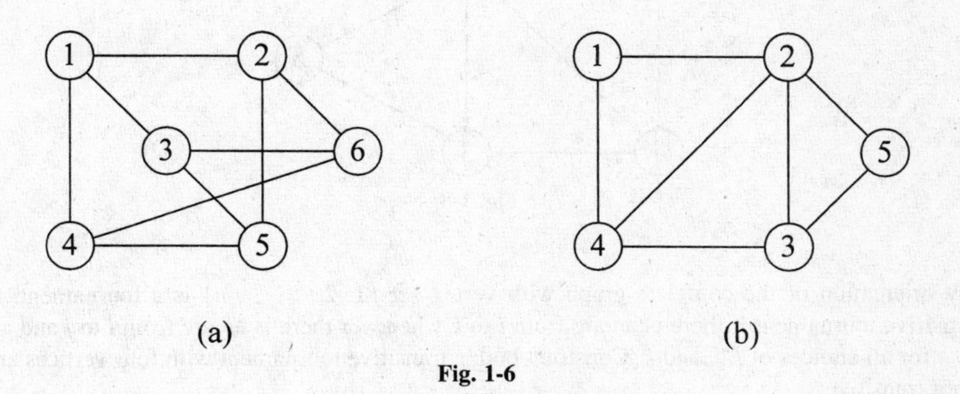

Fig. 1-6

1.3 Identify the simple graphs in the previous two problems. If a simple graph is identified, determine whether it is (*i*) a bipartite graph, (*ii*) a complete graph, (*iii*) a complete bipartite graph, or (*iv*) a complete nonbipartite graph.

Solution. The two graphs in Fig. 1-5 are not simple. The graph in Fig. 1-6(*a*) is a complete bipartite graph (X, Y, E), where $X = \{1, 5, 6\}$ and $Y = \{2, 3, 4\}$. The graph in Fig. 1-6(*b*) is simple and nonbipartite.

1.4 The **complement** of a simple graph $G = (V, E)$ is the simple graph $\overline{G} = (V, F)$ in which there is an edge between two vertices v and w if and only if there is no edge between v and w in G. Obviously, the complement of the complement of $\overline{G}$ is G. Draw the diagrams of the complements of the simple graphs identified in Problems 1.1 and 1.2.

Solution. The complement of Fig. 1-6(*a*) is Fig. 1-7(*a*), and the complement of Fig. 1-6(*b*) is Fig. 1-7(*b*), as shown in Fig. 1-7.

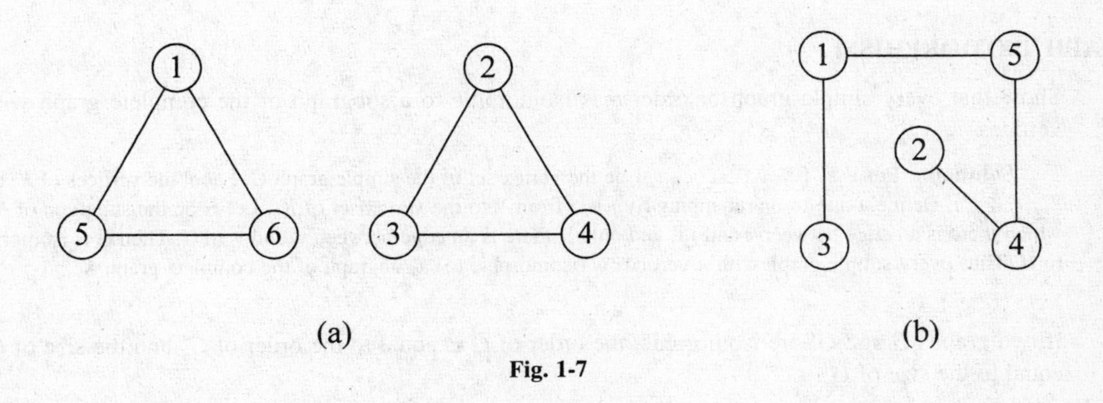

Fig. 1-7

1.5 Show that the complement of a bipartite graph need not be a bipartite graph.

Solution. Figure 1-6(*a*) shows a bipartite graph whose complement, shown in Fig. 1-7(*a*), is not bipartite.

1.6 Draw the diagram of an orientation of the simple nonbipartite graph that is not complete identified from Problems 1.1 and 1.2.

Solution. An orientation of the simple graph in Problem 1.2(*b*) is as shown in Fig. 1-8.

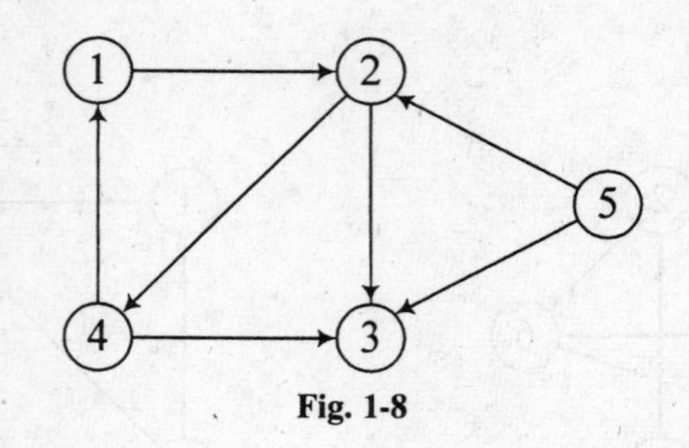

Fig. 1-8

1.7 Any orientation of the complete graph with vertex set $\{1, 2, \ldots, n\}$ is a tournament, and it is a **transitive** tournament if there is an arc from i to k whenever there is an arc from i to j and an arc from j to k for all choices of i, j, and k. Construct both a transitive tournament with four vertices and one that is not transitive.

Solution. A transitive tournament with four vertices is shown in Fig. 1-9. If we replace the arc (1, 3) by the arc (3, 1), the resulting tournament is not transitive.

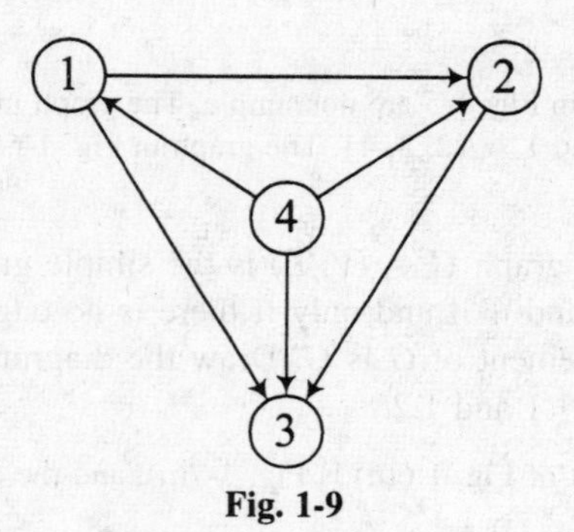

Fig. 1-9

GRAPH ISOMORPHISM

1.8 Show that every simple graph of order n is isomorphic to a subgraph of the complete graph with n vertices.

Solution. Let $V = \{v_1, v_2, \ldots, v_n\}$ be the vertex set of the simple graph G. Label the vertices of K_n as 1, 2, . . . , n. Define a one-to-one mapping $f(v_i) = i$ from V to the vertex set of K_n. Let H be the subgraph of K_n in which there is an edge between i and j if and only if there is an edge between v_i and v_j in G. Then G is isomorphic to H. Thus every simple graph with n vertices is (isomorphic to) a subgraph of the complete graph K_n.

1.9 If two graphs G and G' are isomorphic, the order of G is equal to the order of G' and the size of G is equal to the size of G'.

Solution. Let $G = (V, E)$ and $G' = (V', E')$. If G and G' are isomorphic, there is a bijection f between V and V'. So both G and G' have the same order. Furthermore, the bijection preserves adjacency and nonadjacency: There is an edge between vertex x and vertex y in G if and only if there is an edge between $f(x)$ and $f(y)$ in G' and there is a loop at x if and only if there is a loop at $f(x)$. So both G and G' have the same size.

1.10 Show that two graphs need not be isomorphic even when they both have the same order and same size.

Solution. Let $G = (V, E')$ and $G' = (V', E')$, where $V = V' = \{a, b, c, d\}$, $E = \{\{a, b\}, \{b, c\}, \{c, d\}\}$, and $E' = \{\{a, b\}, \{b, c\}, \{b, d\}\}$, be two graphs. It is impossible to define a bijection between V and V' that will preserve adjacency and nonadjacency even though both G and G' have the same order and same size.

1.11 Show that two simple graphs are isomorphic if and only if their complements are isomorphic.

Solution. Suppose $G = (V, E)$ and $H = (W, F)$ are two simple isomorphic graphs with $|V| = |W| = n$ and $|E| = |F| = m$. Then their complements are also of order n. An isomorphism from V to W that preserves adjacency and nonadjacency between G and H is an isomorphism from V to W that preserves nonadjacency and adjacency between their complements.

1.12 Determine whether the three graphs given in Fig. 1-10 are isomorphic.

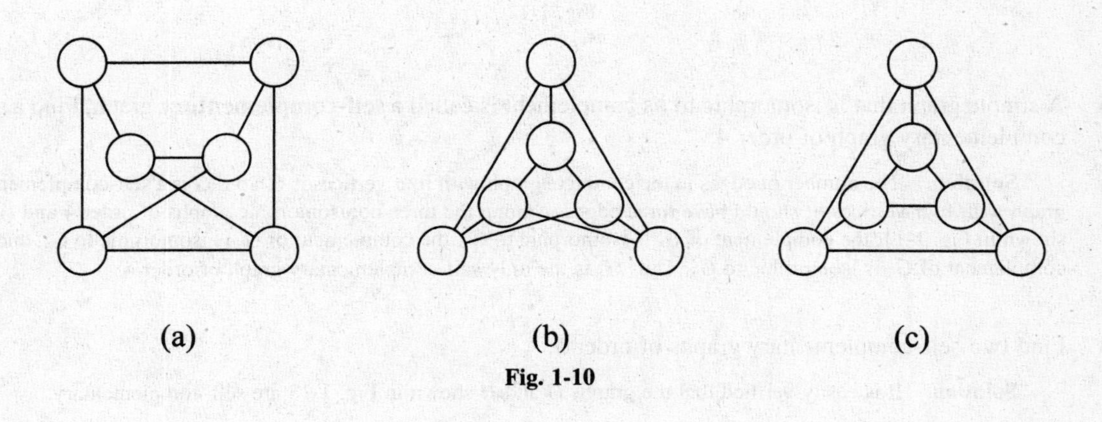

Fig. 1-10

Solution. Both Figure 1-10(*a*) and (*b*) are isomorphic to $K_{3,3}$ and so they are isomorphic to each other. But Figure 1-10(*c*) is not isomorphic to $K_{3,3}$.

1.13 Suppose $N(n, k)$ is the number of nonisomorphic simple graphs with n vertices and k edges. Find $N(4, 3)$.

Solution. There are exactly three graphs in this category, as shown in Fig. 1-11. So $N(4, 3) = 3$.

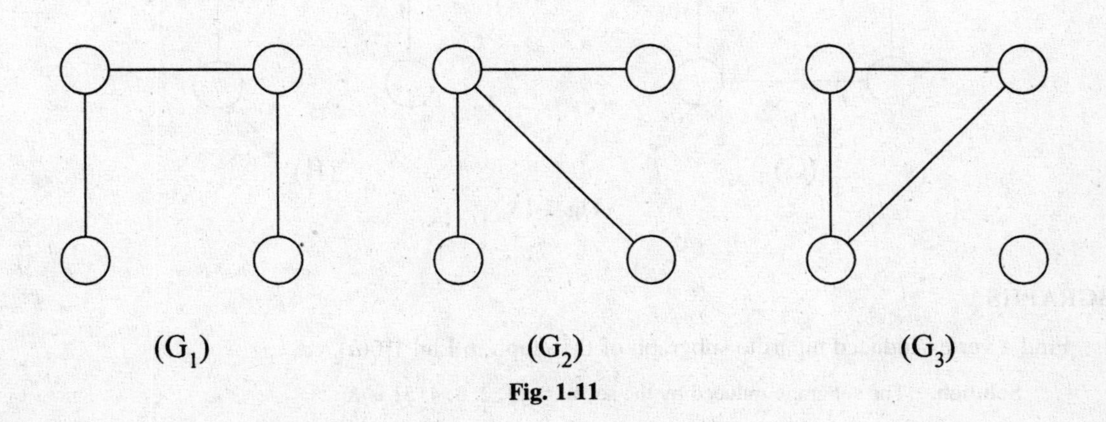

Fig. 1-11

1.14 Find all nonisomorphic simple graphs of order 4.

Solution. The maximum number of edges in a simple graph with four vertices is 6. The complete graph K_4 is the only graph of order 4 and size 6. There is one nonisomorphic graph of order 4 and size 5, and there are two nonisomorphic graphs of order 4 and size 4, as shown in Fig. 1-12. So $N(4, 6) = 1$, $N(4, 5) = 1$, and $N(4, 4) = 2$. In Problem 1.13 it was shown that $N(4, 3) = 3$. It is easy to see that $N(4, 2) = 2$ and $N(4, 1) = N(4, 0) = 1$. So the total number of nonisomorphic simple graphs of order 4 is $N(4, 0) + N(4, 1) + N(4, 2) + N(4, 3) + N(4, 4) + N(4, 5) + N(4, 6) = 1 + 1 + 2 + 3 + 2 + 1 + 1 = 11$.

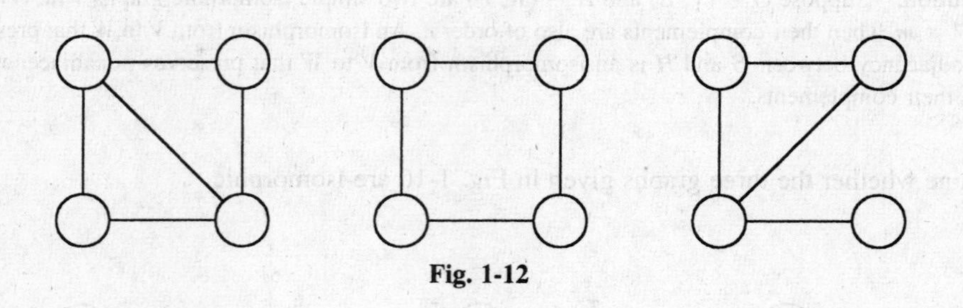

Fig. 1-12

1.15 A simple graph that is isomorphic to its complement is called a **self-complementary** graph. Find a self-complementary graph of order 4.

Solution. The number of edges in the complete graph with four vertices is 6. So if G is a self-complementary graph with four vertices, it should have three edges. Among the three nonisomorphic graphs of order 4 and size 3 shown in Fig. 1-11, the complement of G_1 is isomorphic to G_1, the complement of G_2 is isomorphic to G_3, and the complement of G_3 is isomorphic to G_2. Thus G_1 is the only self-complementary graph of order 4.

1.16 Find two self-complementary graphs of order 5.

Solution. It is easily verified that the graphs G and H shown in Fig. 1-13 are self-complementary.

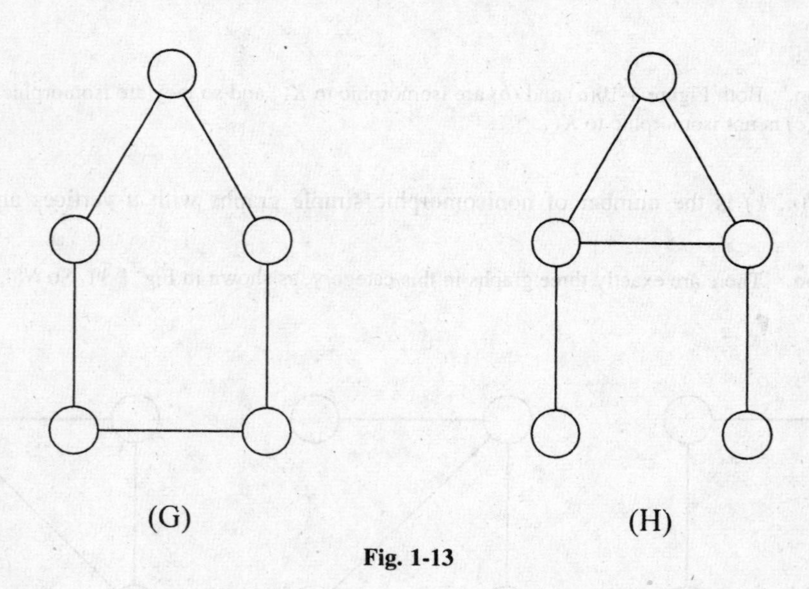

Fig. 1-13

SUBGRAPHS

1.17 Find a vertex-induced bipartite subgraph of the graph in Fig. 1-6(*a*).

Solution. The subgraph induced by the set $W = \{1, 2, 3, 4, 5\}$ is $K_{2,3}$.

1.18 A set I of vertices in a simple graph $G = (V, E)$ is an **independent set** (also known as an **internally stable set**) in G if no two vertices in I are adjacent. A set K of vertices in G is called a **vertex cover** if every edge in the graph is incident to at least one vertex in K. Show that a set K of vertices is a vertex cover if and only if its complement $(V - K)$ is an independent set.

Solution. If K is a vertex cover, no two vertices in $(V - K)$ can be adjacent, so $(V - K)$ is an independent set. On the other hand, if I is any independent set in G, out of the two vertices joined by an edge at least one should be in $(V - I)$. In other words, every edge is adjacent to some vertex in $(V - I)$, so $(V - I)$ is a vertex cover.

1.19 An independent set I in a simple graph G is a **maximum independent set** if there is no independent set I' in G such that $|I'| > |I|$. The number of vertices in a maximum independent set in G is the **independence number** $\alpha(G)$ (also known as the **internal stability number**) of the graph G. A vertex cover K in a graph G is a **minimum vertex cover** if there is no vertex cover K' such that $|K'| < |K|$. The number of vertices in a minimum vertex cover is called the **vertex-covering number** $\beta(G)$ of the graph G. Find the vertex-covering number and the independence number of the graph in Fig. 1-14.

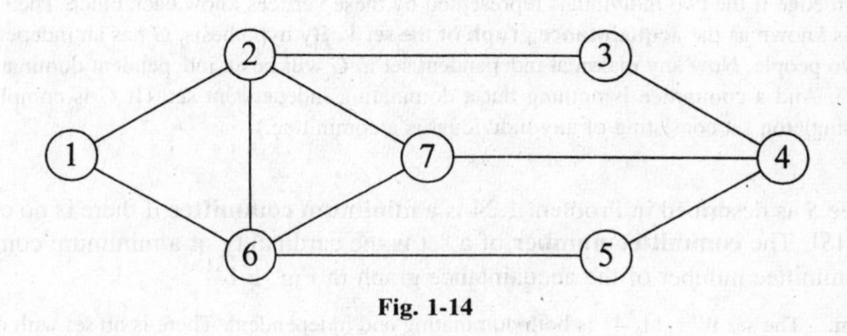

Fig. 1-14

Solution. The set {1, 3, 7, 5} is a maximum independent set, so the vertex-independence number of the graph is 4. The set {2, 4, 6} is a minimum vertex cover, so the vertex-covering number is 3.

1.20 Show that in a simple graph G of order n, $\alpha(G) + \beta(G) = n$.

Solution. Let I be a maximum independent set in G. Hence $|I| = \alpha(G)$ and $|(V - I)| = n - \alpha(G)$. But $(V - I)$ is a vertex cover. So $|(V - I)| \geq \beta(G)$, and consequently, $n \geq \alpha(G) + \beta(G)$. On the other hand, suppose K is a minimum vertex cover in G. Then $|K| = \beta(G)$ and $|(V - K)| = n - \beta(G)$. But $(V - K)$ is an independent vertex set. So $|(V - K)| \leq \alpha(G)$, and consequently, $n \leq \alpha(G) + \beta(G)$. Thus the equality is established.

1.21 A subset D of vertices in a simple graph G is a **dominating vertex set** (also known as an **external dominating set**) if every vertex not in D is adjacent to at least one vertex in D. Find (*a*) a dominating vertex set that is not independent, (*b*) an independent set that is not a dominating vertex set, and (*c*) a set that is an independent set as well as a dominating vertex set in the graph of Fig. 1-14.

Solution. (*a*) {2, 6, 7}. (*b*) {1, 3}. (*c*) {1, 4}.

1.22 A dominating vertex set D is a **minimum dominating vertex set** if there is no dominating set D' such that $|D'| < |D|$. The number of vectors in a minimum dominating vertex set is the **vertex domination number** $\sigma(G)$ (also known as the **external stability number**) of the graph. Show that the vertex domination number of a simple graph cannot exceed its independence number.

Solution. Let X be any minimum dominating set, Y any set that is both dominating and independent, and Z any maximum independent set. Then $\sigma(G) = |X| \leq |Y| \leq |Z| = \alpha(G)$.

1.23 An independent set is a **maximal independent set** if it is not a proper subset of another independent set. Show that an independent set is a dominating vertex set if and only if it is a maximal independent set. (A *maximal* independent set is not the same as a *maximum* independent set defined in Problem 1.19.)

Solution. Let I be a maximal independent set. If this set is not a dominating set, there will be a vertex v (not in I) that is not adjacent to any vertex in I. In that case, $I \cup \{v\}$ is an independent set violating the maximality of I. Conversely, suppose the independent set I is also a dominating set. If I is not a maximal independent set, there exists an independent set J such that I is a proper subset of J. So there is a vertex in $(J - I)$ that is not adjacent to any vertex in I contradicting the assumption that I is a dominating set.

1.24 Show that if there are at least two people who do not know each other among a set of individuals, it is possible to choose people from that set to form a **committee** such that no two individuals in the committee know each other and that every individual in the set not included in the committee is known to at least one person in the committee.

Solution. Construct a graph in which each vertex represents an individual in the set V of people. Join two vertices by an edge if the two individuals represented by these vertices know each other. The simple graph G thus constructed is known as the **acquaintance graph** of the set V. By hypothesis, G has an independent set consisting of at least two people. Now any maximal independent set in G will be an independent dominating set as shown in Problem 1.23. And a committee is nothing but a dominating independent set. (If G is complete, the problem is trivial. The singleton set consisting of any individual is a committee.)

1.25 A committee S as described in Problem 1.24 is a **minimum committee** if there is no committee S' such that $|S'| < |S|$. The **committee number** of a set is the cardinality of a minimum committee in the set. Find the committee number of the acquaintance graph in Fig. 1-14.

Solution. The set $W = \{1, 4\}$ is both dominating and independent. There is no set with only one vertex that is dominating. So the committee number is 2. (The committee number of any complete graph is 1.)

1.26 A set M of edges in a graph is called a **matching** (also known as an **independent edge set**) if no two edges in M have a vertex in common. A set of edges L is an **edge cover** if every vertex of positive degree is a vertex of at least one edge in L. Show that the complement of a matching need not be an edge cover. (Compare this result with that of Problem 1.18.)

Solution. Consider any simple graph G with three vertices and two edges. The set consisting of one of these edges is matching in G. But the complement of that set is not an edge cover in G.

1.27 A matching M in a simple graph is a **maximum matching** (also known as **maximum cardinality matching**) if there is no matching M' such that $|M'| > |M|$. The **edge independence number** $\alpha_1(G)$ of a graph G is the number of edges in a maximum matching. An edge cover L of a simple graph G is a **minimum edge cover** if there is no edge cover L' of G such that $|L'| < |L|$. The **edge-covering number** $\beta_1(G)$ of the graph is the sum of the number of edges in a minimum edge cover and the number of isolated vertices. Find the edge independence number and the edge-covering number of the graph of Fig. 1-14.

Solution. The set $\{\{1, 2\}, \{3, 4\}, \{5, 6\}\}$ is maximum matching, so the edge independence number is 3. The set $\{\{1, 2\}, \{3, 4\}, \{5, 6\}, \{6, 7\}\}$ is a minimum edge cover, so the edge-covering number is 4.

1.28 (*Gallai's Theorem*) Show that in a simple graph, $\alpha_1 + \beta_1 = n$.

Solution. Suppose M is a maximum matching. If M is an edge cover, $\beta_1 \le \alpha_1$, which implies that $\alpha_1 + \beta_1 \le 2\alpha_1 \le n$. If M is not an edge cover, $\beta_1 \le \alpha_1 + (n - 2\alpha_1)$. Thus in any case, the sum of the two numbers cannot exceed n. Next we establish the reverse inequality. Suppose L is a minimum edge cover in G. Let H be the subgraph of G induced by L, and let M be a maximum matching in H. If W is the set of unmatched vertices in this subgraph, the subgraph of H induced by W has no edges since M is a maximum matching. Thus $|L| - |M| = |(L - M)|$, where $(L - M)$ is the relative complement of M in L. Now $|(L - M)| \geq |W| = n - 2|M|$. Hence $|L| + |M| \geq n$. Since M is matching in G, $\alpha_1 \geq |M|$. Thus $\alpha_1 + \beta_1 \geq |M| + |L| \geq n$. This completes the proof.

1.29 A set F of edges in a graph $G = (V, E)$ is a **dominating edge set** if every edge not in F has a vertex in common with an edge in F. The **edge domination number** $\sigma_1(G)$ is the number of edges in a minimum edge domination set. Find the edge domination number of the graph of Fig. 1-14.

Solution. The edge domination number is 3 since the set $\{\{1, 2\}, \{5, 6\}, \{4, 7\}\}$ is a minimum dominating edge set.

1.30 Show that the edge domination number cannot exceed the edge independence number.

Solution. The proof is as in Problem 1.22.

1.31 Find a necessary and sufficient condition to be satisfied by a matching so that it is a dominating edge set.

Solution. A matching is a dominating edge set if and only if it is a maximal (not necessarily a maximum) matching.

DEGREES, INDEGREES, AND OUTDEGREES

1.32 Find the number of edges in the complete graph with n vertices.

Solution. Suppose the vertex set is $V = \{1, 2, \ldots, n\}$. A vertex i can be selected in n ways. There are exactly $(n - 1)$ edges between a selected vertex i and the remaining $(n - 1)$ vertices. The edge joining i and another vertex j is the same as the edge joining j and i. Thus the number of edges in K_n is $n(n - 1)/2$. Equivalently, an edge in K_n is constructed by choosing any two vertices out of a set of n vertices and joining them. The number of ways of choosing any two elements out of a set of n elements is $n(n - 1)/2$.

1.33 Using techniques from graph theory, show that $1 + 2 + \cdots + n = n(n + 1)/2$.

Solution. Consider the complete graph with $(n + 1)$ vertices. By Problem 1.32, it has $n(n + 1)/2$ edges. Now the total number of edges in the graph can be computed as follows. Suppose the vertices are labeled $1, 2, \ldots, (n + 1)$. Joining vertex 1 and the remaining vertices are n edges. Delete these edges. Then joining vertex 2 and the remaining vertices are $(n - 1)$ edges, which are also deleted. Continue the process until all the edges are deleted. The total number of edges deleted is $n + (n - 1) + \cdots + 2 + 1$, which is equal to the total number of edges in the graph.

1.34 Show that the number of vertices in a self-complementary graph is either $4k$ or $4k + 1$, where k is a positive integer.

Solution. Consider a self-complementary graph $G = (V, E)$ with n vertices and m edges. Since G is isomorphic to its complement, both G and its complement have the same number of edges. Now every edge in the complete graph with V as the set of vertices is either an edge in G or an edge in its complement. Thus $m + m = n(n - 1)/2$, showing that $n(n - 1) = 4k$, where k is a positive integer. So $n = 4k$ or $4k + 1$.

1.35 Find the number of edges in the complete bipartite graph $K_{m,n}$.

Solution. Suppose $K_{m,n} = G(X, Y, E)$. There are n edges adjacent to a vertex in X. No vertex in X is joined to a vertex in X. There are m vertices in X. So the total number of edges is mn.

1.36 Prove Theorem 1.1: The sum of the degrees of a graph is twice the number of edges in it.

Solution. An edge that is not a loop contributes to the degrees of two distinct vertices. A loop at a vertex by definition contributes twice to the degree of that vertex. Thus when the degrees of the vertices are added, each edge (whether it is a loop or not) is counted exactly two times. Thus the sum of the degrees is twice the number of edges.

1.37 Use Theorem 1.1 to find the size of K_n and $K_{m,n}$.

Solution. (*i*) Suppose the size of K_n is m. The degree of each vertex is $(n - 1)$. There are n vertices. Thus the sum of the degrees of the n vertices is $n(n - 1)$, which is $2m$ by the theorem. Hence $m = n(n - 1)/2$. (*ii*) Let the number of edges in $K_{m,n} = G(X, Y, E)$ be p. The degree of each vertex in X is n, and the degree of each vertex in Y is m. The sum of the degrees of the m vertices in X is mn. The sum of the degrees of the n vertices in Y is also mn. The sum of the degrees of all the vertices is $2mn$, which is $2p$. Hence $p = mn$.

1.38 Prove Theorem 1.2: Every graph has an even number of odd vertices.

Solution. Suppose the sum of the degrees of the odd vertices is x and the sum of the degrees of the even vertices is y. The number y is even, and the number $x + y$, being twice the number of edges, is also even. So x is necessarily even. If there are p odd vertices, the even number x is the sum of p odd numbers. So p is even.

1.39 Construct two nonisomorphic simple graphs with six vertices with degrees 1, 1, 2, 2, 3, and 3. Find the size of the graph thus constructed.

Solution. Since the sum of the degrees is 12, the size is 6. Two nonisomorphic graphs G and G' with six vertices and six edges are shown in Fig. 1-15.

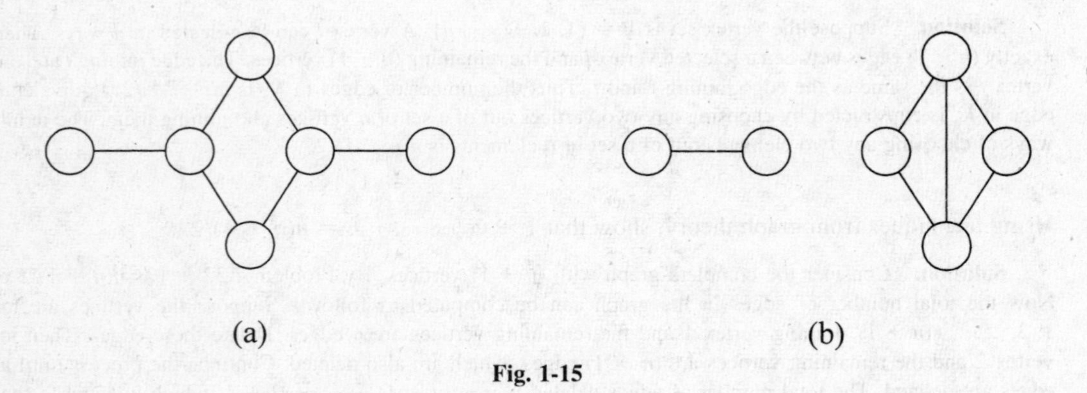

Fig. 1-15

1.40 Show that if G and G' are isomorphic graphs, the degree of each vertex is preserved under the isomorphism.

Solution. Suppose $G = (V, E)$ and $G' = (V', E')$ are two isomorphic graphs with f as the bijection from V to V' that preserves adjacency and nonadjacency. Suppose the degree of v in G is k. Then there are k vertices adjacent to v in G. Let these vertices be v_i $(i = 1, 2, \ldots, k)$. Then the only edges adjacent to $f(v)$ are those edges joining $f(v)$ and $f(v_i)$ for $i = 1, 2, \ldots, k$. So the degree of $f(v)$ is also k.

1.41 Show that two graphs G and G' with the same set $V = \{1, 2, \ldots, n\}$ of vertices such that the degree of vertex i is the same for both the graphs for every i need not be isomorphic.

Solution. The two graphs in Fig. 1-16 are not isomorphic.

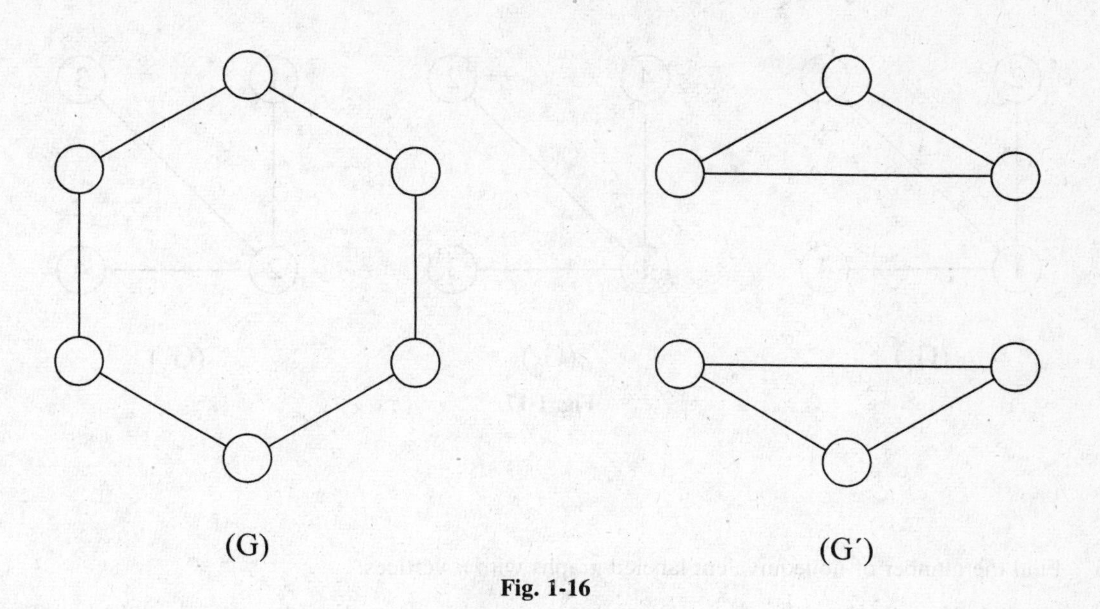

Fig. 1-16

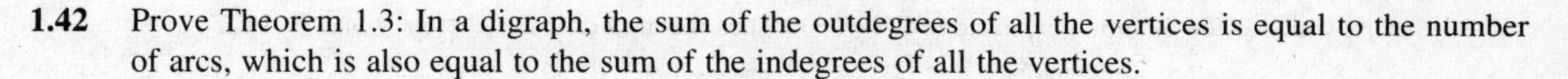

1.42 Prove Theorem 1.3: In a digraph, the sum of the outdegrees of all the vertices is equal to the number of arcs, which is also equal to the sum of the indegrees of all the vertices.

Solution. The outdegree of a vertex is the number of arcs adjacent from that vertex. So when we add all the outdegrees, each arc is counted exactly once. Likewise, when the indegrees are summed, each arc is counted exactly once. Thus the sum of the outdegrees and the sum of the indegrees are both equal to the total number of arcs in the digraph.

1.43 Show that there is no simple graph with 12 vertices and 28 edges in which (*i*) the degree of each vertex is either 3 or 4, and (*ii*) the degree of each vertex is either 3 or 6.

Solution. Suppose there is a graph with p vertices of degree 3 in the graph. (*i*) If the remaining $(12 - p)$ vertices have all degree 4, the equation $3p + 4(12 - p) = 56$ gives a negative value for p. (*ii*) If the remaining $(12 - p)$ vertices have all degree 6, the equation $3p + 6(12 - p) = 56$ gives a noninteger value for p.

1.44 Show that there is no simple graph with four vertices such that three vertices have degree 3 and one vertex has degree 1.

Solution. Suppose there exists a graph G as stipulated. The size of this graph is 5 since the sum of the degrees is 10. So G is a graph with four vertices and five edges. There is only one (up to isomorphism) graph with four vertices and five edges, as shown in Fig. 1.12 with degrees 3, 3, 2, and 2. So there is no simple graph that satisfies the given requirement.

1.45 A **labeled graph with *n* vertices** is obtained by assigning the labels 1, 2, . . . , n to the vertices of a given graph G with n vertices and m edges. Two labeled graphs thus obtained from a given graph G are necessarily isomorphic, but they need not be equivalent. Label the vertices of a simple graph with four vertices of degrees 1, 1, 1, and 3, creating three isomorphic graphs G_1, G_2, and G_3 such that (*i*) G_1 and G_2 are equivalent, and (*ii*) G_1 and G_3 are not equivalent.

Solution. See Fig. 1-17.

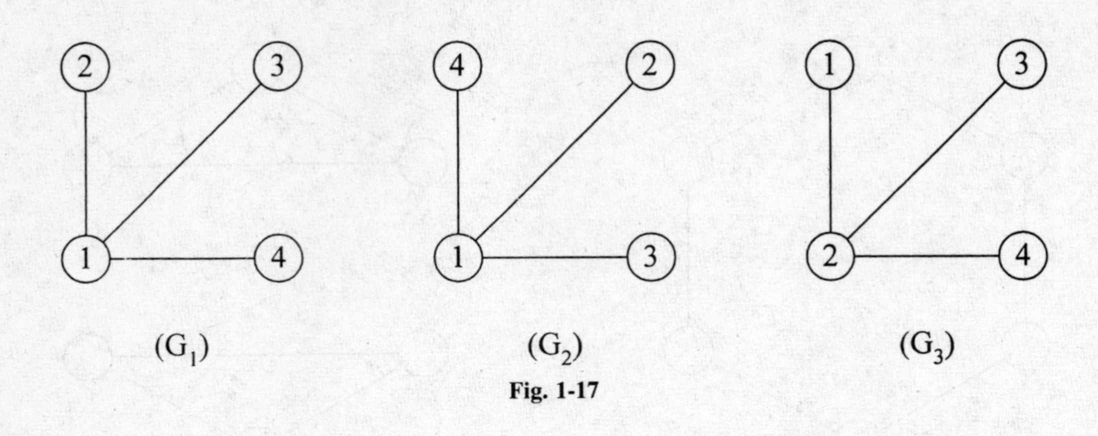

Fig. 1-17

1.46 Find the number of nonequivalent labeled graphs with n vertices.

Solution. Let $L(n, k)$ be the number of nonequivalent labeled graphs with n vertices and k edges. Then the number of nonequivalent graphs of order n is $L(n, 0) + L(n, 1) + \cdots + L(n, r)$, where r is the number of edges in a complete graph with n vertices. Now we can choose k edges out of r edges in $C(r, k)$ ways, where $C(r, k)$ is the binomial coefficient representing the number of ways of choosing k elements from a set of r elements. Thus $L(n, k)$ is equal to $C(r, k)$. So the total number of labeled graphs with n vertices is $C(r, 0) + C(r, 1) + \cdots + C(r, r)$, which is the binomial expansion of $(1 + 1)^r$. Thus the answer is 2^r, where $r = n(n - 1)/2$.

1.47 Show that the number of vertices in a k-regular graph is even if k is odd.

Solution. If the number of vertices n, the product kn is twice the number of edges. Thus n is even if k is odd.

1.48 Show that it is not possible to have a group of seven people such that each person in the group knows exactly three other people in the group.

Solution. Construct an acquaintance graph with seven vertices such that each vertex represents a person. Join two vertices by an edge if the two individuals know each other. If each knows exactly three people, we should have a 3-regular graph with seven vertices, which is a contradiction by Problem 1.47.

1.49 Prove that in any group of six people, there will be either three people who know one another or three people who do not know one another.

Solution. Suppose G is the acquaintance graph involving six people. Then its complementary graph can be considered as the *nonacquaintance* graph in the sense that there will be an edge (in that graph) between two vertices representing two people if and only if they do not know each other. Suppose v is a vertex. Then either in G or in its complement there will be at least three edges incident to v since the degree of every vertex in the complete graph with six vertices is 5. Assume without loss of generality that v is adjacent to vertices p, q, and r in G. If there is an edge joining any two of these three vertices, v and these two vertices form a set of three mutual acquaintances. Otherwise, the three individuals p, q, and r form a set of three nonacquaintances.

1.50 The positive integer n has the **(p, q)-Ramsey property** if for every graph G with n vertices, either K_p is a subgraph of G or K_q is a subgraph of the complement of G. Show that the positive integer 6 has the (3, 3)-Ramsey property whereas the positive integer 5 does not.

Solution. It follows from Problem 1.49 that for any graph G with six vertices, there is a triangle either in G or in its complement. So the number 6 has the (3, 3)-Ramsey property. The cyclic graph with five vertices has no triangle and is isomorphic to its complement. So the number 5 does not have the (3, 3)-Ramsey property.

1.51 Show that the following properties are equivalent: (*i*) the positive integer n has the (p, q)-Ramsey property, (*ii*) every simple graph with n vertices has a clique of p vertices or an independent set of q vertices, and (*iii*) the edges of K_n can be colored using two colors such that there will be either a clique K_p in which all edges are of one color or a clique K_q in which all edges are of the other color.

Solution. These statements are reformulations of the definition given in Problem 1.50.

1.52 The smallest integer n that has the (p, q)-Ramsey property is called a **Ramsey number,** denoted $\boldsymbol{R(p, q)}$. Show that (*a*) $R(p, q) = R(q, p)$, (*b*) $R(p, 2) = p$, and (*c*) $R(3, 3) = 6$.

Solution. (*a*) This is an immediate consequence of the definition. (*b*) For any graph G with n vertices, either G or its complement has an edge. (*c*) This follows from Problems 1.49 through 1.51.

1.53 Show that if a bipartite graph $G = (X, Y, E)$ is regular, both X and Y have the same number of elements.

Solution. Suppose there are x vertices in X and y vertices in Y. If the degree of each vertex is r, the total number of edges is rx, which is also equal to ry. Thus $x = y$.

1.54 A **cubic graph** is simple graph in which the degree of each vertex is 3. Construct two nonisomorphic cubic graphs each with six vertices.

Solution. The bipartite graph $K_{3,3}$ is a cubic graph with six vertices. A nonbipartite cubic graph of order 6 is the graph in Fig. 1-10(*c*).

1.55 Find the maximum number of edges in a bipartite graph.

Solution. Suppose in $G = (X, Y, E)$ there are m vertices in X and n vertices in Y. The number of edges in G cannot exceed mn, which is a maximum when $m = n$. So the maximum number of edges is n^2 when G has $2n$ vertices.

1.56 The ***k*-cube** (also known as the **hypercube**) is the graph $\boldsymbol{Q_k}$ whose vertices are the ordered k-tuples of binary numbers, two vertices being joined by an edge if and only if they differ exactly in one component. Show that a k-cube is a k-regular bipartite graph, and find the number of vertices and edges in a k-cube.

Solution. Let X denote the set of k-tuples consisting of the zero k-tuple and all the k-tuples that differ from the zero k-tuple in an even number of components, and let Y be the set of k-tuples that differ from the zero k-tuple in an odd number of components. Then every edge in the cube is between a vertex in X and a vertex in Y. Moreover, both X and Y have the same number of vertices. Since each component of the k-tuple is either 0 or 1, there are 2^k vertices. For any vertex that corresponds to a fixed k-tuple, there are k vertices with k-tuples that differ from the given k-tuple in exactly one component. Thus the k-cube is a k-regular bipartite graph with $(k)(2^k)/2$ edges.

1.57 Find the fewest vertices needed to construct a complete graph with at least 1000 edges.

Solution. If the number of vertices is n, we have the inequality $n(n - 1) \geq 2000$. So $n \geq 46$.

ADJACENCY MATRICES AND INCIDENCE MATRICES

1.58 Show that the vertices of the bipartite graph $G = (X, Y, E)$ with m vertices in X and n vertices in Y can be enumerated so that the adjacency matrix has the form

$$\begin{bmatrix} 0 & A \\ A^T & 0 \end{bmatrix}$$

where A is an $m \times n$ matrix in which each entry is 0 or 1, A^T is the transpose of A, and 0 is a matrix in which each entry is zero.

Solution. Label the vertices in X from 1 to m, and label the vertices in Y from $(m + 1)$ to $(m + n)$. Then the adjacency matrix is of the form given above.

1.59 Prove Theorem 1.4: (*i*) In the adjacency matrix of a graph, the sum of the entries in a row (or a column) corresponding to a vertex is its degree, and the sum of all the entries of the matrix is twice the number of edges in the graph. (*ii*) In the adjacency matrix of a digraph, the sum of the entries in a row corresponding to a vertex is its outdegree, the sum of the entries in a column corresponding to a vertex is its indegree, and the sum of all the entries of the matrix is equal to the number of arcs in the digraph.

Solution. Suppose the vertex set of G is $\{1, 2, \ldots, n\}$. (*i*) In the adjacency matrix, the (i, j) entry is equal to the number of edges joining vertex i and vertex j. When we add all the entries of the ith row (or the ith column), we count the number of edges adjacent to vertex i and add them. So the sum of the ith row (column) is equal to the degree of vertex i. When all the entries in the matrix are added, we obtain the sum of the degrees of all vertices, which is twice the sum of edges by Theorem 1.1. (*ii*) If the (i, j) entry is positive, there is an arc from i to j. So the sum of the entries in the ith row is the outdegree of i, whereas the sum of the entries in the jth column is the indegree of j. When we add all the entries of the matrix, we get the sum of all outdegrees (which is also the sum of all indegrees), and by Theorem 1.3, this sum is equal to the number of arcs.

1.60 Prove Theorem 1.5: (*i*) The sum of the entries in a row of the incidence matrix of a simple graph corresponding to a vertex is its degree, and the sum of all the entries in the matrix is twice the number of edges. (*ii*) The sum of the entries in a row of the incidence matrix of a digraph is its outdegree minus its indegree, and the sum of all the entries in the matrix is zero.

Solution. Let the vertices be $1, 2, \ldots, n$. (*i*) Suppose the edges are e_i $(i = 1, 2, \ldots, m)$, and let the jth column correspond to e_j. The sum of each column is 2, so the sum of all the entries in the matrix is $2m$, which is twice the number of edges. Moreover, if the (i, j) entry is positive, there is an edge joining i and j. So when all the entries in the ith row are added, all the edges adjacent to i are accounted for. Thus the sum of the ith row is the degree of vertex i. (*ii*) Suppose the arcs are a_i $(i = 1, 2, \ldots, m)$. Each column has two nonzero entries ($+1$ and -1), and therefore each column sum is zero. So the sum of all the entries in the matrix is zero. If the (i, j) entry is -1, it indicates that there is an arc from i, and if it is $+1$, it indicates that there is an arc directed to vertex i. So the sum of all the entries in the ith row is equal to its outdegree minus its indegree.

1.61 A **permutation matrix** is a square binary matrix that has exactly one 1 in each row and column. Two matrices A and A' are **isomorphic** if there is a permutation matrix P such that $A'P = PA$. Show that two graphs are isomorphic if and only if their adjacency matrices are isomorphic.

Solution. Suppose A and A' are the adjacency matrices of two isomorphic graphs. Then one of these matrices can be obtained from the other by rearranging rows and then rearranging the corresponding columns. Now rearranging rows of A is equivalent to premultiplying by a permutation matrix P obtaining the product matrix PA. The subsequent rearrangement of corresponding columns is equivalent to postmultiplying PA by P^{-1}. (Recall that any permutation matrix P is nonsingular.) Thus $A' = PAP^{-1}$. Conversely, if $A'P = PA$, A' can be obtained from A by rearranging columns and then rows, showing that the two graphs are isomorphic.

1.62 Find the adjacency matrices A and A' of the two isomorphic graphs given in Fig. 1-18, and obtain a permutation matrix P such that $A'P = PA$.

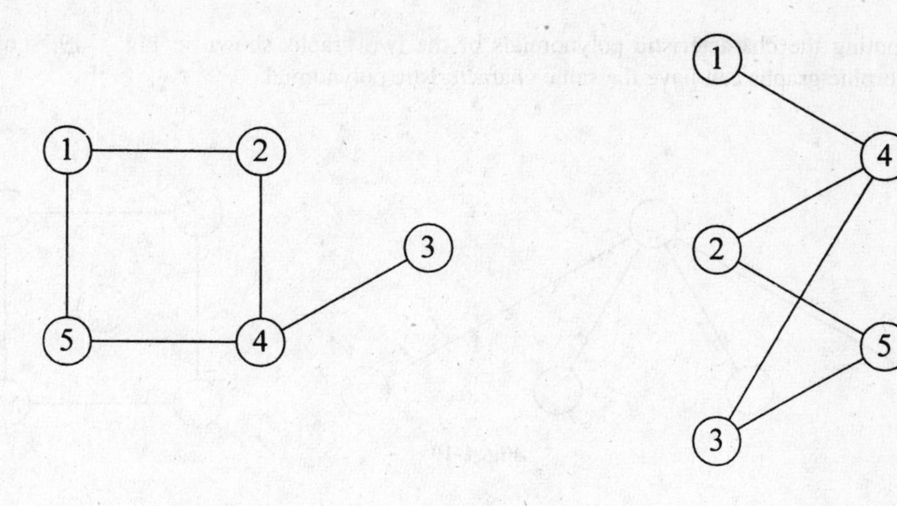

Fig. 1-18

Solution. The adjacency matrices are A and A', where

$$A = \begin{bmatrix} 0 & 1 & 0 & 0 & 1 \\ 1 & 0 & 0 & 1 & 0 \\ 0 & 0 & 0 & 1 & 0 \\ 0 & 1 & 1 & 0 & 1 \\ 1 & 0 & 0 & 1 & 0 \end{bmatrix} \quad \text{and} \quad A' = \begin{bmatrix} 0 & 0 & 0 & 1 & 0 \\ 0 & 0 & 0 & 1 & 1 \\ 0 & 0 & 0 & 1 & 1 \\ 1 & 1 & 1 & 0 & 0 \\ 0 & 1 & 1 & 0 & 0 \end{bmatrix}$$

An isomorphism f from G to G' is $f(1) = 5, f(2) = 2, f(3) = 1, f(4) = 4$, and $f(5) = 3$. To obtain A' from A, we replace the fifth row of A by its first row, the first row of A by its third row, and the third row of A by its fifth row. Suppose the resulting matrix is denoted B. Then replace the fifth column of B by its first column, the first column of B by its third column, and the third column of B by its fifth column. The resulting matrix is A'. To obtain the following permutation matrix P from the 5×5 identity matrix I, we perform the same sequence of operations on I that was performed on A to obtain B:

$$P = \begin{bmatrix} 0 & 0 & 1 & 0 & 0 \\ 0 & 1 & 0 & 0 & 0 \\ 0 & 0 & 0 & 0 & 1 \\ 0 & 0 & 0 & 1 & 0 \\ 1 & 0 & 0 & 0 & 0 \end{bmatrix}$$

It is easy to verify that $PA = B = A'P$. (The permutation matrix P depends on the isomorphism f.)

1.63 The **characteristic polynomial** of a simple graph with n vertices is the determinant of the matrix $(A - \lambda I)$, where A is the adjacency matrix and I is the $n \times n$ identity matrix. Show that if two graphs are isomorphic, their characteristic polynomials are the same. (Note: The determinant of A is written $\det A$.)

Solution. Suppose A and A' are the characteristic polynomials of two isomorphic graphs G and G'. Then $A' = PAP^{-1}$. So $(A' - \lambda I) = (PAP^{-1} - \lambda I) = (PAP^{-1} - \lambda PIP^{-1}) = P(A - \lambda I)P^{-1}$. Thus

$$\begin{aligned} \det(A' - \lambda I) &= \det P(A - \lambda I)P^{-1} \\ &= \det P \det(A - \lambda I) \det P^{-1} \\ &= \det P \det(A - \lambda I)(1/\det P) \\ &= \det(A - \lambda I) \end{aligned}$$

1.64 By computing the characteristic polynomials of the two graphs shown in Fig. 1-19, show that two nonisomorphic graphs can have the same characteristic polynomial.

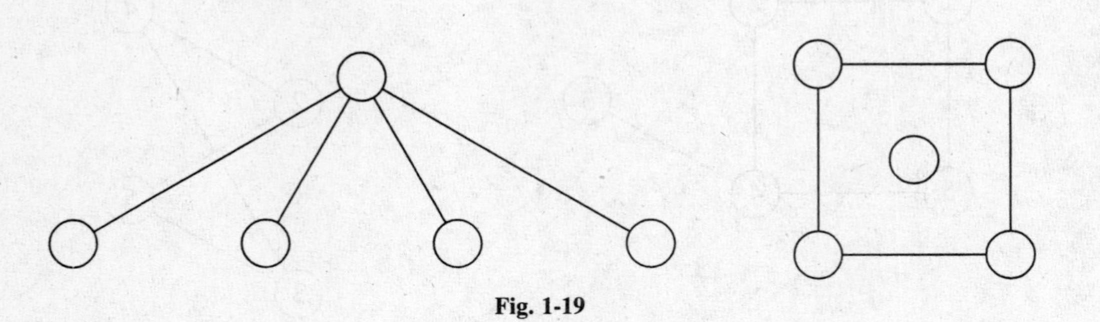

Fig. 1-19

Solution. By expanding the appropriate determinants and simplifying, it is verified that both graphs have the same characteristic polynomial $4\lambda^3 - \lambda^5$. But obviously the two graphs are not isomorphic.

1.65 If $A = [a_{ij}]$ is the adjacency matrix of a simple graph G with n vertices, the **binary code of G with respect to A** is the nonnegative integer $a_{12}2^0 + a_{13}2^1 + \cdots + a_{1n}2^{n-1} + a_{23}2^n + \cdots + a_{2n}2^{2n-3} + \cdots + a_{n-1,n}2^{k-1}$, where $k = n(n-1)/2$. Find the binary code of the adjacency matrix of the graph $G = (V, E)$, where $V = \{1, 2, 3, 4\}$ and $E = \{\{1, 2\}, \{1, 4\}, \{2, 4\}, \{3, 4\}\}$.

Solution. The adjacency matrix is

$$A = \begin{bmatrix} 0 & 1 & 0 & 1 \\ 1 & 0 & 0 & 1 \\ 0 & 0 & 0 & 1 \\ 1 & 1 & 1 & 0 \end{bmatrix}$$

The binary code of A is $1.2^0 + 0.2^1 + 1.2^2 + 0.2^3 + 1.2^4 + 1.2^5 = 53$.

1.66 Show that it is possible to construct a simple graph G if the binary code of one of its adjacency matrices and the order of G are known.

Solution. The given binary code can be uniquely expressed as the sum of terms like $a_i.2^i$, where the coefficient a_i is either 0 or 1. This information gives all the entries above the diagonal of the adjacency matrix provided we know the order of the graph. For example, suppose the binary code is 573, which is equal to $512 + 32 + 16 + 8 + 4 + 1 = 1.2^0 + 0.2^1 + 1.2^2 + 1.2^3 + 1.2^4 + 1.2^5 + 0.2^6 + 0.2^7 + 0.2^8 + 1.2^9$.

(*i*) Suppose A and A' are the adjacency matrices of graphs with 5 and 6 vertices, respectively. Then these two matrices are

$$A = \begin{bmatrix} 0 & 1 & 0 & 1 & 1 \\ 1 & 0 & 1 & 1 & 0 \\ 0 & 1 & 0 & 0 & 0 \\ 1 & 1 & 0 & 0 & 1 \\ 1 & 0 & 0 & 1 & 0 \end{bmatrix} \quad \text{and} \quad A' = \begin{bmatrix} 0 & 1 & 0 & 1 & 1 & 1 \\ 1 & 0 & 1 & 1 & 0 & 0 \\ 0 & 1 & 0 & 1 & 0 & 0 \\ 1 & 0 & 1 & 0 & 0 & 0 \\ 1 & 0 & 0 & 0 & 0 & 0 \\ 1 & 0 & 0 & 0 & 0 & 0 \end{bmatrix}$$

(*ii*) If the order of the graph is 7, it can be easily verified that vertex 3 is an isolated vertex since the third row (and the third column) of the 7×7 adjacency matrix is the zero vector.

1.67 The **minicode of a graph** is its smallest binary code, and the largest binary code is the **maxicode** of the graph. Find the minicode and maxicode of a simple graph with three vertices and two edges.

Solution. The three vertices give rise to 3! adjacency matrices, out of which three are distinct. These matrices give the binary codes 3, 5, and 6. Thus the minicode is 3 and maxicode is 6.

DEGREE VECTORS AND GRAPHICAL VECTORS

1.68 Prove Theorem 1.6: Let $v = [d_1 \quad d_2 \quad d_3 \quad \cdots \quad d_k]$ be a nonincreasing vector of k (where k is at least 2) nonnegative integers such that no component d_i exceeds $(k-1)$. Let v' be the vector obtained from v by deleting d_1 and subtracting 1 from each of the next d_1 components of v. Let v_1 be the nonincreasing vector obtained from v' by rearranging its components if necessary. Then v is a graphical vector if and only if v_1 is a graphical vector.

Solution. (*i*) Suppose v_1 is graphical. So there exists a graph G_1 of order $(k-1)$ for which the degree vector is v_1. Now relabel the vertices of G_1 as $x_2, x_3, \ldots, x_k$ such that the degree of x_i is the ith component of v'. Construct a new vertex x_1 and join x_1 to each of the first d_1 vertices in the ordered set $\{x_2, x_3, \ldots, x_k\}$. The first component of the degree vector of the new graph G thus constructed is d_1, and the next d_1 components are d_j. Thus the first $d_1 + 1$ components of the degree vector of G are the same as the first $d_1 + 1$ components of v. The remaining components of the degree vector are the same as the last $k - (d_1 + 1)$ components of v. Thus the degree vector of G is v, and v is graphical.

(*ii*) Suppose $v = [d_1 \quad d_2 \quad \cdots \quad d_k]$ is a graphical vector. There can be more than one graph with (the ordered) vertex set $V = \{x_1, x_2, \ldots, x_k\}$ such that degree $x_i = d_i$. Choose a graph G with degree vector v such that the sum of the degrees of those vertices adjacent to the first vertex x_1 is as large as possible. Then we prove that x_1 is adjacent to the next d_1 vertices, starting from x_2 in the ordered set V. Suppose this is not the case. So there exist vertices x_j and x_k such that (1) x_j is not adjacent to x_1, (2) x_k is adjacent to x_1, and (3) $d_j > d_k$. Let the sum of the degrees of all vertices adjacent to x_1 in the graph G be $d_k + t$, where $t \geq 0$. Since $d_j > d_k$, there should be a vertex x_i that is adjacent to x_j but not adjacent to x_k. In the graph G, we now delete the edge joining x_1 and x_k and the edge joining x_i and x_j. Then construct an edge joining the nonadjacent vertices x_1 and x_k and another edge joining the nonadjacent vertices x_i and x_j. The degree vector of the newly constructed graph G' is also the same vector v. In G', the sum of the degrees of the vertices adjacent to x_1 is $d_j + t$, which is more than $d_k + t$. This contradicts the assumption that the sum of the degrees of the vertices adjacent to x_1 in G is a maximum. So x_1 is adjacent to the next d_1 vertices in the ordered set V starting from x_2. Thus the vector v_1 constructed from the vector v as in the hypothesis of the theorem is the degree vector of the graph $G - x_1$. So v_1 is also a graphical vector.

1.69 Prove that the algorithm in Section 1.6 determines whether a given vector of nonnegative integers is a graphical vector.

Solution. It is enough if we show that the repetitive process in step 4 of the algorithm eventually results in a zero vector or a vector with at least one negative component. Suppose we start with a vector with n components. Each component is at most $(n-1)$. At the end of the next iteration, we have a vector with $(n-1)$ components. Each component is at most $(n-2)$. At the end of the kth iteration, we have a vector with $(n-k)$ components. Each component is at most $n-k-1$. If step 4 were applied $(n-2)$ times, we have a vector with two components. Each component is at most 1. At this stage we have either the vector $v = [1 \quad 1]$ or $w = [1 \quad 0]$. By iterating once more, we have either the zero vector with one component or the vector with -1 as the only component.

1.70 Test whether $[5 \quad 4 \quad 3 \quad 3 \quad 3 \quad 3 \quad 3 \quad 2]$ is a graphical vector. If it is graphical, draw a simple graph with this vector as the degree vector.

Solution.

Iteration 1: $v = [5 \quad 4 \quad 3 \quad 3 \quad 3 \quad 3 \quad 3 \quad 2]$ and $v_1 = [3 \quad 3 \quad 2 \quad 2 \quad 2 \quad 2 \quad 2]$

Iteration 2: $v = [3 \quad 3 \quad 2 \quad 2 \quad 2 \quad 2 \quad 2]$ and $v_1 = [2 \quad 2 \quad 2 \quad 2 \quad 1 \quad 1]$

Iteration 3: $v = [2 \quad 2 \quad 2 \quad 2 \quad 1 \quad 1]$ and $v_1 = [2 \quad 1 \quad 1 \quad 1 \quad 1]$

Iteration 4: $v = [2 \quad 1 \quad 1 \quad 1 \quad 1]$ and $v_1 = [1 \quad 1 \quad 0 \quad 0]$

Iteration 5: $v = [1 \quad 1 \quad 0 \quad 0]$ and $v_1 = [0 \quad 0 \quad 0]$

So the given vector is graphical. A simple graph with v as degree vector is shown in Fig. 1-20.

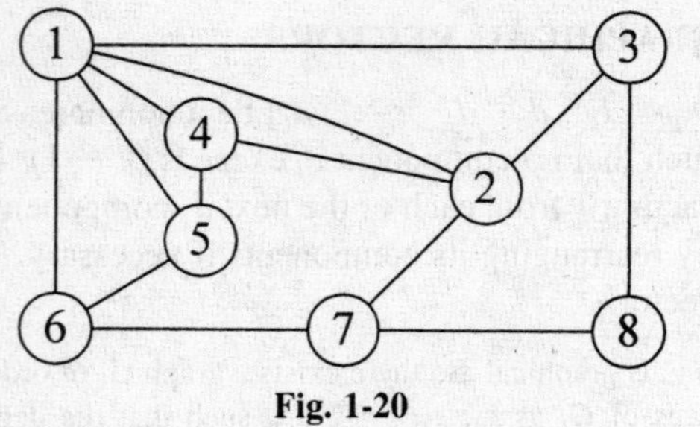

Fig. 1-20

1.71 Test whether [6 6 5 4 3 3 1] is graphical.

Solution.

Iteration 1: $v = [6 \quad 6 \quad 5 \quad 4 \quad 3 \quad 3 \quad 1]$ and $v_1 = [5 \quad 4 \quad 3 \quad 2 \quad 2 \quad 0]$

Iteration 2: $v = [5 \quad 4 \quad 3 \quad 2 \quad 2 \quad 0]$ and $v_1 = [3 \quad 2 \quad 1 \quad 1 \quad -1]$

Since we obtain a vector with a negative component, we conclude that the given vector is not graphical.

1.72 Let $v = [d_1 \quad d_2 \quad \cdots \quad d_n]$ and $w = [w_n \quad w_{n-1} \quad \cdots \quad w_2 \quad w_1]$, where $w_i = n - 1 - d_i$. Show that v is graphical if and only if w is graphical.

Solution. Suppose v is the degree vector of $G = (V, E)$, where $V = \{1, 2, \ldots, n\}$. It is easy to see that w is the degree vector of the complement of G. Thus v is graphical if and only if w is graphical.

1.73 Show that there is no simple graph with six vertices of which the degrees of five vertices are 5, 5, 3, 2, and 1.

Solution. Suppose there is a simple graph, and let x be the degree of the sixth vertex. The sum of the 6 degrees has to be even, and x cannot exceed 5. So $x = 0$ or 2 or 4.

$x = 0$ implies $v = [5 \quad 5 \quad 3 \quad 2 \quad 1 \quad 0]$. At the end of the first iteration we get $[4 \quad 2 \quad 1 \quad 0 \quad -1]$. So x is not 0.

$x = 2$ implies $v = [5 \quad 5 \quad 3 \quad 2 \quad 2 \quad 1]$. At the end of the second iteration we get $[1 \quad 0 \quad 0 \quad -1]$. So x is not 2.

$x = 4$ implies $v = [5 \quad 5 \quad 4 \quad 3 \quad 2 \quad 1]$. At the end of the second iteration we get $[2 \quad 1 \quad 0 \quad -1]$. So x is not 4.

1.74 Find x if $[8 \quad x \quad 7 \quad 6 \quad 6 \quad 5 \quad 4 \quad 3 \quad 3 \quad 1 \quad 1 \quad 1]$ is a graphical vector.

Solution. Obviously x is either 8 or 7.

$x = 8$ eventually leads to $[-1 \quad 0 \quad 0 \quad 0 \quad 0]$. So x is not 8.

$x = 7$ eventually leads us to $v = [0 \quad 0 \quad 0 \quad 0 \quad 0]$. So $x = 7$.

1.75 Show that a finite nonincreasing vector in which no two components are equal cannot be a graphical vector.

Solution. Consider the nonincreasing vector v with k components in which each component is a nonnegative integer. If no two components are equal, $v = [k-1 \quad k-2 \quad \cdots \quad 2 \quad 1 \quad 0]$. If we delete the first component and subtract 1 from each of the remaining components, we get a vector v_1 in which the last component is negative. So v_1 is not graphical, and therefore v is not graphical.

1.76 Show that in a simple graph, there are at least two vertices with equal degrees.

Solution. Suppose no two degrees in a simple graph G are equal. Then no two components of the degree vector of G are equal. This is a contradiction, as established in Problem 1.76.

1.77 Show that there exists a simple graph with 12 vertices and 28 edges such that the degree of each vertex is either 3 or 5. Draw this graph.

Solution. Suppose there are p vertices of degree 3. Then the equation $3p + 5(12 - p) = (2)(28)$ gives the unique solution $p = 2$. So if there exists a graph with two vertices of degree 3 and 10 vertices of degree 5, its degree vector is $v = [5 \quad 5 \quad 5 \quad 5 \quad 5 \quad 5 \quad 5 \quad 5 \quad 5 \quad 5 \quad 3 \quad 3]$. To show that there exists a graph with the desired property, it is enough if we demonstrate that v is a graphical vector. After nine iterations we get the 0 vector:

$$v_1 = [5 \quad 5 \quad 5 \quad 5 \quad 4 \quad 4 \quad 4 \quad 4 \quad 4 \quad 3 \quad 3]$$

$$v_2 = [4 \quad 4 \quad 4 \quad 4 \quad 4 \quad 4 \quad 3 \quad 3 \quad 3 \quad 3]$$

$$v_3 = [4 \quad 3 \quad 3 \quad 3 \quad 3 \quad 3 \quad 3 \quad 3 \quad 3]$$

$$v_4 = [3 \quad 3 \quad 3 \quad 3 \quad 2 \quad 2 \quad 2 \quad 2]$$

$$v_5 = [2 \quad 2 \quad 2 \quad 2 \quad 2 \quad 2 \quad 2]$$

$$v_6 = [2 \quad 2 \quad 2 \quad 2 \quad 1 \quad 1]$$

$$v_7 = [2 \quad 1 \quad 1 \quad 1 \quad 1]$$

$$v_8 = [1 \quad 1 \quad 0 \quad 0]$$

$$v_9 = [0 \quad 0 \quad 0]$$

So v is indeed a graphical vector. A simple graph with v as the degree vector is shown in Fig. 1-21.

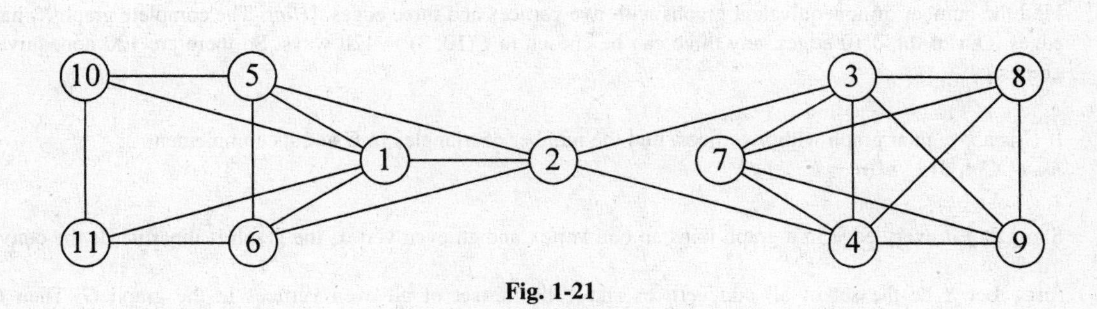

Fig. 1-21

1.78 Show that there exists a simple graph with seven vertices and 12 edges such that the degree of each vertex is 2 or 3 or 4.

Solution. Suppose there are p vectors of degree 2 and q vectors of degree 3. Then the only solution in positive integers of the equation $2p + 3q + 4(7 - p - q) = 24$ is $p = 1$ and $q = 2$. Thus if there is a graph with the desired property, it should have one vertex of degree 2, two vertices of degree 3, and four vertices of degree 4, giving a unique degree vector $v = [4 \quad 4 \quad 4 \quad 4 \quad 3 \quad 3 \quad 2]$. It is easily verified that this is indeed a graphical vector.

Supplementary Problems

1.79 Find the complements of (*a*) K_n and (*b*) $K_{m,n}$.
Ans. (*a*) The graph with n vertices and no edges. (*b*) The graph with $(m + n)$ vertices consisting of two parts (components): a complete graph with m vertices and a complete graph with n vertices.

1.80 Find the number of nonisomorphic graphs with four vertices and at most 3 edges.
Ans. $N(4, 0) + N(4, 1) + N(4, 2) + N(4, 3) = 1 + 1 + 2 + 3 = 7$

1.81 Find the vertex covering number and the independence number of K_n and $K_{m,n}$.
Ans. $\alpha(K_n) = 1$, $\alpha(K_{m,n}) = \max\{m, n\}$, $\beta(K_n) = n - 1$ and $\beta(K_{m,n}) = (m + n) - \max\{m, n\}$

1.82 Find the vertex dominating number of K_n and $K_{m,n}$. *Ans.* $\sigma(K_n) = 1$ and $\sigma(K_{m,n}) = \min\{m, n\}$

1.83 Find the committee number of (*a*) K_n and (*b*) $K_{m,n}$. *Ans.* (*a*) 1; (*b*) minimum of $\{m, n\}$.

1.84 If I is an independent set in a graph, find the subgraph induced by I.
Ans. The graph with I as vertex set and no edges.

1.85 Find the maximum number of edges in (*a*) a simple graph with n vertices and (*b*) a bipartite graph (X, Y, E), where the cardinalities of X and Y are m and n, respectively. [*Hint:* (*a*) Since K_n has exactly $n(n - 1)/2$ edges, the number of edges in a simple graph with n vertices cannot exceed $n(n - 1)/2$.] (*b*) Since $K_{m,n}$ has exactly mn edges, the number of edges in the bipartite graph cannot exceed mn.

1.86 It is known that there exists a simple graph with 12 vertices and 28 edges in which the degree of each vertex is either 3 or 5. Find the number of vertices of degree 3.
Ans. There are two vertices of degree 3 and 10 vertices of degree 5.

1.87 Find the number of nonequivalent graphs with four vertices and three edges. [*Hint:* The complete graph K_4 has six edges, out of which any three can be chosen in $C(6, 3) = 20$ ways. Thus there are 20 nonequivalent graphs with four vertices and three edges, out of which there are exactly three nonisomorphic graphs, as established earlier.]

1.88 Find the number of nonequivalent graphs with five vertices and three edges. [*Hint:* The complete graph K_5 has 10 edges. Out of these 10 edges, any three can be chosen in $C(10, 3) = 120$ ways. So there are 120 nonequivalent graphs.]

1.89 If G is a k-regular graph with n vertices, find the number of triangles in G and its complement.
Ans. $C(n, 3) - nk(n - k - 1)/2$

1.90 Show that if every edge in a graph joins an odd vertex and an even vertex, the graph is bipartite. Is the converse true?
Ans. Let X be the set of all odd vertices and Y be the set of all even vertices in the graph G. Then $G = (X, Y, E)$ is a bipartite graph where E is the set of edges in G. The converse is not true since it is possible to have a bipartite graph in which there is a left vertex (in X) that is odd (or even) and a right vertex (in Y) that is even (or odd).

1.91 Any root of the characteristic polynomial of a graph is an **eigenvalue** of the graph. The **spectrum** of a graph is the collection of all its eigenvalues. Find the spectrum of (*a*) K_3, (*b*) K_4, and (*c*) $K_{2,2}$.
Ans. Since the adjacency matrix is symmetric, every eigenvalue is a real number. (*a*) -1, -1, and 2; (*b*) -1, -1, -1, and 3; (*c*) 0, 0, 2, -2

1.92 Find the spectrum of K_n. *Ans.* The number -1 repeated $(n - 1)$ times and the number $(n - 1)$

1.93 Find the minicode and the maxicode of the simple graph with four vertices such that the degrees of the vertices are 1, 2, 2, and 3.

Ans. The minicode is 15, and the maxicode is 60.

1.94 Find the minicode and maxicode of K_n.

Ans. Since every nondiagonal element of any adjacency matrix is 1, both the minicode and the maxicode are equal to $1 + 2 + 2^2 + \cdots + 2^{k-1}$, where $k = n(n - 1)/2$.

Chapter 2

Connectivity

2.1 PATHS, CIRCUITS, AND CYCLES

Let v and w be two vertices in a graph. A **walk between v and w** in the graph is a finite alternating sequence $v = v_0, e_1, v_1, e_2, v_2, e_3, \ldots, e_n, v_n = w$ of vertices and edges of the graph such that each edge e_i in the sequence joins vertex v_{i-1} and vertex v_i. The vertices and edges in a walk need not be distinct. Two walks $v_0, e_1, v_1, \ldots, e_n, v_n$ and $u_0, f_1, u_1, \ldots, f_m, u_m$ are **equal** if $n = m$, $v_i = u_i$, and $e_i = f_i$ for $0 \le i \le n$. Two walks are said to be **different** if they are not equal. The number of edges in a walk is the **length** of the walk. If the graph is simple, the edges in the sequence defining a walk between v and w need not be explicitly listed; the walk $v = v_0, e_1, v_1, e_2, v_2, e_3, \ldots, e_n, v_n = w$, can be expressed as $v_0 — v_1 — v_2 — \cdots — v_n$ unambiguously. A walk in which no edge is repeated is a **trail.** The walk $v = v_0, e_1, v_1, e_2, v_2, e_3, \ldots, v_{n-1}, e_n, v_n = w$, in which the vertices v_i $(0 < i < n)$ are all distinct is a **path between v and w;** the $(n - 1)$ vertices $v_i(0 < i < n)$ are called the **intermediate vertices** of the path. Obviously, every path is a trail.

If v and w are vertices in a directed graph, a **directed walk from v to w** is a finite sequence $v = v_0, a_1, v_1, a_2, v_2, a_3, \ldots, a_n, v_n = w$, of vertices and arcs of the digraph such that each arc a_i in the sequence is an arc from v_{i-1} to v_i. This is written $v_0 \to v_1 \to v_2 \to \cdots \to v_n$. A directed walk is a **directed trail** if the arcs are distinct. A **directed path from v to w** in a digraph is a directed walk from v to w in which no vertices repeat.

Example 1. In graph G in Fig. 2-1(a), the sequence 2, e_4, 1, e_1, 4, e_2, 1, e_1, 4 is a walk between vertex 2 and vertex 4. The sequence 2, e_3, 1, e_1, 4, e_2, 1, e_4, 2, e_5, 3 is a trail between 2 and 3. The sequence 2, e_5, 3, e_6, 4, e_1, 1 is a path between 2 and 1. In the directed graph shown in Fig. 2-1(b), $2 \to 3 \to 4 \to 1 \to 2 \to 3 \to 6$ is a directed walk from 1 to 6, $1 \to 2 \to 4 \to 1 \to 3$ is a directed trail from 1 to 3, and $1 \to 2 \to 5 \to 7$ is a directed path from 1 to 7.

Theorem 2.1. Every walk in a graph between v and w contains a path between v and w, and every directed walk from v to w in a digraph contains a directed path from v to w. (See Solved Problem 2.1.)

Theorem 2.2. If A is the adjacency matrix of a simple graph $G = (V, E)$, where $V = \{1, 2, \ldots, n\}$, the $(i - j)$ entry in the kth power of A is the number of *different* walks of length k between the vertices i and j. In particular, the $(i - i)$ diagonal entry in A^2 is the degree of vertex i for each i. (See Solved Problem 2.2.)

Example 2. The adjacency matrix A of the simple graph shown in Fig. 2-2, the matrix A^2, and the matrix A^4 are

$$A = \begin{bmatrix} 0 & 1 & 0 & 1 & 0 \\ 1 & 0 & 1 & 0 & 1 \\ 0 & 1 & 0 & 1 & 1 \\ 1 & 0 & 1 & 0 & 0 \\ 0 & 1 & 1 & 0 & 0 \end{bmatrix}, \quad A^2 = \begin{bmatrix} 2 & 0 & 2 & 0 & 1 \\ 0 & 3 & 1 & 2 & 1 \\ 2 & 1 & 3 & 0 & 1 \\ 0 & 2 & 0 & 2 & 1 \\ 1 & 1 & 1 & 1 & 2 \end{bmatrix}, \quad \text{and} \quad A^4 = \begin{bmatrix} 9 & 3 & 11 & 1 & 6 \\ 3 & 15 & 7 & 11 & 8 \\ 11 & 6 & 15 & 3 & 8 \\ 1 & 11 & 3 & 9 & 6 \\ 6 & 8 & 8 & 6 & 8 \end{bmatrix}$$

The degrees of the vertices 1, 2, 3, 4, 5 are 2, 3, 3, 2, 2, respectively, in agreement with the diagonal entries of A^2. The (1, 5) entry in A^4 is 6, indicating that there are six *different* walks of length 4 between 1 and 5. These walks are $1 — 4 — 1 — 2 — 5$, $1 — 2 — 1 — 2 — 5$, $1 — 4 — 3 — 2 — 5$, $1 — 2 — 5 — 2 — 5$, $1 — 2 — 3 — 2 — 5$, and $1 — 2 — 5 — 3 — 5$.

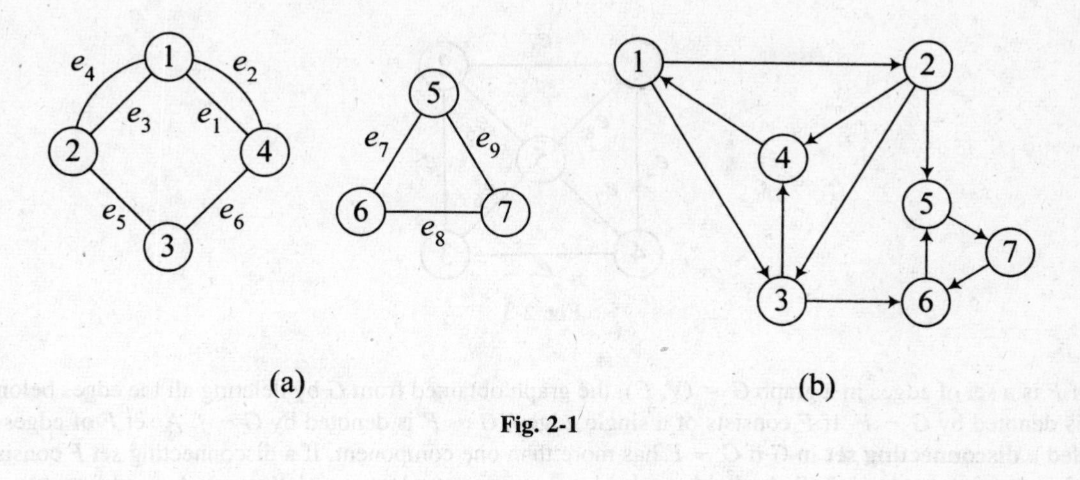

Fig. 2-1

A **closed walk** in a graph is a walk between a vertex and itself. A closed walk in which no edges repeat is a **circuit.** A **cycle** is a circuit with no repeated vertices. Notice that the closed walk v, e_1, w, e_2, v is a cycle, but the closed walk v, e_1, w, e_1, v with no repeated intermediate vertices is not a cycle since it is not a circuit. The subgraph C of a simple graph G is a **cycle in** G if and only if C is a cyclic graph. In a simple graph G, any cycle consisting of k vertices (that is, passing through k vertices) is a k**-cycle** in G; it is an **odd cycle** if k is odd and an **even cycle** if k is even. The terms **directed circuits** and **directed cycles** in the case of digraphs are defined analogously.

Example 3. In the simple graph G shown in Fig. 2-3, the closed walk 1, e_1, 2, e_5, 5, e_6, 3, e_3, 4, e_7, 5, e_8, 1 is a circuit, and the closed walk 1 — 2 — 3 — 4 — 1 is an even cycle.

One way to ascertain whether or not a given graph is bipartite is by identifying its cycles. If there is an odd cycle in a graph G, G is not bipartite. In this context, we have the following theorem, which characterizes bipartite graphs.

Theorem 2.3. A simple graph with three or more vertices is bipartite if and only if it has no odd cycles (See Solved Problem 2.10.)

2.2. CONNECTED GRAPHS AND DIGRAPHS

A pair of vertices in a graph is a **connected pair** if there is a path between them. A graph G is a **connected graph** if every pair of vertices in G is a connected pair; otherwise, it is a **disconnected** graph. A connected subgraph H of a graph G is a **component** of G if $H = H'$ whenever H' is a connected subgraph (of G) that contains H. In other words, a component of a graph is a maximal connected subgraph. A graph is connected if and only if the number of its components is one.

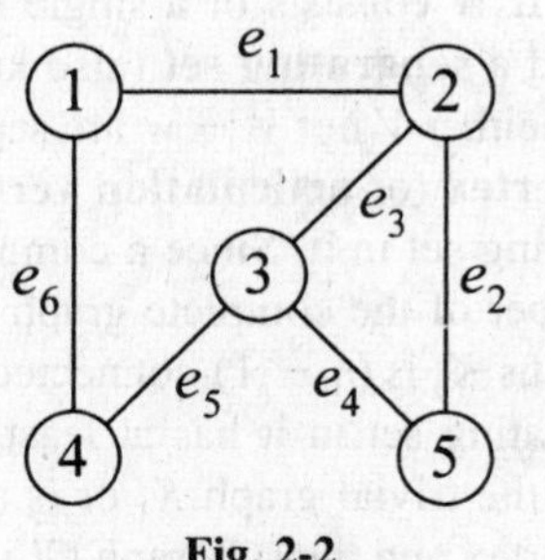

Fig. 2-2

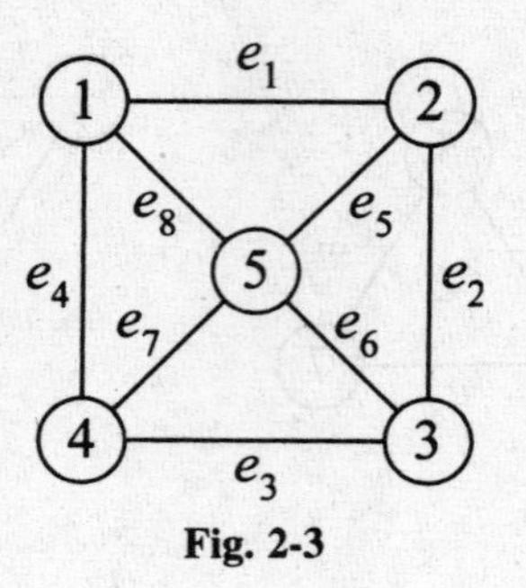

Fig. 2-3

If F is a set of edges in a graph $G = (V, E)$, the graph obtained from G by deleting all the edges belonging to F is denoted by $G - F$. If F consists of a single edge f, $G - F$ is denoted by $G - f$. A set F of edges in G is called a **disconnecting set** in G if $G - F$ has more than one component. If a disconnecting set F consists of a single edge f, that edge is called a **bridge** (also known as a **cut edge** or an **isthmus**). A graph is said to be ***k* edge connected** if every disconnecting set in it has at least k edges. The **edge-connectivity number $\lambda(G)$ of a graph G** is the minimum size of a disconnecting set in G; *by definition*, it is zero when G is the trivial graph K_1. Thus $\lambda(G)$ is zero if and only if G is a disconnected graph or the trivial graph, and it is k edge connected if and only if $\lambda(G)$ is at least k. A disconnecting set F is said to be a **cut set** (also known as a **bond**) if no proper subset of F is a disconnecting set.

Example 4. The graph G of Fig. 2-4 with 13 vertices is not connected since there are several pairs of vertices that are not connected. For example, vertex 5 and vertex 12 do not form a connected pair. The components of G are G^1, G^2, and G^3, as shown in the figure. The set $\{\{1, 6\}, \{2, 6\}\}$ is a disconnecting set but not a cut set. The set $\{\{10, 12\}, \{10, 13\}\}$ is a cut set. The edge $\{2, 6\}$ is a bridge. Since G is disconnected, its edge-connectivity number is zero. The edge-connectivity number of the component G^2 is 2, and it is thus two edge connected.

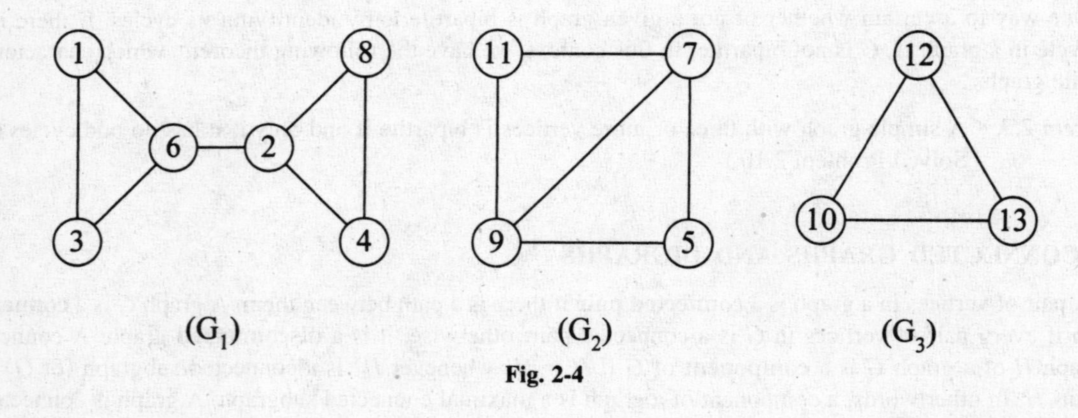

Fig. 2-4

We now define analogous concepts regarding the deletion of vertices. If W is a set of vertices in $G = (V, E)$, the graph obtained from G by deleting all the vertices belonging to W as well as the edges incident to the vertices in W is denoted by $G - W$. If W consists of a single vertex w, the graph $G - W$ is denoted by $G - w$. A set W of vertices in G is called a **separating set** (also known as a **vertex cut**) in G if $G - W$ has more than one component. Observe that neither V nor $V - w$ are separating sets. If a separating set consists of a single vertex w, w is known as a **cut vertex** (or **articulation vertex**). The **connectivity number $\kappa(G)$** of a graph G is the minimum size of a separating set in it. Since a complete graph has no separating set, we adopt the convention that the connectivity number of the complete graph of order n is $(n - 1)$ for all n. A graph G is said to be ***k*-connected** if $\kappa(G) \geq k$. Thus K_n is $(n - 1)$-connected for all n, and a graph that is not complete is k-connected if and only if every separating set in it has at least k vertices. The connectivity number of a graph G is zero if and only if G is either the trivial graph K_1 or is a disconnected graph. A cyclic graph is 2-connected. In Fig. 2-4, vertex 6 is a cut vertex and the subgraph G^1 is 1-connected, whereas both the subgraphs G^2 and G^3 are 2-connected.

Theorem 2.4 **(Whitney's Inequality).** For any graph G, $\kappa(G) \leq \lambda(G) \leq \delta(G)$. (See Solved Problem 2.11.) (Edge connectivity and connectivity are discussed in more detail in Chapter 6 in the context of Ford–Fulkerson theorem and Menger's theorem.)

The number of edges in a path with as few edges as possible between two vertices v and w of a connected graph G is denoted by $d(v, w)$. The **eccentricity** $e(v)$ of the vertex v is the maximum value of $d(v, w)$, where w varies through all the vertices of G. The **radius** $r(G)$ of G is the eccentricity of the vertex of minimum eccentricity. A vertex v is a **central vertex** if its eccentricity is equal to the radius of the graph. The **center** $C(G)$ **of a graph** is the set of all its central vertices.

Example 5. In Fig. 2-5(a), the eccentricities of the vertices A, B, C, D, E, F, and G are 4, 3, 2, 4, 3, 2, and 3, respectively. The central vertices are C and F. The center is the set $\{C, F\}$. In Fig. 2-5(b), the eccentricity of vertex E is 1 and the eccentricities of the other vertices are 2. E is the only central vertex, and the center is the singleton set $\{E\}$.

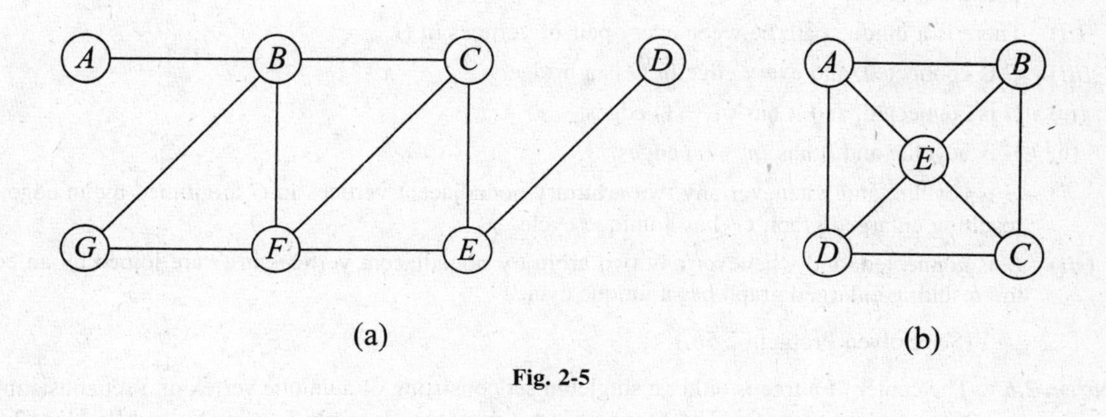

Fig. 2-5

In the case of digraphs, the concept of connectivity becomes more variegated. A digraph G is **strongly connected** if there is a directed path from each vertex to every other vertex. A **strong component of a digraph** is a maximal strongly connected subgraph. A digraph is **unilaterally connected** if for every pair of vertices v and w in G, there is either a path from v to w or from w to v. A digraph is **weakly connected** if its underlying graph is connected.

Example 6. The digraph G' of Fig. 2-1(b) is not strongly connected since there is no directed path from vertex 5 to vertex 3. It is weakly connected because the underlying graph is connected. It is easy to verify that this digraph is unilaterally connected.

A **mixed graph** is a structure $G = (V, E)$, where V is a finite set of vertices and E is a finite set of pairs of vertices in which some pairs are ordered (defining arcs of G) and some pairs are unordered (defining edges of G). The undirected graph $U(G)$ obtained by converting each arc of a mixed graph G into an edge is the underlying graph of G, and G is connected (by definition) if $U(G)$ is connected. The digraph $D(G)$ obtained from G after replacing each edge between v and w by two arcs (v, w) and (w, v) is called the directed graph of G, and G is strongly connected (by definition) if $D(G)$ is strongly connected. An undirected edge e in a connected mixed graph G is a bridge if the deletion of e makes G disconnected.

2.3 TREES AND SPANNING TREES

An **acyclic graph** (also known as a **forest**) is a graph with no cycles. A **tree** is a connected acyclic graph. Thus each component of a forest is a tree, and any tree is a connected forest.

Example 7. The graph in Fig. 2-6 with 13 vertices is a forest consisting of three trees.

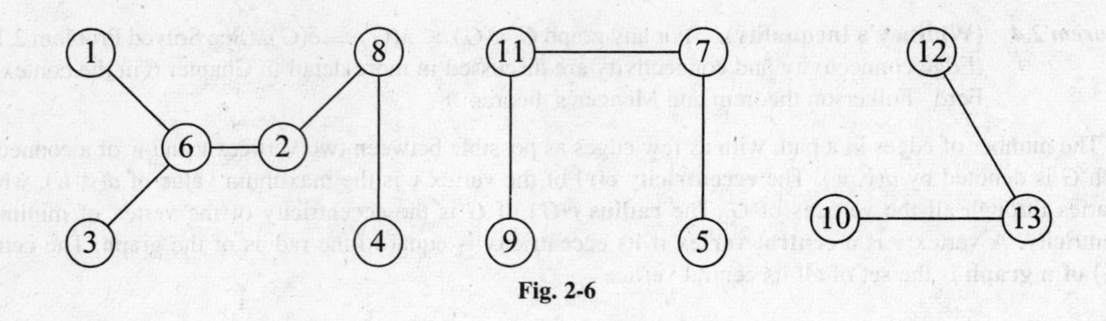

Fig. 2-6

Theorem 2.5. The following are equivalent in a graph G with n vertices.

(*i*) G is a tree.

(*ii*) There is a unique path between every pair of vertices in G.

(*iii*) G is connected, and every edge in G is a bridge.

(*iv*) G is connected, and it has $(n - 1)$ edges.

(*v*) G is acyclic, and it has $(n - 1)$ edges.

(*vi*) G is acyclic, and whenever any two arbitrary nonadjacent vertices in G are joined by an edge, the resulting enlarged graph G' has a unique cycle.

(*vii*) G is connected, and whenever any two arbitrary nonadjacent vertices in G are joined by an edge, the resulting enlarged graph has a unique cycle.

(See Solved Problem 2.56.)

Theorem 2.6. The center of a tree is either a singleton set consisting of a unique vertex or a set consisting of two adjacent vertices. (The converse is not true. See Example 5.) (See Solved Problem 2.62.)

Example 8. In the tree shown in Fig. 2-7(*a*), the center is the set $\{1, 2\}$. In the tree shown in Fig. 2-7(*b*), the center is the singleton set $\{2\}$.

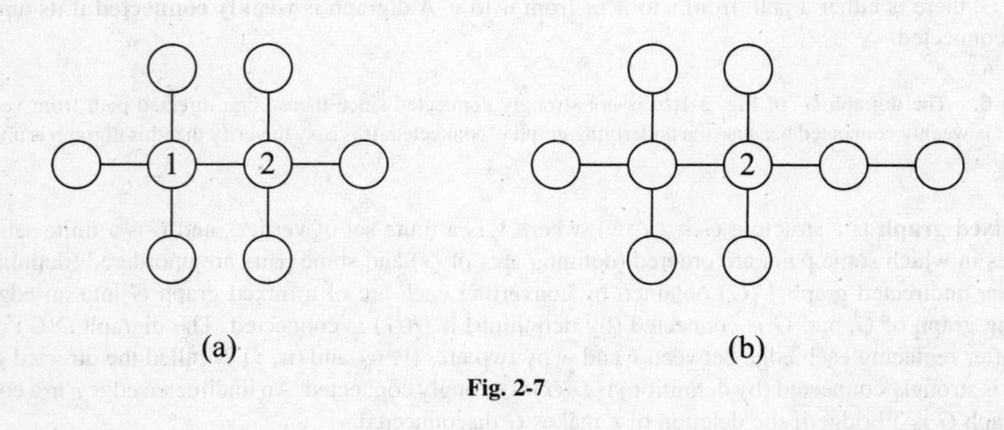

Fig. 2-7

An acyclic spanning subgraph of a graph G is a **spanning forest** in G. A spanning forest in the graph of Fig. 2-4 is the forest shown in Fig. 2-6. An acyclic connected spanning subgraph (if it exists) of G is called a **spanning tree** in G. In the connected graph G of Fig. 2-3, the set $\{e_1, e_2, e_6, e_7\}$ of edges constitutes the edges of a spanning tree in G.

Theorem 2.7. A graph is connected if and only if it has a spanning tree. (See Solved Problem 2.57.)

Theorem 2.8. Let G be a simple graph with n vertices. If a spanning subgraph H satisfies any two of the following three properties, it will satisfy the third property also. (i) H is connected. (ii) H has $(n - 1)$ edges. (iii) H is acyclic. (See Solved Problem 2.58.)

A tree with n vertices is a **labeled tree** if each vertex is assigned a unique positive integer between 1 and n such that no two vertices get the same label. Two labeled trees $T = (V, E)$ and $T' = (V, E')$ are **distinct** if E and E' are not the same set. For example, the labeled tree $T = (V, E)$ with edge set $\{\{1, 2\}$ and $\{2, 3\}\}$ and the labeled tree $T' = (V, E')$ with edge set $\{\{1, 3\}, \{2, 3\}\}$ are distinct even though they are isomorphic. Two nonisomorphic labeled trees are, of course, distinct labeled trees.

Theorem 2.9 **(Cayley's Theorem).** The number of distinct labeled trees with n vertices is n^{n-2}, which is also equal to the number of spanning trees in K_n. (See Solved Problem 2.73.)

Given a graph with n vertices, it is natural to ask whether it is connected. One method to test the connectivity of a graph is by using a recursive procedure known as the **depth first search (DFS) technique** in which we relabel the vertices as follows.

Suppose the vertices of G are $v_1, v_2, \ldots, v_n$. Start the search from any vertex, and relabel that vertex 1. If this vertex has no adjacent vertices, the graph is disconnected. Otherwise, select any vertex adjacent to vertex 1 and relabel it vertex 2, marking the edge $\{1, 2\}$ joining vertex 1 and vertex 2 as a used edge. If vertex 2 has an adjacent vertex other than vertex 1, relabel that vertex 3. If vertex 2 has no other adjacent vertex, revert back to vertex 1 and see whether vertex 1 has an adjacent vertex other than vertex 2. If the answer is no and if $n > 2$, the graph is disconnected. If the answer is yes, relabel the newly located vertex as vertex 3. In either case, mark the edge $\{2, 3\}$ as a used edge. After relabeling a vertex as vertex i, select an arbitrary vertex that is not yet relabeled and that is adjacent to i, relabel that vertex $(i + 1)$. Mark the edge joining i and $(i + 1)$ as a used edge. If a newly relabeled vertex v has no adjacent vertex, go back to vertex w, which is adjacent to v with used edge $\{v, w\}$, and continue the search from w. The procedure continues until all the n vertices are relabeled $1, 2, \ldots, n$, indicating that the graph is connected; or, we are back at vertex 1 with the number of relabeled vertices less than n, showing that the graph is not connected.

If we find that a graph G of order n is a connected graph by using the depth first search, the set of $(n - 1)$ used edges in G constitutes the edges of a spanning tree (in G) known as **DFS spanning tree.**

Example 9. The DFS technique described above is used to test the connectivity of the simple graph with vertex set v_i $(i = 1, 2, \ldots, 8)$, the adjacency matrix of which is

$$\begin{bmatrix} 0 & 1 & 0 & 0 & 0 & 0 & 0 & 1 \\ 1 & 0 & 1 & 0 & 0 & 1 & 1 & 0 \\ 0 & 1 & 0 & 1 & 1 & 1 & 0 & 0 \\ 0 & 0 & 1 & 0 & 0 & 0 & 0 & 0 \\ 0 & 0 & 1 & 0 & 0 & 0 & 0 & 0 \\ 0 & 1 & 1 & 0 & 0 & 0 & 1 & 0 \\ 0 & 1 & 0 & 0 & 0 & 1 & 0 & 1 \\ 1 & 0 & 0 & 0 & 0 & 0 & 1 & 0 \end{bmatrix}$$

Take v_2 and relabel it vertex 1. A vertex adjacent to v_2 is v_7, as can be seen from the adjacency matrix. So v_7 is labeled vertex 2, and edge e_1 joining v_2 and v_7 is marked as a used edge. Then v_6 is labeled vertex 3, and edge e_2 joining v_7 and v_6 is marked as a used edge. At the next stage, v_3 is labeled vertex 4 with used edge $e_3 = \{v_6, v_3\}$, and v_4 is labeled vertex 5 with used edge $e_4 = \{v_3, v_4\}$. Now vertex v_4 (labeled 5) has no adjacent vertex other than v_3 (which is labeled 4). So we revert to vertex 4 and start the search from there. A vertex adjacent to v_3 is v_5, which now gets the label 6, and $e_5 = \{v_3, v_5\}$ becomes a used edge. The search now reverts to vertex v_7 (labeled 2), and v_1 is an adjacent unlabeled vertex that now becomes vertex 7 with $e_6 = \{v_7, v_1\}$ as a used edge. Finally, label vertex v_8 as vertex 8 with used edge $e_7 = \{v_8, v_1\}$. Since all the eight vertices are relabeled, we conclude that G is connected. The seven used edges obtained in this search constitute the edges of a spanning tree, as shown in Fig. 2-8.

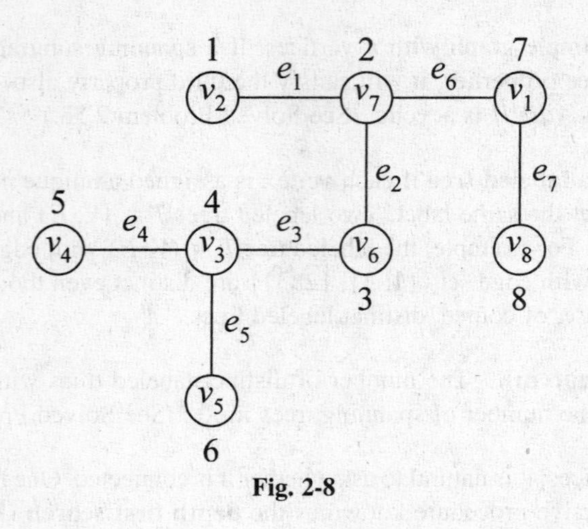

Fig. 2-8

2.4 STRONG ORIENTATIONS OF GRAPHS

If each edge of a simple graph G is replaced by an arc and if the resulting digraph G' is strongly connected, the digraph G' is called a **strong orientation** of G. A simple graph is **strongly orientable** if it has a strong orientation. If each street in a certain region of a city is to be converted into a one-way street, after the conversion one should be able to drive from any street corner to any other street corner in the region. So a conversion of this type is possible only if the streets in the region constitute a strongly orientable graph with street corners as vertices. For a graph to be strongly orientable, it has to be a connected graph in the first place. Furthermore, it should not become disconnected if an edge is deleted from the graph. In other words, a strongly orientable graph is necessarily a connected graph in which no edge is a bridge. The converse of this assertion is also true. Thus we have the following theorem.

Theorem 2.10 **(Robbins's Theorem).** A graph is strongly orientable if and only if it is connected and has no bridges. (See Solved Problem 2.87.)

Example 10. The edge joining vertex 3 and vertex 4 in the connected graph of Fig. 2-9 is a bridge. This graph has no strong orientation.

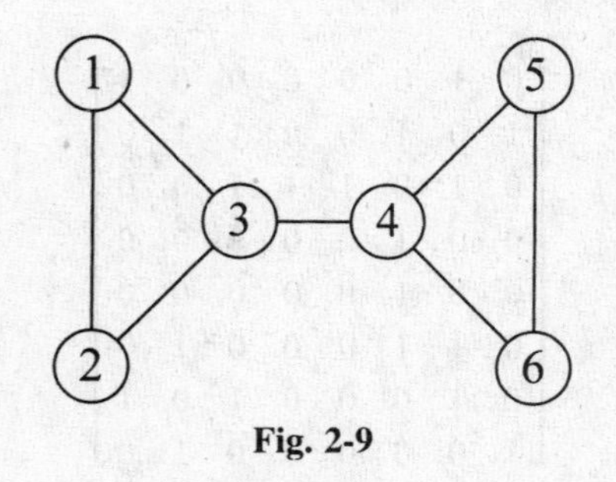

Fig. 2-9

Given a strongly orientable graph with n vertices and m edges, one algorithm mimics the depth first search, enabling us to orient the edges such that the resulting digraph is strongly connected. The procedure is as follows. Suppose the n vertices are relabeled $1, 2, \ldots, n$ with $(n - 1)$ of the m edges specially designated as used edges. A used edge between i and j is converted into an arc from i to j if $i < j$, and an unused edge between i and j is converted into an arc from i to j if $i > j$.

Theorem 2.11 **(Roberts's Theorem).** The orientation procedure using the depth first search in a connected graph with no bridges yields a strongly connected digraph. (See Solved Problem 2.93.)

Example 11. The graph G in Fig. 2-10(a) is connected in which no edge is a bridge. A spanning tree in G obtained by the depth first search is shown in Fig. 2-10(b), with the eight vertices labeled 1, 2, . . . , 8, respectively, and with seven used edges. After orienting the edges according to the rule stipulated in Theorem 2.11, the strongly connected digraph is as shown in Fig. 2-10(c).

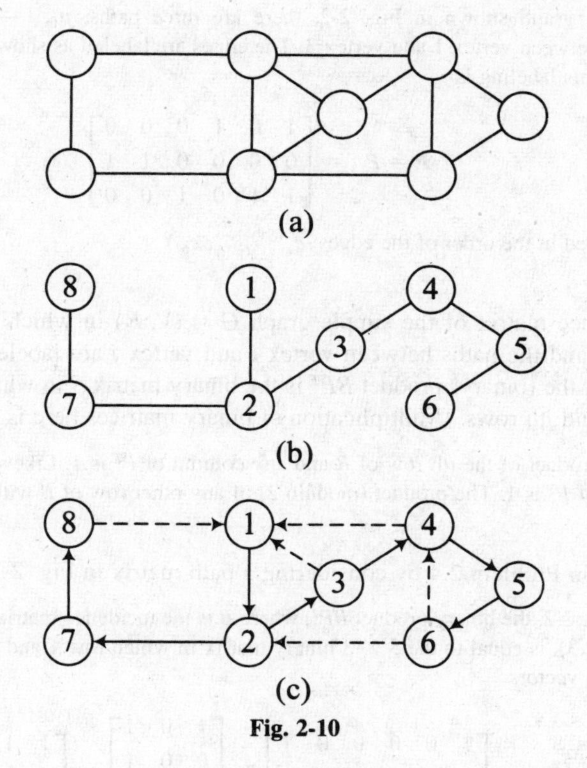

Fig. 2-10

Solved Problems

PATHS, CIRCUITS AND CYCLES

2.1 Prove Theorem 2.1: Every walk in a graph between v and w contains a path between v and w, and every directed walk from v to w in a digraph contains a directed path from v to w.

Solution. Let W be a walk between v and w. If $v = w$, there is the trivial path with no edges. Therefore, assume that v and w are not the same vertex. Suppose W is the walk $v = v_0 — v_1 — \cdots — v_n = w$. It is possible that the same vertex has more than one label in this sequence. If no vertex of the graph appears more than once in the sequence, we have a path between v and w. Otherwise, there will be at least one vertex that appears as v_i and v_j in the sequence with $i < j$. If we remove the terms $v_{i+1}, v_{i+2}, \ldots, v_j$ from the sequence, we still have a walk between v and w that contains fewer edges. We continue this process until each repeated vertex appears only once in the walk; at that stage, we have a path between v and w. The proof in the case of directed walks is similar.

2.2 Prove Theorem 2.2: If A is the adjacency matrix of a simple graph $G = (V, E)$, where $V = \{1, 2, \ldots, n\}$, the $(i - j)$ entry in the kth power of A is the number of *different* walks of length k between the vertices i and j. In particular, the $(i - i)$ diagonal entry in A^2 is the degree of vertex i for each i. (See Solved Problem 2.2.)

Solution. The proof is by induction on k. This is true when $k = 1$. Assume that the result is true for $(k - 1)$. By the induction hypothesis, the (i, j) entry in A^{k-1} is the number of walks between i and j of length $(k - 1)$. The (i, j) entry in A is 1 if and only if i and j are adjacent. Now $A^k = A^{k-1} \cdot A$. So the (i, j) entry in A^k is the number of walks between i and j of length k. Thus it is true for k.

2.3 If the edges in a graph are labeled $e_i (i = 1, 2, \ldots, m)$ and the set of paths between two vertices v and w are labeled p_i $(i = 1, 2, \ldots, k)$, the **v, w path matrix** is the $k \times m$ binary matrix P_{uv} in which the (i, j) entry corresponding to the path p_i is 1 if p_i contains the edge e_j and 0 otherwise. Obtain a path matrix between vertices 1 and 3 in Fig. 2-2.

Solution. In the graph shown in Fig. 2-2, there are three paths, p_1: 1 — 2 — 3, p_2: 1 — 4 — 3, and p_3: 1 — 2 — 5 — 3, between vertex 1 and vertex 3. The edges are labeled as shown. The 1, 3 path matrix of this graph with respect to this labeling is

$$P = P_{13} = \begin{bmatrix} 1 & 0 & 1 & 0 & 0 & 0 \\ 0 & 0 & 0 & 0 & 1 & 1 \\ 1 & 1 & 0 & 1 & 0 & 0 \end{bmatrix}$$

(The columns are labeled in the order of the edges: $e_1, \ldots, e_6$.)

2.4 Let B be the incidence matrix of the simple graph $G = (V, E)$ in which $V = \{1, 2, \ldots, n\}$, $E = \{e_1, e_2, \ldots, e_m\}$, and the paths between vertex i and vertex j are labeled $p_1, p_2, \ldots, p_k$. If P is the (i, j) path matrix, the (binary) product BP^T is the binary matrix S in which the only nonzero entries are those in the ith and jth rows. (Multiplication of binary matrices here is mod 2.)

Solution. The product of the ith row of B and any column of P^T is 1. Likewise, the product of the jth row of B and any column of P^T is 1. The product (modulo 2) of any other row of B with every column of P^T is 0.

2.5 Verify the assertion in Problem 2.4 by considering a path matrix in Fig. 2-2.

Solution. In Fig. 2-2, the binary product BP^T, where B is the incidence matrix and P is the (1, 3) path matrix (obtained in Problem 2.3), is equal to the 5×3 binary matrix in which row 1 and row 3 are nonzero vectors and the other rows are zero vectors:

$$BP^T = \begin{bmatrix} 1 & 0 & 0 & 0 & 0 & 1 \\ 1 & 1 & 1 & 0 & 0 & 0 \\ 0 & 0 & 1 & 1 & 1 & 0 \\ 0 & 0 & 0 & 0 & 1 & 1 \\ 0 & 1 & 0 & 1 & 0 & 0 \end{bmatrix} \times \begin{bmatrix} 1 & 0 & 1 \\ 0 & 0 & 1 \\ 1 & 0 & 0 \\ 0 & 0 & 1 \\ 0 & 1 & 0 \\ 0 & 1 & 0 \end{bmatrix} = \begin{bmatrix} 1 & 1 & 1 \\ 0 & 0 & 0 \\ 1 & 1 & 1 \\ 0 & 0 & 0 \\ 0 & 0 & 0 \end{bmatrix}$$

2.6 If the edge set in a simple graph is labeled $E = \{e_1, e_2, \ldots, e_m\}$ and the set of cycles is labeled $\{C_1, C_2, \ldots, C_k\}$, the **cycle matrix** of the graph is the $k \times m$ binary matrix C defined as follows. In the row corresponding to the ith cycle C_i, the (i, j) entry is 1 if and only if e_j is an edge in C_i. Obtain a cycle matrix of the graph of Fig. 2-3.

Solution. The five cycles in Fig. 2-3 are $C_1 = \{e_1, e_5, e_8\}$, $C_2 = \{e_2, e_5, e_6\}$, $C_3 = \{e_3, e_6, e_7\}$, $C_4 = \{e_4, e_7, e_8\}$, and $C_5 = \{e_1, e_2, e_3, e_4\}$. The cycle matrix C is the following 5×8 matrix:

$$C = \begin{bmatrix} 1 & 0 & 0 & 0 & 1 & 0 & 0 & 1 \\ 0 & 1 & 0 & 0 & 1 & 1 & 0 & 0 \\ 0 & 0 & 1 & 0 & 0 & 1 & 1 & 0 \\ 0 & 0 & 0 & 1 & 0 & 0 & 1 & 1 \\ 1 & 1 & 1 & 1 & 0 & 0 & 0 & 0 \end{bmatrix}$$

(The columns are labeled in the order of the edges: $e_1, \ldots, e_8$.)

2.7 Let B be the incidence matrix of the simple graph $G = (V, E)$, where $V = \{1, 2, \ldots, n\}$, $E = \{e_1, e_2, \ldots, e_m\}$, and the cycles in G are labeled $C_1, C_2, \ldots, C_k$. If C is the $k \times m$ cycle matrix, the (binary) matrix product BC^T and the (binary) matrix product CB^T are zero matrices. (Multiplication of binary matrices here is mod 2.)

Solution. The product of any row of the incidence matrix and any column of the transpose of the cycle matrix is 0 (mod 2). Similarly, the product of any row of the cycle matrix with any column of the transpose of the incidence matrix is 0 (mod 2).

2.8 Verify the assertion of Problem 2.7 by considering the cycle matrix in Fig. 2-3.

Solution. In Fig. 2-3, the cycle matrix C is as in Problem 2.6, and the incidence matrix B is

$$B = \begin{bmatrix} 1 & 0 & 0 & 1 & 0 & 0 & 0 & 1 \\ 1 & 1 & 0 & 0 & 1 & 0 & 0 & 0 \\ 0 & 1 & 1 & 0 & 0 & 1 & 0 & 0 \\ 0 & 0 & 1 & 1 & 0 & 0 & 1 & 0 \\ 0 & 0 & 0 & 0 & 1 & 1 & 1 & 1 \end{bmatrix}$$

It is easily verified that the binary matrix product BC^T and the binary matrix product CB^T are both zero matrices.

CONNECTED GRAPHS AND DIGRAPHS

2.9 Show that a graph G is bipartite if and only if every component of G is bipartite.

Solution. Every subgraph of a bipartite graph G is bipartite. In particular, every component of G is bipartite. Conversely, suppose that the components of $G = (V, E)$ are $G_i = (X_i, Y_i, F_i)$, where $i = 1, 2, \ldots, k$. Let $X = \cup X_i$, $Y = \cup Y_i$, and $F = \cup F_i$. Then $G = (X, Y, F)$.

2.10 Prove Theorem 2.3: A simple graph with three or more vertices is bipartite if and only if it has no odd cycles.

Solution. Let $G = (X, Y, E)$ be a bipartite graph. It is easy to see that if there is a cycle C in G, C should have an even number of vertices (and therefore an even number of edges) since the vertices of C are alternatively from X and from Y. Thus C is an even cycle.

On the other hand, suppose $G = (V, E)$ has no odd cycles. Assume without loss of generality that G is connected as in Problem 2.9. Let $d(u, v)$ be the length of a path between u and v of minimum length. A u, v path of length $d(u, v)$ is called a shortest path between u and v. Let u be any vertex in G.

Define $X = \{x \in V: d(u, x) \text{ is even}\}$ and $Y = V - X$. We have to establish that whenever v and w are two vertices in X (or in Y), there is no edge joining v and w.

Case (i): Let v be any vertex in X other than u. There is a path of even length between u and v. If there is an edge between u and v, we will get an odd cycle. So u is not adjacent to any vertex in X.

Case (ii): Let v and w be two vertices in X other than u. Suppose there is an edge e joining v and w. Let P be a shortest u, v path of length $2m$, and let Q be a shortest u, w path of length $2n$. If these two shortest paths have no common vertex other than u, these two paths and the edge together will form an odd cycle. If the two paths have common vertices, let u' be that common vertex such that the subpath P' between u' and v and the subpath Q' between u' and w have no vertex in common. Since P and Q are shortest paths, the subpath of P between u and u' is a shortest (u, u') path. The subpath of Q between u and u' is also a shortest (u, u') path. Thus both these subpaths have equal number of edges. Let the length of the shortest path between u and u' be k. Then the length of the subpath of P between u' and v is $2m - k$, and the length of the subpath of Q between u' and w is $2n - k$. If there is an edge between v and w, we will have a cycle of length $(2m - k) + (2n - k) + 1$, which will be an odd cycle. So no two vertices in X are adjacent.

Case (iii): Suppose v and w are in Y. Then, as in *(ii)*, the subpath from u' to v of length $(2m - 1) - k$, the subpath from u' to w of length $(2n - 1)$, and the edge joining v and w will form an odd cycle. Thus no two vertices in Y are adjacent.

Case (iv): Suppose u is the only vertex in X. The remaining vertices are all in Y. Every edge is from u to some vertex in Y.

This completes the proof.

2.11 Prove Theorem 2.4: For any graph G, $\kappa(G) \leq \lambda(G) \leq \delta(G)$.

Solution. The set of edges incident to a vertex of minimum degree is a disconnecting set, so $\lambda(G) \leq \delta(G)$. If $\lambda(G)$ is zero, the graph is either trivial or disconnected, implying that $\kappa(G)$ is also zero. If $\lambda(G) = 1$, the graph G has a bridge, implying that the graph has a cut vertex; therefore, $\kappa(G) = 1$. Let us assume that $\lambda(G) > 1$. If we delete $\lambda(G) - 1$ edges from the graph, we get a connected subgraph with a bridge joining two vertices v and w. For each deleted edge, we can choose a vertex incident to it other than v and w. Let W be the set of vertices thus chosen. Suppose the deletion of all the vertices in W results in a graph G'. If G' is not connected, it follows that $\kappa(G) < \lambda(G)$. If G' is connected, it has a vertex u whose deletion results in a trivial graph or a disconnected graph. Thus the set of vertices consisting of u and the chosen vertices [at most $\lambda(G) - 1$ in number] constitutes a separating set for G, implying that $\kappa(G) \leq \lambda(G)$.

2.12 Exhibit a graph for which the inequality established in Problem 2.11 is strict.

Solution. For graph G in Fig. 2-11, vertex 3 with degree 4 is a vertex of minimum degree. So $\delta(G) = 4$. Edges {3, 6}, {5, 6}, and {5, 8} constitute a cut set. Thus $\lambda(G) = 3$. Vertices 3 and 5 together constitute a separating set of minimum cardinality. Hence $\kappa(G) = 2$.

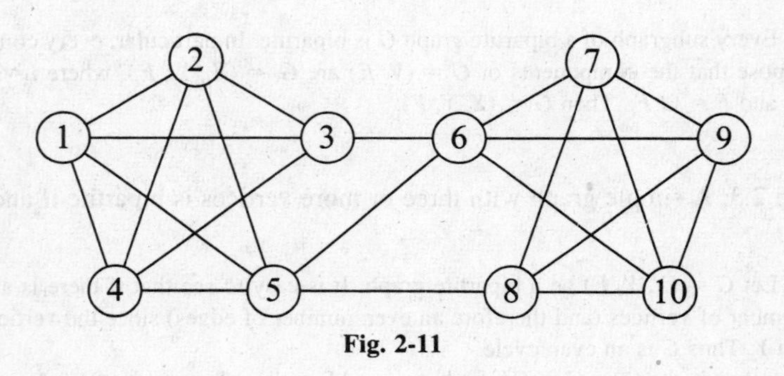

Fig. 2-11

2.13 A matrix is called a **totally unimodular (TU) matrix** if the determinant of every square submatrix is either -1 or 0 or 1. Show that each entry in a TU matrix is -1 or 0 or 1 but that the converse is not true.

Solution. Each entry of a TU matrix is the determinant of a 1×1 matrix; therefore, it is -1 or 0 or 1. On the other hand, the matrix with first row [1 -1] and second row [1 1] is not a TU matrix since its determinant is 2.

2.14 Show that the matrix $A = [a_{ij}]$ in which each element is -1 or 0 or 1 is a TU matrix if it satisfies the following two conditions: (*i*) no column can have more than two nonzero elements, and (*ii*) it is possible to partition the set I of rows of the matrix into sets I_1 and I_2 such that if a_{ij} and a_{kj} are the two nonzero elements in column j, row i and row k belong to the same subset of the partition if and only if they are of opposite sign.

Solution. Let C be any $k \times k$ submatrix of A. The proof is by induction on k. If $k = 1$, the theorem is true. Suppose it holds for $(k - 1)$. Let C' be any $(k - 1) \times (k - 1)$ submatrix of A. By hypothesis, the determinant of C' is -1 or 0 or 1. There are three different possibilities:

1. C has a column in which every entry is 0. Then the determinant of C is 0.

2. C has a column with exactly one nonzero entry that could be -1 or 1. By expanding the determinant of C along this column, we find that that determinant of C is -1 or 0 or 1.
3. Every column of C has exactly two nonzero entries.

Suppose $E = \{r_1, r_2, \ldots, r_k\}$ is the set of rows of C. By hypothesis, the set of k rows is partitioned into two subsets I_1 and I_2. Without loss of generality, let us assume that I_1 is the set of the first p rows and I_2 is the set of the remaining $(k - p)$ rows. It is possible that $p = 0$.

By the way these two subsets are constructed, it is easy to see that $r_1 + r_2 + \cdots + r_p = r_{p+1} + \cdots + r_k$, showing that E is a linearly dependent set. Thus the determinant of C is 0 in this case.

2.15 Show that (*a*) the incidence matrix of a digraph and (*b*) the incidence matrix of a bipartite graph are both TU matrices.

Solution. A matrix in which each element is -1 or 0 or 1 is totally unimodular if in each column there is at most one $+1$ and at most one -1 as a consequence of the result established in Problem 2.14. So the incidence matrix of a digraph and the incidence matrix of a bipartite graph are totally unimodular.

2.16 Show that a graph is bipartite if and only if its incidence matrix is a totally unimodular matrix.

Solution. If G is bipartite, its incidence matrix is a TU matrix, as shown in Problem 2.15. Suppose the incidence matrix of $G = (V, E)$ is a TU matrix. If G is not bipartite, there is at least one odd cycle in G. Let $V = \{1, 2, 3, \ldots, n\}$, and assume that the first $(2k + 1)$ vertices constitute an odd cycle as $1 — 2 — 3 — \cdots — (2k + 1) — 1$.

Let A be the incidence matrix such that the first $(2k + 1)$ rows correspond to the first $(2k + 1)$ vertices and the first $(2k + 1)$ columns correspond to edges $\{1, 2\}$, $\{2, 3\}$, $\ldots$, $\{2k, 2k + 1\}$, and $\{2k + 1, 1\}$. Then the determinant of the submatrix formed by the first $(2k + 1)$ rows and the first $(2k + 1)$ columns is 2. (In fact, the determinant of the incidence matrix of an odd cycle is always -2 or 2.) This contradicts that A is a TU matrix.

2.17 The smallest number of colors needed to color the vertices of a graph such that each vertex gets a unique color and no two adjacent vertices get the same color is called the **chromatic number** (or **vertex chromatic number**) of the graph. Show that a graph is bipartite if and only if its chromatic number is two.

Solution. In the bipartite graph $G = (X, Y, E)$, assign the same color (say red) to each vertex in X. Then assign a unique color other than red (say blue) to each vertex in Y. Thus the chromatic number of G is 2. On the other hand, suppose the chromatic number of $G = (V, E)$ is two. Let X be the set of vertices such that each vertex in X has the same color. Let $Y = (V - X)$. Then every edge in G is between a vertex in X and a vertex in Y. So $G = (X, Y, E)$.

2.18 Show that the following are equivalent in a simple graph G: (*a*) G is bipartite, (*b*) G has no odd cycles, (*c*) the incidence matrix of G is a totally unimodular matrix, and (*d*) the chromatic number of G is two.

Solution. This is a consequence of Problems 2.10, 2.16, and 2.17.

2.19 Suppose each set in a family of subsets of a finite set is represented as a vertex. Two vertices representing two distinct subsets belonging to the family are joined by an edge if they have at least one element in common. The simple graph thus constructed is called the **intersection graph** of the family of subsets of the given set. Construct the intersection graph of the family of subsets of the set $X = \{1, 2, \ldots, 10\}$ with the family $\{A, B, C, D, E, F\}$, where $A = \{1, 3, 5, 7, 9\}$, $B = \{2, 4, 6, 8, 10\}$, $C = \{1, 2, 3\}$, $D = \{4, 5, 6, 8, 9\}$, $E = \{5, 6, 7, 9\}$, and $F = \{4, 6, 10\}$.

Solution. The only nonempty intersections between pairs of distinct sets are $A \cap C$, $A \cap D$, $A \cap E$, $B \cap C$, $B \cap D$, $B \cap F$, $D \cap E$, $D \cap F$, and $E \cap F$. Thus we join A and C, A and D, A and E, B and C, B and D, B and F, D and E, D and F, and, finally, E and F by edges. The intersection graph thus constructed has six vertices and nine edges, as shown in Fig. 2-12.

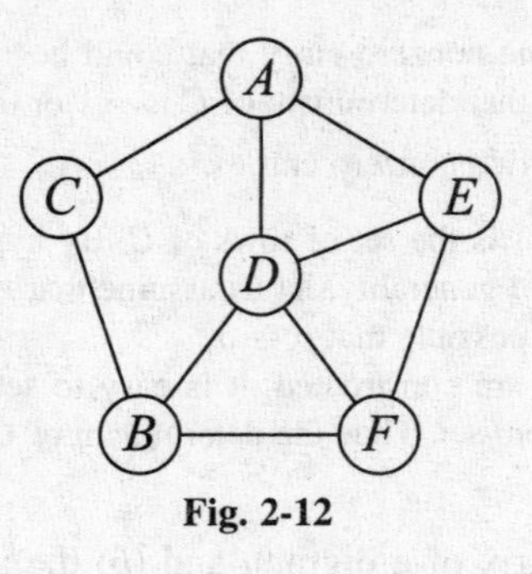

Fig. 2-12

2.20 Show that every simple graph is (isomorphic to) the intersection graph of a family of subsets of a finite set.

Solution. Let $G = (V, E)$, where V is the set $\{1, 2, \ldots, n\}$, be a simple graph. We construct a unique intersection graph corresponding to this graph as follows. For each vertex i in G, we define the set $X(i)$ as the union of the set $\{i\}$ and the set of all edges in G adjacent to i. Thus we have a family of n sets, each representing a vertex. It is easy to see that the intersection of $X(i)$ and $X(j)$ is nonempty if and only if there is an edge between i and j in G. The graph G is isomorphic to the intersection graph of the family $\{X(1), X(2), \ldots, X(n)\}$.

2.21 The **intersection number** $\omega(G)$ of a graph G is the minimum number of elements in a set X such that G is an intersection graph of a family of subsets of X. Show that the intersection number of a connected graph cannot exceed its size.

Solution. Suppose the connected graph is $G = (V, E)$, where $V = \{1, 2, \ldots, n\}$. Let $X(i)$ be the set of edges adjacent to vertex i. Then the union of the family $\{X(1), X(2), \ldots, X(n)\}$ is the set E. Thus G is the intersection graph of the family. So the intersection number cannot exceed the size of the graph.

2.22 Construct (*a*) a connected graph such that its intersection number is equal to its size and (*b*) a connected graph such that its intersection number is less than its size.

Solution.

(*a*) Consider the cyclic graph C_4 with edges e_i $(i = 1, 2, 3, 4)$. The intersection number of this graph is four since it is isomorphic to the intersection graph defined by the family $\{\{e_1, e_2\}, \{e_1, e_3\}, \{e_2, e_4\}, \{e_3, e_4\}\}$.

(*b*) Construct the graph G of size 5 by joining two nonadjacent vertices of the graph C_4 by an edge denoted by e_5. The intersection number of this graph of size 5 is four since it is isomorphic to the intersection graph defined by the family $\{\{e_1, e_2\}, \{e_1, e_3\}, \{e_2, e_4\}, \{e_1, e_4\}\}$.

2.23 Show that the intersection number of a connected graph with at least four vertices is equal to its size if and only if it has no triangles.

Solution. Consider any connected graph $G = (V, E)$ with n vertices and m edges, where n is at least four with no triangles. If $\omega(G) = k$, there exists a set X of cardinality k such that G is isomorphic to an intersection graph of a family of subsets of X. Specifically, there exists a family $\{X_1, X_2, \ldots, X_n\}$ of nonempty subsets of X such that (1) each vertex of G corresponds to a unique set in this family and (2) each edge of G corresponds to a pair of unique sets in the family with a common element that does not appear in any other subset in the family. So the cardinality of X should not be less than the number of edges of G. In other words, $k \geq m$. But $k \leq m$, as established in Problem 2.21. Thus $k = m$.

On the other hand, consider a connected graph $G = (V, E)$ with n vertices, m edges, and $\omega(G) = m$. We now prove that G has no triangle. Suppose G has a triangle. Let $G_1 = (V, E_1)$ be a maximal triangle-free spanning subgraph of G with m_1 edges. Then $\omega(G) = m_1$. Thus there exists a set X of cardinality m_1 such that G_1 is the intersection graph of a family of subsets X. Specifically, there exists a family $\{X_1, X_2, \ldots, X_n\}$ of subsets of X

such that vertex v_i of G_1 corresponds to the subset X_i for each i. Let e be any edge of G that is not in G_1, and let $G_2 = (V, E_2)$ be the spanning subgraph obtained by adding e to the graph G_1. This addition creates a unique triangle in G_2. Assume that this triangle is formed by vertices v_1, v_2, and v_3, and assume that edge e is the edge that joins v_1 and v_2. Notice that the intersection of X_1 and X_2 is not empty since $\{v_1, v_2\}$ is an edge in the graph G_1. Let $t \in (X_1 \cap X_2)$.

Two mutually exclusive cases need to be examined. (*i*) In G_1, the degree of $v_2 = 2$. In this case, we replace X_2 by $\{t\}$ and X_3 by $X_3 \cup \{t\}$ in the family $\{X_1, X_2, X_3, \ldots, X_n\}$ of subsets of X. (*ii*) In G_1, the degree of $v_2 > 2$. In this case, we replace X_3 by $X_3 \cup \{t\}$ in the family. In either case, we have a family of subsets of X defining the intersection graph of G_2. If G is isomorphic to G_2, graph G will have a triangle that contradicts the hypothesis. So assume that this is not the case. In that case, let $m - (m_1 + 1) = m_0$, where $m_0 > 0$. So the intersection number of G is $m_1 + m_0 = m - 1 < m$, which is a contradiction.

2.24 Find the intersection number of K_n, where $n > 1$.

Solution. The complete graph with two vertices is the intersection graph of the family $\{\{1\}, \{1, 2\}\}$, so its intersection number is two. The complete graph with three vertices is the intersection graph of the family $\{\{1, 2\}, \{1, 3\}, \{2, 3\}\}$, so its intersection number is three. The intersection number of any complete graph with n vertices (where $n > 3$) is less than its size as established in Problem 2.23. It is easy to see that K_4 is the intersection graph of the family $\{\{1, 2, 3\}, \{1, 2\}, \{1, 3\}, \{2, 3\}\}$, so its intersection number is three. To obtain a family of subsets for K_5, we introduce a new element 4 and adjoin it to each of the sets in the family for K_4. The family consisting of the four enlarged sets and the singleton set $\{4\}$ defines the intersection graph corresponding to K_5. Thus the intersection number for K_5 is four. By a simple inductive argument, it follows that the intersection number of a complete graph with n vertices (where $n > 3$) is $n - 1$.

2.25 The intersection graph of a finite family of open intervals of the real line is called an **interval graph.** Show that the cyclic graph with n vertices is (isomorphic to) an interval graph only when $n = 3$.

Solution.

(*i*) Suppose the vertices of a cyclic graph with three vertices are A, B, and C. Give an assignment of open intervals to these three vertices as $I(A) = (a, a')$, $I(B) = (b, b')$, and $I(C) = (c, c')$, where $a < b < c < a' < b' < c'$. The vertices of C_3 correspond to these three open intervals.

(*ii*) We must show that such an interval assignment is not possible for a cyclic graph when $n > 3$. It is enough if we show this when $n = 4$. Suppose the vertices are A, B, C, and D such that there is no edge between A and C and no edge between B and D. So any interval assignment $\{I(A), I(B), I(C), \text{and } I(D)\}$ should satisfy the requirement that the intervals $I(B)$ and $I(D)$ are disjoint and the intervals $I(A) \cap I(B)$ and $I(A) \cap I(D)$ are nonempty. Once these assignments are made for A, B, and D, we have to make an assignment for $I(C)$ such that $I(C) \cap I(A)$ is empty and, at the same time, both sets $I(C) \cap I(B)$ and $I(C) \cap I(D)$ are nonempty. It is simply impossible to make an assignment $I(C)$ without violating the earlier assignments.

Thus every interval graph is an intersection graph, but an intersection graph need not be an interval graph in general.

2.26 Show that (*a*) any induced subgraph of an interval graph is an interval graph, and (*b*) an arbitrary subgraph of an interval graph need not be an interval graph.

Solution.

(*a*) Let $H = (W, F)$ be an induced subgraph of the interval graph $G = (V, E)$. The interval assignments in G for the vertices in W will serve as the interval assignments for the vertices in W for the graph H as well.

(*b*) The graph G obtained by joining any two nonadjacent vertices of the cyclic graph C_4 is an interval graph in which the subgraph C_4 is not an interval graph.

2.27 A graph G is called a **chordal graph** if in every cycle C in G there is an edge (belonging to G) joining two nonadjacent (in C) vertices. Show that every interval graph is a chordal graph.

Solution. Suppose an interval graph G is not a chordal graph. Thus it has a cycle C with four or more vertices such that there is no edge in G between any pair of nonadjacent (in C) vertices. In other words, G has a cyclic subgraph C_n as an induced subgraph, where n is more than three, which is not an interval graph; this contradicts that an induced subgraph of an interval graph is an interval graph.

2.28 Show that a graph G is chordal if and only if C_n is not an induced subgraph of G for any $n > 3$.

Solution. This is a consequence of the definition and that C_n is not chordal when $n > 3$.

2.29 Give an example of a chordal graph that is not an interval graph.

Solution. Each of the two graphs in Fig. 2-13 is a chordal graph but not an interval graph.

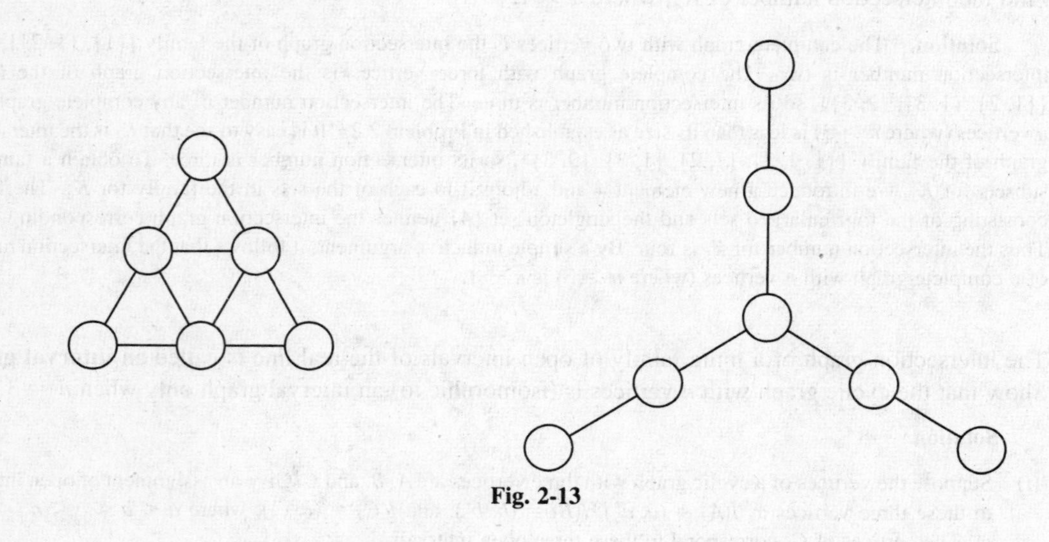

Fig. 2-13

2.30 The graph $G = (V, E)$ is called an **indifference graph** if for every positive number δ, there exists a mapping f from V to the set of real numbers such that $|f(v) - f(w)| < \delta$ if and only if v and w are adjacent. Show that the graph in Fig. 2-14 is an indifference graph.

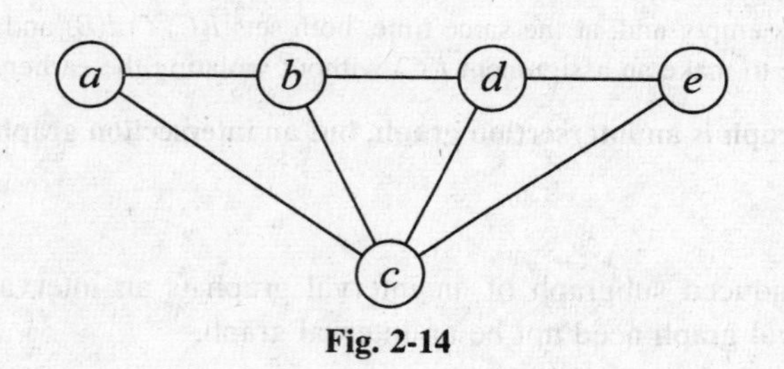

Fig. 2-14

Solution. First notice that without loss of generality, we may take $\delta = 1$. By assigning the values $f(a) = 0.2$, $f(b) = 0.7$, $f(c) = 0.9$, $f(d) = 1.3$, and $f(e) = 1.8$, we see that the graph is indeed an indifference graph.

2.31 Show that every indifference graph is an interval graph.

Solution. Let $G = (V, E)$ be an indifference graph that assigns the value $f(v)$ for each vertex v in the graph. Corresponding to each vertex v, we define the open interval $I(v) = (f(v) - \delta/2, f(v) + \delta/2)$. Then G is an interval graph.

2.32 By constructing an example, show that an interval graph need not be an indifference graph.

Solution. In graph $K_{1,3}$ shown in Fig. 2-15, let $I(a) = (2, 6)$, $I(b) = (8, 10)$, $I(c) = (13, 17)$, and $I(d) =$ 4, 15). Thus G is an interval graph. Suppose G is an indifference graph with weight function f. Assume without loss of generality that $f(a) < f(b) < f(c)$. If $f(c) - f(a) < 2\delta$, either $f(b) - f(c) < \delta$ or $f(a) - f(b) < \delta$. Thus $f(c) - f(a) \geq 2\delta$. Then $f(d)$ cannot be within δ units of both $f(a)$ and $f(c)$. Thus $K_{1,3}$ is an interval graph but not an indifference graph.

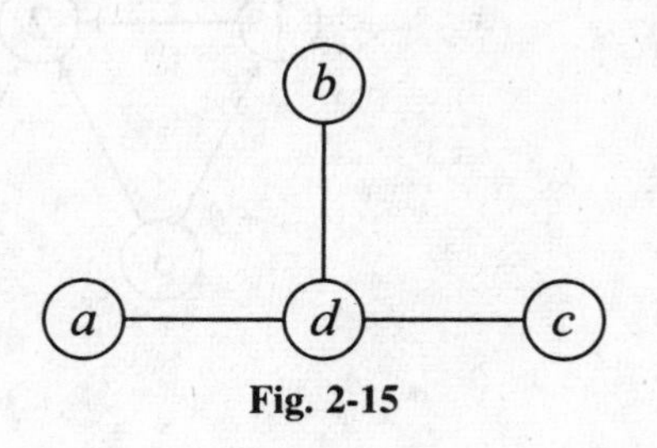

Fig. 2-15

2.33 Show that every indifference graph is an interval graph, every interval graph is a chordal graph, and every chordal graph is an intersection graph.

Solution. This follows from Problems 2.20, 2.27, and 2.31.

2.34 An interval graph is called a **unit interval graph** if the length of the open interval that corresponds to a vertex is the same for every vertex. Show that a graph is an indifference graph if and only if it is a unit interval graph.

Solution. Suppose $G = (V, E)$ is an indifference graph with a weight function f defined on V. Let $\delta = 1$. For each vertex v of G, define the open interval $I(v) = (f(v) - \frac{1}{2}f(v) + \frac{1}{2})$. Now v and w are adjacent in the indifference graph G if and only if $|f(v) - f(w)| < 1$. This is true if and only if the intervals $I(v)$ and $I(w)$ have a nonempty intersection. Thus G is a unit interval graph. Conversely, suppose G is a unit interval graph with $\delta \doteq 1$. If vertex v corresponds to the unit interval (a, b), define $f(v) = (a + b)/2$. Thus G is an indifference graph.

2.35 (*Ghouila–Houri Theorem*) A digraph D is a **transitive digraph** if there is an arc from u to v whenever there is an arc from u to w and there is an arc from w to v for any set of three distinct vertices u, v, and w in D. A simple graph G is a **transitively orientable graph** (or a **comparability graph**) if G has an orientation D that is a transitive digraph. Show that the complement of an interval graph is a transitively orientable graph.

Solution. Let $\{I(v): v \in V\}$ be the interval assignment for the vertices in the interval graph $G = (V, E)$. Suppose v and w are two nonadjacent vertices in G. Then $I(v)$ and $I(w)$ are disjoint. So either $I(v)$ is completely on the left of $I(w)$ or completely on the right. Let us write $I(v) < I(w)$ if $I(v)$ is on the left of $I(w)$. Since v and w are not adjacent in G, there is an edge between v and w in the complement of G. If $I(v) < I(w)$, let the edge between v and w become an arc from v to w. Otherwise, the edge is an arc from w to v. Thus we have an orientation of the complement of G. If $I(u) < I(v)$ and $I(v) < I(w)$, $I(u) < I(w)$. Thus the digraph thus constructed is transitively orientable.

2.36 Obtain a transitive orientation for the complement of the interval graph $K_{1,3}$.

Solution. The complement of $K_{1,3}$ is the graph consisting of an isolated vertex and a triangle, the edges of which can be oriented such that it becomes transitively orientable.

2.37 Show that if the complement of G is transitively orientable, it is not necessary that G is an interval graph by constructing an example.

Solution. The bipartite graph G in Fig. 2-16(*a*) has a cycle $1 — 5 — 3 — 4 — 1$ with no edge joining 1 and 3 or 4 and 5. So G is not chordal; therefore, it is not an interval graph. But its complement is transitively orientable, as shown in Fig. 2-16(*b*).

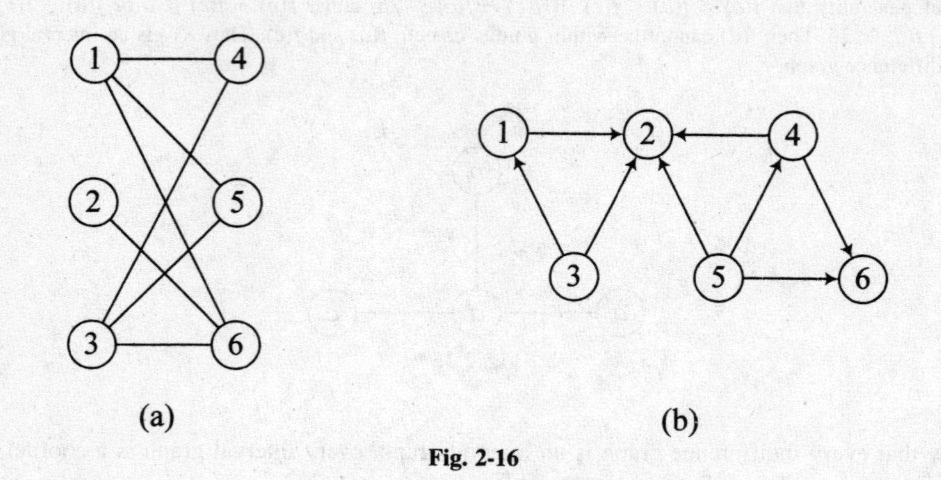

Fig. 2-16

2.38 Suppose $G = (V, E)$ is a simple graph with at least one edge. The **line graph** $L(G)$ (also known as the **interchange graph, adjoint graph, derived graph,** or **edge graph**) of G is the graph (W, F), where there is a one-to-one correspondence ϕ from E to W such that there is an edge between $\phi(e)$ and $\phi(e')$ if and only if the edges e and e' have a vertex in common. Construct the line graph of the graph of K_4.

Solution. The line graph of K_4 is as shown in Fig. 2-17.

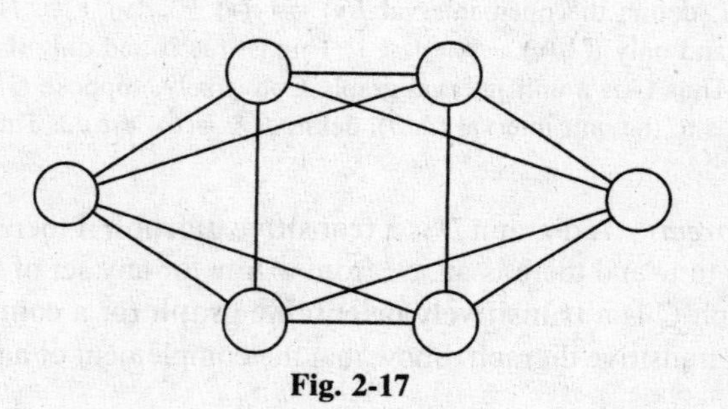

Fig. 2-17

2.39 Find the line graph of the graph shown in Fig. 2-18(*a*).

Solution. The line graph is shown in Fig. 2-18(*b*).

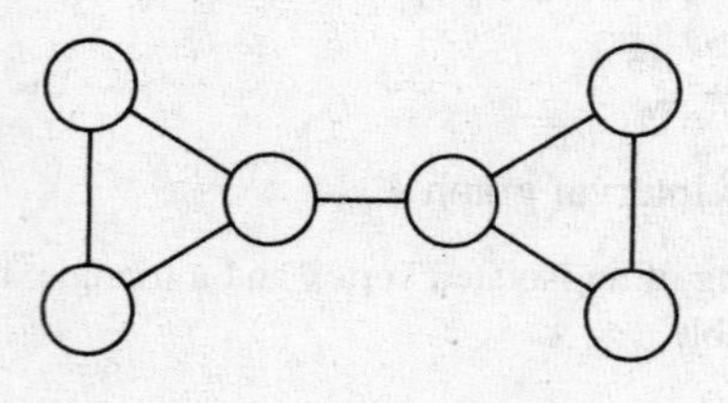

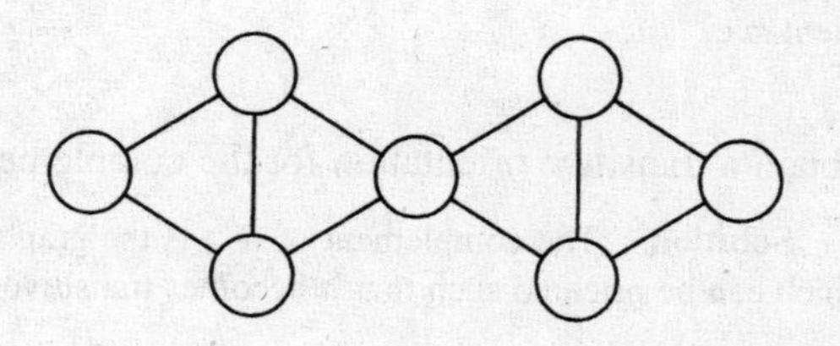

(a) (b)

Fig. 2-18

2.40 Show that the line graph of $G = (V, E)$ is the intersection graph of a family of subsets of the product set $V \times V$.

Solution. Each edge in G defines a two-element set in the product set. Then $L(G)$ is the intersection graph of the set S of all two-element sets that correspond to the edges of G.

2.41 If v is the vertex in the line graph $L(G)$ that corresponds to the edge joining vertex x and vertex y in G, find the degree of v in $L(G)$.

Solution. The degree of v in $L(G)$ = (degree of x in G + degree of y in G) − 2.

2.42 Find the order and size of $L(K_n)$.

Solution. The order of $L(K_n)$ is $n(n-1)/2$, which is the size of K_n. An edge joining two vertices in K_n is adjacent to $2(n-2)$ edges. So the degree of each vertex in $L(K_n)$ is $2(n-2)$. Hence, the sum of the degrees of all the vertices in $L(K_n)$ is $n(n-1)(n-2)$, which is twice the size of $L(K_n)$.

2.43 Find the order and size of the line graph $L(G)$ of a simple graph G with n vertices and m edges.

Solution. By definition, the order of $L(G)$ is m. In counting the number of edges in $L(G)$, we have to examine only those vertices in G with degree more than 1. Each edge of $L(G)$ corresponds to a pair of vertices. Let $V = \{1, 2, \ldots, n\}$ be the vertex set of G, and let d_i be the degree of vertex i. If $d_i > 1$, any two of the d_i edges that are incident at vertex i can be chosen in $C(d_i, 2)$ ways. Two edges e and e' of G that are incident at vertex i correspond to two adjacent vertices in $L(G)$ joined by an edge. Let the degrees of the vertices of G be $d_1, d_2, \ldots, d_n$. Thus the total number of edges in $L(G)$ is

$$r = C(d_1, 2) + C(d_2, 2) + \cdots + C(d_n, 2)$$

Then

$$r = \frac{d_1(d_1 - 1)}{2} + \cdots + \frac{d_n(d_n - 1)}{2} =$$

$$\frac{\{[(d_1)^2 + \cdots + (d_n)^2] - [d_1 + \cdots + d_n)]\}}{2} = \frac{\{[(d_1)^2 + \cdots + (d_n)^2] - 2m\}}{2}$$

2.44 Suppose G is a simple graph with five vertices with degrees 1, 2, 3, 3, and 3. Find the number of vertices and edges in $L(G)$.

Solution. The sum of degrees is 12, so G has six edges. Thus $L(G)$ has six vertices. The sum of the squares of the degrees is 32. Hence, the number of edges in $L(G)$ is $(32 - 12)/2 = 10$.

2.45 Find the line graph of (*a*) a simple path with k edges, where $k > 1$, and (*b*) the cyclic graph C_n with n edges, where $n > 2$.

Solution.

(*a*) The line graph of a path with k edges is a path with $(k - 1)$ edges.

(*b*) The line graph of C_n is C_n.

2.46 Show that there is no graph G such that $L(G) = K_{1,3}$.

Solution. Since $K_{1,3}$ has four vertices, if there is a graph G it should have four edges. Suppose these four edges are a, b, c, and d. Assume that the left vertex of $K_{1,3}$ corresponds to the side a and that the other three correspond to the right vertices. There is no graph G with four sides a, b, c, and d such that a has a vertex in common with the three edges and at the same time no two of these three edges have a vertex in common.

2.47 Construct an example to show that if $L(G)$ and $L(H)$ are isomorphic, it is not necessary that G and H are isomorphic.

Solution. It is easy to see that $L(K_3) = K_3 = L(K_{1,3})$.

2.48 Show that (*a*) a graph G is isomorphic to its line graph if and only if the degree of each vertex is 2, (*b*) a connected graph is isomorphic to its line graph if and only if it is a cyclic graph, and (*c*) the line graph of a connected graph G is (isomorphic to) K_n if and only if G is (isomorphic to) $K_{1,n}$ when $n > 3$.

Solution.

(*a*) If the degree of each vertex of G is 2, the degree of each vertex of $L(G)$ is also 2, and G and $L(G)$ both have the same number of vertices and the same number edges. So G and $L(G)$ are isomorphic. Conversely, if both G and $L(G)$ are isomorphic, they both have the same number of vertices and same number of edges, and the degree of each vertex is 2.

(*b*) This is a special case of (*a*).

(*c*) If $n > 3$, $L(K_{1,n}) = K_n$. Conversely, if $L(G) = K_n$, G has n edges, and all these edges have exactly one vertex in common since the degree of each vertex in K_n is $(n - 1)$.

2.49 Let X be the set of all the vertices and edges of a graph G with n vertices and m edges. Construct a graph $T(G) = (X, F)$ in which two elements x and y are joined by an edge if (*a*) x and y are adjacent vertices in G, (*b*) x is a vertex and y is an edge such that one of the vertices of y is x, or (*c*) if both x and y are edges, they have a vertex in common. The graph thus constructed is called the **total graph** of the graph. Construct the total graph of the complete graph with three vertices.

Solution. The total graph is a four-regular graph with six vertices.

TREES AND SPANNING TREES

2.50 Show that a graph is a tree if and only if there is a unique path between every pair of vertices in the graph.

Solution. Suppose the graph G is a tree. Let v and w be any two vertices in G. Since G is connected, there is a path P between v and w. If Q is another path between these two vertices, let $e = \{v_i, v_{i+1}\}$ be the first edge in P that is not in Q as we go from v to w along P. Let W and W' be the set of intermediate vertices between v_i and w in P and Q, respectively. If W and W' have no vertices in common, we have a cycle in the graph that is acyclic by assumption. If the intersection of W and W' is not empty, let u be the first common vertex as we go from v_i to w either along P or along Q. In this case, we also locate a cycle in the graph. Hence, there is a unique path between every pair of vertices in the tree. Conversely, let G be a graph in which there is a unique path between every pair of vertices. Then G is connected. Suppose G is not a tree. Then there is a cycle C in G. Obviously, there are two paths between any pair of vertices in C, which contradicts the hypothesis.

2.51 Show that a graph is a tree if and only if it is connected and every edge in it is a bridge.

Solution. If G is a tree, it is connected by definition. By Problem 2.50, there is a unique path between every pair of vertices in G. In particular, edge e joining two vertices v and w is the path P: v, e, w. If e is deleted, there is no path between v and w. Thus every edge in a tree is a bridge.

To establish the converse, suppose the connected graph in which every edge is a bridge is not a tree. Let G' be the subgraph of G obtained from G after deleting edge $e = \{v, w\}$ belonging to some cycle C in G. This graph G' is not connected since e is a bridge in G. Let p and q be any two arbitrary vertices. Since G is connected, there is a path P between p and q in G. If e is not an edge in this path P, P is a path in G' between p and q. Suppose e is an edge in P. Let P_1 be the subpath of P between p and v, and let P_2 be the subpath of P between w and q.

Moreover, let P' be the unique path between v and w in the cycle that does not contain e. Suppose Q is the union of these three paths. Then Q is a path in G' between p and q. Thus there is a path between every pair of vertices in G'. But G' is not a connected graph. This is a contradiction.

2.52 Show that a graph with n vertices is a tree if and only if it is connected and has $(n - 1)$ edges.

Solution. Suppose G is a tree with n vertices. It is a connected graph. We prove that it has $(n - 1)$ edges by induction on n. If $n = 1$, it is certainly true. Assume that it is true for $n = 1, 2, \ldots, (n - 1)$. Since every edge is a bridge (as established in Problem 2.51), the subgraph G' obtained from G after deleting an edge will have two components G_1 and G_2 with n_1 and n_2 vertices, respectively, where $n_1 + n_2 = n$. By the induction hypothesis, the number of edges in both the components together is $(n_1 - 1) + (n_2 - 1) = (n - 2)$. Thus the number of edges in G will be $(n - 2) + 1 = (n - 1)$.

Suppose the connected graph G with n vertices and $(n - 1)$ edges is not a tree. Then it has an edge e that is not a bridge. If e is deleted, the resulting subgraph is still a connected graph with n edges and $(n - 2)$ edges. We continue this process of locating edges that are not bridges and deleting them until we get a connected subgraph G' with n vertices and $(n - k)$ edges (where $k > 1$) in which every edge is a bridge. But G' is a tree, so it should have $(n - 1)$ edges. Thus $n - 1 = n - k$, where $k > 1$. This is a contradiction.

2.53 Show that a graph with n vertices is a tree if and only if it is acyclic and has $(n - 1)$ edges.

Solution. If G is a tree with n edges, it is acyclic by definition and has $(n - 1)$ edges, as established in Problem 2.52. On the other hand, consider an acyclic graph G with n vertices and $(n - 1)$ edges. Suppose G is not connected. Let the components of G be $G_i (i = 1, 2, \ldots, k)$ such that G_i has n_i vertices, where $n_1 + n_2 + \cdots + n_k = n$. Notice that each component G_i is a tree with $n_i - 1$ edges. Thus the total number of edges in G is $n - k$, where $k > 1$. This contradiction establishes that G is connected. Thus G is a tree.

2.54 Show that a graph G is a tree if and only if it is acyclic and whenever any arbitrary two vertices in G are joined by an edge, the resulting enlarged graph G' has exactly one cycle.

Solution. If G is a tree, it is connected and acyclic. Let u and v be any two nonadjacent vertices in G. There is a unique path between u and v. If we join u and v by an edge, this edge and path P create a unique cycle in the enlarged graph G'. On the other hand, suppose G is an acyclic graph in which u and v are two any arbitrary nonadjacent vertices such that the linking of the two by a new edge creates a unique cycle in G'. This implies that there is a path in G between u and v. So G is connected and hence is a tree.

2.55 Show that a graph G is a tree if and only if it is connected and whenever any two arbitrary vertices in G are joined by an edge, the resulting enlarged graph G' has exactly one cycle.

Solution. If G is a tree, it is connected and acyclic. If two nonadjacent vertices are joined by an edge, the unique path G between the two vertices and the edge together form a unique cycle. On the other hand, suppose G is connected. There cannot be a cycle in G since the enlarged graph G' obtained by joining two nonadjacent vertices has a unique cycle. So G is a tree.

2.56 Prove Theorem 2.5.

Solution. The proof follows from Problems 2.50 through 2.55.

2.57 Prove Theorem 2.7: A graph is connected if and only if it has a spanning tree.

Solution. Let G be a connected graph. Delete edges from G that are not bridges until we get a connected subgraph H in which each edge is a bridge. Then H is a spanning tree. On the other hand, if there is a spanning tree in G, there is a path between every pair of vertices in G; thus G is connected.

2.58 Prove Theorem 2.8: Let G be a simple graph with n vertices. If a spanning subgraph H satisfies any two of the following three properties, it will satisfy the third property also. (*i*) H is connected. (*ii*) H has $(n - 1)$ edges. (*iii*) H is acyclic.

Solution. The proof follows from Problems 2.54 and 2.55.

2.59 Show that if a graph is disconnected, its complement is connected.

Solution. If a graph G is not connected, it will have at least two components. Suppose u and v are two vertices belonging to two different components of G. Then these two vertices are adjacent in the complement of the graph. In other words, G and its complement cannot both be disconnected graphs. So whenever G is a disconnected graph, its complement is necessarily a connected graph.

2.60 A vertex of degree 1 in a graph is called a **terminal vertex** (or **pendant vertex** or **end-vertex**). Show that every tree of order two or more has at least two terminal vertices.

Solution. Suppose the degrees of the n vertices of a tree are d_i, where $i = 1, 2, \ldots, n$. Then $d_1 + d_2 + \cdots + d_n = 2n - 2$. If each degree is more than 1, the sum of the n degrees is at least $2n$. So there is at least one vertex (say vertex 1) with degree 1. Then $d_2 + d_3 + \cdots + d_n = 2n - 1$. At least one of these $(n - 1)$ positive numbers is necessarily 1. So there is one more vertex of degree 1. Thus at least two of the degrees must be 1.

2.61 Show that the vector $d = [d_1 \quad d_2 \quad \ldots \quad d_n]$ of positive integers, where $d_1 \leq d_2 \leq \ldots \leq d_n$ is the degree vector of a tree with n vertices if and only if $d_1 + d_2 + \cdots + d_n = 2(n - 1)$.

Solution. The necessity is obvious. We prove the sufficiency by induction on n. The property holds for $n = 1$ and $n = 2$. Assume that the property holds for $(n - 1)$ integers, where $n \geq 3$. Let $0 < d_1 \leq d_2 \leq \cdots \leq d_n$ and $d_1 + d_2 + \cdots + d_n = 2(n - 1)$. At least one of these numbers is 1. So $d_1 = 1$. Also $d_n > 1$. Let $d' = d_n - 1$. Then $d_2 + \cdots + d_{n-1} + d' = 2(n - 2)$. So by the induction hypothesis, there exists a tree T with $(n - 1)$ vertices and degrees $d_2, d_3, \ldots, d_{n-1}$ and d'. Construct a new vertex x and join that to the vertex of degree d'. Now we have a tree with n vertices with degrees $1, d_2, \ldots, d_n$. Thus the property hold for n. (Notice that an arbitrary vector satisfying the property stipulated in the problem could also be the degree vector of a graph that is not a tree.)

2.62 Prove Theorem 2.6: The center of a tree is either a singleton set consisting of a unique vertex or a set consisting of two adjacent vertices.

Solution. If a tree has two vertices, the center is the set of those two vertices. If there are three vertices in a tree, the center is the set consisting of the nonterminal vertex. A tree with four vertices is either $K_{1,3}$ (with three terminal vertices) or a path with two terminal vertices. In the former case, the cardinality of the center is 1; in the latter case, the center is the set of two adjacent nonterminal vertices. More generally, let T be a tree with five or more vertices, and let T' be the tree obtained from T by deleting all terminal vertices of T simultaneously. Observe that the eccentricity of any vertex in T' is one less than the eccentricity of that vertex in T. Thus the center of T is equal to the center of T'. If the process of deleting terminal vertices is carried out successively, we finally have a tree with four or fewer vertices.

2.63 A path P between two distinct vertices in a connected graph G is a **diametral** path if there is no other path in G whose length is more than the length of P. Show that (*a*) every diametral path in a tree will pass through its central vertices, and (*b*) the center of a tree can be located once a diametral path in the tree is discerned.

Solution. Let t be the length of any diametral path in a tree, and let P be a fixed diametral path joining the vertices v and w.

(*a*) If t is even, there exists a unique vertex c in P that is equidistant from either v or w. In this case, c is a central vertex. Suppose Q is another diametral path. Since the graph is connected, the two diametral paths should have a vertex in common. If c is not a common vertex, it is possible to obtain path whose length is more than t. So if the length of a diametral path is even, there exists a unique central vertex on that path through which every diametral path passes.

(*b*) If t is odd, there exist two vertices c' and c'' in P such that the number of edges in the path between v and c' is equal to the number of edges between w and c''. In this case, both c' and c'' are central vertices. Suppose Q is another diametral path. Then both P and Q share the edge joining c' and c'' as a common edge. Thus once a diametral path in a tree is located, it is easy to find the center of the tree.

2.64 A tree with exactly one vertex v of degree 2 in which the degree of every nonterminal vertex (other than v) is 3 is called a **binary tree,** and the **root** of the binary tree is the unique vertex of degree 2. Show that the number of vertices in a binary tree is odd.

Solution. Every vertex other than the root is an odd vertex. The number of odd vertices is even. If we now include the root also, the total number of vertices is odd.

2.65 Show that the number of terminal vertices in a binary tree with n vertices is $(n + 1)/2$.

Solution. Suppose there are k terminal vertices. Then the sum of the degrees of the n vertices is $k + 2 + 3(n - k - 1)$, which is equal to $2(n - 1)$ since the graph is a tree. Thus $k = (n + 1)/2$.

2.66 Show that if T is a tree with n vertices and G is a graph with $\delta(G) \geq (n - 1)$, T is isomorphic to a subgraph of G.

Solution. The proof is by induction on n. This is true when the tree has two vertices. The induction hypothesis is that if T' is any tree with $(n - 1)$ vertices and G' is any graph with $\delta(G') \geq (n - 2)$, then T' is isomorphic to a subgraph of G'. Let T be any tree with n vertices, and let G be any graph with $\delta(G) \geq (n - 1)$. Let v be any terminal vertex in T, and let u be the vertex adjacent to v in T. Then $T - v$ is a tree with $(n - 1)$ vertices. Moreover, $\delta(G) \geq (n - 1) > (n - 2)$. So by the induction hypothesis, the tree $T - v$ is isomorphic to a subgraph of G. Let u' be the vertex in G that corresponds (for this isomorphism) to vertex u. Then $\delta(u') \geq (n - 1)$ in G. The graph $T - v$ has only $n - 2$ vertices in addition to vertex u. So there should be a vertex w in G that is adjacent to u' such that w does not correspond to any vertex in $T - v$. By identifying v with vertex w, we see that T is isomorphic to a subgraph of G. Thus the theorem is true for n as well.

2.67 Show that a tree with n vertices is isomorphic to a subgraph of the complement of the cyclic graph with $(n + 2)$ vertices.

Solution. The complement of the cyclic graph with $(n + 2)$ vertices is an r-regular graph G, where $r = n > n - 1$. So by Problem 2.66, any tree with n vertices is isomorphic to a subgraph of G.

2.68 A graph is said to be a **unicyclic graph** if it has exactly one cyclic subgraph. Show that if any two of the following conditions are satisfied in a graph with n vertices and m vertices, the third condition also is satisfied: (*a*) G is connected, (*b*) G is unicyclic, and (*c*) $n = m$.

Solution.

(*a*) Suppose G is connected and unicyclic. Let the vertices in the cycle be $v_1, v_2, \ldots, v_k$, and let T_i be a tree with root at v_i containing $n_i - 1$ vertices, excluding the root and the two roots adjacent to the root in the cycle. Then the total number of vertices in the graph is $(n_1 - 1) + (n_2 - 1) + \cdots + (n_k - 1) + k$, which is equal to the total number of edges in G.

(*b*) Suppose G is connected and $n = m$. Since $m > (n - 1)$, there is at least one cycle. Suppose there is more than one cycle. Then $m > n$.

(*c*) Suppose G is unicyclic and $m = n$. If G is not connected, $m < n$.

2.69 Show that each labeled tree with n vertices corresponds to a unique vector $s = [s_1 \quad s_2 \cdots s_{n-2}]$, where $s_i \in N = \{1, 2, \ldots, n\}$ for $i = 1, 2, \ldots, (n-2)$.

Solution. Let T be any labeled tree, and let W be the set of terminal vertices in T. Arrange the vertices in W such that their labels are in increasing order. If w_1 is the first element in W, find the label s_1 of the unique vertex adjacent to w_1. Then delete w_1 from T to obtain a tree T'. Let W' be the set of all terminal vertices in T'. Arrange the vertices in the set W' such that their labels are in increasing order. If w'_1 is the first element in W', find the label s_2 of the unique vertex adjacent to w'_1 in T'. The operation is repeated until s_{n-3} has been defined, leaving behind a tree with exactly two vertices with labels p and q, where $p < q$. We take $s_{n-2} = p$. Thus each spanning tree defines a unique vector with $(n-2)$ components.

2.70 Find the unique vector corresponding to the labeled tree shown in Fig. 2-19.

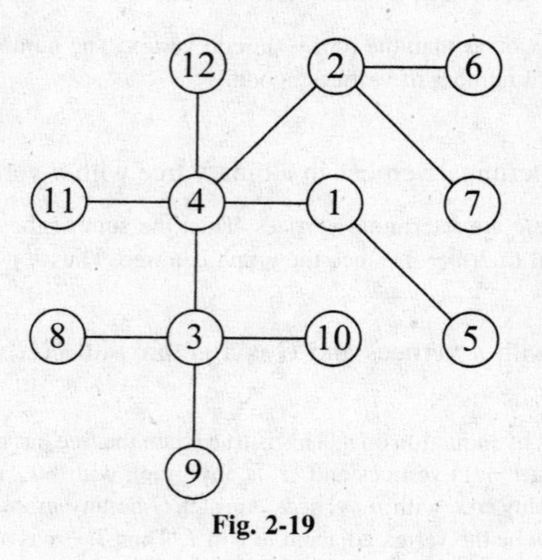

Fig. 2-19

Solution. Since there are 12 vertices in T, we are looking for a vector s with 10 components, each component being an integer between 1 and 12. Here $W = \{5, 6, 7, 8, 9, 10, 11, 12\}$ is the set of all terminal vertices in T, with labels arranged in increasing order. The vertex adjacent to 5 is 1, so $s_1 = 5$. Deleting vertex 5 from T, we get the subtree (the current tree, again denoted by T) in which the set of terminal vertices (the current set, again denoted by W) is $W = \{1, 5, 6, 7, 8, 9, 10, 11, 12\}$. The vertex adjacent to 1 is 4, so $s_2 = 4$.

Deleting 1 from the current tree, we get $W = \{6, 7, 8, 9, 10, 11, 12\}$ and $s_3 = 2$. In the next iteration, $W = \{7, 8, 9, 10, 11, 12\}$ and $s_4 = 2$. In the next iteration, $W = \{2, 8, 9, 10, 11, 12\}$ and $s_5 = 4$. In the next iteration, $W = \{8, 9, 10, 11, 12\}$ and $s_6 = 3$. In the next iteration, $W = \{9, 10, 11, 12\}$ and $s_7 = 3$. In the next iteration, $W = \{10, 11, 12\}$ and $s_8 = 3$. In the next iteration, $W = \{11, 12\}$ and $s_9 = 4$. At the final stage, we have the tree consisting of two vertices 4 and 12, so $s_{10} = 4$. Thus $s = [1 \quad 4 \quad 2 \quad 2 \quad 4 \quad 3 \quad 3 \quad 3 \quad 4 \quad 4]$ is the unique vector defined by the given labeled tree.

2.71 Show that every vector s with $(n-2)$ components, where each component is an element of $N = \{1, 2, \ldots, n\}$, corresponds to a unique labeled tree with n vertices.

Solution. Observe that in Problem 2.69, when we construct the vector s for the given labeled tree, the vertex i (with degree d_i) of T occurs $d_i - 1$ times in s. In particular, terminal vertices do not appear in s. We exploit this idea to prove the reverse implication.

Let v_1 = the first element in N that is not in s, v_2 = the first element in $N - \{v_1\}$ that is not in the subvector $s - s_1$ obtained by deleting s_1 from s, v_3 = the first element in $N - \{v_1, v_2\}$ that is not in the subvector $s - s_1 - s_2$ obtained by deleting s_2 from $s - s_1$, and so on. This process is repeated until we get the set $\{v_1, v_2, \ldots, v_{n-2}\}$. The two remaining vertices are denoted by x and y. Now join s_i and v_i for each i. Also join x and y. The graph thus defined with n vertices has $n - 1$ edges and acyclic-giving a spanning tree that is unique.

2.72 Obtain the unique labeled tree corresponding to the vector $s = [1 \ 4 \ 2 \ 2 \ 4 \ 3 \ 3 \ 3 \ 4 \ 4]$.

Solution. Since there are 10 components in s, we are looking for a labeled tree with 12 vertices. So $N = \{1, 2, 3, 4, 5, 6, 7, 8, 9, 10, 11, 12\}$. Since the elements in s are 1, 2, 3, and 4, the first element in N that is not in s is 5. So $v_1 = 5$, and we join s_1 and v_1 by an edge. Thus the first edge in T is obtained by joining 1 and 5.

At this stage, $N - \{v_1\} = \{1, 2, 3, 4, 6, 7, 8, 9, 10, 11, 12\}$, and the elements in the subvector $s - s_1 = [4 \ 2 \ 2 \ 4 \ 3 \ 3 \ 3 \ 4 \ 4]$ are 2, 3, and 4. The first element in $N - \{v_1\}$ that does not appear in the subvector is 1. So $v_2 = 1$. We now join 4 and 1.

At the next iteration, $N - \{v_1, v_2\} = \{2, 3, 4, 6, 7, 8, 9, 10, 11, 12\}$, and the corresponding subvector $s - s_1 - s_2$ is $[2 \ 2 \ 4 \ 3 \ 3 \ 3 \ 4 \ 4]$. Thus $v_3 = 6$. Join 2 and 6.

Continuing this process, we get $v_4 = 7$, $v_5 = 2$, $v_6 = 8$, $v_7 = 9$, $v_8 = 10$, $v_9 = 3$, and $v_{10} = 11$. We then join each of these vertices to the corresponding components of the subvector $[2 \ 4 \ 3 \ 3 \ 3 \ 4 \ 4]$, which is obtained by deleting the first 3 components of s.

At the end, we have the set $N - \{5, 1, 6, 7, 2, 8, 9, 10, 3, 11\}$ consisting of exactly two vertices, 4 and 12. At this stage we join 4 and 12. The unique labeled tree is the tree in Fig. 2-19, as expected.

2.73 Prove Theorem 2.9 (Cayley's theorem): The number of distinct labeled trees with n vertices is n^{n-2}, which is also equal to the number of spanning trees in K_n. (*a*) Show that the number of distinct labeled trees with n vertices is n^{n-2}. (*b*) Show that the number of spanning trees in K_n is also n^{n-2}.

Solution.

(*a*) A vector $s = [s_1 \ s_2 \ \cdots \ s_{n-2}]$ of $(n - 2)$ components from the set $N = \{1, 2, \ldots, n\}$ can be formed in n^{n-2} ways. From Problems 2.69 and 2.71, we see that there is a bijection between the set of all distinct labeled trees with n vertices and the set of these vectors.

(*b*) Every labeled tree with n vertices corresponds to a unique spanning tree in K_n. On the other hand, every spanning tree in K_n is a uniquely labeled tree. Thus the total number of spanning trees in K_n is n^{n-2}.

2.74 A directed tree T such that there is a unique vertex v of indegree zero and that the indegree of every other vertex is 1 is called an **arborescence** rooted at v. Find the number of distinct labeled arborescences with n vertices.

Solution. If T is a tree with n vertices, we can choose a root v to construct an arborescence in n ways. Once a root v is selected, a unique arborescence rooted at v is obtained by orienting each edge so that any vertex w can be reached from v by a unique directed path. By Cayley's theorem, for each choice of a root there are n^{n-2} labeled arborescences. Thus there are $n.n^{n-2} = n^{n-1}$ labeled arborescences with n vertices.

2.75 (*Palmer's Generalization of Cayley's Theorem*) A tree with n vertices is an **edge-labeled tree** if each edge is assigned a unique positive integer between 1 and $(n - 1)$. Show that the number of distinct edge-labeled trees with $(n - 1)$ labeled edges and n unlabeled vertices is n^{n-3}.

Solution. Suppose v is a fixed vertex in a labeled tree with n vertices. There is a unique path P from any vertex i to the vertex v. Define $f_v(i) = j$ if j is the first vertex in the path P after the vertex i. Thus $f_v(i) = j$ for i = 1, 2, . . . , $(n - 1)$. The function f_v: $\{1, 2, \ldots, (n - 1)\} \to \{1, 2, \ldots, n\}$ is the **tree function with respect to** v of the vertex labeled tree. Assign the label i to the edge joining i and j if $f_v(i) = j$. Thus each edge is assigned a label from the set $\{1, 2, \ldots, (n - 1)\}$.

Let X be the set of all vertex labeled trees, and let Y be the set of all edge labeled trees with unlabeled vertices. The cardinality of X is n^{n-2} by Cayley's theorem.

We have a mapping of X into Y via the tree function. If $n \geq 3$, each edge-labeled tree is the image of n vertex-labeled trees since the vertex v can be chosen in n ways and the labels of the other vertices are uniquely determined by the labels of these edges. Thus there are $n^{n-2}/n = n^{n-3}$ edge-labeled trees with n unlabeled vertices.

2.76 Let G be an undirected graph with n labeled vertices and m labeled edges. Assign each edge an arbitrary orientation, and let A be the incidence matrix of the resulting digraph. Show that (*a*) if G is connected, the rank of A is $(n - 1)$; and (*b*) the determinant of any nonsingular submatrix of A is either -1 or 1.

Solution.

(*a*) Each column has $(n - 2)$ zeros. The other two entries are 1 and -1. So the sum of all n rows is a zero vector with m components. Thus the rank of A is less than n. If $k < n$, the sum of any k rows should not be zero; otherwise, G would not be connected. So any set of k rows is linearly independent if $k < n$. Thus the rank of A is at least $(n - 1)$.

(*b*) If B is any nonsingular $r \times r$ submatrix, no column in B can be the zero vector. If each column of B contains two nonzero entries, B will become nonsingular. So there is at least one column in B with exactly one nonzero entry. We then expand the determinant of B along this column. Thus the determinant of B is the product of the nonzero entry in that column and the determinant of B', where B' is a $(k - 1) \times (k - 1)$ nonsingular submatrix. By a simple inductive argument, we conclude that the determinant of B is either -1 or 1.

2.77 Suppose an arbitrary incidence matrix A of a connected graph $G = (V, E)$ with n vertices is defined as in Problem 2.77. The **reduced incidence matrix A_r** is the matrix obtained from A by deleting a row, say the nth row. Show that any $(n - 1) \times (n - 1)$ submatrix B of matrix A_r is nonsingular if and only if the edges corresponding to the columns of B constitute the edges of a spanning tree in G.

Solution. Let B be an $(n - 1) \times (n - 1)$ submatrix of the reduced incidence matrix. Let $F = (V, E)$ be the spanning subgraph of G defined by the columns of B. Then B is the reduced incidence matrix of F and is nonsingular if and only if its rank is $(n - 1)$. Hence, B is nonsingular if and only if F is connected. But F has n vertices and $(n - 1)$ edges. So B is nonsingular if and only if F is a spanning tree.

2.78 (*Matrix Tree Theorem*) Show that if A_r is the reduced incidence matrix (as defined in Problem 2.77) of a connected graph G and if $(A_r)^T$ is its transpose, the number of spanning trees in G is equal to the determinant of $A_r(A_r)^T$.

Solution. Suppose P is a $p \times q$ matrix and Q is a $q \times p$ matrix, where $p \leq q$. Then it is a known result (known as **Cauchy–Binet formula** in matrix theory) that $\det PQ = \Sigma (\det B)(\det C)$, where the sum is taken over all $p \times p$ matrices B and C of P and Q such that the columns of P in B are numbered the same as the rows of Q in C. Let $P = A_r$ and $Q = (A_r)^T$. Then $(\det A_r)(\det (A_r)^T) = \Sigma (\det B)(\det B^T) = \Sigma (\det B)^2 = \Sigma 1$, where the last summation is over all $(n - 1) \times (n - 1)$ nonsingular submatrices of A_r. According to Problem 2.77, each such matrix corresponds to a spanning tree.

2.79 If e is an edge in graph G, $G - e$ is the subgraph of G obtained from G by deleting e from G. After edge e joining the vertices v and w is deleted, suppose the vertices v and w are merged to constitute a single vertex. The resulting graph G' is called the **contracted graph** obtained by contracting edge e and is denoted by $G.e$. If $\tau(G)$ is the number of spanning trees in G, show that $\tau(G) = \tau(G - e) + \tau(G.e)$.

Solution. Every spanning tree in G that does not contain edge e corresponds to a spanning tree in $G - e$. Every spanning tree in G that contains edge e corresponds to a spanning tree in the contracted graph $G.e$.

2.80 Show that if $T_i = (V_i, E_i)$, where $i = 1, 2, \ldots, k$ are subtrees of $T = (V, E)$ such that every pair of subtrees have at least one vertex in common, the entire set of subtrees have a vertex in common.

Solution. Let n be the number of vertices of T. The proof is by induction on n. The desired property holds if $n = 2$. Assume that the property holds for all trees with at most n vertices.

Let T be a tree with $(n + 1)$ vertices in which x is a terminal vertex adjacent to a vertex y. Suppose the subtrees $T_1, T_2, \ldots, T_k$ of T are such that every pair of them has at least one vertex in common. If x is not a vertex in any of these trees, the trees are subtrees of a tree with n vertices; thus the property holds for the graph T with $(n + 1)$ vertices. If one of these trees is the tree with just one vertex x, x is common to all the vertices; thus the property holds in this case.

We now examine the remaining case. Let $T_i(x)$ be the subtree of T_i obtained by deleting x from T_i. If x is a vertex common to T_i and T_j, y is also a vertex common to T_i and T_j. Therefore, y is a common vertex for $T_i(x)$ and $T_j(x)$. Thus by the induction hypothesis, the subtrees $T_i(x)$ have a common vertex. Therefore, the entire collection $\{T_i\}$ has a vertex in common.

2.81 Obtain a DFS spanning tree starting the search from vertex 2 in the graph shown in Fig. 2-20.

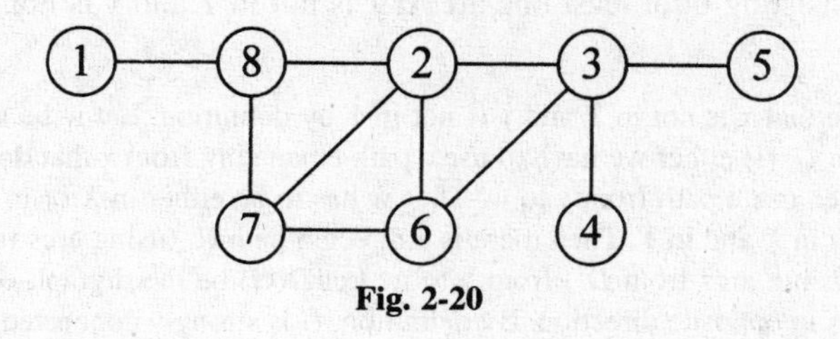

Fig. 2-20

Solution. From 2, we go to 3 and then from 3 to 4. Return to 3 and from 3 to 5. Return to 3 and then 3 to 6, 6 to 7, 7 to 8, and finally 8 to 1. If the search starts from 2, the edges of a DFS spanning tree are {2, 3}, {3, 4}, {3, 5}, {3, 6}, {6, 7}, {7, 8}, and {8, 1}.

STRONG ORIENTATIONS OF GRAPHS

2.82 Show that a digraph $G = (V, E)$ is strongly connected if and only if the following property is satisfied: For every nonempty subset X of vertices, there exists an arc from a vertex x in X to a vertex y in the complement of X.

Solution. Suppose $G = (V, E)$ is strongly connected and X is an arbitrary nonempty subset of V. Let u be an arbitrary vertex in X and y be an arbitrary vertex in $Y = (V - X)$. Then there is at least one directed path P from u to v that will have an arc of the form (x, y), where $x \in X$ and $y \in Y$. So a strongly connected graph satisfies the property. To prove the sufficiency part, assume that G is a digraph that satisfies the property. Suppose G is not strongly connected. Suppose u and v are two vertices such that there is no directed path from u to v in the digraph. Let X be the set of all vertices that are terminal vertices of directed paths that originate from the vertex u. By assumption, X is a proper subset of V. So there exists an arc e from vertex $x \in X$ to vertex $y \in (V - X)$. Now x is the terminal vertex of a directed path P from u, and this path can be enlarged into path P' from u to y using arc e. Thus the vertex y cannot be in $(V - X)$, which is a contradiction.

2.83 Show that if a tournament has a directed circuit, it has a directed triangle.

Solution. If a tournament $G = (V, E)$ has a directed circuit, it will have a directed cycle C. Suppose C: $v_1 \to v_2 \to \cdots \to v_k \to v_1$ is a directed cycle of minimum length in G. We have to show that $k = 3$. Suppose $k > 3$. There are two cases: (*i*) There exists an arc from v_1 to v_{k-1} creating a directed cycle passing through three vertices. This is a contradiction. (*ii*) There is an arc from v_{k-1} to v_1 creating a cycle of length $(k - 1)$. This also is a contradiction. So $k = 3$.

2.84 If G is a tournament with n vertices, the vector whose components are the n outdegrees arranged in a nondecreasing order is called the **score vector** of the tournament. A tournament is a **transitive tournament** if whenever (u, v) and (v, w) are arcs, (u, w) is also an arc. Show that a tournament G with n vertices is transitive if and only if its score vector is $[0 \quad 1 \quad 2 \quad \cdots \quad (n - 1)]$.

Solution. Let G be a transitive tournament. In any tournament, the outdegree is at most $(n - 1)$. Suppose there are two vertices u and v with equal outdegrees. If there is an arc from u to v, the outdegree of u is more than the outdegree of v. If there is an arc from v to u, the outdegree of v is more than that of u. In other words, in a transitive tournament, the outdegrees are all distinct. Now suppose $G = (V, E)$ is a tournament with score vector $s = [0 \quad 1 \quad 2 \quad \cdots \quad (n - 1)]$. If $V = \{v_1, v_2, \ldots, v_n\}$, E will be the set $\{(v_i, v_j)\colon 1 \le j < i \le n\}$ because of the way the outdegrees are defined. Then G is obviously transitive.

2.85 Suppose D' is the digraph obtained from a *strongly connected mixed* graph G (see Section 2.2) by deleting an undirected edge $e = \{x, y\}$ from G and replacing every other edge in G by two arcs in opposite directions. Let X be the set of all vertices v such that there is a directed path in D' from x to v consisting of at least one arc. Likewise, let Y be the set of all vertices v such that there is a direct path in D' from y to v consisting of at least one arc. If x is not in Y and y is not in X, $e = \{x, y\}$ is a bridge in the mixed graph.

Solution. Notice that x is not in X and y is not in Y by definition. Let w be any vertex that is neither x nor y. To go from x to w in G, (*i*) either we have to use a path emanating from x that does not go through y or (*ii*) first go to y using e, and then use a path from y to w. Thus w has to be either in X or in Y.

Suppose w is both in X and in Y. Then there is a directed path P_1 (using arcs from D') from x to w, and there is a directed path P_2 (using arcs from D') from y to w. Let $D(G)$ be the digraph obtained from G by converting each edge into two arcs in opposite direction. By definition, G is strongly connected if and only if $D(G)$ is strongly connected. By hypothesis, there is a directed path P_3 [consisting of arcs from $D(G)$] from w to x and a directed path P_4 [consisting of arcs from $D(G)$] from w to y.

There are three cases to be examined:

(*i*) Both P_3 and P_4 did not use e in either direction. In that case, there is a path in D' from x to y that will imply that y is in X, which is a contradiction, and there is a path in D' from y to x that will imply that x is in Y, which also is a contradiction.

(*ii*) Path P_3 is from w to y, using arcs from D' and then using the arc (y, x), which will imply that there is a directed path in D' from x to y. In that case, y will be in X, which is a contradiction.

(*iii*) Path P_4 is from w to y, using edges from D' and then using the arc (x, y). This will imply that x is in Y, which is also a contradiction.

Thus sets X and Y are disjoint, consequently, there is no edge or arc connecting a vertex in X and a vertex in Y. So the only reason that the mixed graph is connected is because of the existence of edge e joining x and y. In other words, e is a bridge.

2.86 Show that if G is a strongly connected mixed graph, it is possible to convert any edge that is not a bridge into an arc such that the resulting mixed graph is also strongly connected.

Solution. Let $e = \{x, y\}$ be any edge in a strongly connected mixed graph G. Construct X and Y as in Problem 2.87. If e is not a bridge, either x is in Y or y is in X. If x is in Y, edge e is converted into an arc from x to y. If y is in X, edge e is converted into an arc from y to x. In either case, the resulting mixed graph is strongly connected.

2.87 Prove Theorem 2.10 (Robbins's theorem): A graph is strongly orientable if and only if it is connected and has no bridges.

Solution. If an undirected graph G is strongly orientable, it is necessarily connected; thus no edge in it can be a bridge. To prove the converse, start from an undirected connected graph with no bridges. By Problem 2.86, we can orient one edge at a time. Notice at each stage that if $e = \{x, y\}$ is not a bridge, it will not become a bridge after the orientation.

2.88 Suppose G is a connected graph in which no edge is a bridge, and let T be the DFS spanning (directed) tree obtained by starting the search from a fixed vertex r called the root of the tree. Show that there is a directed path in T to every vertex from root r.

Solution. As a result of the search, each vertex v gets a label $f(v)$ that is a positive integer between 1 and n. The vertex r with label $f(r) = 1$ is the root of the DFS spanning tree T. We can use an inductive argument (induction on the label) to establish that there is a directed path from the root to every other vertex. If $f(r) = 1$, there is a directed path from 1 to 1. Suppose there is a directed path from the root to any vertex x, where $f(x) < k$. Assume there is a vertex y such that $f(y) = k$. Now there is some vertex z in T [with $f(z) < k$] such that there is an arc from z to y. By the induction hypothesis, there is a directed path from root r to z and an arc from z to y. So there is a directed path from r to y. Thus there is a directed path from the root to every other vertex.

2.89 Let G be a connected simple graph in which no edge is a bridge, and let T be a DFS spanning tree in the graph. Assume that $f(x)$ is the unique label of vertex x assigned to it during the search, as in Problem 2.88. If $\{v, u\}$ is any edge that is not used as an arc in T and if w is any vertex such that $f(v) < f(w) \leq f(u)$, there is a directed path in T from v to w.

Solution. The proof is by induction on $f(w)$. Initially, let $f(w) = f(v) + 1$. Now $\{v, u\}$ is an edge with $f(u) > f(v)$. So arc (v, w) is in T. In this case, there is a directed path from v to w. Suppose the result is true for vertices with labels less than t, and suppose $f(w) = t$. Then there is a vertex x with $f(x) < f(w)$ so that arc (x, w) is in T; consequently, $f(x) \geq f(v)$. Now $f(x) > f(v)$; otherwise, (v, w) will be in T and there will be a path in T from v to w. Thus there is a path in T from v to x. The arc (x, w) is in T. So there is a path from v to w in T.

2.90 Suppose G and T are as in Problem 2.89. If there is a directed path P in T from vertex a to vertex x and a directed path Q in T from another vertex b to x, either there is a directed path from a to b or from b to a in tree T.

Solution. The proof again is by induction on the length of the path P. Observe that the indegree of each vertex (other than the root) is 1. If (a, x) is in T, path Q uses this arc since (a, x) is the only arc pointing to x in T. So there is a path from b to a. By inductive reasoning, there is an arc (w, x) in T for some w. Now both P and Q must use this arc. Thus there are paths in T from both a and b to w. So there is either a path from a to b or from b to a.

2.91 Suppose G and T are as in Problem 2.89 with the same labeling as before. If $\{u, v\}$ is an edge not in T and if $f(u) < f(v)$, the edge is converted into an arc from v to u. Thus each edge in G is now converted into an arc, giving an orientation G' of graph G based on the depth first search. The length of the (unique) path in T from the root to vertex x is denoted by $d(x)$. If $d(x) \geq 1$ and y is any vertex with $d(y) < d(x)$, there is a directed path in G' from x to y.

Solution. Since $d(x) > 0$, there is a vertex w such that there is an arc in T from w to x. Let $R = \{u$: there is a path in T from x to $u\}$. Let $\bar{R}$ be the (complement) set of vertices not in R. Both these sets are nonempty. So there exist u in R and v in the complement such that there is an edge in G between these two vertices. While orienting this edge, there are four possibilities:

(*i*) There is an arc from u to v in T. This will imply that v is in R, which is a contradiction.

(*ii*) There is an arc from v to u in T. Since there is a path from x to u in T, by Problem 2.90, there is a path from v to x or from x to v in T. Since v is not in R, there cannot be a path in T from x to v. Thus there is a path from v to x and then a path from x to u. This implies that $v = w$ and $x = u$.

(*iii*) There is an arc from v to u in G' but not in T. So $f(v) > f(u)$, implying that (Problem 2.89) that there is a path in T from u to v showing that v is in R.

(*iv*) There is an arc from u to v in G' but not in T.

Since the first three alternatives are ruled out, we conclude that there exist u in R and y in $\bar{R}$ such that there is an arc in G' (but not in T) from u to y. So $f(u) > f(y)$, implying that there is a path in T from y to u. Since there is a path in T from x to u, by Problem 2.90, there has to be a path in T from y to x. Hence $d(y) < d(x)$. Thus there is a path in T from x to u and the arc (u, y) in G'. So there is a directed path in G' from x to y. An inductive argument as before completes the proof.

2.92 Let G and G' be as in Problem 2.91. Show that there is a directed path in G' from every vertex to the root.

Solution. This is proved by induction on $d(x)$. If $d(x) = 0$, x is r. The result is true. Suppose the result is true for all vertices v with $d(v) < t$. Let $d(x) = t$. By Problem 2.91, there is a path from x to vertex y with $d(y) < t$. So by the induction hypothesis, there is a path in G' from y to r.

2.93 Prove Theorem 2.11 (Roberts's theorem): The orientation procedure using the depth first search in a connected graph with no bridges yields a strongly connected digraph. (This is another way of establishing Robbins's theorem. Thus the proof given here may be called Roberts's proof of Robbins's theorem.)

Solution. Let G, T, and G' be as in Problem 2.91. There is a directed path in T (and hence in G') from the root to every other vertex, as established in Problem 2.88. There is a directed path from every vertex to the root, as established in Problem 2.92. So G' is strongly connected.

Cut Vertices, Bridges, and Blocks

2.94 Show that a vertex v of a connected graph is a cut vertex if and only if there exist two distinct vertices u and w such that every path between these two vertices passes through v.

Solution. Let v be a cut vertex in a connected graph G. Then $G - v$ has at least two components. If we choose u from one component and w from another component, any path in G between u and w has to pass through v. On the other hand, suppose there are two vertices in a connected graph such that every path between these two vertices passes through vertex v. If this vertex is deleted, there cannot be a path between these two vertices in the resulting graph. In other words, this deletion makes G disconnected. So v is a cut vertex of G.

2.95 Show that any nontrivial graph has at least two vertices that are not cut vertices.

Solution. We may assume without loss of generality that graph G is connected. The distance $d(u, v)$ between two vertices u and v is the number of edges in a path with as few edges as possible between u and v. Let x and y be two vertices such that $d(x, y)$ is a maximum. Suppose x is a cut vertex. Then $G - x$ has at least two components. Let z be a vertex in a component that does not contain y. In G, there is a path between y and z. Since x is a cut vertex, this path has to pass through x, implying that $d(z, y) > d(x, y)$, which contradicts the maximality of $d(x, y)$. So x is not a cut vertex. Similarly, y is also not a cut vertex.

2.96 Show that an edge of a connected graph is a bridge if and only if there exist vertices v and w such that every path between these two vertices contains this edge.

Solution. The deletion of a bridge from a connected graph creates two connected components of the graph, and any path in the original graph joining a vertex in one component and a vertex in the other component definitely contains the bridge. On the other hand, if there are two vertices such that every path between these two vertices contains an edge of the graph, the deletion of this edge will no doubt disconnect the graph.

2.97 Show that an edge is a bridge if and only if no cycle contains that edge.

Solution. Assume without loss of generality that the graph under consideration is connected. If e is a bridge joining two vertices x and y and if there is a cycle that contains this edge, there is a path in the graph between these two vertices other than the edge. Since the graph is connected, every vertex in the graph is connected to every vertex in the cycle. So the deletion of e will not disconnect the graph, which contradicts that e is a bridge. Conversely, let G be a connected graph and e be an edge joining x and y such that no cycle contains this edge. Suppose e is not a bridge. Then $G - e$ is still a connected graph that has a path joining x and y. This path and the edge e together constitute a cycle containing e in G, contradicting the hypothesis.

2.98 A nontrivial connected graph is called a **nonseparable graph** if it has no cut vertices. A subgraph H of a graph G is a **block in graph G** if H is nonseparable and maximal with respect to this property: If there is a nonseparable subgraph H' such that H is a subgraph of H', $H = H'$. Obtain the blocks of the graph in Fig. 2-21.

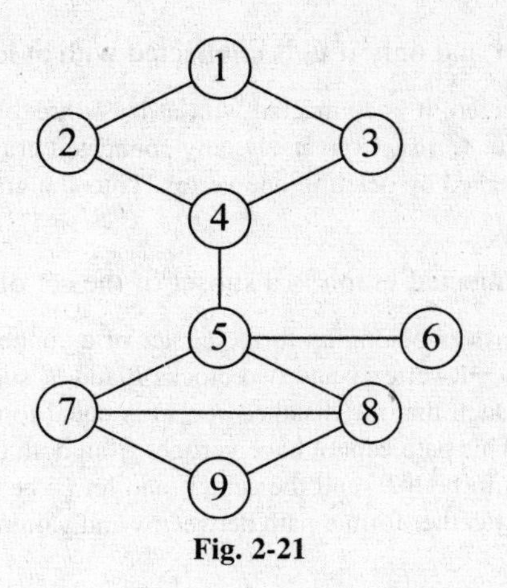

Fig. 2-21

Solution. By definition, each block in a graph is an induced subgraph, and the blocks collectively constitute a partition of the edges of the graph. The graph in Fig. 2-21 has four blocks, as shown in Fig. 2-22.

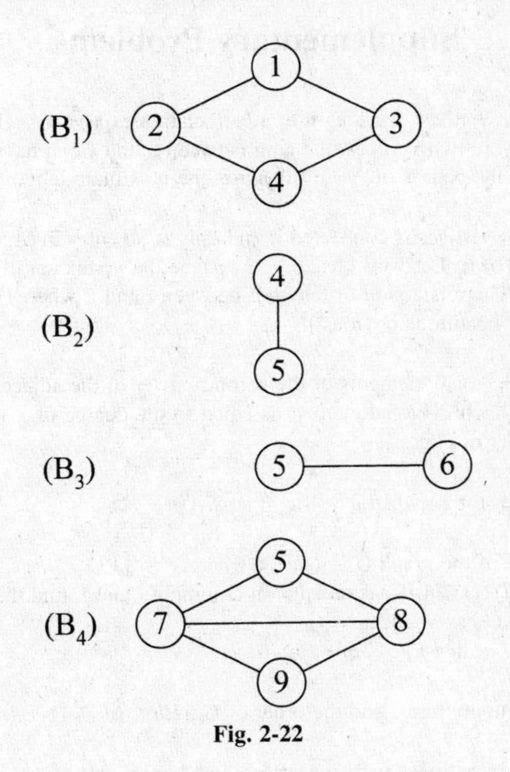

Fig. 2-22

2.99 Show that two blocks have at most one vertex in common. What distinguishing feature does a vertex common to two blocks have?

Solution. We may assume that the graph G under consideration is connected. If a vertex belonging to a block of G is deleted, the graph does not get disconnected. If two blocks B_1 and B_2 have two vertices x and y in common, let $G' = G - x$. In graph B_1, there is a path between every vertex in B_1 to every vertex in $(B_1 \cap B_2) - x$. Similarly, in graph B_2, there is a path between every vertex in B_2 and every vertex in $(B_1 \cap B_2) - x$. Hence, the union of these two blocks is a block violating the maximality of these blocks. If two blocks have a vertex in common, that vertex is necessarily a cut vertex.

2.100 Show that G is 2-connected if and only if G is connected with at least three vertices but no cut vertices.

Solution. If G is 2-connected, it is connected with at least three vertices, and the deletion of a vertex will not disconnect it. So it has no cut vertex. Conversely, any connected graph with at least three vertices and no cut vertices cannot become disconnected by deleting one vertex. Thus the graph is 2-connected.

2.101 Show that the center of a connected graph is a subset of the set of vertices of a block.

Solution. Suppose the vertices belonging to the center of a connected graph are not vertices belonging to the same block. So there exists a cut vertex v and two blocks B and B' such that each block contains vertices from the center. Let x be any vertex such that the distance $d(x, v)$ is equal to the eccentricity $e(v)$, and let P be a path between x and v of length $e(v)$. This path cannot have vertices from both the components. Suppose P has no vertex in B. Let y be a vertex common to both B' and the center, and let P' be a path of minimum length joining y and v. Then the two paths P and P' together form a path between x and y, implying that $e(y) > e(v)$, which contradicts that y is in the center.

Supplementary Problems

2.102 Show that in a graph with n vertices, the length of a path cannot exceed $(n - 1)$ and the length of a cycle cannot exceed n. [*Hint:* If u and v are two vertices, the path between u and v can have at most $(n - 2)$ distinct vertices. So the maximum length of the path is $(n - 1)$. Likewise, the maximum length of a cycle is n.]

2.103 Show that a graph G with n vertices is connected if and only if no entry in $(A + A^2 + \cdots + A^{n-1})$ is zero, where A is its adjacency matrix. [*Hint:* Let $V = \{1, 2, \ldots, n\}$ be the vertex set of G. Suppose G is connected. Let i and j be any two vertices. There is a path of length p between i and j, where $0 < p < n$. So the (i, j) entry in A^p is nonzero. The reverse implication is obvious.]

2.104 Show that the sum of the diagonal elements of the second power of the adjacency matrix is twice the number of edges of the graph. [*Hint:* Each diagonal element is equal to the degree of a vertex. So the sum of the diagonal elements is twice the number of edges.]

2.105 Find the intersection number for K_n when $n > 4$. *Ans.* $(n - 1)$

2.106 Let $T(G)$ be the total graph of the graph $G = (V, E)$, where $V = \{1, 2, \ldots, n\}$. Let d_i be the degree of i in G. (*a*) Find the degree of i in $T(G)$. (*b*) If e is an edge in G joining i and j, find the degree of e in $T(G)$. (*c*) Find the number of edges in $T(G)$ if G has m edges.
Ans. (*a*) $2(\deg i)$, (*b*) $\deg i + \deg j$, (*c*) $2m + \frac{1}{2}\Sigma (d_i)^2$.

2.107 If both G and its complement are trees, find the order of G. [*Hint:* $(n - 1) + (n - 1) = n(n - 1)/2$. So $n = 4$.]

2.108 Find the number of edges in a forest with n vertices and k trees. [*Hint:* $m = (n_1 - 1) + (n_2 - 1) + \cdots + (n_k - 1) = (n - k)$.]

2.109 Show that a tree is a bipartite graph. [*Hint:* A tree has no cycles. Specifically, it has no odd cycles. So it is bipartite.]

2.110 Show that if a tree has exactly two terminal vertices, the degree of every other vertex is 2 and thus it is a path. [*Hint:* $d_1 + d_2 + \cdots + d_{n-2} + 1 + 1 = (2n - 2)$, where each $d_i > 1$. Hence, each $d_i = 2$.]

2.111 Find the number of terminal vertices in a tree with n vertices. [*Hint:* Let $v_1, v_2, \ldots, v_k$ be the vertices of degree 3 or more in a tree with n vertices. Then the number of vertices of degree 1 is $2 + (d_1 + d_2 + \cdots + d_k) - 2k$, where d_i is the degree of v_i.]

2.112 Show that if the degree of every nonterminal vertex in a tree is 3, the number of vertices in the tree is even. [*Hint:* Let k be number of terminal vertices. Solve $k + 3(n - k) = 2(n - 1)$ for k.]

2.113 In a tree with 14 terminal vertices, the degree of every nonterminal vertex is either 4 or 5. Find the number of vertices of degree 4 and of degree 5.
Ans. There are three vertices of degree 4 and two vertices of degree 5.

2.114 Find the number of distinct labeled trees with five vertices. *Ans.* $5^{5-2} = 125$

2.115 If G is the subgraph obtained by deleting an edge from K_n, find the number of spanning trees in G.
Ans. $(n - 2)^{n-3}$

2.116 Show that a bipartite graph is a transitively orientable graph but not conversely. [*Hint:* A bipartite graph has no odd cycle and therefore no triangle. Thus the condition for transitive orientability is vacuously satisfied. On the other hand, consider an orientation C' (of a cycle G) that is not a directed cycle. Here C is not bipartite, but C' is a transitive orientation of C.]

Chapter 3

Eulerian and Hamiltonian Graphs

3.1 EULERIAN GRAPHS AND DIGRAPHS

A trail between two distinct vertices in a connected graph G (it need not be a simple graph) is an **Eulerian trail** if it contains all the edges of G. A circuit that contains all the edges is an **Eulerian circuit.** A graph is said to be **semi-Eulerian** (or **unicursal**) if it has an Eulerian trail. An **Eulerian graph** is a graph that has an Eulerian circuit. In Fig. 3-1(*a*), the trail $\{e_1, e_2, \ldots, e_{11}\}$ is an Eulerian trail between the vertices 1 and 4, and this graph is unicursal but not Eulerian. The graph in Fig. 3-1(*b*), however, is an Eulerian graph with an Eulerian circuit $\{e_1, e_2, \ldots, e_{10}\}$.

Theorem 3.1. In graph $G = (V, E)$, the following four statements are equivalent: (*i*) G is Eulerian, (*ii*) G is connected and each vertex in it is even, (*iii*) G is connected and there exists a partition of E such that the edges in each subset of the partition constitutes a cycle in G, and (*iv*) G is connected and each edge in G is an edge in an odd number of cycles in G. (Since we are considering graphs that are not necessarily simple, the number of edges in a cycle could be less than three.) (See Solved Problems 3.2, 3.5, and 3.6.)

Example 1. In the connected graph of Fig. 3-1(*b*), the degree of each vertex is even. The set of edges can be partitioned into $E_1 = \{e_1, e_4, e_8\}$, $E_2 = \{e_2, e_3\}$, $E_3 = \{e_5, e_6, e_7\}$, and $E_4 = \{e_9, e_{10}\}$ such that the edges in each subset constitute a cycle in the Eulerian graph. The edge e_1 joining vertices 1 and 2 is an edge that belongs to the following 15 cycles in the graph:

1. 1—2—3—1
2. 1—2—4—3—1 (2—4 is e_2)
3. 1—2—4—3—1 (2—4 is e_3)
4. 1—2—4—5—1 (2—4 is e_2, and 5—1 is e_9)
5. 1—2—4—5—1 (2—4 is e_3, and 5—1 is e_9)
6. 1—2—4—5—1 (2—4 is e_2, and 5—1 is e_{10})
7. 1—2—4—5—1 (2—4 is e_3, and 5—1 is e_{10})
8. 1—2—3—5—1 (5—1 is e_9)
9. 1—2—3—5—1 (5—1 is e_{10})
10. 1—2—3—4—5—1 (5—1 is e_9)
11. 1—2—3—4—5—1 (5—1 is e_{10})
12. 1—2—4—3—5—1 (2—4 is e_2, and 5—1 is e_9)
13. 1—2—4—3—5—1 (2—4 is e_2, and 5—1 is e_{10})
14. 1—2—4—3—5—1 (2—4 is e_3, and 5—1 is e_9)
15. 1—2—4—3—5—1 (2—4 is e_3, and 5—1 is e_{10})

We can obtain an Eulerian circuit in G because the set of edges in an Eulerian graph G is the union of edge-disjoint cycles. The algorithm is as follows:

Step 1. Start from any vertex v and construct a cycle C.

Step 2. If C contains all the edges of the graph, stop. If not, select a vertex w common to C and the subgraph G' obtained from G by deleting all the edges of C from it.

Step 3. Starting from w, construct a cycle in G', say C'.

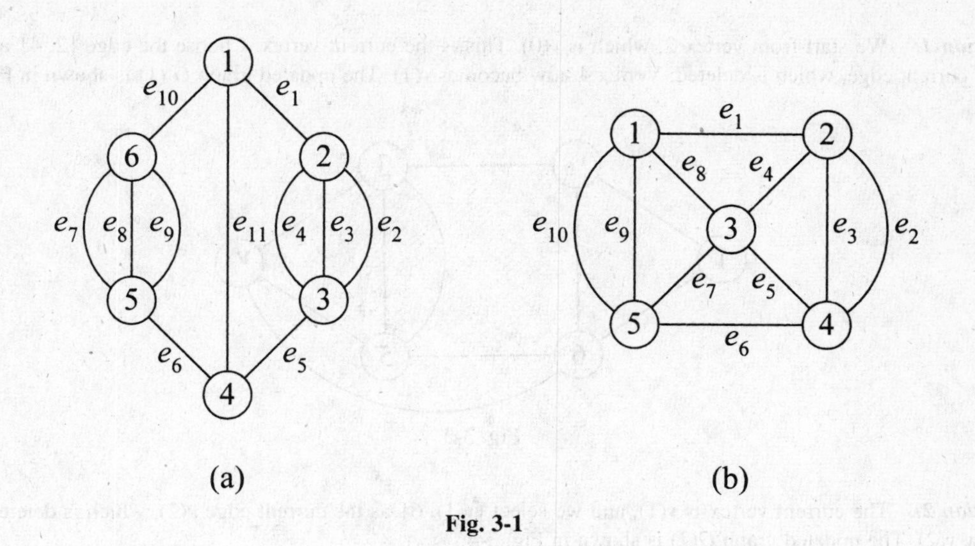

Fig. 3-1

Step 4. Combine the edges of C and C' to obtain a circuit in G. The new circuit is C. Return to Step 2.

(See Solved Problem 3.5.)

We now present another method (known as **Fleury's algorithm**) to obtain an Eulerian circuit starting from any vertex in a connected graph $G = (V, E)$ in which each vertex is even.

Step 1. Initially, $i = 0$. Start from vertex v_0 and define $T_0 : v_0$.

Step 2. Let $T_i = v_0, e_1, v_1, e_2, \ldots, e_i, v_i$ be the trail between v_0 and v_i at stage i. Select an edge e_{i+1} joining v_i and v_{i+1} from the set $E_i = E - \{e_1, e_2, \ldots, e_i\}$. If edge e_{i+1} is a bridge in the subgraph obtained from G after deleting the edges belonging to E_i from E, select it for inclusion in the updated trail $T_{i+1} = v_0, e_1, v_1, e_2, \ldots, e_i, v_i, e_{i+1}, v_{i+1}$ only if there is no other choice. If there is no such edge, stop.

Step 3. Replace i by $i + 1$ and go to step 2.

(See Solved Problem 3.7.)

Example 2. In the connected graph G in Fig. 3-2 (with six vertices and 11 edges), the degree of each vertex is even. Using Fleury's algorithm, an Eulerian circuit in this graph is obtained as follows. There will be 11 iterations, and at the end, we will have the subgraph with six vertices in which the degree of each vertex is zero.

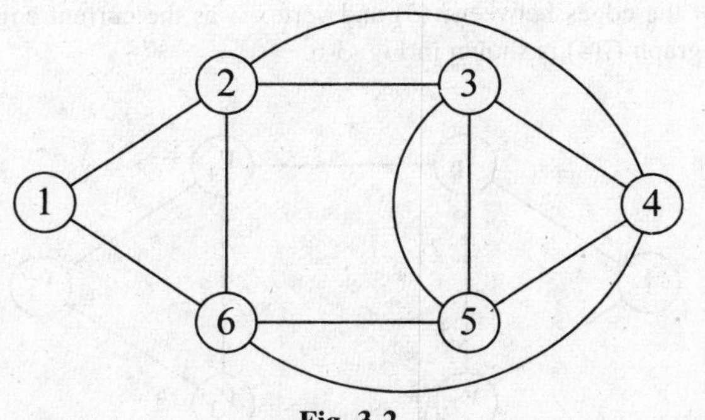

Fig. 3-2

Iteration 1: We start from vertex 2, which is $v(0)$. This is the current vertex. Choose the edge $\{2, 4\}$ as $e(1)$. This is the current edge, which is deleted. Vertex 4 now becomes $v(1)$. The updated graph $G(1)$ is shown in Fig. 3-3.

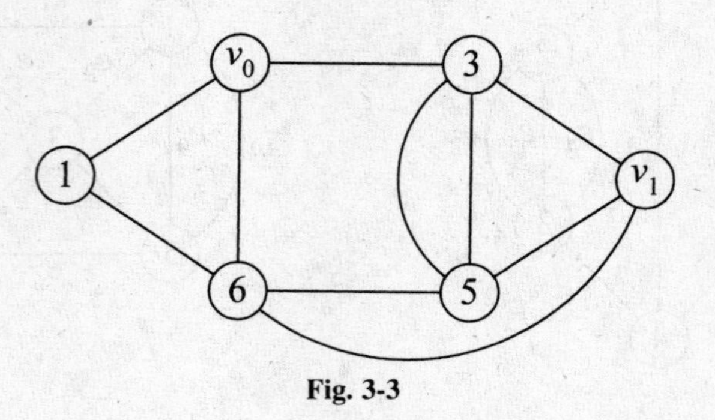

Fig. 3-3

Iteration 2: The current vertex is $v(1)$, and we select $\{v(1), 6\}$ as the current edge $e(2)$, which is deleted. Vertex 6 is now $v(2)$. The updated graph $G(2)$ is shown in Fig. 3-4.

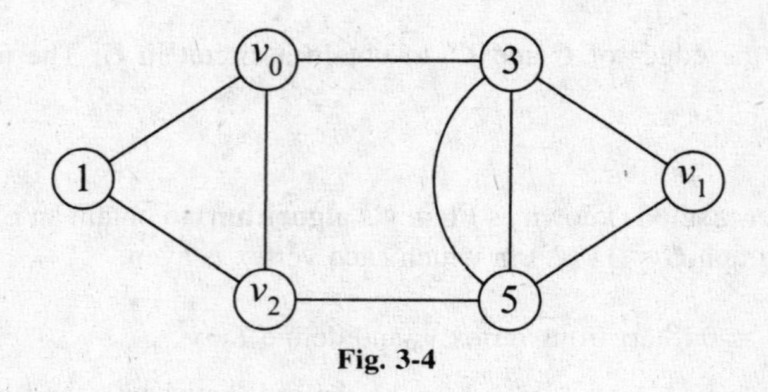

Fig. 3-4

Iteration 3: Select $\{v(2), 5\}$ as $e(3)$, and delete it. Vertex 5 is $v(3)$. The updated graph $G(3)$ is shown in Fig. 3-5.

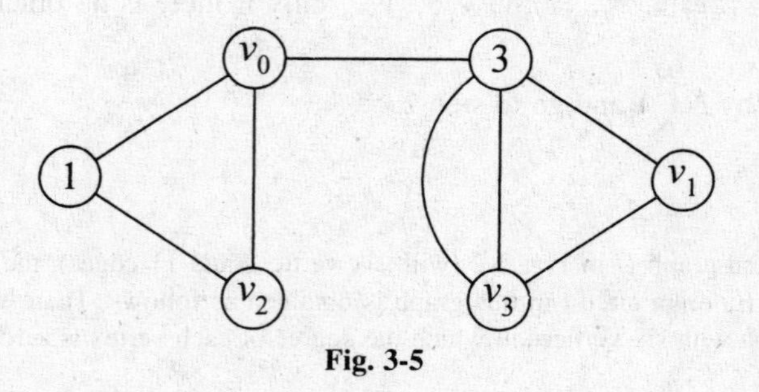

Fig. 3-5

Iteration 4: Select one of the edges between $v(3)$ and vertex 3 as the current edge $e(4)$, and delete it. Vertex 3 is now $v(4)$, and the updated graph $G(4)$ is shown in Fig. 3-6.

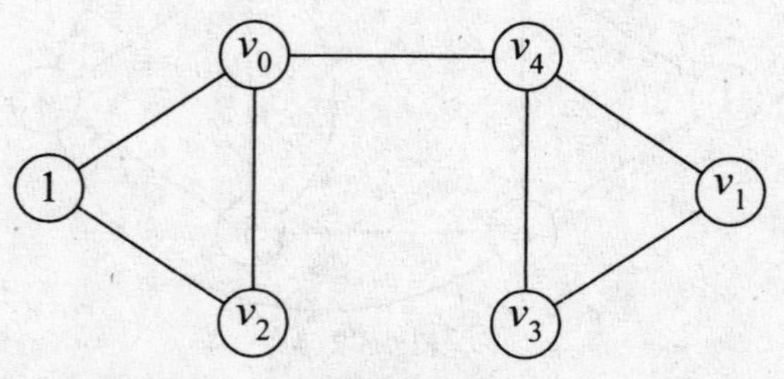

Fig. 3-6

Iteration 5: The current vertex is $v(4)$, and we have to select an edge incident with this vertex. **The edge joining $v(4)$ and $v(0)$ is not selected since it is a bridge.** Instead, we select the edge joining $v(4)$ and $v(1)$ as the edge $e(5)$ to be deleted. At this stage, vertex $v(1)$ gets the updated label $v(5)$. The updated graph $G(5)$ is shown in Fig. 3-7.

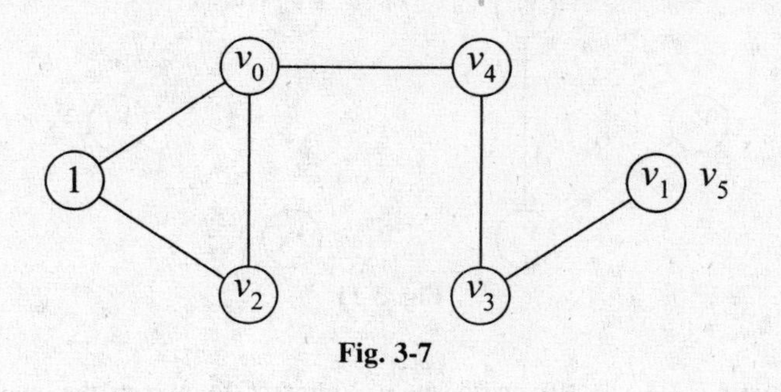

Fig. 3-7

Iteration 6: The current vertex is $v(5)$, and the **only edge** incident to this vertex is the **bridge** that joins this vertex and $v(3)$, which now gets the label $v(6)$. Bridge $e(6)$ is deleted, making $v(5)$ as an isolated vertex, as shown in Fig. 3-8.

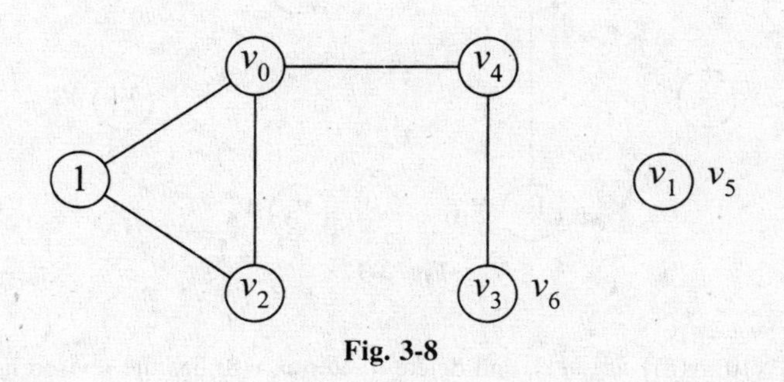

Fig. 3-8

Iteration 7: Select the edge joining $v(6)$ and $v(4)$ as $e(7)$, and delete it. The new label for $v(4)$ is $v(7)$, as shown in Fig. 3-9.

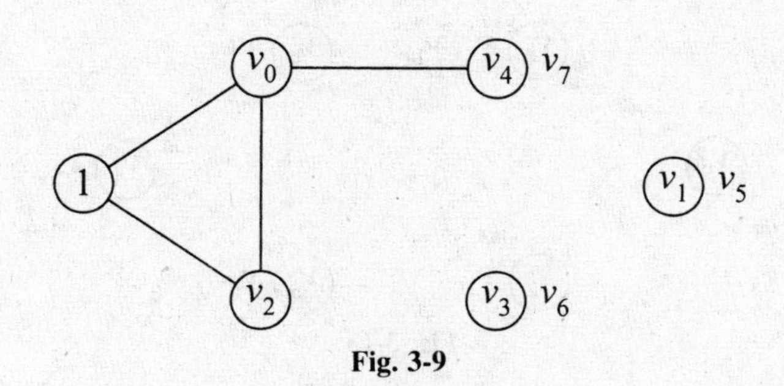

Fig. 3-9

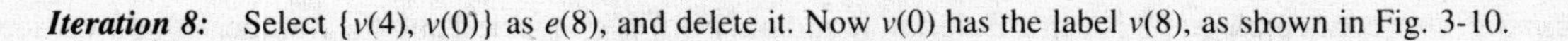

Iteration 8: Select $\{v(4), v(0)\}$ as $e(8)$, and delete it. Now $v(0)$ has the label $v(8)$, as shown in Fig. 3-10.

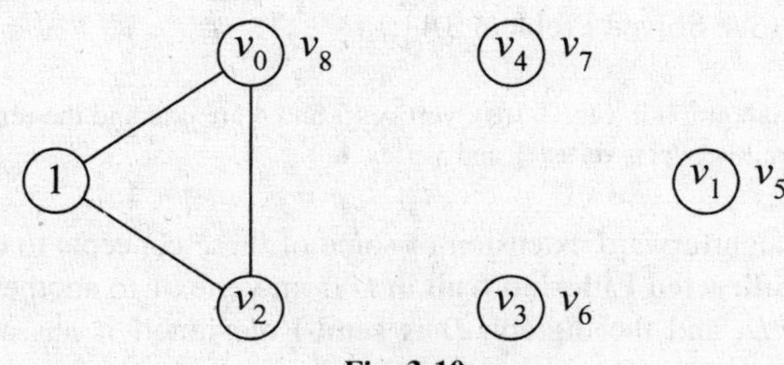

Fig. 3-10

Iteration 9: Select $\{v(8), 1\}$ as $e(9)$, and delete it. Vertex 1 has the label $v(9)$, as shown in Fig. 3-11.

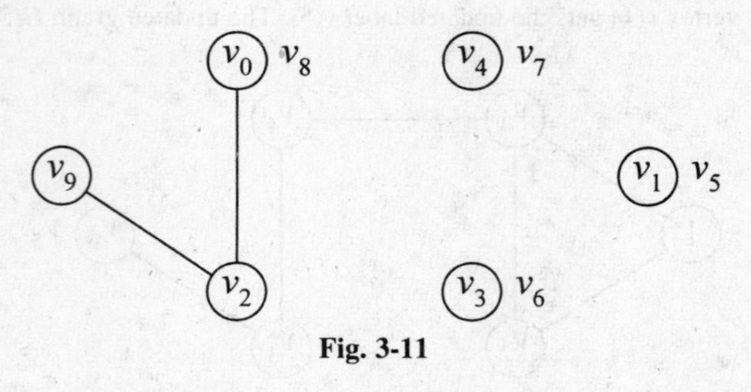

Fig. 3-11

Iteration 10: Select $\{v(9), v(2)\}$ as $e(10)$, and delete it. Vertex $v(2)$ now gets the revised label $v(10)$, as shown in Fig. 3-12.

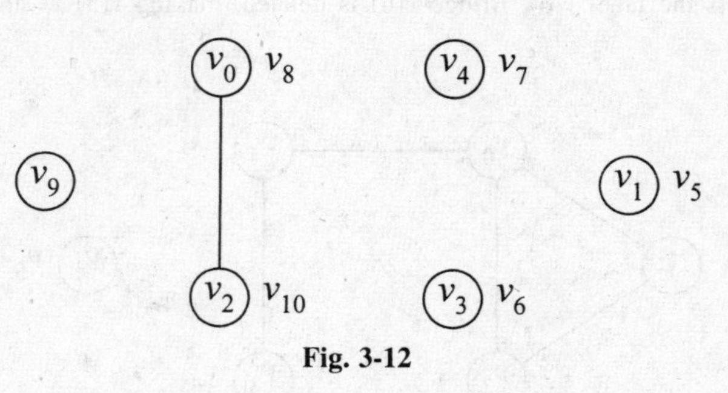

Fig. 3-12

Iteration 11: Select $\{v(10), v(8)\}$ as $e(11)$, and delete it. Vertex $v(8)$ has the revised label $v(11)$, as shown in Fig. 3-13. At this stage, we have a graph with no edges, and the algorithm terminates. The sequence of edges $e(1), e(2), \ldots, e(3)$ gives an Eulerian circuit in the graph.

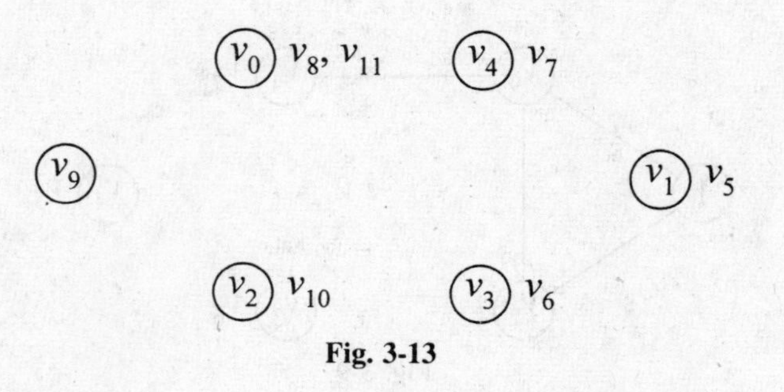

Fig. 3-13

Theorem 3.2. A connected graph is semi-Eulerian if and only if it is connected and the number of odd vertices in it is exactly two. Furthermore, in a semi-Eulerian graph, any Eulerian trail is between its two odd vertices. (See Solved Problem 3.11.)

Example 3. In the semi-Eulerian graph in Fig. 3-1(*a*), vertices 1 and 4 are odd and the remaining vertices are even. Any Eulerian trail in this graph is a trail between vertex 1 and vertex 4.

We now consider the straightforward extension of some of these concepts to directed graphs. Let D be a weakly connected digraph. A **directed Eulerian trail** in D from vertex v to another vertex w is a directed trail that contains all the arcs of D, and the digraph D is **semi-Eulerian** if it has a directed Eulerian trail. A

directed Eulerian circuit in D is a directed circuit that contains all its arcs, and the digraph is **Eulerian** if it contains all its arcs. Figure 3-14(a) shows an Eulerian trail $1 \to 5 \to 1 \to 4 \to 5 \to 4 \to 3 \to 2 \to 5 \to 2$ from vertex 1 to vertex 2, so this digraph is semi-Eulerian. Figure 3-14(b) shows a digraph with an Eulerian circuit $1 \to 2 \to 3 \to 2 \to 5 \to 3 \to 4 \to 1 \to 4 \to 5 \to 1$, and so this digraph is Eulerian.

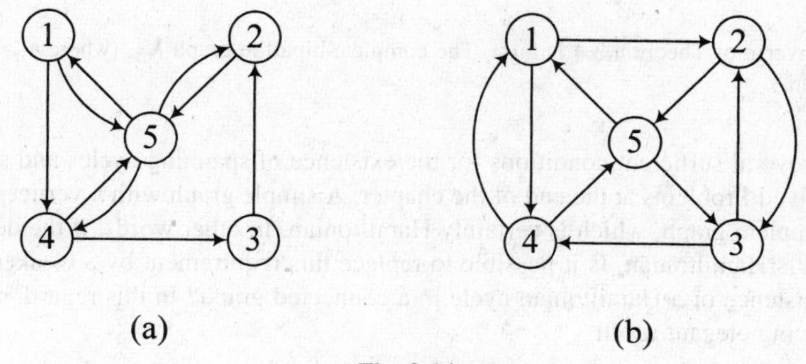

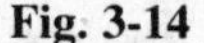

Fig. 3-14

The following theorem characterizes Eulerian and semi-Eulerian digraphs.

Theorem 3.3. (i) A weakly connected digraph is Eulerian if and only if the indegree of each vertex is equal to its outdegree. (ii) A weakly connected digraph is semi-Eulerian if and only if there are two vertices v and w such that (a) (outdegree of v) = (indegree of v) + 1, (b) (outdegree of w) = (indegree of w) − 1, and (c) the outdegree of every other vertex is equal to its indegree. In a weakly connected digraph satisfying these three properties, any directed Eulerian trail is from v to w. (See Solved Problems 3.12 and 3.13.)

Example 4. In the digraph of Fig. 3-14(a), (outdegree of vertex 1) = (its indegree) + 1 and (outdegree of vertex 2) = (its indegree) − 1, and for every other vertex, both the outdegree and indegree are equal. Any directed Eulerian trail is from vertex 1 to vertex 2. In the digraph of Fig. 3-14(b), the outdegree and indegree are the same for every vertex, and it has a directed Eulerian circuit.

3.2 HAMILTONIAN GRAPHS AND DIGRAPHS

A path between two vertices in a graph is a **Hamiltonian path** if it passes through every vertex of the graph. A closed Hamiltonian path is called a **Hamiltonian cycle** in the graph. In other words, a Hamiltonian path is a spanning path, and a Hamiltonian cycle is a spanning cycle. A **Hamiltonian graph** is a graph that has a Hamiltonian cycle. Every Hamiltonian graph has a Hamiltonian path, but the converse is not true; consider the graph that is a path. In our discussion of Hamiltonian graphs, we restrict our attention to simple graphs because the trails under consideration pass through a vertex at most once. In the case of directed graphs, the definitions are analogous.

Unlike Eulerian graphs, there is no elegant (easily testable) characterization of Hamiltonian graphs, even though many necessary conditions and many sufficient conditions are known for the existence of spanning cycles and spanning paths in connected graphs and weakly connected digraphs. Of course, any Hamiltonian graph is necessarily 2-connected since the deletion of a vertex from the graph results in a connected graph that has a Hamiltonian path. Consequently, no graph with a cut vertex is Hamiltonian. Every cyclic graph is Hamiltonian, as is every complete graph with three or more vertices. A complete bipartite graph $K_{m,n}$ is Hamiltonian if and only if $m = n$.

The following necessary condition is a straightforward generalization of the fact that no vertex of a Hamiltonian graph is a cut vertex.

Theorem 3.4. **(A necessary condition for a graph to be Hamiltonian).** If $G = (V, E)$ is Hamiltonian and if W is any nonempty proper subset of V, the graph $G - W$ has at most $|W|$ components.

Proof. Let G_i ($i = 1, 2, \ldots, k$) be the distinct components of $G - W$, and let C be any Hamiltonian cycle in graph G. If the last (considered clockwise) vertex in C that belongs to G_i is denoted by u_i, the vertex in the cycle immediately after u_i is necessarily in W for each i. So each component defines a unique vertex in W. Hence, the number of distinct components cannot exceed the cardinality of W.

Example 5. The converse of Theorem 3.4 is false. The complete bipartite graph $K_{2,n}$ (where $n > 2$) has no cut vertex, but it is not Hamiltonian.

We will obtain several sufficient conditions for the existence of spanning cycles and spanning paths in an arbitrary graph in Solved Problems at the end of the chapter. A simple graph with n vertices and as many edges as possible is the complete graph, which is certainly Hamiltonian. In other words, if the degree of each vertex is $(n - 1)$, the graph is Hamiltonian. Is it possible to replace this requirement by a weaker condition that will still guarantee the existence of a Hamiltonian cycle in a connected graph? In this regard, one such sufficiency criterion is the following elegant result.

Theorem 3.5 **(Ore's Theorem: A sufficient condition for a graph to be Hamiltonian).** A simple graph with n vertices (where $n > 2$) is Hamiltonian if the sum of the degrees of every pair of non-adjacent vertices is at least n.

Proof. Suppose a graph G with n vertices satisfying the given inequality condition is not Hamiltonian. So it is a subgraph of the complete graph K_n with fewer edges. We recursively add edges to the graph by joining nonadjacent vertices until we obtain a graph H such that the addition of one more edge joining two nonadjacent vertices in H will produce a Hamiltonian graph with n vertices. Let x and y be two nonadjacent vertices in H. Thus they are nonadjacent in G also. Since the sum of the degree of x and the degree of y is at least n in G, this sum is at least n in H as well. If we join the nonadjacent vertices x and y, the resulting graph is Hamiltonian. Hence, in graph H, there is a Hamiltonian path between the vertices x and y. If we write $x = v_1$ and $y = v_n$, this Hamiltonian path can be written as $v_1—v_2—\cdots—v_{i-1}—v_i—v_{i+1}—\cdots—v_{n-1}—v_n$. Suppose the degree of v_1 is r in graph H. If there is an edge between v_1 and v_i in this graph, the existence of an edge between v_{i-1} and v_n will imply that H is Hamiltonian. So whenever vertices v_1 and v_i are adjacent in H, vertices v_n and v_{i-1} are not adjacent. This is true for $i = 2, 3, \ldots, (n - 1)$. Hence, the degree of v_n cannot exceed $(n - 1) - r$ since the degree of v_1 is r. This implies that the sum of the degrees of the two nonadjacent (in G) vertices is less than n, which contradicts the hypothesis. So any connected graph satisfying the given condition is Hamiltonian.

Theorem 3.6 **(Dirac's Theorem: A sufficient condition for a graph to be Hamiltonian).** A simple graph with n vertices (where $n > 2$) is Hamiltonian if the degree of every vertex is at least $n/2$.

Proof. If each degree is at least $n/2$, the sum of every pair of vertices is at least n. In particular, the sum of every pair of nonadjacent vertices is at least n. So by Ore's theorem, the graph is Hamiltonian.

Example 6. The converses of both Theorem 3.5 and Theorem 3.6 are false. The cyclic graph with five or more vertices is Hamiltonian, but the degree of every vertex in that graph is only 2. The sum of the degrees of every pair of vertices is only 4.

A graph is said to be **Hamilton-connected** if there is a Hamiltonian path in that graph between every pair of vertices. Obviously, every Hamiltonian-connected graph is Hamiltonian. But a Hamiltonian graph need not be Hamiltonian-connected; the cyclic graph with more than three vertices is Hamiltonian but not Hamiltonian-connected. The following sufficient condition for a graph to be Hamilton-connected is also due to O. Ore (1963).

Theorem 3.7. A graph G with n vertices (n is at least three) is Hamilton-connected if the sum of the degrees of any two nonadjacent vertices is more than n. In particular, it is Hamilton-connected if the degree of each vertex is more than $n/2$. (See Solved Problem 3.60.)

Example 7. Figure 3-15 shows a graph that satisfies the sufficiency criterion (first part) stated in Theorem 3.7.

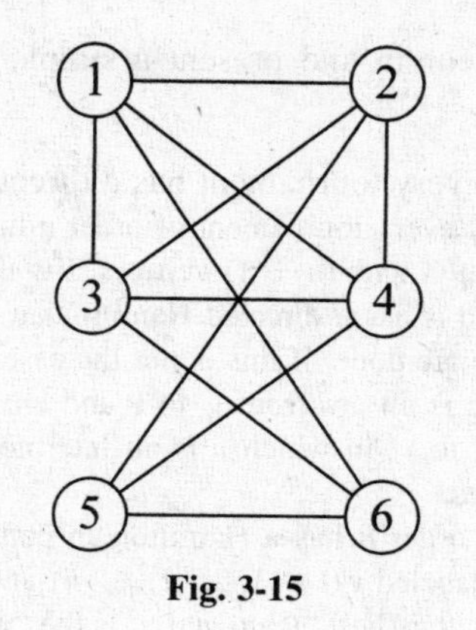

Fig. 3-15

We conclude this section with a brief discussion on Hamiltonian digraphs. Again there is no nice characterization of these graphs. We state two sets of theorems stating sufficient conditions—one set for the existence of directed spanning cycles and the other for the existence of directed spanning paths—in an arbitrary digraph. The interested reader may refer to *Graphs and Digraphs* by Mehdi Behzad, Gary Chartrand, and Linda Lesniak-Foster (1979) for proofs of these assertions.

***Theorem 3.8*(a)** **(Meyniel's Theorem).** A strongly connected digraph with n vertices is Hamiltonian if the sum of the degrees of every pair of nonadjacent vertices is at least $(2n - 1)$.

Theorem 3.8(b) **(Woodall's Theorem).** A nontrivial digraph with n vertices is Hamiltonian if (outdegree of u) + (indegree of v) is at least n whenever u and v are vertices such that there is no arc from u to v.

Theorem 3.8(c) **(Ghoula–Houri Theorem).** A strongly connected digraph with n vertices is Hamiltonian if the degree of each vertex is at least n.

Theorem 3.8(d). A digraph with n vertices is Hamiltonian if both the indegree and outdegree of each vertex are at least $n/2$.

***Theorem 3.9*(a).** A digraph with n vertices has a directed Hamiltonian path if the sum of the degrees of every pair of nonadjacent vertices is at least $(2n - 3)$.

Theorem 3.9(b). A digraph with n vertices has a directed Hamiltonian path if (outdegree of u) + (indegree of v) is at least $(n - 1)$ whenever u and v are vertices such that there is no arc from u to v.

Theorem 3.9(c). A digraph with n vertices has a directed Hamiltonian path if the degree of each vertex is at least $(n - 1)$.

Theorem 3.9(d). A digraph with n vertices has a directed Hamiltonian path if both the indegree and outdegree of each vertex are at least $(n - 1)/2$.

3.3 TOURNAMENTS

The digraph obtained by converting each edge of the complete graph K_n into an arc is known as a **tournament** with n vertices. In other words, any tournament is an orientation of a complete graph. A digraph of this kind can be used to record the results of games in a tournament in which each team (player) plays against every other team (player) in a match such that no match ends in a draw. If the match between two teams ends in a draw, they continue to play against each other until one of them becomes the winner. A typical vertex v represents a team, and the presence of an arc from v to another vertex w indicates that in the match between v and w, the winner is v. The outdegree of a vertex v is the number of matches in which v is the winner, and its indegree is the number of matches v lost during the tournament. The sum of the outdegree and the indegree of each vertex is the order of the digraph, hence, by Theorem 3.9(c), we notice that every tournament has a directed

Hamiltonian path. We state this as a theorem and present a simple proof by induction on the order of the digraph.

Theorem 3.10 **(Redei's Theorem).** Every tournament has a directed Hamiltonian path.

Proof. The induction hypothesis is that every tournament of order n has a Hamiltonian path. If $n = 2$, the theorem is true. Consider any tournament G with $(n + 1)$ vertices. If v is a vertex of this tournament, graph $G - v$ is a tournament with n vertices, so it has a directed Hamiltonian path: $v_1 \rightarrow v_2 \rightarrow \cdots \rightarrow v_n$. If there is an arc from v to v_1 or from v_n to v, we are done. If this is not the case, let i be the largest integer such that there is no arc from v to v_i. Thus there is an arc from v_i to v and an arc from v_{i+1} to v. Then we have a directed Hamiltonian path in G from v_1 to v_n in which v is an intermediate vertex. So the theorem is true for any tournament with $(n + 1)$ vertices.

That every tournament of order n has a Hamiltonian path implies that once a Hamiltonian path P is discerned, the vertices can be labeled $v_i (i = 1, 2, \ldots, n)$ and ranked such that v_i defeats v_{i+1}. Then, according to this choice of P, v_1 is the "best" team and v_n is the "worst" team. But a tournament can have more than one Hamiltonian path or a unique Hamiltonian path. Figure 3-16(*a*) shows four Hamiltonian paths, suggesting that each team could be the "best" as well as the "worst," whereas Fig. 3-16(*b*) shows a unique Hamiltonian path from vertex 1 to vertex 4.

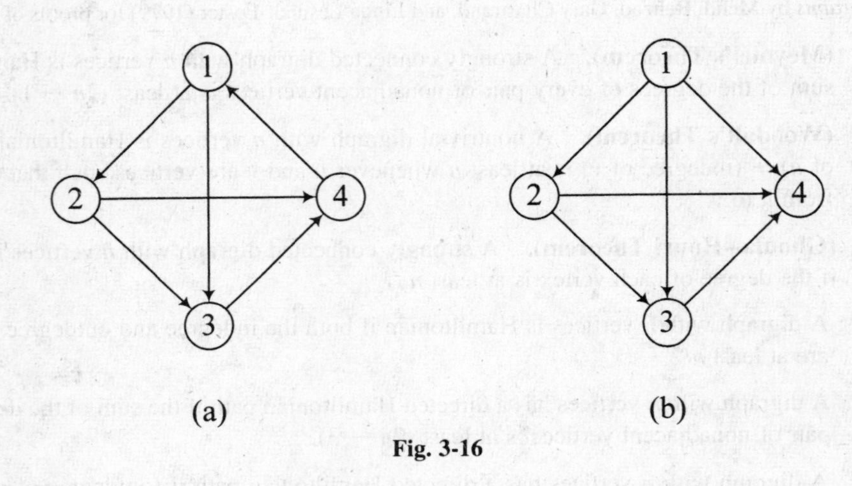

Fig. 3-16

Since every tournament has a Hamiltonian path, it is natural to find the conditions under which a tournament will have a unique Hamiltonian path. A tournament is called a **transitive tournament** if u defeats w whenever u defeats v and v defeats w. Observe that this definition implies that a tournament is transitive if and only if it has no directed cycles consisting of three arcs. A transitive tournament can be characterized in more than one way, as can be seen from the following two theorems. Theorem 3.12 settles the uniqueness problem.

Theorem 3.11. A tournament is transitive if and only if it is acyclic.

Proof. Let G be an acyclic tournament. Suppose u, v, and w are three vertices in the graph. If there is an arc from u to v and an arc from v to w, there has to be an arc from u to w since the graph is acyclic. So an acyclic tournament is necessarily transitive. On the other hand, consider a transitive tournament. By definition, it cannot have a cycle consisting of three arcs. If it has a cycle with more than three arcs, it should, again by transitivity, have a cycle with three arcs, which is an impossibility. Thus a transitive tournament is acyclic.

Theorem 3.12. A tournament has a unique Hamiltonian path if and only if it is a transitive tournament.

Proof. If there are two Hamiltonian paths in a transitive tournament G, there will be two vertices u and v such that in one path u precedes v and in the other v precedes u. Then, due to transitivity, there is an arc from u to v as well as an arc from v to u, which is a contradiction. So if a tournament is transitive, it has a unique Hamiltonian path. To prove the converse, consider a tournament G with a unique Hamiltonian path P: $v_1 \rightarrow v_2 \rightarrow \cdots v_n$. If $i < j$ implies that there is an arc from v_i to v_j, the tournament is obviously transitive. Suppose this is not the case. Select the smallest i such that for some j with $i < j$, there is an arc from v_j to v_i. Once i is selected, choose j as large as possible. If $1 < i$ and $j < n$, there is an arc from v_{i-1} to v_{i+1}. There is also an arc from v_i to v_{j+1}. So we have another Hamiltonian path:

$$v_1 \to v_2 \to \cdots \to v_{i-1} \to v_{i+1} \to \cdots \to v_j \to v_i \to v_{j+1} \to \cdots \to v_n$$

which is not the same as the path P. If $i = 1$ and $j < n$, we have the spanning path $v_2 \to \cdots \to v_j \to v_1 \to v_{j+1} \to v_n$. If $1 < i$ and $j = n$, there is the additional spanning path $v_1 \to v_{i-1} \to v_{i+1} \to v_n \to v_i$, which is a contradiction. Finally, if $i = 1$ and $j = n$, we get a Hamiltonian cycle in the graph. Thus, if we assume that the tournament is not transitive, the uniqueness requirement is violated. So any tournament with a unique spanning path is transitive.

We conclude this section with a nice characterization of Hamiltonian tournaments. For a proof of this theorem and many related results in tournaments, see the section on tournaments in the Solved Problems at the end of the chapter.

Theorem 3.13 **(Camion's Theorem).** A tournament is Hamiltonian if and only if it is strongly connected. (See Solved Problem 3.80.)

Solved Problems

EULERIAN GRAPHS AND DIGRAPHS

3.1 If the degree of each vertex in a graph is at least two, show that there is a cycle in the graph.

Solution. If there is a loop at a vertex, that loop can be considered a cycle. If there is more than one edge between two vertices, any two edges joining two vertices will form a cycle. Suppose the graph is simple. Let v_0 be any vertex in the graph, and let e_1 be the edge joining this vertex and vertex v_1. Now there exists a third vertex v_2 and edge e_2 joining v_1 and v_2. This process of finding new vertices and edges is repeated, and at the kth stage, we have edge e_k joining vertices v_{k-1} and v_k and a path from v_0 to v_k consisting of k edges. Since the number of vertices in the graph is finite, we must ultimately choose a vertex that has been chosen before. Suppose v_i is the first repeated vertex in this process. Then the path between the two occurrences of this repeated vertex is a cycle.

3.2 (*Euler–Hierholzer Theorem*) Show that a connected graph is Eulerian if and only if the degree of each vertex of the graph is even.

Solution. Whenever a circuit passes through an arbitrary vertex of the graph, two distinct edges are used up; as such, each such passing results in a contribution of two to the degree of that vertex. So if there is a circuit that contains all the edges of the graph, the degree of each vertex is necessarily even. To prove the converse, we use induction on the size of a connected graph. If there are no edges, there is only one vertex in the graph since the graph is connected; thus the problem is trivial. Suppose the degree of each vertex is even. Since the graph is connected, each degree is at least 2. So by Problem 3.1, there is a cycle C in the graph. If this cycle contains all the edges, we are done. Suppose this is not the case. Then all the edges belonging to this cycle are deleted from the graph, resulting in a spanning subgraph H (with fewer edges) that need not be connected. But each vertex in H is even. The induction hypothesis is that every graph with fewer edges in which each degree is even is Eulerian. So each component of H is Eulerian. Furthermore, each component has a vertex in common with the cycle C. We can thus obtain an Eulerian circuit in G as follows. Start from any vertex of C and traverse the edges of this cycle sequentially until we reach vertex v_1 that is also a vertex in a (nontrivial) component that is Eulerian. Then we traverse all the edges of this component sequentially starting from v_1 and return to it, and then continue along the edges of C until we find another vertex v_2 that is a vertex in another component. We repeat this process and eventually return to the starting vertex in C, thereby obtaining an Eulerian circuit in the graph.

3.3 (*The Königsberg Bridge Problem*) The two islands (say the East Island and the West Island) in the river Pregel (known as Pregolya these days) that flows from east to west through the city of Königsberg (now known as Kaliningrad) in eastern Prussia (now a part of Russia) were connected by a bridge. Two bridges connected the west island (W) to the north shore (N), and two bridges connected it to the south shore (S). A bridge connected the east island (E) and the north shore, and another bridge connected it to the south shore. Show that the following problem posed to Leonard Euler (in 1736) by the citizens of Königsberg is not solvable: Start from one of these four land masses in the city and return to that point after walking along each bridge exactly once.

Solution. Consider the graph (Fig. 3-17) with vertices denoted by N, S, E, and W. Each bridge is represented by an edge joining the corresponding vertices. We thus have a connected graph with four vertices and seven edges in which the degree of each vertex is odd. So by Euler's theorem, this problem has no solution.

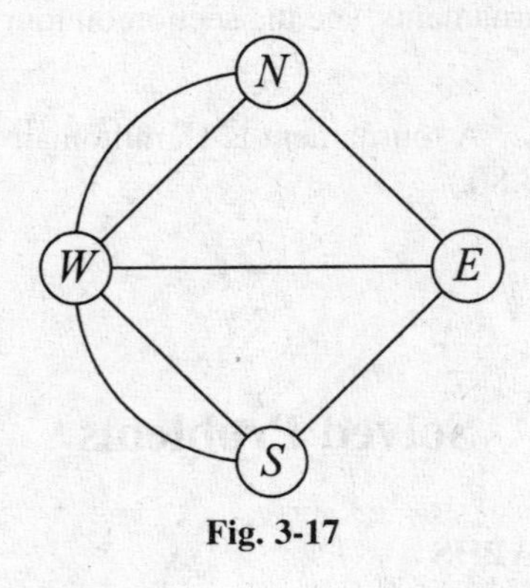

Fig. 3-17

3.4 Solve the modified Königsberg bridge problem by (*a*) deleting two edges from the graph, (*b*) constructing two new edges, and (*c*) deleting an edge and constructing an edge.

Solution.

(*a*) Delete the edges $\{N, W\}$ and $\{S, E\}$ in Fig. 3-17.

(*b*) Join N and E by an edge, and then join S and W by another edge.

(*c*) Delete the edge joining W and E, and then join N and S by an edge. In each case, we end up with a connected graph in which the degree of each vertex is even.

3.5 Show that a graph is Eulerian if and only if it is connected and if the set of its edges can be partitioned into a disjoint union of cycles.

Solution. Suppose G is an Eulerian graph. Then each vertex in G is even and of degree at least 2. So it is not a tree since it does not have a vertex of degree 1. (See Solved Problem 2.60.) Thus there is at least one cycle C_1 in the graph. If G is not this cycle, let G_1 be the subgraph (possibly disconnected) obtained from G by deleting all the edges belonging to the cycle. Since every vertex in a cycle is of degree 2, every vertex in G_1 is also even and, as before, has cycle C_2. Let $G_2 = G_1 - C_2 = G - C_1 - C_2$. We repeat this process of identifying cycles until we get the graph $G_k = G - C_1 - C_2 - \ldots - C_k$ with no edges. Thus the set of edges is the disjoint union of these k cycles. Conversely, suppose the set of edges in a connected graph G is the disjoint union of k cycles. Consider any one of these cycles, say cycle C_1. Since the graph is connected, there is a cycle (say C_2) such that the two cycles have a vertex v_1 in common. Let Q_{12} be the circuit that consists of all the edges in these two cycles. As before, there is a cycle C_3 such that this cycle and circuit Q_{12} have no edge in common but do have vertex v_2 in common. Let Q_{123} be the circuit that contains all the edges of these three edge-disjoint cycles. We repeat this procedure until we get a circuit that contains all the edges of the graph. Thus the graph is Eulerian.

3.6 (*Toida-McKee Theorem*) Show that a connected graph is Eulerian if and only if every edge in it belongs to an odd number of cycles.

Solution. Let e be an edge joining two vertices x and y in an Eulerian graph G. The subgraph $G' = G - e$ is necessarily a connected graph. Let X be the set of trails in G' between x and y in which y is not a repeated vertex. Since the degree of x in G' is odd, the number of ways of choosing the *first* edge in the trail (considering it as a trail *from x to y*) is odd. Once the first edge is chosen, the number of ways of choosing a *second* edge in the trail is also odd. So $|X|$ is odd. Let Y be the set of trails in X with repeated vertices. Suppose v is a repeated vertex in a trail T, and let Q be the circuit consisting of the edges in T between any two occurrences of vertex v. Then, by reversing the order in which the edges of circuit Q appear in T, we can construct a new trail T'. So $|Y|$ is an even number. Thus the number of paths between x and y is $|X| - |Y|$, which is an odd number. So edge e belongs to an odd number of cycles in the graph. Conversely, let G be a connected graph with the property that every edge belongs to an odd number of cycles. Suppose G is not Eulerian. Let v be an odd vertex in G. For each edge e incident with v, let $r(e)$ be the number of cycles that contain edge e. Any cycle passing through v contains two edges incident at v. So the sum $\Sigma r(e)$ (where the summation is over all the edges e incident at v) is even. But by hypothesis, $r(e)$ is odd, so the number of terms in the summation is odd. This contradiction shows that G is Eulerian.

3.7 Show that any trail constructed by Fleury's algorithm in an Eulerian graph is an Eulerian circuit.

Solution. Suppose $T(0)$: $v(0), e(1), v(1), e(2), v(2), \ldots, v(i-1), e(i), v(i)$ is the trail at stage i, and let $G(i)$ be the subgraph $G - \{e(1), e(2), \ldots, e(i)\}$. If the trail is not closed, both the *first vertex* $v(0)$ and the *last vertex* $v(i)$ are odd and every other vertex is even.

Case (i): The degree of $v(i)$ is 1. So there is a unique edge e in $G(i)$ joining $v(i)$ and a vertex w, and this e has to be selected for upgrading the trail. Once this choice is made, $v(i)$ becomes an isolated vertex in $G(i+1)$. In other words, e is a bridge in $G(i)$, and the upgrading of the trail using this bridge not only disconnects $G(i)$ but makes the last vertex become an isolated vertex in $G(i+1)$.

Case (ii): If the degree of $v(i)$ is more than 1, we claim that there is at most one bridge incident at $v(i)$. Suppose $e = \{v(i), p\}$ and $f = \{v(i), q\}$ are two bridges in $G(i)$. If we delete these two bridges from $G(i)$, we get a disconnected graph G' with three components: one component with $v(i)$, another one with p, and a third one with q. In G', we see that the degree of $v(i)$ is still odd since two edges incident with it are deleted. Since p and q are even vertices in $G(i)$, they now become odd vertices in G'. The deletion of these two bridges does not affect the degrees of the other vertices. Thus G' is a disconnected graph with three components and exactly four odd vertices. So there are at least two components in which the number of odd vertices is odd. But the number of odd vertices in any graph is even. Thus there is at most one bridge incident at $v(i)$. So whenever there are three or more edges incident at $v(i)$, we can always select one of them (which is not a bridge) and call it $e(i+1)$ in upgrading the trail.

On the other hand, if the current trail $T(i)$ is a closed trail, the degree of each of its vertices is even in $G(i)$. If the degree of $v(i)$ is positive, no edge incident with $v(i)$ is a bridge; otherwise, we will have a component with one odd vertex. These edges are used in upgrading the trail, and we return to this vertex later.

Thus we conclude that once we start constructing a trail using Fleury's algorithm in an Eulerian graph, the only stage at which we find it impossible to execute step 2 of the algorithm is when we reach a stage in which the current trail is a circuit containing all the edges of the graph, since any selection of a bridge (for inclusion in the trail in the absence of any other edge incident to the current last vertex) always results in an isolated vertex.

3.8 Using Fleury's algorithm, obtain an Eulerian circuit in the graph of Fig. 3-18.

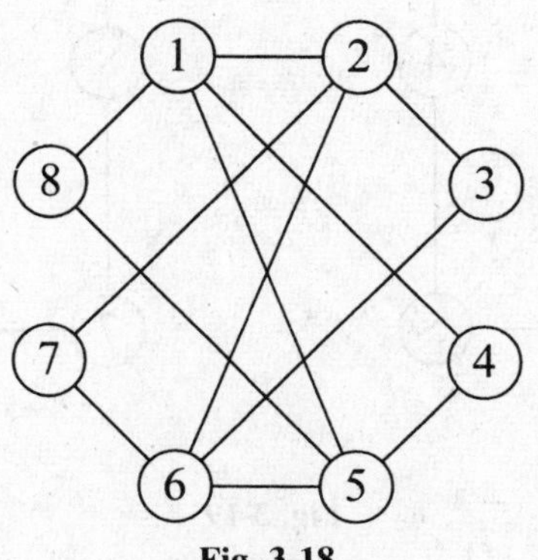

Fig. 3-18

Solution. The starting vertex is 1, and we let $T(0) = 1$. (For convenience, the edge joining i and j is written as ij.)

$T(1)$: $T(0)$, 12, 2
$T(2)$: $T(1)$, 23, 3
$T(3)$: $T(2)$, 36, 6
$T(4)$: $T(3)$, 67, 7
$T(5)$: $T(4)$, 72, 2
$T(6)$: $T(5)$, 26, 6
$T(7)$: $T(6)$, 65, 5
$T(8)$: $T(7)$, 58, 8
$T(9)$: $T(8)$, 81, 1
$T(10)$: $T(9)$, 15, 5
$T(11)$: $T(10)$, 54, 4
$T(12)$: $T(11)$, 41, 1

An Eulerian circuit obtained by applying Fleury's algorithm in this simple graph (Fig. 3-18) is 1—2 — 3 — 6 — 7 — 2 — 6 — 5 — 8 — 1 — 5 — 4 — 1.

3.9 If the number of odd vertices in a connected graph $G = (V, E)$ is $2k$, show that the set E can be partitioned into k subsets such that the edges in each subset constitute a trail between two odd vertices.

Solution. Suppose the odd vertices are v_i $(i = 1, 2, \ldots, k)$ and w_i $(i = 1, 2, \ldots, k)$. Construct k new vertices x_i $(i = 1, 2, \ldots, k)$ and $2k$ new edges $\{x_i, v_i\}$ and $\{x_i, w_i\}$ for $i = 1, 2, \ldots, k$. In the graph G' thus constructed, each vertex is even, so G' is Eulerian. Construct an Eulerian circuit Q in G'. Observe that in this circuit, the new edges adjacent to the new vertex x_i appear consecutively for each i. Now delete from this circuit all the new vertices (and, of course, all the new edges). The remaining edges in Q precisely constitute k pairwise disjoint sets, forming a partition of E such that the edges in each subset of the partition constitute a trail between two distinct odd vertices.

3.10 Locate the odd vertices in the graph of Fig. 3-19 and then partition the set of edges in the graph into subsets such that the edges in each subset constitute a trail.

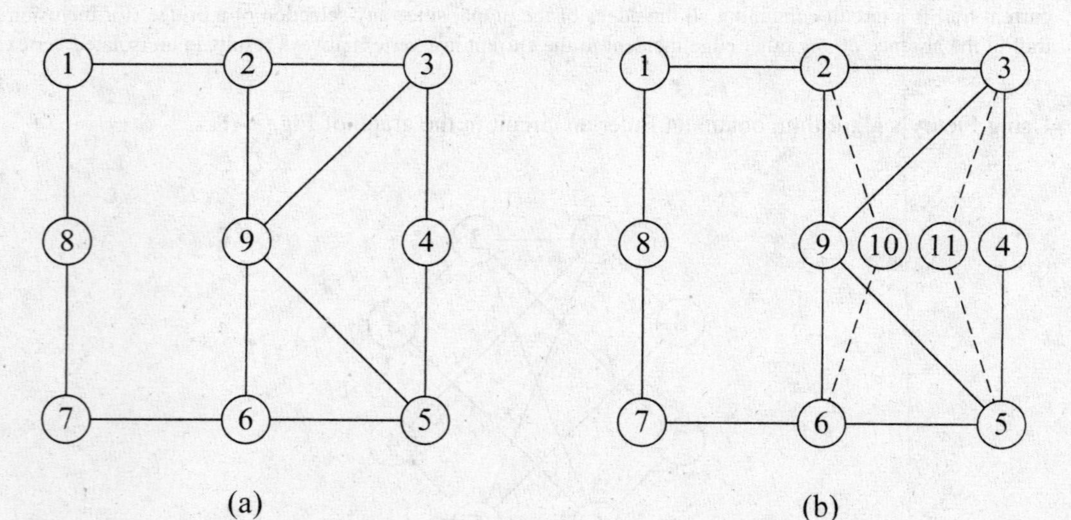

Fig. 3-19

Solution. The odd vertices are 2, 3, 5, and 6. So the set of edges can be partitioned into a family consisting of two subsets. We construct two new vertices 10 and 11 and new edges {2, 10}, {6, 10}, {3, 11}, and {5, 11} to obtain an Eulerian graph as in Figure 3-19(*b*). An Eulerian circuit in the new graph is constructed, and the four new edges are deleted from this circuit. We then obtain a trail 3 — 9 — 5 — 6 between the odd vertices 3 and 6. We also obtain the trail 5 — 4 — 3 — 2 — 1 — 8 — 7 — 6 — 9 — 2 between the odd vertices 5 and 2.

3.11 Prove Theorem 3.2: A connected graph is semi-Eulerian if and only if the number of odd vertices in it is exactly two. Furthermore, in a semi-Eulerian graph, any Eulerian trail is between its two odd vertices.

Solution. Suppose there is a trail between two distinct vertices u and v that contains all the edges of the graph. Any vertex in the trail other than these two is an intermediate vertex, the degree of which in the graph (and in the circuit) is necessarily even because whenever the trail passes through an intermediate vertex, it uses two edges: one to enter and one to exit. By the same argument, both the starting vertex and the terminal vertex are odd. So if the graph is semi-Eulerian, there exist two unique odd vertices and an Eulerian trail between these two vertices such that the degree of every other intermediate vertex in the trail is even. Conversely, according to Problem 3.10, if there are exactly two odd vertices, there is a trail between these two vertices consisting of all the edges of the graph.

3.12 Prove that a weakly connected digraph is Eulerian if and only if the indegree of each vertex is equal to its outdegree.

Solution. Whenever a circuit passes through a vertex, two distinct arcs are used: one to the vertex and one from the vertex. So each such passing results in a contribution of one to the outdegree and one to the indegree. Thus the existence of a directed circuit containing all the arcs implies that the outdegree of each vertex equals its indegree. Conversely, let G be a weakly connected digraph with m arcs in which the indegree of each vertex equals its outdegree. We prove by induction on m that the digraph is Eulerian. The result is obviously true when $m = 2$. Assume that the theorem is true for all weakly connected digraphs with $(m - 1)$ or fewer arcs in which outdegree and indegree are the same for every vertex, and let D be one such digraph. Observe that the outdegree (and therefore the indegree) of each vertex in D is positive. Let T be any trail from an arbitrary vertex u to another vertex v in this digraph. There will be at least one arc incident from v that is not an arc in T. So it is possibe to extend the trail and end up with a trail that terminates at u. If the closed trail T' thus obtained contains all the arcs of D, we are done. Otherwise, delete from D all the arcs belonging to T' as well as the vertices that become isolated as a result of this deletion. Each component of the resulting digraph D' is weakly connected with fewer arcs in which the outdegree and the indegree are the same. So by the induction hypothesis, each component has a directed Eulerian circuit. Since D is weakly connected, each weak component of D has a vertex in common with the closed circuit T'. An Eulerian circuit for D can now be constructed by inserting an Eulerian circuit of each weak component F of D' at a vertex common to both F and T'.

3.13 A weakly connected digraph is semi-Eulerian if and only if there are two vertices v and w in it such that (1) (outdegree of v) = (indegree of v) + 1, (2) (outdegree of w) = (indegree of w) − 1, and (3) the outdegree of every other vertex is equal to its indegree. Show that in a weakly connected digraph satisfying these three properties, any directed Eulerian trail is from v to w.

Solution. If the digraph G has an Eulerian trail T from vertex v to vertex w, the indegree and outdegree of each intermediate vertex in T are necessarily the same. Since v is the initial vertex, it contributes 1 to the outdegree in the beginning. Thereafter, whenever T passes through v, it contributes 1 to the outdegree and to the indegree. Thus the outdegree of v is its indegree plus 1. Likewise, the outdegree of w is its indegree minus 1. To prove the converse, let G be a weakly connected digraph satisfying the given conditions. Construct a new vertex x and two new arcs: one from the terminal vertex w to the new vertex x and another from x to the initial vertex v. The resulting digraph G' is weakly connected in which the outdegree of each vertex is equal to its indegree, so it is Eulerian. Let C be an Eulerian circuit in G' that starts from v and ends in v. The last two arcs in this circuit are (w, x) and (x, v). If we delete these two arcs and the new vertex from C, we have an Eulerian trail from v to w.

3.14 If every vertex in a graph G is even, no edge in that graph is a bridge.

Solution. Every vertex in any component $H = (W, F)$ of G is also even. So H is an Eulerian graph. The set F is the union of edge-disjoint cycles. So no edge is F; therefore, no edge in G is a bridge.

3.15 Find all positive integers n such that K_n is (*a*) Eulerian and (*b*) semi-Eulerian.

Solution.

(*a*) The degree of any vertex in a complete graph with n vertices is $(n - 1)$, so the graph is Eulerian if and only if n is odd.

(*b*) Since a complete graph is regular, it cannot be semi-Eulerian.

3.16 If $L(G)$ is the line graph of a simple graph G, show that $L(G)$ is Eulerian whenever G is Eulerian.

Solution. Let e be an edge in G joining two vertices x and y. By definition, the degree of the vertex in $L(G)$ that corresponds to e is equal to (degree of x in G) + (degree of y in G) $-$ 2. Since G is Eulerian, every vertex in it is even. So the degree of every vertex in the line graph is also even.

3.17 Show that if the line graph of a simple graph G is Eulerian, it is not necessary that G is Eulerian.

Solution. The complete graph with four vertices is not Eulerian since it has odd vertices. But the degree of each vertex of its line graph is even, so the line graph is Eulerian.

3.18 Show that a digraph that has an Eulerian circuit is a strongly connected digraph. Is the converse true?

Solution. Since the digraph is Eulerian, there is a closed directed trail emanating from every vertex that returns to it after traversing through each arc exactly once. Such a closed trail no doubt passes through each vertex of the graph at least once. So there is a directed trail (and therefore a directed path) from every vertex to every other vertex, establishing the strong connectivity of the digraph. The converse is not true; a strongly connected digraph need not be an Eulerian graph. As a counterexample, consider the digraph G' obtained by introducing a new arc from a vertex to a nonadjacent vertex in the cyclic digraph $1 \to 2 \to 3 \to 4 \to 1$.

3.19 Show that a digraph that has an Eulerian trail is a unilaterally connected digraph. Is the converse true?

Solution. If a digraph has an Eulerian trail, there is a directed path from a unique vertex in the digraph to another unique vertex in the digraph such that every other vertex is an intermediate vertex in this path. So if u and v are any two arbitrary vertices in the digraph, there is either a path from u to v or from v to u. Thus the digraph is unilaterally connected. The converse is not true. The digraph G', obtained by introducing a new vertex labeled 4 and the new arc (3, 4) to the cyclic graph $1 \to 2 \to 3 \to 1$, is unilaterally connected but has no Eulerian trail.

3.20 (*a*) Is there an Eulerian graph of even order and odd size? (*b*) Is there an Eulerian graph of odd order and even size?

Solution.

(*a*) Yes. Suppose C is a cycle of even order in which v is a vertex. Now consider a cycle C' of odd order passing through v such that the two cycles have no edge in common. The circuit G that consists of the edges of these two cycles is a graph in which each vertex is even.

(*b*) Yes. In part (*a*), suppose both C and C' are odd cycles. Then every vertex in the circuit G that has an odd number of vertices and even number of edges is even.

3.21 Show that if the degree of every vertex of a connected multigraph $G = (V, E)$ is 4, the graph has two spanning subgraphs such that (1) the degree of each vertex in these two subgraphs is 2, (2) the two subgraphs have no edge in common, and (3) E is the union of the sets of edges of these two subgraphs.

Solution. Since each vertex is even, G is Eulerian. If the graph has n vertices and m edges, $4n = 2m$, implying that the number of edges is even. We construct an Eulerian circuit starting from an arbitrary vertex and alternately color the edges of this circuit red and green. Then the red edges constitute a spanning subgraph in which the degree of each vertex is 2, and the blue edges constitute another spanning subgraph in which the degree of each vertex is also 2. Construct an Eulerain circuit. (Any k-regular spanning a subgraph of a graph is k-factor. We have shown that **every 4-regular graph can be decomposed into two factors.**)

3.22 Obtain a 2-factor decomposition of the 4-regular graph shown in Fig. 3-20.

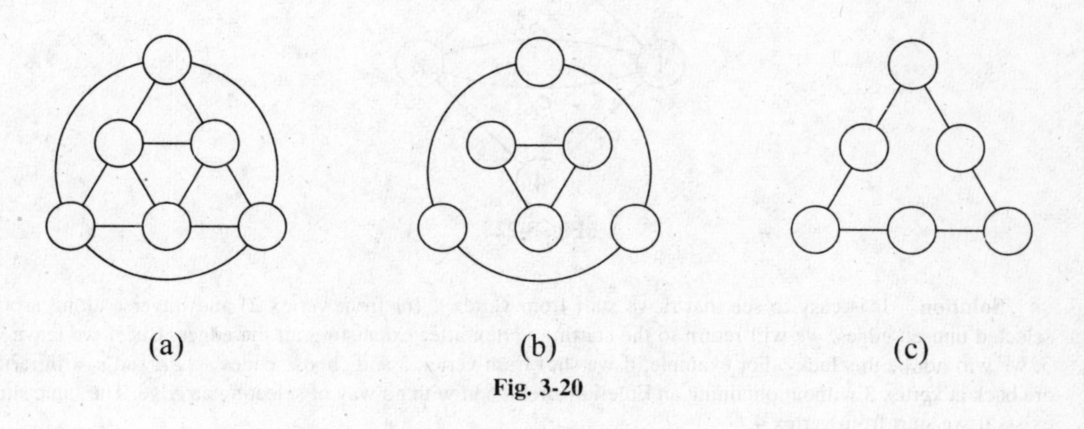

Fig. 3-20

Solution. A 2-factor decomposition of the connected graph in Fig. 3-20(a) is as shown in Fig. 3-20(b) and (c).

3.23 Obtain a 2-factor decomposition of the complete bipartite graph $K_{4,4}$.

Solution. A decomposition is shown in Fig. 3-21.

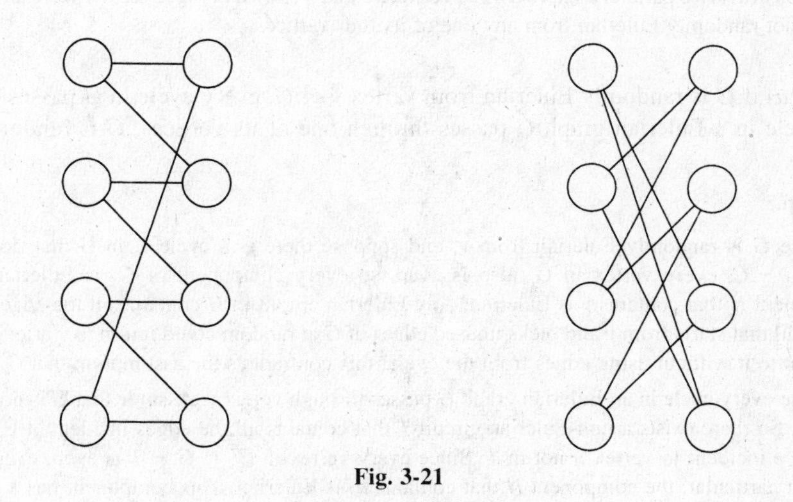

Fig. 3-21

3.24 A graph is said to be an **even graph** if the degree of every vertex in the graph is even. Find the number of nonequivalent labeled even graphs with $(n + 1)$ vertices with labels 1, 2, . . . , $(n + 1)$.

Solution. Consider any labeled graph with n vertices with labels 1, 2, . . . , n. Construct a new vertex with label $(n + 1)$, and join this vertex to every odd vertex in G. The resulting graph G' is even and is labeled with $(n + 1)$ vertices. This correspondence, which is clearly 1 to 1, implies that the number of labeled even graphs of order $(n + 1)$ is equal to the number of labeled graphs of order n.

Randomly Eulerian Graphs

3.25 A graph G is **randomly Eulerian from a vertex** v if every trail in the graph starting from v can be extended to a circuit terminating at v that consists of all the edges of the graph. Show that the graph in Fig. 3-22 is randomly Eulerian only from vertex 1.

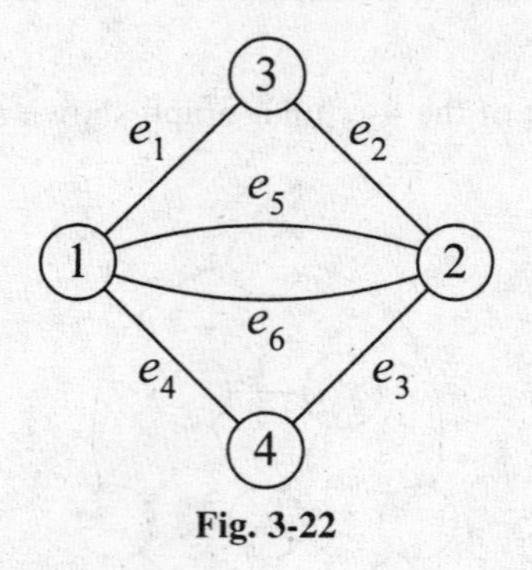

Fig. 3-22

Solution. It is easy to see that if we start from vertex 1 (or from vertex 2) and traverse along arbitrarily selected unused edges, we will return to the starting vertex after exhausting all the edges. But if we leave vertex 3, we will not be this lucky. For example, if we start from vertex 3 and choose edges e_2, e_5, and e_1 arbitrarily, we are back at vertex 3 without obtaining an Eulerian circuit and with no way of selecting an edge. The same situation exists if we start from vertex 4.

3.26 (*a*) Give an example of a graph that is randomly Eulerian from every vertex in it. (*b*) Give an example of an Eulerian graph that is not randomly Eulerian from any of its vertices.

Solution.

(*a*) Any cyclic graph is randomly Eulerian from any of its vertices.

(*b*) If we construct two parallel edges between vertices 3 and 4 shown in Fig. 3-22, we have an Eulerian multigraph that is not randomly Eulerian from any one of its four vertices.

3.27 Prove that (*a*) if G is randomly Eulerian from vertex v of G, every cycle in G passes through v; and (*b*) if every cycle in a Eulerian graph G passes through one of its vertices, G is randomly Eulerian from that vertex.

Solution.

(*a*) Suppose G is randomly Eulerian from v, and suppose there is a cycle C in G that does not contain v. If $G' = G - C$, every vertex in G' also is even, so every component of G' is Eulerian. In particular, the component H that contains v is Eulerian. Any Eulerian circuit in H contains all the edges of G incident at v. So a trail that starts from v and picks unused edges in G at random could return to v after exhausting all edges incident to it without using edges from the cycle; this contradicts the assumption that G is Eulerian from v.

(*b*) Suppose every cycle in an Eulerian graph G passes through vertex v. Assume that G is not randomly Eulerian from v. So there exists a non-Eulerian circuit T that contains all the edges incident at v, implying that there is edge e incident to vertex u not in T. Since every vertex in $G' = G - T$ is even, every component of G' (and, in particular, the component H that contains u) is Eulerian. This component has a circuit that does not pass through v; therefore, it has a cycle that does not pass through v, which contradicts our assumption.

3.28 Show that if G is randomly Eulerian from v, $G - v$ is acyclic. Also show that if v is a vertex in an Eulerian graph G such that $G - v$ is acyclic, G is randomly Eulerian from v.

Solution. If G is randomly Eulerian from v, every cycle in G passes through v, as established in Problem 3.27. So $G - v$ is acyclic. On the other hand, if G is Eulerian and if $G - v$ is acyclic, every cycle in G passes through v. Consequently, G is Eulerian, as shown in Problem 3.27.

3.29 If graph G is randomly Eulerian from a vertex, show that the degree of that vertex is equal to $\Delta(G)$, which is the maximum among the degrees of all its vertices. Show that an arbitrary Eulerian graph need not be randomly Eulerian from a vertex of maximum degree.

Solution. Suppose a graph is Eulerian from v, and let w be any other vertex. Since the graph is randomly Eulerian from v, every cycle in the graph passes through v. Thus between any two consecutive occurrences of the vertex w in an Eulerian circuit in the graph there is a cycle that contains the vertex v. So the degree of w cannot exceed the degree of v. The converse is not true. Consider the cyclic graph 1 — 2 — 3 — 4 — 1 and the Eulerian multigraph G obtained by constructing two new edges between vertices 1 and 3 and four new edges between vertices 2 and 4. Vertex 2 is a vertex with maximum degree, but the graph is not randomly Eulerian from that vertex.

3.30 Show that if G is randomly Eulerian from v and if w is another vertex such that both v and w have the same degree, G is randomly Eulerian from w as well.

Solution. Between any two consecutive occurrences of w in an Eulerian circuit randomly obtained by starting from v, there is an occurrence of v. In the same circuit, there is an occurrence of w between any two consecutive occurrences of v since the degrees of the two vertices are the same. So every cycle in the graph passes through the vertex w, implying that the graph is randomly Eulerian from w also.

3.31 A graph is **randomly Eulerian** if it is randomly Eulerian from every vertex in the graph. Obtain a necessary and sufficient condition for a graph to be randomly Eulerian.

Solution. Notice that a cyclic graph is randomly Eulerian. If an arbitrary graph is randomly Eulerian from a vertex, every cycle in that graph passes through that vertex. Thus if a graph is randomly Eulerian, it has to be a cyclic graph.

3.32 Show that if a graph is not randomly Eulerian, it is randomly Eulerian from at most two of its vertices.

Solution. If a graph is randomly Eulerian from two distinct vertices v and w, in any two consecutive occurrences of v in an Eulerian circuit there is an occurrence of w and vice versa. If there is a third vertex u from which the graph is also randomly Eulerian, we obviously have a cyclic graph, which is a contradiction.

3.33 A digraph is a **randomly Eulerian digraph from a vertex v** if every directed trail of the digraph starting from v can be extended to an Eulerian circuit. Prove that (*a*) an Eulerian digraph is randomly Eulerian from a vertex v if and only if every directed cycle in the digraph passes through v; (*b*) if a digraph is randomly Eulerian from vertex v, the maximum outdegree among its vertices is equal to the outdegree of v; (*c*) if a digraph is randomly Eulerian from v and if w is another vertex such that both v and w have the same outdegree, the digraph is randomly Eulerian from w also; and (*d*) if an Eulerian digraph of order n is not randomly Eulerian from every vertex, it is randomly Eulerian from at most $n/2$ vertices.

Solution. These are the direct generalizations of Problems 3.27 through 3.32.

De Bruijn Digraphs and Sequences

3.34 If W is the set $\{0, 1, 2, \ldots, p - 1\}$, any linear arrangement (repetitions are allowed) using some of or all these numbers is called a **word** in the **alphabet** W. Any word with n numbers from W is called an **n-letter** word in W. The set of all n-letter words in the alphabet W with p numbers is denoted by $W(p, n)$. The **de Bruijn digraph $G(p, n)$** is constructed by the following inductive procedure. Suppose all the words in $W(p, k)$ are known for $k = 1, 2, \ldots, (n - 1)$. The vertex set of $G(p, n)$ is $W(p, n - 1)$. If t is any $(n - 2)$-letter word, and if i and j are numbers (not necessarily distinct) in W, draw arcs (for each i) from vertex it to vertex tj as j varies from 0 to $p - 1$. Then the arc from xt to ty represents the n-letter word xty. Find the order and size of the de Bruijn digraph $G(p, n)$, and show that it is an Eulerian digraph.

Solution. In a word with $(n-1)$ numbers, each number can be chosen in p ways. So the order of $G(p, n)$ is $p^{(n-1)}$. The outdegree of each vertex by our construction is p. The indegree is also p. Thus the graph is Eulerian, and its size is p^n.

3.35 Construct the de Bruijn digraph $G(3, 2)$.

Solution. Here $W = \{0, 1, 2\}$. Each vertex is a one-letter word, and each arc is a two-letter word. There are three vertices (representing the three one-letter words) and eight arcs (representing the eight two-letter words) in the digraph, as shown in Fig. 3-23.

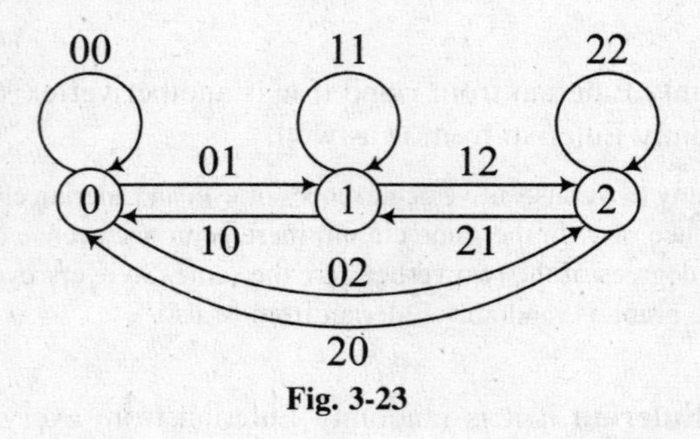

Fig. 3-23

3.36 If $x = 21134$ is a word in $W(4, 5)$, construct the arcs (a) incident from the vertex x and (b) incident to x in $G(4, 5)$.

Solution.

(a) Take $t = 1134$. We draw arcs from x to $t0$, $t1$, $t2$, $t3$, and $t4$ to denote the five six-letter words 211340, 211341, 211342, 211343, and 211344, respectively.

(b) Take $t = 2113$. We draw arcs from $0t$, $1t$, $2t$, $3t$, and $4t$ to x to denote the five six-letter words 021134, 121134, 221134, 321134, and 421134, respectively.

3.37 If p and n are two positive integers and if $p^n = r$, a sequence $\langle a(0), a(1), \ldots, a(r-1)\rangle$, where each $a(i)$ is in $W = \{0, 1, 2, \ldots, p-1\}$, is called a **de Bruijn sequence,** denoted by $B(p, n)$, if and only if any n-letter in W is of the form $a(i)a(i+1)\cdots a(i+n-1)$, where i is at most r and where addition of the subscripts is modulo r. (Equivalently, these r numbers in the sequence form a circular arrangement such that any choice of n consecutive (in the clockwise direction) numbers in this arrangement gives a unique word.) Show that a de Bruijn sequence exists for every choice of p and n.

Solution. Construct the de Bruijn digraph $G(p, n)$. If the sequence of arcs that appear in an Eulerian circuit of this digraph is $\langle e(0), e(1), \ldots, e(r-1)\rangle$, where $r = p^n$ defines the sequence $\langle a(0), a(1), \ldots, a(r-1)\rangle$, where $a(i)$ is the first number that appears in $e(i)$, any word of the form $a(i)a(i+1)\cdots a(i+n-1)$, where i is at most r and where addition of the subscripts is modulo r, is obviously an n-letter word. Conversely, suppose $w = d(1)d(2)\cdots d(n)$ is any word. Now the arcs in the digraph are defined such that a Eulerian circuit in this graph must go from vertex $d(1)d(2)\cdots d(n-1)$ to vertex $d(2)d(3)\cdots d(n)$ and then to $d(3)d(4)\cdots d(n+1)$, and so on. Thus the first letters of the arcs in this sequence will constitute the word w.

3.38 Obtain a de Bruijn sequence such that any three-letter word using 0, 1, and 2 can be obtained from this sequence.

Solution. There are nine two-letter words and 27 three-letter words using the numbers 0, 1, and 2. So the digraph $G(3, 2)$ has nine vertices (representing the two-letter words) and 27 arcs (representing the three-letter words). Both the indegree and outdegree of each vertex is three. The vertices are 00, 01, 02, 10, 11, 12, 20, 21, and 22. From 00, we draw arcs to 00, 01, and 02 representing 000, 001, and 002. Continue like this until all the arcs are drawn and labeled. An Eulerian circuit starting from 00 and ending at 00 is the following sequence of 27 words: ⟨000, 001, 010, 101, 011, 111, 112, 122, 222, 221, 212, 121, 210, 102, 021, 211, 110, 100, 002, 022, 220, 202, 020, 201, 012, 120, 200⟩. If we select the first number from each arc in this Eulerian sequence, we get the following sequence: ⟨0 0 0 1 0 1 1 1 2 2 2 1 2 1 0 2 1 1 0 0 2 2 0 2 0 1 2⟩, which is a $B(3, 3)$ sequence for this problem. Any three-letter words using the numbers 0, 1, and 2 can be obtained from this sequence by choosing any three "consecutive" numbers from this sequence. The three consecutive numbers starting from the last 1 in the sequence are 1, 2, and the first element in the sequence, which is 0. So this choice gives the word 120. If we start from the last element, we choose 2 and the first and second numbers from the sequence. This choice gives the word 200. The other 25 words can be easily obtained by starting from any number and selecting that number and the two numbers following it. For example, if we start from the 1 that appears for the first time in the sequence, the word 101 is obtained.

3.39 (*Rotating Drum Problem*) A rotating drum has 2^p sectors. The problem is to assign each sector the label 0 or 1 such that no two sequences of p consecutive labels are the same. Solve this problem when (*a*) $k = 3$ and (*b*) $k = 4$.

Solution. To solve this problem, we construct the de Bruijn graph $G(2, p - 1)$ and find an Eulerian circuit in this graph. Then the corresponding de Bruijn sequence will give an assignment of labels to the sectors.

(*a*) The graph $G(2, 2)$ is shown in Fig. 3-24(*a*) in which ⟨000 001 010 101 011 111 110 100⟩ is an Eulerian circuit. By choosing the first letter from each arc, we obtain the corresponding de Bruijn sequence, which assigns the following eight labels consecutively: 0, 0, 0, 1, 0, 1, 1, and 1.

(*b*) The graph $G(2, 3)$ is shown in Fig. 3-24(*b*). An Eulerian circuit in this digraph has the following 16 arcs consecutively: 0000, 0001, 0011, 0111, 1111, 1110, 1101, 1010, 0101, 1011, 0110, 1100, 1001, 0010, 0100, and 1000. A de Bruijn sequence corresponding to this Eulerian circuit gives the following assignment of labels to the 16 consecutive sectors: 0, 0, 0, 0, 1, 1, 1, 1, 0, 1, 0, 1, 1, 0, 0, 1.

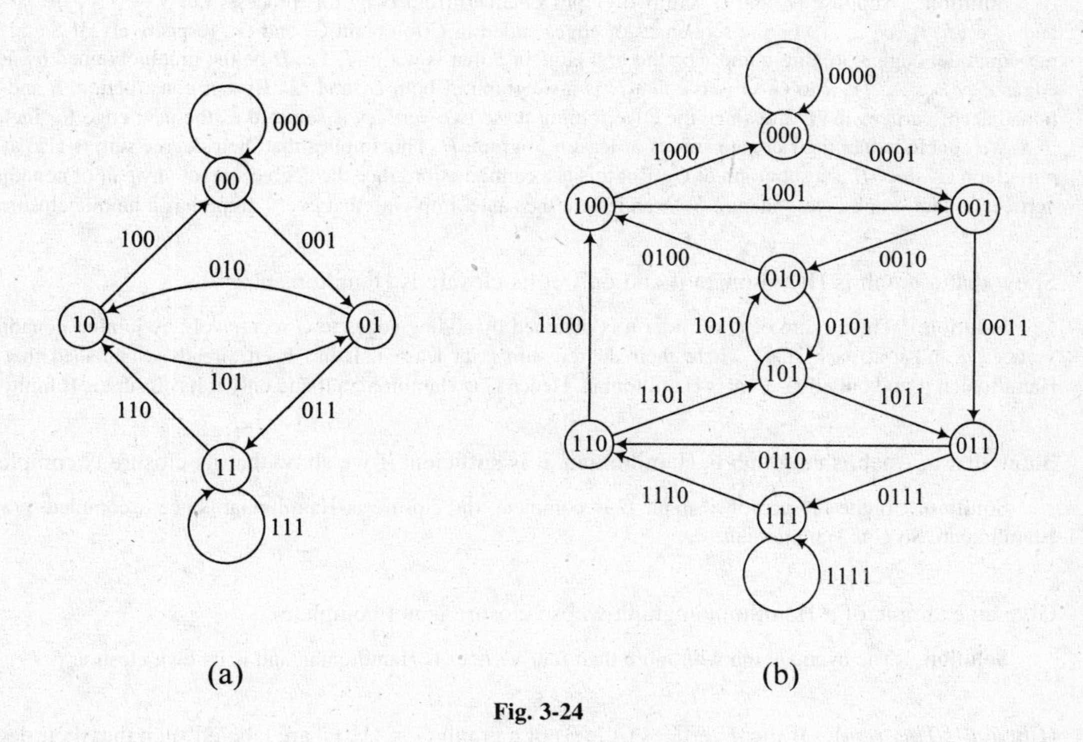

Fig. 3-24

3.40 Show that there are two unique four-letter binary words that can be deleted from the set of all such words so that any of the remaining words can be obtained by selecting four consecutive elements in a binary sequence placed in a circular arrangement. Obtain such a binary sequence.

Solution. The set of arcs in $G(2, 3)$ is in 1-to-1 correspondence with the set of all four-digit binary words. The only loops in the digraph are the arcs 0000 and 1111. If we delete these two arcs, we still have an Eulerian digraph. So the words that can be deleted are 0000 and 1111. Once they are deleted, we have the Eulerian sequence 0001, 0011, 0111, 1110, 1101, 1010, 0101, 1011, 0110, 1100, 1001, 0010, 0100, and 1000, consisting of 14 arcs. If we choose the first element of each of these 14 words and make a circular arrangement of these elements, any four-letter word (other than 0000 and 1111) can be obtained by selecting any four consecutive (clockwise) elements from this circular arrangement. Notice that any such arrangement will have seven zeros and seven ones.

HAMILTONIAN GRAPHS AND DIGRAPHS

3.41 (*Bondy and Chvátal*) Let u and v be two nonadjacent vertices in a simple graph $G = (V, E)$ of order n such that the sum of their degrees is at least n, and let $G + uv$ be the graph obtained from G by joining these two nonadjacent vertices. Then G is Hamiltonian if and only if $G + uv$ is Hamiltonian.

Solution. If G is Hamiltonian, $G + uv$ is Hamiltonian. To prove the converse, assume that $G + uv$ is Hamiltonian but G is not. Then there is a Hamiltonian path in G between u and v. Let us consider this undirected path as a path from u to v. For each vertex adjacent to vertex u, the vertex in the path immediately preceding that vertex cannot be adjacent to v; otherwise, there will be a Hamiltonian cycle in G. So the degree sum of u and v cannot exceed $(n - 1)$ as established in the proof of Theorem 3.5. In other words, the degree sum of these two nonadjacent vertices is less than n, which is a contradiction.

3.42 The **closure** $c(G)$ of a graph G of order n is obtained from G by recursively joining pairs of nonadjacent vertices whose degree sum is at least n until no such pair exists. Show that every graph has a unique closure.

Solution. Suppose G_1 and G_2 are two graphs obtained from G by this process. Let $S = \langle e_1, e_2, \ldots, e_k \rangle$ and $T = \langle f_1, f_2, \ldots, f_r \rangle$ be the sequences of edges added to G to obtain G_1 and G_2, respectively. If S and T are not equal, let edge e_i joining u and v be the first edge in S that is not in T. Let H be the graph obtained by adding edges $e_1, e_2, \ldots, e_{i-1}$ to G. Observe that H is a subgraph of both G_1 and G_2. By our construction, u and v are nonadjacent vertices in H, and since the edge joining these two vertices is selected as the next edge for inclusion in S, we conclude that their degree sum is at least n in graph H. This implies that their degree sum is also at least n in graph G_2 since H is a subgraph of G_2. But this is a contradiction since the degree sum of any pair of nonadjacent vertices in that graph is less than n. So S and T are the same, implying that every graph has a unique closure.

3.43 Show that a graph is Hamiltonian if and only if its closure is Hamiltonian.

Solution. The closure of G of order n is obtained by adding edges to G recursively by joining nonadjacent vertices u and v at each stage where their degree sum is at least n. It has been already established that G is Hamiltonian if and only if $G + uv$ is Hamiltonian. Hence G is Hamiltonian if and only if its closure is Hamiltonian.

3.44 Show that to establish a graph is Hamiltonian, it is sufficient if we show that its closure is complete.

Solution. If the closure of a graph G is complete, the closure is Hamiltonian since a complete graph is Hamiltonian. So G is Hamiltonian.

3.45 Give an example of a Hamiltonian graph whose closure is not complete.

Solution. The cyclic graph with more than four vertices is Hamiltonian and is its own closure.

3.46 (*Chvátal's Theorem*) If the n vertices $(n \geq 3)$ of a graph $G = (V, E)$ are labeled such that their degrees d_i $(i = 1, 2, \ldots, n)$ can be arranged as a sequence $d_1 \leq d_2 \leq \cdots \leq d_n$, and if $d_{n-k} \geq (n - k)$ whenever $d_k \leq k < n/2$, G is Hamiltonian.

Solution. It is enough if we prove that the closure $c(G)$ is complete. We denote the degree of vertex v in the closure by $d'(v)$. Suppose the closure is not complete. We select two nonadjacent vertices u and v in the closure such that $d'(u) \leq d'(v)$ and the degree sum $d'(u) + d'(v)$ is as large as possible. Since these two vertices are nonadjacent, their degree sum is, of course, less than n.

Let $S = \{x \in V - v$: x and v are not adjacent in $c(G)\}$ and $T = \{x \in V - u$: x and u are not adjacent in $c(G)\}$. Now $x \in S \Rightarrow d'(x) + d'(v) \leq n \Rightarrow d'(x) + d'(v) \leq d'(u) + d'(v) \Rightarrow d'(x) \leq d'(u)$. Similarly, $x \in T \Rightarrow d'(x) \leq d'(v)$. Obviously, $|S| = (n - 1) - d'(v)$ and $|T| = (n - 1) - d'(u)$.

Now denote $d'(u)$ by the integer k. Then $k < n/2$ since $k \leq d'(v)$ and $k + d'(v) < n$. Now $d'(u) = k \Rightarrow d'(v) < (n - k) \Rightarrow |S| > (n - 1) - (n - k) \Rightarrow |S| \geq k$. Again, $d'(u) = k \Rightarrow |T| = (n - 1) - k \Rightarrow |T| = (n - 1) - k \Rightarrow |T| < (n - k)$. Thus the closure $c(G)$ has at least k vertices of degree at most k and at least $(n - k)$ vertices of degree less than $(n - k)$. Since G is a spanning subgraph of the closure, the same degree conditions prevail in G. We have thus found an integer k satisfying the inequalities $d_k \leq k < n/2$ and $d_{n-k} < (n - k)$, contradicting the hypothesis. So the closure is complete.

3.47 (*Bondy's Theorem*) If the n vertices ($n \geq 3$) of a graph $G = (V, E)$ are labeled such that their degrees d_i ($i = 1, 2, \ldots, n$) can be arranged as a sequence $d_1 \leq d_2 \leq \cdots \leq d_n$, and if $d_j + d_k \geq n$ whenever $d_j \leq j < k$ and $d_k \leq (k - 1)$, G is Hamiltonian.

Solution. Suppose the hypothesis of this theorem does not imply the hypothesis of Chvátal's theorem. So there is an integer p such that $d_p \leq p < n/2$ and $d_{n-p} < (n - p)$. Then $d_p + d_{n-p} < n$. Also $p < n/2$ implies $p < (n - p)$. If $n - p = q$, we have the inequalities $d_p + d_q < n$, $d_p \leq p < q$, and $d_q \leq (q - 1)$, which will contradict the hypothesis of Bondy's theorem. So Bondy's hypothesis implies Chvátal's hypothesis.

3.48 (*Posa's Theorem*) If the n vertices ($n \geq 3$) of a graph $G = (V, E)$ are labeled such that their degrees d_i ($i = 1, 2, \ldots, n$) can be arranged as a sequence S: $d_1 \leq d_2 \leq \cdots \leq d_n$, and if $d_k > k$ whenever $1 \leq k < n/2$, G is Hamiltonian.

Solution. Suppose the hypothesis of this theorem does not imply the hypothesis of Bondy's theorem. So there exist j and k such that $d_j \leq j < k$, $d_k \leq (k - 1)$ and $d_j + d_k < n$. So $k + (k - 1) < n$, which implies that $k < (n + i)/2$. Thus $d_j < n/2$.

Let d_j be denoted by p. Then $d_j \leq j \Rightarrow p \leq j \Rightarrow d_p \leq d_j \Rightarrow d_p \leq p$. But $p < n/2$. Thus there exists p such that $1 \leq p < n/2$ and $d_p \leq p$, which contradicts the hypothesis of Posa's theorem. So Bondy's hypothesis implies Posa's hypothesis.

3.49 Prove Ore's theorem (Theorem 3.5) using Posa's theorem.

Solution. The hypothesis in Ore's theorem is that the degree sum of any pair of nonadjacent vertices in a graph of order n is at least n. Suppose this hypothesis holds, and suppose there exists a positive integer p such that $p < n/2$ and $d_p \leq p$. If $q < p$, $d_q + d_q \leq 2p < n$. Thus $(q < p) \Rightarrow$ the vertices v_p and v_q are adjacent because of the hypothesis of Ore's theorem. So the subgraph induced by $W = \{v_i: i = 1, 2, \ldots, p\}$ is a clique. Since $d_p \leq p \Rightarrow d_i \leq p$ for every vertex in W, and since each v_i in W is adjacent to the remaining $(p - 1)$ vertices in W, a vertex in W can be adjacent to at most one vertex in the complement of W. Now $n - p > p$ since $p < n/2$. So the complement has more vertices than W; consequently, there is vertex v_i in W and vertex v_j in the complement such that there is no edge joining them. Also, $d_j \leq (n - p) - 1$. So $d_i + d_j \leq p + (n - p) - 1 < n$. But vertices v_i and v_j are not adjacent. This contradicts Ore's hypothesis. So Ore's hypothesis implies Posa's hypothesis.

3.50 Point out the implications regarding sufficient conditions (established in this sequel) for the existence of a spanning cycle in a graph.

Solution. The theorems are due to Chvátal (1972), Bondy (1969), Posa (1962), Ore (1963), and Dirac (1952), in reverse chronological order. Chvátal's theorem is proved in Problem 3.46, and it implies Bondy's theorem as shown in Problem 3.47. This theorem in turn implies Posa's theorem as proved in Problem 3.48. It was shown in Problem 3.49 that Posa's theorem implies Ore's theorem. That Ore's theorem implies Dirac's theorem was established in Theorem 3.6.

3.51 Show that Chvátal's condition is not a necessary condition for the existence of a spanning cycle in a graph.

Solution. Consider the graph in Fig. 3-25. It is obviously Hamiltonian. Its degree sequence is [2, 2, 2, 3, 3, 4]. Here $n = 6$ and $d_2 = 2 < 3 = n/2$. But $d_{n-2} = d_4 = 3$, violating the requirement that $d_4 \geq (6 - 2)$.

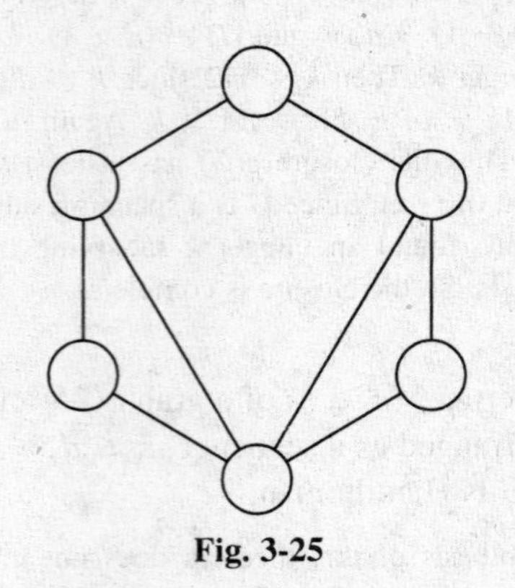

Fig. 3-25

3.52 Show that Chvátal's theorem is stronger than Bondy's theorem.

Solution. The degree sequence in the graph of Fig. 3-26 is [2, 2, 2, 4, 4, 4]. Here $d_2 = 2 \leq 2$ and $d_3 = 2 \leq 3$. But their sum is only 4, which is less than the order of the graph. So this Hamiltonian graph does not satisfy Bondy's hypothesis. It can be easily verified that it satisfies Chvátal's hypothesis, however.

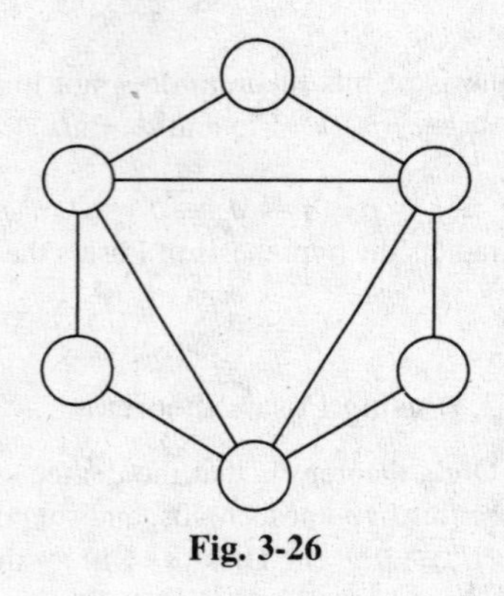

Fig. 3-26

3.53 Show that Bondy's theorem is stronger than Posa's theorem.

Solution. The degree sequence in the graph of Fig. 3-27 of order 7 is [2 2 4 5 5 5 5] in which d_2 is not more than 2 even through 2 is less than 7/2. Thus for this Hamiltonian graph, Posa's hypothesis is not satisfied. Bondy's hypothesis is satisfied, however.

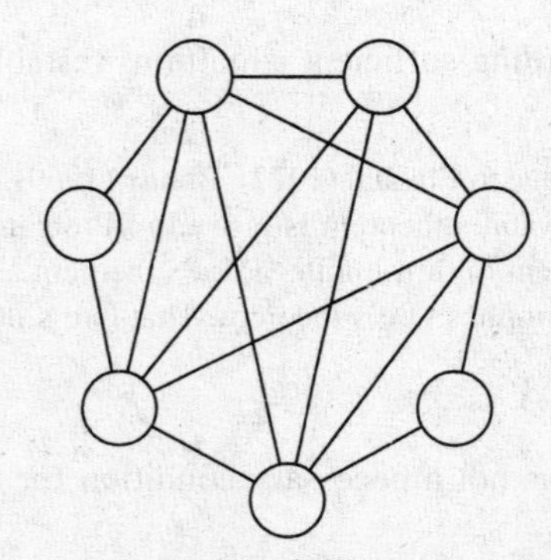

Fig. 3-27

3.54 Show that Posa's theorem is stronger than Ore's theorem.

Solution. In the graph of Fig. 3-28 of order 6, it is easy to locate two nonadjacent vertices whose degree sum is less than 6. This Hamiltonian graph satisfies Posa's hypothesis, however.

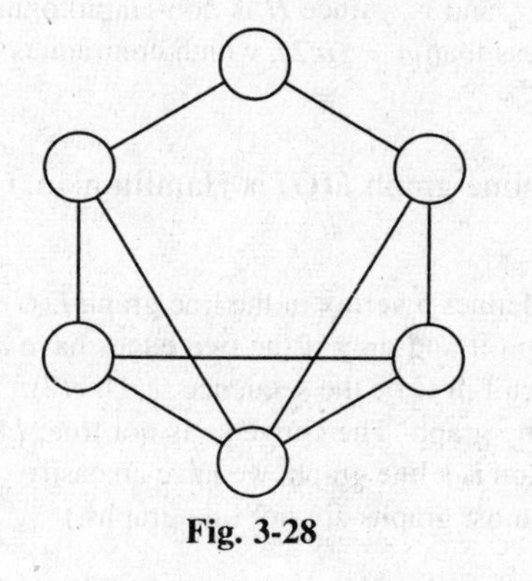

Fig. 3-28

3.55 Show that Ore's theorem is stronger than Dirac's theorem.

Solution. In the Hamiltonian graph of Fig. 3-29 of order 5, there is a vertex with degree less than 5/2. The degree sum of every pair of nonadjacent vertices is at least 5, however. Thus Ore's hypothesis is satisfied.

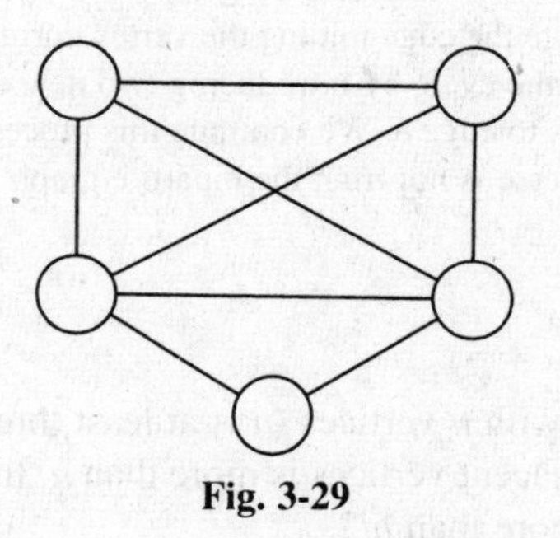

Fig. 3-29

3.56 (*a*) If $G = (V, E)$ is a graph with n vertices and m edges (where m is at least three), and if $m \geq [(n-1)(n-2)/2] + 2$, G is Hamiltonian. (*b*) Show that the converse is not true by exhibiting a counterexample. (*c*) Show that the inequality with this bound on the size of the graph is "sharp" in the sense that there is a non-Hamiltonian graph whose size is one less than this bound.

Solution.

(*a*) If G is complete, it is Hamiltonian. Otherwise, let u and v be two nonadjacent vertices in G with degrees x and y, respectively. If we delete these two vertices from G, we get a subgraph with $(n-2)$ vertices and q edges, where $q \leq (n-2)(n-3)/2$. Now $m = q + x + y$ since u and v are nonadjacent. So $x + y = m - q \geq \{[(n-1)(n-2)/2] + 2\} - \{[(n-2)(n-3)/2]\} = n$. So by Ore's theorem, G is complete.

(*b*) The converse is not true. Any cyclic graph with four or more vertices is Hamiltonian. But the inequality does not hold.

(*c*) Consider the graph G obtained by creating a new vertex and joining it to exactly one vertex of the cyclic graph with three vertices.

3.57 If $G = (V, W, E)$ is a bipartite graph with $|V| = |W| = n$, and if the degree of each vertex is more than $n/2$, G is Hamiltonian.

Solution. Suppose G is not Hamiltonian. Add as many edges as possible joining vertices in V and W until we obtain a graph H that will become Hamiltonian if one more such edge is added. H cannot be the complete bipartite graph $K_{n,n}$. If the degree in G of each vertex is more than $n/2$, the degree of each vertex in H is also more than $n/2$. Let $u \in V$ and $v \in V$ be two nonadjacent vertices H. Obviously, there is a Hamiltonian path $u = v_1, v_2, \ldots, v_{2n} = v$ in H from u to v where v_i is in V if and only if i is odd. If there is an edge joining v_1 and v_i, there cannot be an edge joining v_{2n} and v_{i-i} since H is non-Hamiltonian. Since the degree of v_1 is more than $n/2$, we find that the degree of v_{2n} is less than $n - (n/2)$, which contradicts the hypothesis.

3.58 Show that if G is Eulerian, its line graph $L(G)$ is Hamiltonian. Give a counterexample to show that the converse is not true.

Solution. Each edge in G defines a vertex in the line graph $L(G)$ such that the vertices corresponding to two edges are adjacent in the line graph if and only if the two edges have a vertex in common. Suppose G is Eulerian with m edges. Let an Eulerian circuit in G be the sequence $\langle e(1), e(2), \ldots, e(m)\rangle$. Then this sequence of vertices defines a spanning cycle in the line graph. The converse is not true; $L(K_4)$ is Hamiltonian, but K_4 is not Eulerian. (So if a graph G under consideration is a line graph, we have an easily verifiable sufficient condition to test whether it is Hamiltonian. Unfortunately, most graphs are not line graphs.)

3.59 Show that if a graph G is Hamiltonian, its line graph $L(G)$ is Hamiltonian. Give a counterexample to show that the converse is not true.

Solution. Let C be a Hamiltonian cycle in G. If $L(C)$ is a Hamiltonian cycle in $L(G)$, we are done. If that is not the case, let e be an edge in G that is not in C, and let the edges in C adjacent to e be e_p and e_q. Then there is an edge f_p in $L(G)$ joining the vertices corresponding to E and e_p and an edge f_q in $L(G)$ joining the vertices corresponding to e and e_q. Now delete the edge joining the vertex corresponding to e_p and the vertex corresponding to e_q in the cycle $L(C)$, and enlarge the cycle by introducing two new consecutive edges f_p and f_q together with the intermediate vertex that corresponds to edge e. We continue this process until every edge not in C is taken care of. This completes the proof. The converse is not true; the bipartite graph $K_{1,3}$ is not Hamiltonian, but its line graph is Hamiltonian.

Hamiltonian-Connected Graphs

3.60 Prove Theorem 3.7: A graph G with n vertices (n is at least three) is Hamiltonian-connected if the sum of the degrees of any two nonadjacent vertices is more than n. In particular, it is Hamiltonian-connected if the degree of each vertex is more than $n/2$.

Solution. The hypothesis implies that G is Hamiltonian by Ore's theorem. Relabel the vertices (if necessary) so that C: $v_1, v_2, \ldots, v_n, v_1$ is a Hamiltonian cycle in the graph. If the graph is not Hamiltonian-connected, there will be two vertices v_i and v_j (with $i < j$) that are not joined by a Hamiltonian path. This implies that the vertices v_{i+i} and v_{j+1} (the addition of subscripts is modulo n) are not adjacent; otherwise, the path $v_i, v_{i-1}, \ldots, v_1, v_n, v_{n-1}, \ldots, v_{j+1}, v_{i+1}, v_{i+2}, v_j$ will be a Hamiltonian path between v_i and v_j. We now seek a lower bound on the number of vertices that are adjacent to v_{i+1}. Suppose there are p vertices $v_r (i + 1 < r < j + 1)$ and q vertices v_s ($s < i + 1$ or $s > j + 1$) that are adjacent to v_{i+1}. Thus the degree of v_{i+1} is $(p+q)$. That there is no Hamiltonian path between v_i and v_j implies that there is no edge between v_{r-1} and v_{j+1} in the first category and no edge between v_{s+1} and v_{j+1} in the second category. Observe that vertex v_{i+1} is common to both these sets. So there are at least $p + q - 1$ vertices to which v_{j+1} is not adjacent. Hence, (degree v_{j+1}) $\le (n - 1) - (p + q - 1) = n -$ (degree v_{i+1}). This contradicts the hypothesis. So G is Hamiltonian-connected. In particular, if the degree of each vertex is more than $n/2$, the graph is Hamiltonian-connected. Thus the theorems of Ore and Dirac are generalized.

3.61 Give a counterexample to show that the sufficient conditions established in Problem 3.60 for a graph to be Hamiltonian-connected are not necessary conditions.

Solution. Figure 3-30 shows a Hamiltonian-connected graph with eight vertices. The degree of each vertex is only 3.

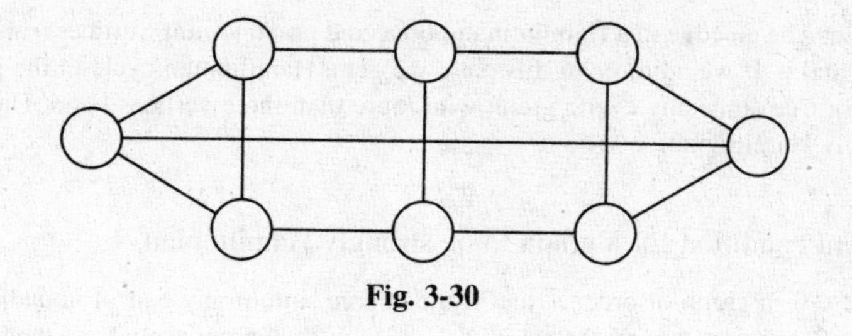

Fig. 3-30

3.62 Define the concept of closure in the context of graphs that are Hamiltonian-connected. Show that the closure of G thus defined is unique. State and prove the appropriate theorem in this context.

Solution. It is easy to show that if u and v are nonadjacent vertices of a graph of order n such that the degree sum of these two vertices is at least $(n + 1)$, G is Hamiltonian-connected if and only if $G + uv$ is Hamiltonian-connected, as in Problem 3.41. A closure of a graph G is obtained from G by recursively joining pairs of nonadjacent vertices whose degree sum is at least $(n + 1)$. As in Problem 3.42, it can be shown that the closure that may be denoted by $c'(G)$ is unique. Thus the appropriate theorem is the assertion that G is Hamiltonian-connected if and only if $c'(G)$ is Hamiltonian-connected. This assertion can be verified as done in Problem 3.43. Consequently, a graph is Hamiltonian-connected if its closure $c'(G)$ is complete.

3.63 Give an example of a Hamiltonian-connected graph whose closure as defined in Problem 3.62 is not complete.

Solution. The graph in Fig. 3-30 is Hamiltonian-connected and is its own closure. But it is not complete.

3.64 (*a*) If $G = (V, E)$ is a graph with n vertices and m edges (where m is at least three), and if $m \geq [(n - 1)(n - 2)/2] + 3$, G is Hamiltonian-connected. (*b*) Show that the condition is sufficient but not necessary by exhibiting a counterexample.

Solution.

(*a*) The proof is similar to that of Problem 3.56 when we use the sufficient condition in Theorem 3.7.

(*b*) Consider the graph in Fig. 3-30, which is Hamiltonian-connected. Here $n = 8$ and $m = 12$. Thus $[(n - 1)(n - 2)/2] + 3 = 24$.

3.65 Show that if a graph with n vertices ($n > 3$) and m edges is Hamilton-connected, it is necessary that $m \geq (3n)/2$. Show that this condition is by no means sufficient for a graph to be Hamiltonian-connected.

Solution. Observe that if a graph with more than three vertices is Hamiltonian-connected, the degree of each vertex has to be at least three. Thus the sum of all the degrees is at least $3n$. So $2m \geq 3n$. The converse is not true; the inequality is certainly satisfied for the complete bipartite graph $K_{3,4}$, but it is not Hamiltonian-connected.

3.66 A Hamiltonian graph is said to be **strongly Hamiltonian** if every edge in the graph belongs to a Hamiltonian cycle. Give an example of (*a*) a strongly Hamiltonian graph and (*b*) a Hamiltonian graph that is not strongly Hamiltonian.

Solution.

(*a*) Every cyclic graph is obviously strongly Hamiltonian.

(*b*) The graph obtained by joining two nonadjacent vertices of a cyclic graph with more than three vertices is not strongly Hamiltonian.

3.67 Show that a Hamiltonian-connected graph is a strongly Hamiltonian graph. Is the converse true?

Solution. Let e be an edge in a Hamiltonian-connected graph joining vertices u and v. There is a Hamiltonian path P between u and v. If we adjoin e to this path, we get a Hamiltonian cycle in the graph that contains edge e. The converse is not true since any cyclic graph with more than three vertices is not Hamiltonian-connected, even though it is strongly Hamiltonian.

3.68 Obtain a sufficient condition for a graph to be strongly Hamiltonian.

Solution. If G is a graph of order n, and if the degree sum of any pair of nonadjacent vertices is more than n (or if the degree of each vertex is more than $n/2$), G is Hamiltonian-connected and therefore strongly Hamiltonian.

3.69 Show that if the degree sum of every pair of nonadjacent vertices in a non–Hamiltonian-connected graph with three or more vertices is at least k (where k is some positive integer), the graph contains a path of length k.

Solution. Let P: $v_0, v_1, v_2, \ldots, v_p$ be a longest path in the graph. In other words, the number of edges in any path in the graph cannot exceed p. Since P is a longest path, neither of its terminal vertices can be adjacent to a vertex not in P. Let terminal vertex v_0 be adjacent to intermediate vertex v_i. Then we claim that the other terminal vertex v_p cannot be adjacent to the vertex v_{i-1}. Suppose this were the case. Then the p vertices in the path constitute the cycle C: $v_0, v_i, v_{i+1}, \ldots, v_p, v_{i-1}, v_{i-2}, \ldots, v_1, v_0$, which cannot contain all the vertices of the non-Hamiltonian graph. This implies that there is a vertex w that is not a vertex in the cycle but that is adjacent to one of its vertices. So there is a path of length $(p + 1)$ in the graph violating the maximality of P. Since v_p is adjacent to v_{p-1}, we conclude that the two terminal vertices are nonadjacent. Hence, (degree of v_p) $\leq p -$ (degree of v_0). Thus $p \geq$ (degree of v_p) + (degree of v_0). But (degree of v_p) + (degree of v_0) $\geq k$ since the two terminal vertices of the longest path are nonadjacent. Since there is a path with p edges and since $p \geq k$, there exists a path with k edges.

3.70 If the degree sum of every pair of nonadjacent vertices in a graph of order n is at least $(n - 1)$, the graph has a Hamiltonian path. In particular, if the degree of each vertex is at least $(n - 1)/2$, the graph has a Hamiltonian path.

Solution. The hypothesis implies that the graph is connected. The result follows from Problem 3.69.

Randomly Hamiltonian Graphs (Chartrand–Kronk Theory)

3.71 A graph G is **randomly traceable** if a Hamiltonian path results upon starting from any vertex and successively proceeding to any adjacent vertex that is not yet in the path. Furthermore, if G has at least three vertices and the terminal vertex of each such path is adjacent to the initial vertex, the graph is called **randomly Hamiltonian.** Obviously, cyclic graphs and complete graphs with three or more vertices are randomly traceable and randomly Hamiltonian. Show that (*a*) G is randomly traceable if and only if it is randomly Hamiltonian, and (*b*) every randomly traceable graph is Hamiltonian.

Solution.

(*a*) Obviously, any randomly Hamiltonian graph is randomly traceable. To prove the converse, assume that G is randomly traceable, and let P be any Hamiltonian path $\langle v_1, v_2, \ldots, v_n \rangle$. It is enough if we show that there is an edge in the graph joining the first vertex and last vertex of this path. Since the graph is randomly traceable, once we have path $P' = \langle v_2, v_3, \ldots, v_n \rangle$, the next vertex has to be v_1 to complete this path into a Hamiltonian path. This implies that the vertices v_n and v_1 are adjacent.

(*b*) Every randomly traceable graph is randomly Hamiltonian; therefore, it is Hamiltonian.

3.72 Show that if $C = \langle v_1, v_2, \ldots, v_n, v_1 \rangle$ is a Hamiltonian cycle in a randomly Hamiltonian graph and if there is an edge in the graph between v_j and v_k, there is an edge between v_{j+i} and v_{k+i} for $i = 1, 2, \ldots, n - 1$. (Here the addition of the subscripts is modulo n.)

Solution. The path $P = \langle v_{j+2}, v_{j+3}, \ldots, v_k, v_j, v_{j-1}, v_{j-2}, \ldots, v_1, v_n, v_{n-1}, \ldots, v_{k+1} \rangle$, which started from v_{j+2} and ended in v_{k+1}, passes through every vertex of the graph except v_{j+1}. Since the graph is randomly

Hamiltonian, there has to be an edge between v_{j+1} and v_{k+1}. Now consider the path $P' = \langle v_{j+3}, v_{j+4}, \ldots, v_k, v_{k+1}, v_{j+1}, v_j, \ldots, v_1, v_n, v_{n-1}, \ldots, v_{k+2}\rangle$, which started from v_{j+3} and ended in v_{k+2} and which did not pass through v_{j+2}. So there is an edge between v_{k+2} and v_{j+2}. By repeating the process, we get the desired result.

3.73 Show that if $C = \langle v_1, v_2, \ldots, v_n, v_1\rangle$ is a Hamiltonian cycle in a randomly Hamiltonian graph G and if there is an edge between v_p and v_{p+2} (for some p), G is a complete graph.

Solution. Select any two arbitrary vertices v_j and v_k in G. By Problem 3.72, there is an edge between v_{j-1} and v_{j+1}. Now consider the path $\langle v_{k+1}, v_{k+2}, \ldots, v_{j-1}, v_{j+1}, v_{j+2}, \ldots, v_k\rangle$, which starts from v_{k+1} and terminates in v_k, that contains all the vertices except v_j. So there has to be an edge between v_j and v_k. In other words, G is complete.

3.74 Let G be a randomly Hamiltonian graph of order n, and let C be any fixed Hamiltonian cycle in this graph. The edges in C are called **cycle edges,** and the edges not in C are called **diagonal edges.** Any cycle C' consisting of $(k - 1)$ cycle edges and exactly one diagonal edge is called an **outer k-cycle.** Show that the minimum value of k such that G has an outer k-cycle is either 3 or 4. Furthermore, if $k = 3$, the graph is a complete graph, and if $k = 4$, the number of vertices in the graph is even.

Solution. Let $C = \langle v_1, v_2, \ldots, v_n, v_1\rangle$ be a Hamiltonian cycle in G. Suppose there is no outer cycle of length 4, and suppose $C' = \langle v_1, v_2, \ldots, v_k, v_1\rangle$ is an outer k-cycle, where $k > 4$. Since the edge joining v_1 and v_k is a diagonal edge, the edge joining v_2 and v_{k+1} is also a diagonal by Problem 3.72. Now consider path P, which starts from v_4 and ends in v_n, defined by the following sequence $\langle v_4, \ldots, v_k, v_1, v_2, v_{k+1}, \ldots, v_n\rangle$. This path goes through every vertex other than v_3. So there is an edge joining v_n and v_3. Hence, we have an outer cycle $\langle v_1, v_2, v_3, v_n\rangle$ consisting of four edges, which contradicts the hypothesis. Thus $k = 3$ or $k = 4$. In the former case, there is an edge between every pair of vertices in cycle C (see Problem 3.72); therefore, the graph is complete. Suppose n is odd in the latter case when the smallest outer cycle has four edges. So G contains all the diagonal edges joining v_i and v_{i+3} ($i = 1, 2, \ldots, n$). Consider the path starting from v_5 defined by the sequence $\langle v_5, v_4, v_7, v_6, v_9, v_8, \ldots, v_{n-1}, v_{n-2}, v_2, v_1\rangle$. This path does not pass through v_3. Since the graph is randomly Hamiltonian, there is an edge between v_1 and v_3, creating an outer cycle with less than four edges. This contradiction establishes that the number of vertices is even.

3.75 Show that a graph with three or more vertices is randomly Hamiltonian if and only if it is a cyclic graph or a complete graph or a complete bipartite graph with equal number of vertices in each part.

Solution. If a graph is a cyclic, a complete, or a complete bipartite graph with equal number of vertices in each part, the graph is obviously randomly Hamiltonian. Conversely, let G be a randomly Hamiltonian graph with n vertices, and let C be a Hamiltonian cycle in G. If every edge in G is an edge of C, G is a cyclic graph. So for the remaining part of the theorem we assume that G is not a cyclic graph. The smallest value of k such that G has an outer k-cycle is either 3 or 4. If $k = 3$, by Problem 3.72, the graph is complete. If $k = 4$, the order of the graph is even, as established in Problem 3.74. Observe that there is a diagonal edge joining vertex v_i and vertex v_{i+3} for $i = 1, 2, \ldots, n$. Let v_r be any vertex where r is even. Consider the path from v_3 to v_1 defined by the following sequence: $\langle v_3, v_2, v_5, v_4, v_7, v_6, \ldots, v_{r-1}, v_{r-2}, v_{r+1}, v_{r+2}, \ldots, v_n, v_1\rangle$. This path contains all the vertices except v_r. So there is an edge joining v_1 and v_r where r is even. By Problem 3.72, there is an edge joining v_2 and v_r where r is odd. More generally, there is an edge between v_i and v_j whenever i and j are of opposite parity. Next, we show that there is no edge in the graph joining v_i and v_j when i and j are of the same parity. It is enough if we show that there is no edge between v_1 and v_s when s is odd. Suppose there is an edge joining v_1 and v_s when s is odd. Now we construct the path $\langle v_{s+2}, v_{s+1}, v_{s+4}, v_{s+3}, v_{s+6}, v_{s+5}, \ldots, v_{n-1}, v_{n-2}, v_1, v_s, v_{s-1}, v_{s-2}, \ldots, v_3, v_2\rangle$, which passes through every vertex other than v_n. So there is an edge joining v_2 and v_n, implying that we have an outer cycle consisting of three vertices, which contradicts the assumption.

3.76 Show that the cyclic graph is the only graph that is both randomly Eulerian and randomly Hamiltonian.

Solution. A graph is randomly Eulerian if and only if it is a cyclic graph. This assertion due to O. Ore was established in Problem 3.31. In conjunction with the Chartrand–Kronk theory, this assertion leads us to the conclusion that the only graph that is both randomly Eulerian and randomly Hamiltonian is the cyclic graph.

3.77 A graph G is **strongly randomly traceable** if for every two distinct vertices u and v, a Hamiltonian path between u and v exists whenever we start from u and successively proceed to another vertex not yet encountered, with the restriction that v will be chosen only when there is no other alternative. Show that a graph G of order n is a strongly randomly traceable graph if and only if it is the complete graph K_n.

Solution. Any complete graph is strongly randomly traceable. On the other hand, suppose G is strongly randomly traceable. If we delete a vertex from G, the resulting graph should be randomly Hamiltonian. So G can be neither a cyclic graph with more than three vertices nor a complete bipartite graph with equal number of vertices on its two parts. Thus the only alternative is that G is the complete graph with n vertices.

TOURNAMENTS

3.78 (*Landau's Theorem*) The outdegree of a vertex in a tournament can be considered as the score of the player represented by that vertex. A player with maximum score is a winner. If u is a player who has defeated a winner w in the tournament, show that w has defeated some player who has defeated u.

Solution. Let G be a tournament of order n. Suppose the outdegree of winner w is p. Thus w has defeated p players who constitute the set X, and at the same time, w has been defeated by $(n - 1 - p)$ players who constitute the set Y. Let u be any player in Y. If u has defeated every player in X, the outdegree of u will at least $(p + 1)$, which is a contradiction since the outdegree of no vertex can exceed p. So there is a vertex v in X such that there is an arc from v to u. This completes the proof.

[A set D of vertices in a graph (digraph) is called a ***k*-dominating set** if there is a path (directed path) from every vertex in D to every other vertex in the graph (digraph) such that the number of edges (arcs) in the path does not exceed k. If $k = 1$, D is a dominating set as defined in Solved Problem 1.21. Any player with the maximum outdegree is called a winner in the tournament. Landau's theorem asserts that the winners in a tournament constitute a 2-dominating set.]

3.79 (*Moon–Moser Theorem*) A graph (digraph) of order n $(n \geq 3)$ is **vertex-pancyclic** if every vertex is contained in a cycle (directed cycle) of length p for every p $(3 \leq p \leq n)$. Show that a strongly connected tournament is vertex-pancyclic.

Solution. Let v be any vertex in the graph $G = (V, E)$. The proof is by induction on p. Since the graph is strongly connected, both the outdegree and indegree of each vertex is necessarily positive.

Let $X = \{u \in V$: there is an arc from v to u in the tournament$\}$ and $Y = \{u \in V$: there is an arc from u to v in the tournament$\}$. Since the graph is strongly connected, there should be an arc from a vertex in X to a vertex in Y. So there is a cycle of length 3 passing through v. Hence, the theorem is true when $p = 3$.

Suppose there is a directed cycle $C = \langle v_1, v_2, \ldots, v_k, v_1\rangle$, where $k < n$ passing through vertex v_1. Let v be any vertex not in this cycle. We consider two cases.

Case (i): There is an arc from v to one of the vertices in C, and there is an arc from one of the vertices in C to v. Then there must be two adjacent vertices in the cycle such that there is an arc from one of them to v_1 and an arc from v_1 to the other. Then using these two vertices and v, we can construct a cycle of length $(k + 1)$ passing through v_1.

Case (ii): No vertex exists, as in case (*i*). Let $X = \{x \in V$: there is an arc from each vertex in C to $x\}$ and $Y = \{y \in V$: there is an arc to y to each vertex in $C\}$. Since the graph is strongly connected, both X and Y are nonempty. The strong connectivity also implies that there has to be an arc from vertex u in X to vertex v in Y. Then the length of the cycle $C' = \langle v_1, v_2, v_3, \ldots, v_{k-2}, u, v, v_1\rangle$ that passes through v_1 is $(k + 1)$, which completes the induction argument.

3.80 (*Camion's Theorem*) Prove Theorem 3.13: A tournament is Hamiltonian if and only if it is strongly connected.

Solution. If the tournament is strongly connected, it is vertex-pancyclic (as proved in Problem 3.79); hence, it is Hamiltonian. On the other hand, if it is Hamiltonian, there is a path from every vertex to every other vertex. So the graph is strongly connected.

3.81 Obtain a necessary and sufficient condition to be satisfied by a vertex in a tournament so that there is a Hamiltonian path starting from that vertex.

Solution. In a tournament $G = (V, E)$, any subset W of vertices is an **outclassed group** (as defined by O. Ore) if there is no arc in the tournament from a vertex in W to a vertex in $(V - W)$. So the initial vertex of any Hamiltonian path in a tournament cannot be a vertex in any outclassed group. We now claim that there is a Hamiltonian path from any vertex v_1 that does not belong to any outclassed group. Suppose we start from v_1 and get the path $P = \langle v_1, v_2, \ldots, v_k \rangle$. If this path has all the vertices, we are done. Let v_{k+1} be a vertex not in this path. If there is an arc from any vertex in P to v_{k+1}, we can enlarge the path into path P' (with $k + 1$ vertices) that includes this vertex either as an intermediate vertex or as the terminal vertex. Thus the path starting from v_1 is extended as much as possible. If w is a vertex not in the path thus extended, there is an arc from w to every vertex in the path that implies that the vertices in the path (including v_1) form an outclassed set, which is a contradiction. Thus the extended path contains all the vertices.

3.82 A tournament $G = (V, E)$ is **irreducible** whenever W is a subset of V, there should be an arc from a vertex in W to a vertex in $V - W$. Show that in a tournament $G = (V, E)$, the following concepts are equivalent: (*a*) G is Hamiltonian, (*b*) G is strongly connected, (*c*) G is irreducible.

Solution. The equivalence of (*a*) and (*b*) was already established in Problem 3.80. If a tournament is Hamiltonian, and if W is any set of vertices, there should be an arc from a vertex in W to a vertex in $(V - W)$. So any Hamiltonian tournament is irreducible. Now consider an irreducible tournament G. In any Hamiltonian path in the tournament, there should be an arc from the terminal vertex to some vertex v in the path. If v is the starting vertex of the path, we are done. Otherwise, we have cycle $C = \langle v_1, v_2, \ldots, v_k, v_1 \rangle$ through k vertices, where k is less than the order n of the tournament. Let X be the set of those vertices x not in C such that there is an arc from x to a vertex in C, and let Y be the set of those vertices y not in C such that there is an arc from y to a vertex in C. If X is empty, there is no arc from any vertex in Y to a vertex in $V - Y$, which is against the hypothesis. If Y is empty, there is no arc from C to $(V - C)$, which contradicts the hypothesis. Thus both X and Y are not empty. By the same reasoning, there is a vertex y in Y and a vertex x in X such that there is an arc from y to x. Then we can locate two vertices v_i and v_{i+1} in the cycle such that there is an arc from the former to y and an arc from the latter to x. Thus cycle C becomes enlarged with two intermediate vertices between v_i and v_{i+1}. We continue this process until we get a cycle that includes all the vertices.

3.83 Show that every tournament is either a strongly connected digraph or a digraph that can be converted into a strongly connected digraph by reversing the orientation of exactly one arc.

Solution. Suppose G is a tournament. Let u and v be any two vertices in G. So there is a directed Hamiltonian path from u to v. Either there is an arc from v to u or there is an arc from u to v. In the former case, the graph is strongly connected. In the latter case, change the orientation of the arc; the resulting digraph then becomes strongly connected.

3.84 The sequence $\langle s_1, s_2, \ldots, s_n \rangle$ of nonnegative integers is called a **score sequence** (of a tournament) if there exists a tournament of order n whose vertices can be labeled v_i so that the outdegree of each v_i is s_i for each $i = 1, 2, \ldots, n$. Show that a nondecreasing sequence of n nonnegative integers is the score sequence of a transitive tournament if and only if the sequence is $\langle 0, 1, 2, \ldots, n - 1 \rangle$.

Solution. If the score sequence of a tournament G is $\langle 0, 1, 2, \ldots, n \rangle$, the n vertices of G can be labeled $v_1, v_2, \ldots, v_n$ such that every arc is from vertex v_i to vertex v_j, where $i < j$. Obviously, this is a transitive tournament. On the other hand, suppose G is a transitive tournament. Thus it has a unique Hamiltonian path. Label the vertices v_i ($i = 1, 2, \ldots, n$) such that this unique path is $v_1 \to v_2 \to \cdots \to v_n$. Since G is transitive, every arc is from vertex v_i to vertex v_j, where $i < j$. So the outdegree of v_k is $(k - i)$ for each k. [It follows from this problem that there is exactly one (up to isomorphism) transitive tournament of order n for every positive integer n.]

3.85 Show that the sequence $\langle s_1, s_2, \ldots, s_n \rangle$ of nondecreasing nonnegative integers is a score sequence of a tournament if and only if $s_1 + s_2 + \cdots + s_k \geq k(k-1)/2$ for $1 \leq k \leq n$ with equality holding when $k = n$.

Solution. Consider the subgraph H induced by the set $W = \{v_1, v_2, \ldots, v_k\}$ of k vertices. H is a subtournament. The sum of the outdegrees (in H) of these k vertices is $k(k-1)/2$. The sum $\Sigma_{i=1}^{k} s_i$ has to be at least equal to $k(k-1)/2$ since there could be arcs in the tournament from vertices in W to vertices in $V - W$. Thus the inequality holds for all $k \leq n$. Obviously, the inequality becomes an equality when $k = n$.

The converse is proved by contradiction. If there is a sequence satisfying the given conditions that is not the score sequence of a tournament, there should be a sequence with n elements, where n is as small as possible. Among such sequences, choose $S = \langle s_1, s_2, \ldots, s_n \rangle$, where the first component s_1 is as small as possible. We consider two cases.

Case (i): Suppose there exists k ($0 < k < n$) such that $s_1 + s_2 + \cdots + s_k = k(k-1)/2$. Then $S_1 = \langle s_1, s_2, \ldots, s_k \rangle$ is the score sequence of some tournament $H_1 = (V_1, E_1)$ of order k because of the minimality assumption on n.

Consider the sequence $S_2 = \langle t_1, t_2, \ldots, t_{n-k} \rangle$, where $t_i = s_{k+i} - k$, $i = 1, 2, \ldots, (n-k)$. This is a nondecreasing sequence. Suppose m is any positive integer where $m \leq (n-k)$. Then $t_1 + t_2 + \cdots + t_m = (s_{k+1} + s_{k+2} + \cdots + s_{k+m}) - mk = (s_1 + s_2 + \cdots + s_{m+k}) - (s_1 + s_2 + \cdots + s_k) - mk \geq [(m+k)(m+k-1)/2] - [k(k-1)/2] - mk = m(m-1)/2$ and $t_1 + t_2 + \cdots + t_{n-k} = (n-k)(n-k-1)/2$. So, by hypothesis, S_2 is the score sequence of some tournament $H_2 = (V_2, E_2)$ of order $(n-k)$. Let $H = (V, E)$ be a digraph where $V = V_1 \cup V_2$ and $E = E_1 \cup E_2 \cup F$, where F is the set of arcs obtained by drawing an arc from each vertex in V_2 to each vertex in V_1. Then it is easy to see that H is a tournament with the score sequence S, which contradicts the assumption that there is no tournament with S as the score sequence.

Case (ii): $s_1 + s_2 + \cdots + s_k > k(k-1)/2$ for every $k < (n-1)$. In particular, s_1 is positive. Then $S' = \langle s_1 - 1, s_2, \ldots, s_{n-1}, s_n + 1 \rangle$ is a another nondecreasing sequence satisfying the requirements of the hypothesis. Due to the minimality condition on s_1, we conclude that there is a tournament H' for which S' is the score sequence.

Let x, y be the vertices in H' with respective outdegrees $s_n + 1$ and $s_1 - 1$. Then $(s_1 - 1) + 2 \leq (s_n + 1)$, which implies that there is a vertex v in H' such that there is an arc from x to v and an arc from v to x. If we reverse the directions of these two arcs, we get a new tournament with S as a score sequence, which is again a contradiction. (The proof presented here is due to C. Thomassen.)

3.86 Show that the sequence $\langle s_1, s_2, \ldots, s_n \rangle$ of nondecreasing nonnegative integers is a score of a strongly connected tournament if and only if $s_1 + s_2 + \cdots + s_k > k(k-1)/2$ for $1 \leq k \leq (n-1)$ and $s_1 + s_2 + \cdots + s_n = n(n-1)/2$.

Solution. Let $S = \langle s_1, s_2, \ldots, s_n \rangle$ be the score sequence of a strongly connected tournament G, where s_i is the score (outdegree) of vertex v_i. Then, obviously, $s_1 + s_2 + \cdots + s_n = n(n-1)/2$. Now consider the subgraph H induced by $W = \{v_1, v_2, \ldots, v_k\}$, where $k < n$. The subgraph H is a strongly connected tournament, and the sum of the outdegrees (in H) of these k vertices is $k(k-1)/2$. As we saw in Problem 3.85, this sum is at most equal to $s_1 + s_2 + \cdots + s_k$. The strong connectivity implies that there is at least one vertex v_i in W such that the outdegree of v_i in G is more than the outdegree of the same vertex in H. Thus $s_1 + s_2 + \cdots + s_k > k(k-1)/2$ for $k = 1, 2, \ldots, (n-1)$.

To prove the converse, let the sequence $S = \langle s_1, s_2, \ldots, s_n \rangle$ satisfy the hypothesis of the theorem. By Problem 3.85, there is a tournament $G = (V, E)$ with S as the score sequence. We have to show that G is strongly connected. Suppose it is not. Then there is a set W of k vertices ($k < n$) such that there is a directed path between every pair of vertices in W. The subgraph H induced by W is called a strong component of G. If u is any vertex in $(V - W)$ and w is any vertex in W, the arc between these two vertices has to be an arc from u to w. Consequently, the outdegree of every vertex in W is the same for both G and H. So the sum of the outdegrees (in G) of these k vertices in W is $k(k-1)/2$.

Now, $s_1 + s_2 + \cdots + s_k) \le$ (sum of outdegrees in G of the k vertices in W) since the sequence S is nondecreasing. So $(s_1 + s_2 + \cdots + s_k) \le k(k-1)/2$, which contradicts the hypothesis. Thus the tournament is strongly connected.

Supplementary Problems

3.87 If G is a connected graph with k odd vertices, find the minimum number of trails in G such that every edge in the graph is an edge in exactly one of these trails. *Ans.* $(k/2)$

3.88 Show that if a multidigraph has an Eulerian walk but not an Eulerian circuit, exactly one vertex has an excess of one indegree and exactly one vertex has an excess of one outdegree. [*Hint:* Generalize Theorem 3.3.]

3.89 Show that if a graph has a circuit of odd length, it has a cycle of odd length. [*Hint:* Every circuit is a disjoint union of cycles.]

3.90 Suppose in a group of n people ($n > 3$), any two of them together know all the other people in the group. Show that these n people can be seated around a circular table so that each person is seated between two acquaintances. [*Hint:* Consider the acquaintance graph and use Problem 3.69.]

3.91 Show that a k-regular graph with $(2k - 1)$ vertices is Hamiltonian. [*Hint:* Use Dirac's theorem.]

3.92 Show that any k-regular simple graph with $(2k - 1)$ vertices is Hamiltonian. [*Hint:* Use Dirac's theorem.]

3.93 A nontrivial connected graph is Eulerian if and only if every block in the graph is Eulerian. [*Hint:* This follows from the definition of a block in a graph.]

3.94 Find the number of Hamiltonian graphs in K_n. *Ans.* $\frac{1}{2}[(n-1)!]$

3.95 Find the number of Hamiltonian cycles in $K_{n,n}$. *Ans.* $\frac{1}{2}[(n-1)!](n!)$

3.96 Show that if n is odd, the set of edges of K_n can be partitioned into $\frac{1}{2}(n-1)$ disjoint Hamiltonian cycles. [*Hint:* If the vertices are $1, 2, 3, \ldots, n$, arrange these vertices in a cycle in that order. The different cycles are $1 — (1 + i) — (1 + i + i) — \cdots — 1$, where addition is modulo n and $1 \le i \le \frac{1}{2}(n-1)$.]

3.97 Thirteen mathematicians are attending a six-day conference. Each night they sit around a circular table for dinner so that no two persons sit next to each other more than once and so that each person gets a chance to sit to next to each other person exactly once during those six nights.
Ans. List the six distinct Hamiltonian cycles in the complete graph with 13 vertices (as indicated in Problem 3.96), where the vertices are labeled 1, 2, 3, . . . , 13, representing the 13 individuals. The six cycles (giving the seating arrangements around the circular table) are

(*a*) 1 — 2 — 3 — 4 — 5 — 6 — 7 — 8 — 9 — 10 — 11 — 12 — 13 — 1

(*b*) 1 — 3 — 5 — 7 — 9 — 11 — 13 — 2 — 4 — 6 — 8 — 10 — 12 — 1

(*c*) 1 — 4 — 7 — 10 — 13 — 3 — 6 — 9 — 12 — 2 — 5 — 8 — 11 — 1

(*d*) 1 — 5 — 9 — 13 — 4 — 8 — 12 — 3 — 7 — 11 — 2 — 6 — 10 — 1

(*e*) 1 — 6 — 11 — 3 — 8 — 13 — 5 — 10 — 2 — 7 — 12 — 4 — 9 — 1

(*f*) 1 — 7 — 13 — 6 — 12 — 5 — 11 — 4 — 10 — 3 — 9 — 2 — 8 — 1

3.98 (*a*) It is known that $\langle 1, 1, 2, 3, 3, t \rangle$ is the score sequence of a tournament. Is it a score sequence of a strong tournament? (*b*) It is known that $\langle 1, 2, 2, 3, 3, t \rangle$ is the score sequence of a tournament. Is it a score sequence of a strong tournament?
Ans. (*a*) No. (Here $t = 5$.) (*b*) Yes. (Here $t = 4$.)

3.99 Show that the sum of the squares of the terms in the score sequence of a tournament is equal to the sum of the squares of the indegrees of the vertices. [*Hint:* If n is the order of the graph, the sum of the outdegree and indegree at each vertex is $(n - 1)$, and the sum of all the indegrees is equal to the sum of all the outdegrees.]

3.100 Show that a regular tournament is strongly connected. [*Hint:* If the outdegree of each vertex is d in a tournament with n vertices, nd is $n(n - 1)/2$. Use the condition established in Problem 3.86.]

3.101 Show that if $\langle s_1, s_2, \ldots, s_n \rangle$ is the score sequence of a tournament, $\langle t_1, t_2, \ldots, t_n \rangle$ is also a score sequence of a tournament, where $t_i = (n - 1) - s_i$ for $i = 1, 2, \ldots, n$. [*Hint:* Use the condition established in Problem 3.85.]

3.102 Examine the case when the inequalities in the statement of Problem 3.85 are all equalities.
Ans. In this case, the tournament is transitive.

3.103 Show that in a tournament involving n players, the number of players with the score $(n - 1)$ is at most 1. [*Hint:* Otherwise, the inequality condition established in Problem (85) will not hold.]

3.104 Show that if n is an odd number other than 1, there is a tournament of order n in which every vertex is a winner. [*Hint:* Each term in the score sequence is $(n - 1)/2$.]

3.105 Show that the k-cube (see Solved Problem 1.56) is a Hamiltonian graph. [*Hint:* The k-cube is a k-regular bipartite graph $G = (X, Y, E)$, where both X and Y have 2^{k-1} vertices. If $k = 3$, the path 000 010 110 100 101 111 011 001 is a Hamiltonian path from 000 to 001 that can be completed into a Hamiltonian cycle. Notice that the first four vertices have 0 as the last digit and that the last four have 1 as the last digit. To obtain a Hamiltonian path (when $k = 4$) from 0000 to 0001, we take these eight vertices and attach a 0 to them as a last digit, then take these new vertices in reverse order and replace the 0 in the last digit by 1. Finally, join 0001 and 0000 to get the Hamiltonian cycle. The proof is completed by this induction argument.]

3.106 For what value of k is the k-cube graph an Eulerian graph?
Ans. The k-cube is Eulerian if and only if k is even.

3.107 Show that the de Bruijn digraph is a Hamiltonian graph. [*Hint:* See the definition of the de Bruijn graph.]

Chapter 4

Optimization Involving Trees

4.1 MINIMUM WEIGHT SPANNING TREES

If each edge e of a graph G is assigned a real number $w(e)$ as its weight, the graph equipped with this allocation of weights is known as a **weighted graph.** The definition of a weighted digraph is analogous. A **network** is a weighted graph or a weighted digraph. In some situations with weights defined on the edges and arcs of a mixed graph, we have a network in which there are weighted edges as well as weighted arcs. In a figure representing a network, the weight of an edge or arc is usually written as a number on the edge or arc, as the case may be. If H is a subgraph of a weighted graph G, the weight of H is the sum of the weights of all the edges in H; this is denoted as $w(H)$. A spanning tree T in a weighted digraph G is a **minimum weight spanning tree** in G (or **minimum spanning tree** or **minimum connector**) if $w(T) \leq w(T')$ for every spanning tree T' in G.

In this section, we discuss two algorithms to obtain a minimum spanning tree (M.S.T.) in a connected weighted graph. Both the procedures are "greedy" in the sense that at every stage, a decision is to be made to select the best possible edge from the graph for inclusion as an edge in the M.S.T. after making sure that the selection of an edge does not create a cycle in the subgraph already constructed.

Kruskal's Algorithm

The input is the set of edges in a weighted graph with n vertices. The output is either a report that the graph is not connected or the set of edges in a minimum spanning tree T in the network. There are four steps in this algorithm.

Step 1. Arrange the edges of the weighted graph in nondecreasing order of their weights as a list L, and set T to be the empty set.

Step 2. Select the first edge from L, and include that in T.

Step 3. If every edge in L is examined for possible inclusion in T, stop and report that G is not connected. Otherwise, take the first unexamined edge in L. If it does not form a cycle in T, select it for inclusion in T and go to step 4. Otherwise, discard that edge as an examined (but unselected) edge and repeat step 3.

Step 4. Stop if T has $(n - 1)$ edges. Otherwise, go to step 3.

Theorem 4.1. Kruskal's algorithm solves the M.S.T. problem in a network. (See Solved Problem 4.14.)

Example 1. Show by using Kruskal's algorithm that the network in Fig. 4-1 is a disconnected weighted graph.

Here $n = 7$. The list L of edges with weights in nondecreasing order is $\{1, 2\}$, $\{1, 3\}$, $\{5, 6\}$, $\{6, 7\}$, $\{2, 3\}$, $\{5, 7\}$, $\{1, 4\}$, and $\{3, 4\}$. As specified in the algorithm, we select edges $\{1, 2\}$, $\{1, 3\}$, $\{5, 6\}$, and $\{6, 7\}$. Then edge $\{2, 3\}$ is examined but not selected. Subsequently, edge $\{5, 7\}$ is examined but not selected. Then edge $\{1, 4\}$ is examined and selected. Finally, edge $\{3, 4\}$ is examined but not selected. At this stage all the edges in L are examined. So the graph is not connected.

Example 2. Obtain an M.S.T. in the network shown in Fig. 4-2 using Kruskal's algorithm.

Here $n = 8$. We take edges $\{1, 2\}$, $\{3, 4\}$, $\{1, 8\}$, $\{4, 5\}$, and $\{7, 8\}$, in that order, using the greedy procedure. We disregard the next entry, $\{2, 7\}$, and select $\{5, 6\}$. Then we ignore $\{3, 6\}$ and selected $\{6, 7\}$. At this stage, the number of

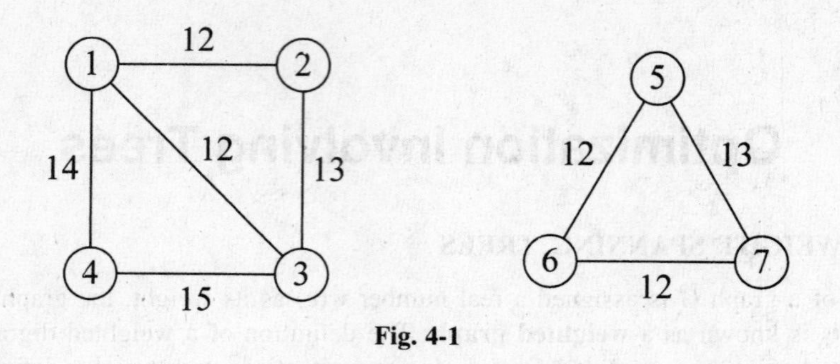

Fig. 4-1

edges selected for inclusion in the tree is seven, which is the number of trees in a spanning tree. The seven edges thus selected constitute the edges of a minimum spanning tree T in the given network with $w(T) = 93$, as shown in Fig. 4-3. The positive integer displayed on an edge in this figure indicates the order in which the edge was chosen for inclusion in T, and it should not be confused with the weight of the edge.

Prim's Algorithm

The input and the output are the same as in Kruskal's algorithm. There are three steps in this algorithm.

Step 1. Select an arbitrary vertex of the weighted graph, and include it as a vertex in T.

Step 2. Let W be the set of vertices in T. If the disconnecting set $(W, V - W)$ is empty, report that G is not connected. Otherwise, find an edge of minimum weight in the disconnecting set $(W, V - W)$. Tiebreaking is arbitrary. If this edge does not create a cycle in T, select it for inclusion in T and go to step 3. Otherwise, discard that edge and repeat step 2.

Step 3. Stop if T has $(n - 1)$ edges. Otherwise, go to step 2.

Theorem 4.2. Prim's algorithm solves the M.S.T. problem. (See Solved Problem 4.17.)

Example 3. Use Prim's algorithm to show that the network shown in Fig. 4-1 is not a connected graph.

Since $n = 7$, there should be six edges in the tree. Initially, $W = \{1\}$. The algorithm selects edges $\{1, 2\}$, $\{1, 3\}$, and $\{1, 4\}$. At this stage, the number of selected edges is less than six, and $W = \{1, 2, 3, 4\}$. The disconnecting set $(W, V - W)$ is empty. So the graph is not connected.

Example 4. Obtain an M.S.T. in the network shown in Fig. 4-2 using Prim's algorithm.

We start from vertex 1 and select edge $\{1, 2\}$. At this stage, W consists of two vertices 1 and 2. An edge of minimum weight in the disconnecting set $(W, V - W)$ is edge $\{1, 8\}$. Currently, edges $\{1, 2\}$ and $\{1, 8\}$ are in T, and $W = \{1, 2, 8\}$. An edge of smallest weight in $(W, V - W)$ is either $\{2, 7\}$ or $\{7, 8\}$. The tie is broken by selecting edge $\{2, 7\}$.

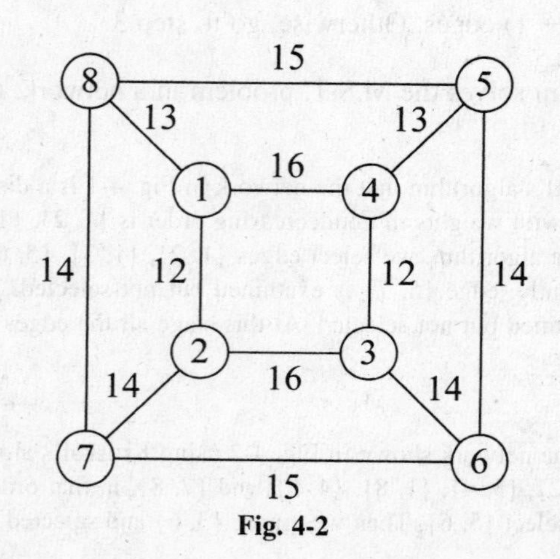

Fig. 4-2

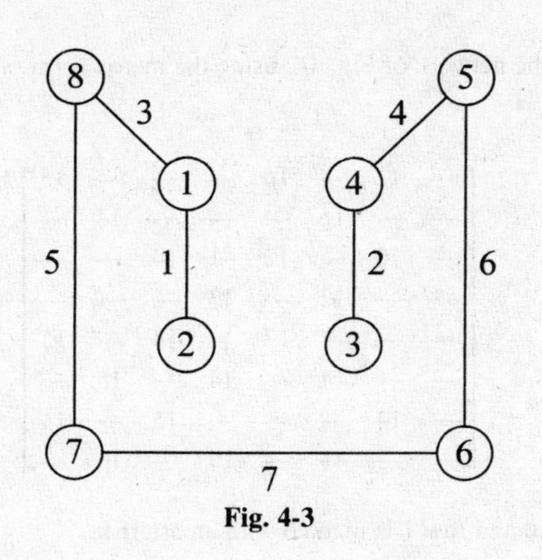

Fig. 4-3

At this stage, $W = \{1, 2, 7, 8\}$, and the edges in T are $\{1, 2\}$, $\{1, 8\}$, and $\{2, 7\}$. An edge of minimum weight in $(W, V - W)$ is $\{7, 8\}$, which forms a cycle in T. So this edge is discarded. Then *an* edge of minimum weight in $(W, V - W)$ is $\{5, 8\}$. We continue like this and select $\{8, 5\}$, $\{5, 4\}$, $\{4, 3\}$, and $\{3, 6\}$, in that order. At this stage, T has seven edges and we stop. In Fig. 4-3, the graph has seven edges and we stop. Figure 4-4 shows the graph of T in which the integers marked on the edges indicate the order in which they are selected.

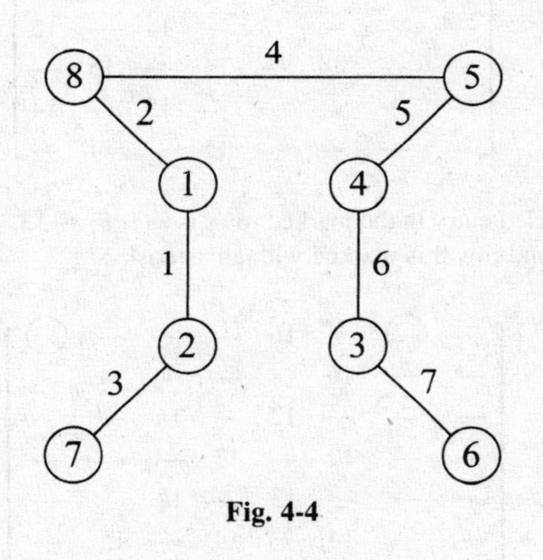

Fig. 4-4

Prim's Algorithm (Matrix Version)

For the sake of simplicity, we assume that the network is connected. Let $V = \{1, 2, \ldots, n\}$ be the vertices of a connected network. If there is an edge joining vertices i and j, let $w(i, j)$ be the weight of this edge, and let $D = [w(i, j)]$ be the matrix in which the (i, j) entry is blank if there is no edge joining i and j. Delete all elements of column 1, and mark row 1 with an asterisk. This means that we are starting with vertex 1. So $W = \{1\}$. Initially, all numerical entries in the matrix are uncircled. Each iteration has two steps, as follows.

Step 1. Select a smallest element from the uncircled numerical entries in the rows that are marked with an asterisk. Stop if no such element exists. The edges that correspond to the circled entries constitute an M.S.T.

Step 2. If $w(i, j)$ is selected, circle the (i, j) entry in D and mark row j with an asterisk. Delete the remaining uncircled elements in column j. Go to step 1.

Example 5. Obtain an M.S.T. in the network of Fig. 4-2 using the matrix form of Prim's algorithm.

The initial matrix is

$$\begin{bmatrix} - & 12 & - & 16 & - & - & - & 13 \\ - & - & 16 & - & - & - & 14 & - \\ - & 16 & - & 12 & - & 14 & - & - \\ - & - & 12 & - & 13 & - & - & - \\ - & - & - & 13 & - & 14 & - & 15 \\ - & - & 14 & - & 14 & - & 15 & - \\ - & 14 & - & - & - & 15 & - & 14 \\ - & - & - & - & 15 & - & 14 & - \end{bmatrix}\begin{matrix} * \\ \\ \\ \\ \\ \\ \\ \\ \end{matrix}$$

All the entries in column 1 are deleted and row 1 is marked with an asterisk.

Iteration 1: The smallest uncircled entry in the marked row is $w(1, 2) = 12$. This entry is circled, the remaining entries in column 2 are deleted, and row 2 is marked with an asterisk:

$$\begin{bmatrix} - & \textcircled{12} & - & 16 & - & - & - & 13 \\ - & - & 16 & - & - & - & 14 & - \\ - & - & - & 12 & - & 14 & - & - \\ - & - & 12 & - & 13 & - & - & - \\ - & - & - & 13 & - & 14 & - & 15 \\ - & - & 14 & - & 14 & - & 15 & - \\ - & - & - & - & - & 15 & - & 14 \\ - & - & - & - & 15 & - & 14 & - \end{bmatrix}\begin{matrix} * \\ * \\ \\ \\ \\ \\ \\ \\ \end{matrix}$$

Iteration 2: The smallest uncircled entry in the marked rows is $w(1, 8) = 13$. This element is circled, the remaining entries in column 8 are deleted, and row 8 is marked with an asterisk:

$$\begin{bmatrix} - & \textcircled{12} & - & 16 & - & - & - & \textcircled{13} \\ - & - & 16 & - & - & - & 14 & - \\ - & - & - & 12 & - & 14 & - & - \\ - & - & 12 & - & 13 & - & - & - \\ - & - & - & 13 & - & 14 & - & - \\ - & - & 14 & - & 14 & - & 15 & - \\ - & - & - & - & - & 15 & - & - \\ - & - & - & - & 15 & - & 14 & - \end{bmatrix}\begin{matrix} * \\ * \\ \\ \\ \\ \\ \\ * \end{matrix}$$

Iteration 3: The smallest uncircled entry in the marked rows is $w(2, 7) = 14$. This element is circled, the remaining entries in column 7 are deleted, and row 7 is marked with an asterisk:

$$\begin{bmatrix} - & \textcircled{12} & - & 16 & - & - & - & \textcircled{13} \\ - & - & 16 & - & - & - & \textcircled{14} & - \\ - & - & - & 12 & - & 14 & - & - \\ - & - & 12 & - & 13 & - & - & - \\ - & - & - & 13 & - & 14 & - & - \\ - & - & 14 & - & 14 & - & - & - \\ - & - & - & - & - & 15 & - & - \\ - & - & - & - & 15 & - & - & - \end{bmatrix}\begin{matrix} * \\ * \\ \\ \\ \\ \\ * \\ * \end{matrix}$$

Iteration 4: The smallest uncircled entry in the marked rows is $w(7, 6) = 15$. This element is circled, the remaining entries in column 6 are deleted, and row 6 is marked with an asterisk:

$$\begin{bmatrix} - & \textcircled{12} & - & 16 & - & - & - & \textcircled{13} \\ - & - & 16 & - & - & - & \textcircled{14} & - \\ - & - & - & 12 & - & - & - & - \\ - & - & 12 & - & 13 & - & - & - \\ - & - & - & 13 & - & - & - & - \\ - & - & 14 & - & 14 & - & - & - \\ - & - & - & - & - & \textcircled{15} & - & - \\ - & - & - & - & 15 & - & - & - \end{bmatrix}\begin{matrix} * \\ * \\ \\ \\ \\ * \\ * \\ * \end{matrix}$$

Iteration 5: The smallest uncircled entry in the marked rows is $w(6, 5) = 14$. This entry is circled, the remaining entries in column 5 are deleted, and row 5 is marked with an asterisk:

$$\begin{bmatrix} - & \textcircled{12} & - & 16 & - & - & - & \textcircled{13} \\ - & - & 16 & - & - & - & \textcircled{14} & - \\ - & - & - & 12 & - & - & - & - \\ - & - & 12 & - & - & - & - & - \\ - & - & - & 13 & - & - & - & - \\ - & - & 14 & - & \textcircled{14} & - & - & - \\ - & - & - & - & - & \textcircled{15} & - & - \\ - & - & - & - & - & - & - & - \end{bmatrix}\begin{matrix} * \\ * \\ \\ \\ * \\ * \\ * \\ * \end{matrix}$$

Iteration 6: The smallest uncircled entry in the marked rows is $w(5, 4) = 13$. This entry is circled, the remaining entries in column 4 are deleted, and row 4 is marked with an asterisk:

$$\begin{bmatrix} - & \textcircled{12} & - & - & - & - & - & \textcircled{13} \\ - & - & 16 & - & - & - & \textcircled{14} & - \\ - & - & - & - & - & - & - & - \\ - & - & 12 & - & - & - & - & - \\ - & - & - & \textcircled{13} & - & - & - & - \\ - & - & 14 & - & \textcircled{14} & - & - & - \\ - & - & - & - & - & \textcircled{15} & - & - \\ - & - & - & - & - & - & - & - \end{bmatrix}\begin{matrix} * \\ * \\ \\ * \\ * \\ * \\ * \\ * \end{matrix}$$

Iteration 7: The smallest uncircled entry in the marked rows is $w(4, 3) = 12$. This entry is circled, the remaining vertices in column 3 are deleted, and row 3 is marked with an asterisk:

$$\begin{bmatrix} - & \textcircled{12} & - & - & - & - & - & \textcircled{13} \\ - & - & - & - & - & - & \textcircled{14} & - \\ - & - & - & - & - & - & - & - \\ - & - & \textcircled{12} & - & - & - & - & - \\ - & - & - & \textcircled{13} & - & - & - & - \\ - & - & - & - & \textcircled{14} & - & - & - \\ - & - & - & - & - & \textcircled{15} & - & - \\ - & - & - & - & - & - & - & - \end{bmatrix}\begin{matrix} * \\ * \\ * \\ * \\ * \\ * \\ * \\ * \end{matrix}$$

At this stage, all the rows of the matrix are marked, giving $(n - 1)$ circled entries corresponding to the edges of an M.S.T. The weight of the M.S.T. is the sum of the circled entries in this last matrix.

4.2 MAXIMUM WEIGHT BRANCHINGS

A **directed forest** is a digraph whose underlying graph is a forest. A **branching B in a digraph G** is a subgraph of G such that B is a directed forest and that the indegree of each vertex in B is either 0 or 1. The problem of finding a branching of maximum weight in a weighted digraph is known as the **maximum weight branching problem.**

It is easy to see that the greedy approach may not produce a maximum weight branching in a weighted digraph in general. For example, consider the digraph $G = (V, E)$ and $V = \{1, 2, 3, 4\}$ and E the set of arcs $\{(1, 2,)\ (2, 3), (3, 4), (4, 3)\}$ with weights 9, 8, 5, and 10, respectively. The greedy method will select arcs (4, 3) and (1, 2) with a total weight of 19, whereas the optimal solution consists of arcs (1, 2), (2, 3), and (3, 4) with a total weight of 22.

If e is an arc from i to j, the **source** $s(e)$ of e is i and the **terminal** $t(e)$ of e is j. Arc e from i to j is a **critical arc** if its weight is not less than the weight of any other arc whose terminal is also vertex j. If there is more than one critical arc directed to a vertex, we select any one of them. The subgraph consisting of the set of all critical arcs thus chosen in a weighted digraph is a **critical graph** in the digraph. If a critical graph H in a digraph G is acyclic, H is obviously a maximum weight branching in G. In this case, the optimization problem is readily solvable. What is now needed is a procedure to obtain a maximum weight branching in a digraph when it does not have an acyclic critical graph.

Condensing a Weighted Digraph

Let H be a critical graph in G with weight function w, and let the cycles in H be $C_i (i = 1, 2, \ldots, k)$. Also let W be the set of those vertices in G that do not belong to any of the cycles in H. Replace each cycle C_i in H by a single vertex X_i. Let $V_1 = \{X_1, X_2, \ldots, X_k\} \cup W$. If e is an arc in G that is not an arc of C_i and if $t(e)$ is a vertex of C_i, define $w_1(e) = w(e) - w(f) + w(e_i)$, where f is the unique arc in C_i such that $t(e) = t(f)$ and e_i is an arc of minimum weight among all the arcs in that cycle. If $t(e)$ is not a vertex of any of these k cycles, $w_1(e) = w(e)$. The weighted multigraph G_1 thus constructed with V_1 as the set of vertices with the revised weight function w_1 is the **condensed graph** of G. If H_1 is a critical graph of G_1, we have moved from the pair (G, H) to the (G_1, H_1) after condensing G by using H. We continue this condensation process until we reach the pair (G_m, H_m), where H_m is acyclic.

Example 6. Carry out the condensation process in the weighted digraph shown in Fig. 4-5.

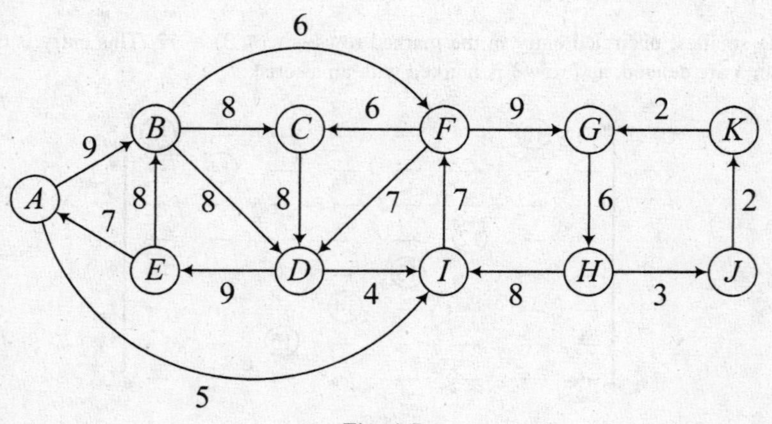

Fig. 4-5

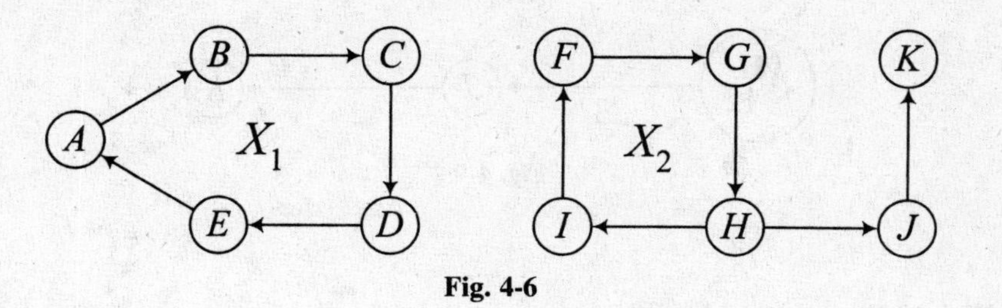

Fig. 4-6

A critical graph H in G is shown in Fig. 4-6 with two disjoint cycles marked X_1 and X_2. If the two cycles are shrunk into two vertices and the weight function is updated, we get the condensed graph G_1 shown in Fig. 4-7.

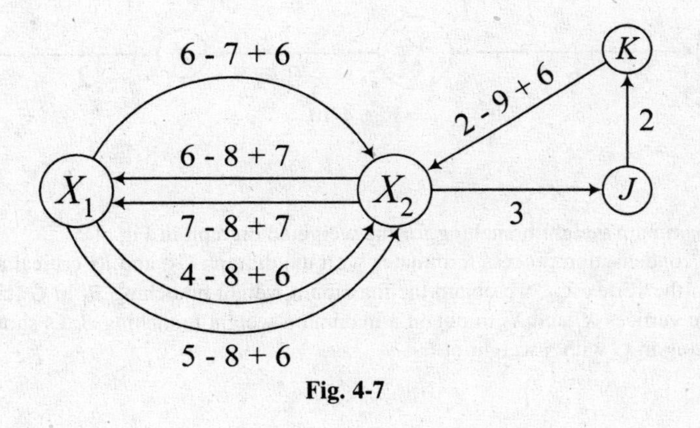

Fig. 4-7

A critical graph H_1 in G_1 is shown in Fig. 4-8. The cycle in H_1 is condensed into a single vertex denoted by X_{12}, as shown in Fig. 4-9, defining the acyclic digraph G_2 in which the critical graph H_2 is acyclic and is the same as G_2.

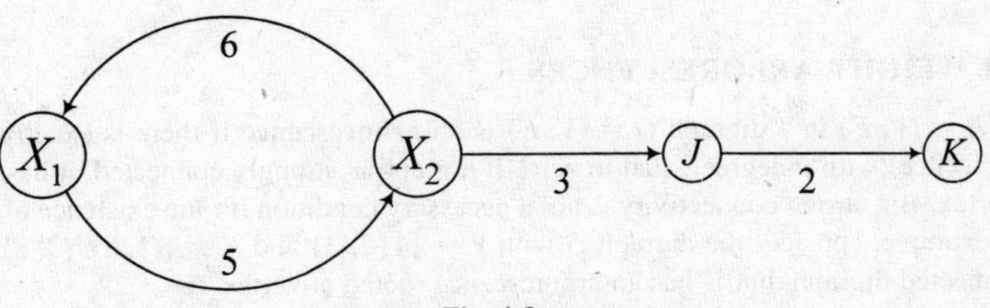

Fig. 4-8

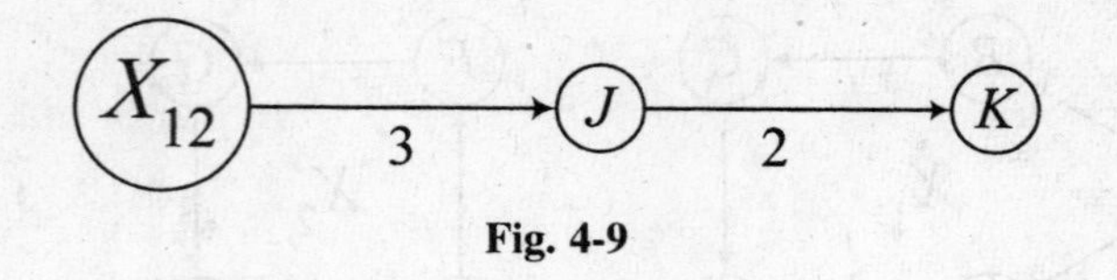

Fig. 4-9

The Maximum Weight Branching Algorithm

Step 1. (The condensation process) The input is the weighted digraph $G = G_0$. Construct G_1 from G_{i-1} by condensing the cycles in its critical digraph. G_k is the first digraph in the sequence for which the critical graph H_k is acyclic.

Step 2. (The unraveling process) The graph H_k is a maximum weight branching in G_k. Let $B_k = H_k$. Construct B_{i-1} from B_i by expanding the condensed cycles. B_i is a maximum weight branching in G_i for $i = k, k-1, k-2, \ldots, 1, 0$. The output is $B = B_0$. (See Solved Problem 4.25.)

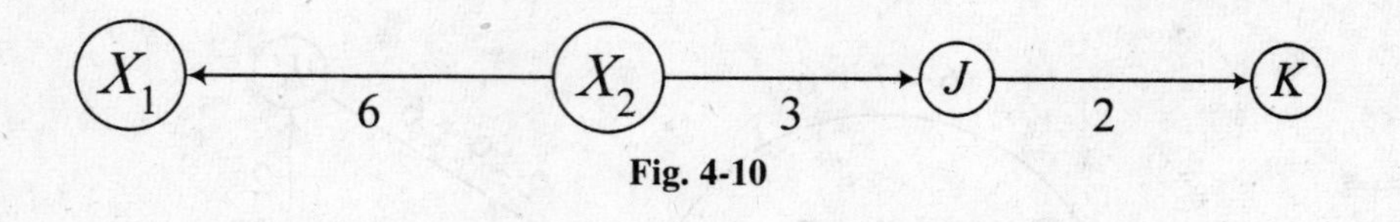

Fig. 4-10

Example 7. Obtain a maximum weight branching for the weighted digraph in Fig. 4-5.

For this example, the condensation process terminates with the digraph G_2, and its critical acyclic graph is as shown in Fig. 4-9. When we open the vertex X_{12}, we obtain the maximum weight branching B_1 in G_1, as shown in Fig. 4-10. At the next stage, we open the vertices X_1 and X_2 to obtain a maximum weight branching B_0 as shown in Fig. 4-11, which is a maximum weight branching in G with a weight of 69.

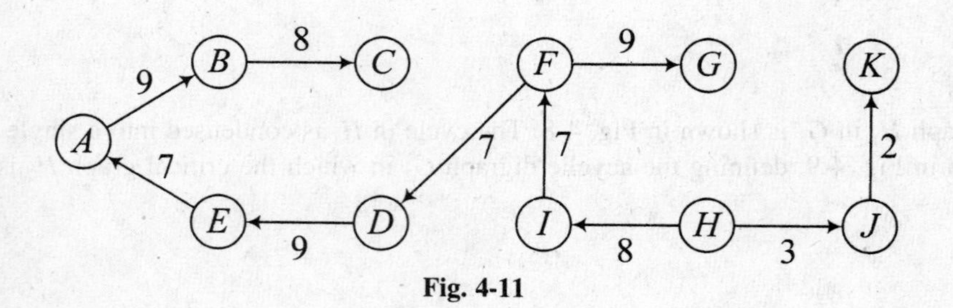

Fig. 4-11

4.3 MINIMUM WEIGHT ARBORESCENCES

A branching $B = (V, F)$ in a digraph $G = (V, E)$ is an **arborescence** if there is exactly one vertex (the **root** of the arborescence) with indegree equal to zero. If a graph is strongly connected, it has an arborescence rooted at every vertex. But strong connectivity is not a necessary condition for the existence of an arborescence in a digraph. For example, consider the digraph G with $V = \{1, 2, 3\}$ and $E = \{(1, 2), (2, 3), (2, 1)\}$. This is not a strongly connected digraph, but is has an arborescence rooted at vertex 1.

A digraph is **quasi-strongly connected** if for every pair of vertices u and v in the digraph there exists a vertex w such that there are directed paths from w to u and from w to v. Strong connectivity implies quasi-strong connectivity. But the converse is not true, as was seen in Example 7. Obviously, any graph that has an arborescence is quasi-strongly connected. The converse also is true. Thus we have the following theorem.

Theorem 4.3. A digraph has an arborescence if and only if it is quasi-strongly connected. (See Solved Problem 4.29.)

Suppose it is known that a weighted digraph has an arborescence with root at vertex r. If we adopt a greedy procedure and choose an arc of minimum weight directed to each vertex other than the root and if we can obtain a subgraph H that is acyclic by this method, H is indeed a minimum weight arborescence in the digraph. Observe that H is acyclic if and only if it is connected. For example, if the weighted digraph is $G = (V, E)$, where $V = \{1, 2, 3, 4\}$ and $E = \{(1, 2), (1, 3), (2, 4), (3, 2), (4, 3)\}$ with weights 3, 5, 4, 6, and 7, respectively, the greedy algorithm will select (1, 2), (2, 4), and (1, 3), giving a minimum weight arborescence of weight 12. If, on the other hand, the weights are 6, 7, 4, 3, and 5, respectively, the resulting digraph is not acyclic and not connected. So we need a procedure to solve the minimum weight arborescence problem when the digraph chosen by the greedy procedure is not acyclic.

Theorem 4.4. Let $T^* = (V, E^*)$ be a minimum weight arborescence rooted at vertex r in a weighted digraph $G = (V, E)$ with distinct arc weights, and let $H = (V, E')$ be the subgraph obtained from G by choosing the arc of minimum weight directed to each vertex other than r. Then $|(C - E^*)| = 1$ for every cycle C in H. (See Solved Problem 4.30.)

Condensing a Weighted Digraph

Let $G = (V, E)$, T^*, and $H = (V, E')$ be as in Theorem 4.4. Suppose H is not acyclic and W is the set of vertices in a cycle C of H. Consider the digraph G_1 (probably with multiple arcs) obtained from G by condensing the vertices in W to a single vertex X. The vertex set of this digraph is $V_1 = (V - W) \cup X$, and E_1 is the set of arcs. Each arc in E_1 corresponds to a unique arc in E; therefore, E_1 can be considered a subset of E. Then $T_1^* = (V_1, E^* \cap E_1)$ is an arborescence in G_1, and $w(T^*) = w(T_1^*) - w(C) + w(e')$. We now define a weight function w_1 on set E_1 as follows. If e is an arc that is not directed to a vertex in the cycle C, $w_1(e) = w(e)$. Otherwise, $w_1(e) = w(e) - w(e')$. If there is more than one arc from a vertex to another vertex in this condensed graph, replace each by a single arc of minimum weight where tiebreaking is arbitrary. Then $w_1(T_1^*) = w(T_1^*) - w(e')$. So $w(T^*) = w(T_1^*) + w(C) - w(e') = w_1(T_1^*) + w(C)$. Thus T^* is a minimum weight arborescence in G if and only if T_1^* is a minimum weight arborescence in the condensed graph G/C.

The Minimum Weight Arborescence Algorithm

(*a*) Step $i = 0$. $G^0 = G$, $w^0(e) = w(e)$ for each i in G.

(*b*) At step i, using the weight function w^i, construct the subgraph H^i of G^i by selecting the arc of smallest weight directed to every vertex other than the root of G^i.

(*c*) If there is no cycle in H^i, it is a minimum weight arborescence in G^i from which a minimum weight arborescence in G^0 can be derived. Otherwise, go to (*d*).

(*d*) If H^i has a cycle C, define $G^{i+1} = G^i/C$ and $w^{i+1}(e) = w^i(e)$ when e is not directed to a vertex in the cycle and $w^{i+1}(e) = w^i(e) - w^i(e')$ when $e = (i, j)$ is directed to a vertex j in C and $e' = (k, j)$ is an arc of the cycle. If H^i contains cycles $C_1, C_2, \ldots$, first condense G^i with respect to C_1, then condense G^i/C_1 with respect to C_2, and so on. Set $i = i + 1$ and return to (*b*).

[If we assume that at each step of the algorithm, all arcs of the current digraph have different weights, the minimum weight arborescence is unique. When there are arcs of equal weight, the algorithm remains valid since the difference $w(T^*) - w(T_1^*)$ is always equal to the weight of the cycle C. If C has two arcs $e' = (p, q)$ and $f' = (r, s)$ that are not in E^*, arcs e and f are directed to q and s, respectively. Then $w_1(e) = w(e) - w(e')$ and $w_1(f) = w(f) - w(f')$.]

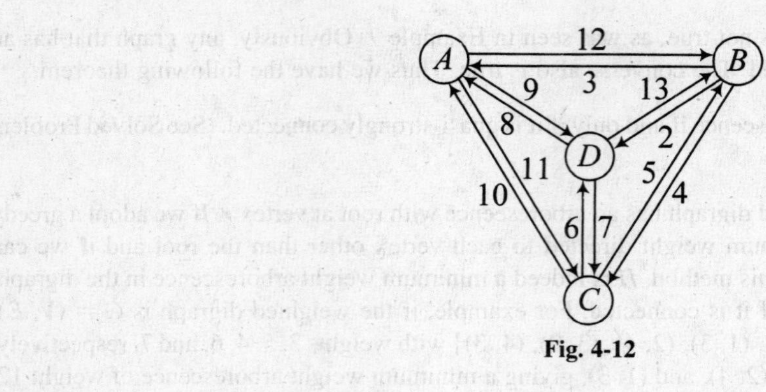

Fig. 4-12

Example 8. Obtain a minimum weight arborescence with root at vertex A in the weighted digraph of Fig. 4-12.

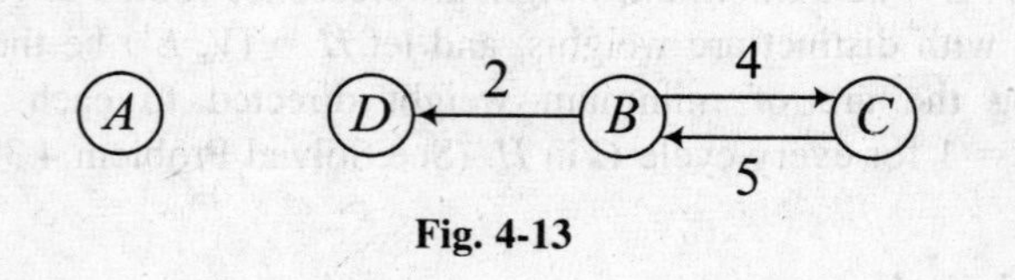

Fig. 4-13

Step 0. The arcs of H^0 are (B, D), (B, C), and (C, B), as shown in Fig. 4-13. Since (B, C) and (C, B) form a cycle, the two vertices are condensed into a single vertex BC. At this stage, we construct the condensed graph G^1 with vertices as shown in Fig. 4-14.

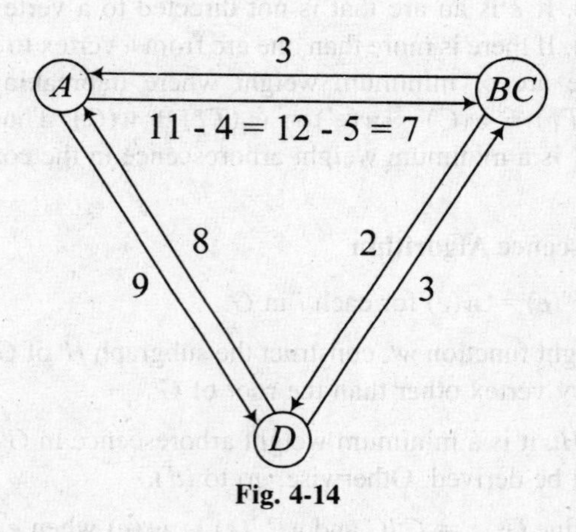

Fig. 4-14

Step 1. The digraph H^1 is shown in Fig. 4-15. Vertices D and BC constitute a cycle, and this cycle is shrunk into a single vertex BCD, forming the digraph G^2 as shown in Fig. 4-16.

Fig. 4-15

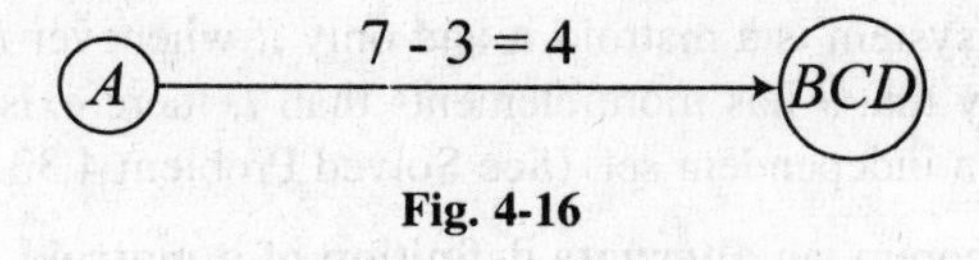

Fig. 4-16

Step 2. At this stage, we have a minimum weight arborescence in the condensed graph. We now derive a minimum weight arborescence in G by working backward from the current graph G^2. Since the weight of arc (A, BCD) is $7 - 3 = 4$, we locate the arc from A with weight 7 in G^1. Thus the minimum weight arborescence in G^1 is $A \rightarrow BC \rightarrow D$. The weight of arc (A, BC) is 7, which is either $11 - 4$ or $12 - 5$, indicating a tie. In the former case, we take (A, C) and (C, B). In the latter case, we take (A, B) and (B, C). In G^1, the weight of the arc from BC to D is 2, which is the weight of the arc from B to D in G. So the arc (B, D) is in the minimum weight arborescence.

Thus there are two minimum weight arborescences with root at A, as shown in Fig. 4-17.

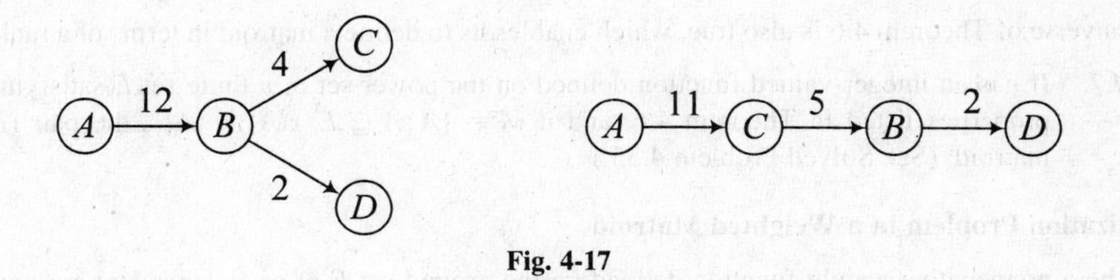

Fig. 4-17

Arborescence in a Quasi-connected Graph

Let $G = (V, E)$ be a quasi-connected weighted digraph. Construct a new vertex r, and draw arcs from r to each vertex in G. Let each of these new arcs be assigned a weight W that is some number larger than the weight of the maximum weight arc in G. The enlarged digraph G' is quasi-strongly connected and has an arborescence rooted at the vertex r. We can obtain a minimum weight arborescence T' in G' with root at r using the algorithm. By deleting the new vertex from T', we obtain a minimum weight arborescence T in G.

4.4 MATROIDS AND THE GREEDY ALGORITHM

If E is a finite set and $\mathcal{A}$ is a class of subsets of E, the pair $(E, \mathcal{A})$ is called an **independent system** if $A \in \mathcal{A}$ whenever $A \subset B$ and $B \in \mathcal{A}$. The set E is the **ground set** of the independent system. The sets in $\mathcal{A}$ are the **independent sets,** and a subset of E that is not independent is called a **dependent set** of the system. A dependent set C is called a **circuit** if every proper subset of C is independent. An independent set I that is a subset of $A \subset E$ is **maximal in** A if $I \subset J \subset A$ implies that $I = J$ whenever J is an independent set. An independent system is a **matroid** if any two sets maximal in a set have the same cardinality. An independent set in a matroid $(E, \mathcal{A})$ that is maximal in E is a **base** of the matroid. The matroid in which the only independent sets are the ground set and the empty set is the **trivial matroid,** whereas the matroid in which every subset of the ground set is an independent set is the **discrete matroid.** If E is a set with n elements, the ***k*-uniform matroid** with E as the ground set is the matroid in which a subset I of E is independent if and only if it has at most k elements, where $k \leq n$.

If H is a subgraph of an undirected graph $G = (V, E)$, any two spanning forests in H will have the same number of edges. So if $\mathcal{A} = \{A : A \subset E, A \text{ is the set of edges of an acyclic subgraph of } G\}$, the pair $(E, \mathcal{A})$ is a matroid and is the **circuit matroid** defined by G. Thus every undirected graph can be considered a matroid. The set of edges belonging to any spanning forest of graph G is a base of this matroid. Furthermore, a circuit in this matroid is precisely the set of edges belonging to a cycle in the graph.

Theorem 4.5. An independent system is a matroid if and only if whenever I and J are two independent sets with the property that J has more elements than I, there exists an element $e \in (J - I)$ such that $I \cup \{e\}$ is an independent set. (See Solved Problem 4.33.)

As a consequence of this theorem, an **alternate definition of a matroid** can be given as follows. If $\mathcal{A}$ is a class of subsets of a finite set E, the pair $(E, \mathcal{A})$ is called a matroid if (*i*) $I \in \mathcal{A}$ whenever $I \subset J$ and $J \in \mathcal{A}$ and (*ii*) I and J are any two sets in $\mathcal{A}$ such that J has more elements than I. Then there exists an element $e \in (J - I)$ such that $I \cup \{e\}$ is an independent set.

The cardinality of an independent set in a matroid that is maximal in a set A is the **rank** of A, denoted by $\boldsymbol{r(A)}$. Consequently, the **rank function $\boldsymbol{r}$ of a matroid** is a function whose domain is the power set of E and whose range is a subset of the set of nonnegative integers. The **rank of the matroid** is the rank of its ground set.

Theorem 4.6. The rank function r of a matroid satisfies the following four properties: (*i*) the rank of the empty set is 0, (*ii*) $r(A) \le r(B)$ if $A \subset B$, (*iii*) the function r is **submodular:** $r(A \cup B) + r(A \cap B) \le r(A) + r(B)$, and (*iv*) the rank of a singleton set is either 0 or 1. (See Solved Probiem 4.34.)

The converse of Theorem 4.6 is also true, which enables us to define a matroid in terms of a rank function.

Theorem 4.7. If r is an integer-valued function defined on the power set of a finite set E satisfying the four properties listed in Theorem 4.6, and if $\mathcal{A} = \{A : A \subseteq E, r(A) = |A|\}$, the pair $(E, \mathcal{A})$ is a matroid. (See Solved Problem 4.35.)

An Optimization Problem in a Weighted Matroid

Let w be a nonnegative weight function defined on the ground set E of an independent system. If A is a subset of E, the **weight of $\boldsymbol{A}$**, denoted by $w(A)$, is the sum of the weights of all the elements in A. An optimization problem associated with the independent system is the problem of finding an independent set with maximum weight. A greedy approach to solve this optimization problem (reminiscent of the algorithms of Kruskal and Prim) will have the following two steps at iteration k:

Step 1. Choose $x(k)$ distinct from $x(1), x(2), \ldots, x(k-1)$ such that (*i*) the set $\{x(1), x(2), \ldots, x(k-1), x(k)\}$ is an independent set; and (*ii*) if $\{x(1), x(2), \ldots, x(k-1), x\}$ is an independent set, the weight of x does not exceed the weight of $x(k)$.

Step 2. Stop if no such $x(k)$ exists.

In the case of an arbitrary independent system, this greedy approach may not produce an optimal solution, as in the following simple example. Let $E = \{a, b, c, d\}$ with weights 2, 6, 7, and 5, respectively. Suppose the independent sets are $\{a\}$, $\{b\}$, $\{c\}$, $\{a, b\}$, and the empty set. The greedy algorithm will choose the independent set $\{c\}$, whereas the optimal solution is the set $\{a, b\}$. On the other hand, if the independent sets are $\{a\}$, $\{c\}$, $\{a, c\}$, and the empty set, the greedy algorithm will pick the optimal solution. Observe that in the former case (when the greedy method failed), the independent system is not a matroid, whereas in the latter case, the independent system is a matroid.

We are now ready to establish the connection between the greedy algorithm and the structure of matroids.

Theorem 4.8. A solution of the problem of finding a maximum weight independent set in an independent system can be obtained by using the greedy algorithm for every nonnegative weight function defined on its ground set if and only if the independent system is a matroid. (See Solved Problem 4.36.)

As a consequence of Theorem 4.8, we once again conclude that the greedy methods of Kruskal and Prim to obtain a minimum spanning tree in a connected graph will correctly solve the optimization problem because (1) the underlying structure is a matroid, and (2) the problem of finding a minimum weight spanning tree is equivalent to that of finding a maximum weight spanning tree.

Solved Problems

MINIMUM WEIGHT SPANNING TREES

4.1 Show that if the vertex set of a connected graph $G = (V, E)$ is partitioned into two nonempty sets X and Y, the disconnecting set $F = (X, Y)$ consisting of all edges of G joining vertices in X and vertices in Y is a cut set if the subgraph $G' = (V, E - F)$ has exactly two components.

Solution. Suppose $F = (X, Y)$ is not a cut set. Then there is a proper subset F' of F that is a cut set. Let $e = \{x, y\}$ in an edge in $F - F'$ joining vertex x in X and vertex y in Y. If v is any vertex in X, there is a path joining v and x consisting of vertices from X only. Likewise, if w is any vertex in Y, there is a path joining w and y consisting of vertices from Y only. Thus the deletion of all the edges belonging to F' from the graph will not make it disconnected. This is a contradiction since F' is a disconnecting set.

4.2 The deletion of an edge belonging to a spanning tree $T = (V, F)$ of a graph $G = (V, E)$ from F defines a partition of V into two subsets X and Y, creating a disconnecting set $D = \{X, Y\}$ of the graph. Show that this disconnecting set is a cut set of G.

Solution. Since every edge in a tree is a bridge, the deletion of an edge from a spanning tree gives two sets X and Y that constitute a partition of V. So $D = \{X, Y\}$ is a disconnecting set in G. It is a cut set since $G' = (V, E - D)$ has exactly two components.

4.3 If T is a spanning tree in a graph G, the cut set of G formed by deleting an edge e of T, denoted by $D_T(e)$, is called the **fundamental cut set** of G with respect to T relative to edge e. Find the number of cut sets of a connected graph with n vertices with respect to a spanning tree.

Solution. Since the graph has n vertices, any spanning tree in the graph will have $(n - 1)$ edges. The deletion of each edge defines a fundamental cut set. So there are $(n - 1)$ fundamental cut sets of the graph with respect to a spanning tree.

4.4 If T is a spanning tree in G, any edge $e = \{x, y\}$ in G that is not an edge in T is called a **chord** of T. The unique cycle in G formed by the path in T joining x and y and edge e, denoted by $C_T(e)$, is called the **fundamental cycle** of G with respect to T relative to edge e. Find the number of fundamental cycles of a connected graph with respect to a spanning tree.

Solution. If G is a connected graph of order n and size m, it has $r = m - (n - 1)$ chords with respect to any spanning tree T, so it has r fundamental cycles with respect to T.

4.5 Let G be a connected graph in which C is a cycle, D is a cut set, and T is a spanning tree. Show that (*a*) the number of edges common to C and D is even, (*b*) at least one edge of C is a chord of T, and (*c*) at least one edge of D is an edge of T.

Solution.

(*a*) Let $D = \{X, Y\}$. If all the vertices of C are exclusively in X (or in Y), C and D have no edges in common. Suppose C has two vertices x and y, where x is in X and y is in Y. Then the cycle C that starts from x and ends in x will contain the edges from D an even number of times.

(*b*) If no edge of C is a chord of T, then every edge of C is an edge of T, which is a contradiction since T is acyclic.

(*c*) If no edge of D is an edge of T, the deletion of all the edges belonging to D will not affect T, implying that G continues to be a connected even after all the edges belonging to D are deleted. This is a contradiction since D is a cut set.

4.6 Let e be an edge of a spanning tree T in a graph G. Show that (*a*) if f is any edge (other than e) in the fundamental cut set $D_T(e)$, f is a chord of T and e is an edge of the fundamental cycle $C_T(f)$; and (*b*) e is not an edge of the fundamental cycle $C_T(e')$ relative to any chord e' that is not an edge in $D_T(e)$.

Solution.

(*a*) An edge f in the cut set cannot be an edge of the spanning tree T since T is acyclic. So f is necessarily a chord of T. Now $D_T(e) = \{e, f\} \cup A$ and $C_T(f) = \{f\} \cup B$, where A is a set of chords of T and B is a set of edges of T. Since the intersection of A and B is empty and since the intersection of $D_T(e)$ and $C_T(f)$ has an even number of elements, edge e should be an element of B. Hence, e is an edge of $C_T(f)$.

(*b*) The set of edges of $C_T(e')$ is the union of $\{e'\}$ and a set X of edges of T. If e is an edge of $C_T(e')$, e is an element of X. The set of edges of $D_T(e)$ is the union of $\{e\}$ and a set Y of chords of T. Since the intersection of $D_T(e)$ and $C_T(e')$ has an even number of edges and since X and Y have no elements in common, edge e' is an element of $D_T(e)$, which contradicts the hypothesis.

4.7 Let e be a chord of a spanning tree T in a graph G. Show that (*a*) if f is any edge (other than e) in the fundamental cycle $C_T(e)$, f is an edge of T and e is an edge of the fundamental cut set $D_T(f)$; and (*b*) e is not an edge of the fundamental cut set $D_T(e')$ relative to any edge e' of T that is not an edge in $C_T(e)$.

Solution.

(*a*) By definition, any edge f of $C_T(e)$ other than e is an edge of T. Now $D_T(f)$ is the union of the set $\{f\}$ and a set A of chords of T. At the same time, $C_T(e)$ is the union of $\{e, f\}$ and a set B of edges of T. So e is an element of A; therefore, it is an element of $D_T(f)$.

(*b*) $C_T(e)$ is the union of $\{e\}$ and a set X of edges of T. If e' is in $D_T(e')$, $D_T(e')$ is the union of $\{e, e'\}$ and a set Y of chords of T. This implies that e' is an element of $C_T(e)$, which contradicts the hypothesis.

4.8 Show that if T and T' are two spanning trees in G and if e is an edge of T and a chord of T', there exists an edge e' in T' such that e' is a chord of T and $T - e + e'$ is a spanning tree in G.

Solution. If $e = \{x, y\}$ is deleted from T, the set of vertices of G is partitioned into two sets X and Y such that x is in X and y is in Y. Since e is a chord of T', there will be a path in T' between x and y that will have edge $e' = \{p, q\}$ joining vertex p in X and vertex q in Y. This edge e' is necessarily a chord of T. Then $T - e + e'$ is obviously a spanning tree in G. At the same time, $T' - e' + e$ is also a spanning tree.

4.9 If no two edge weights of a connected graph G are equal, show that G has a unique minimum spanning tree.

Solution. Suppose there are two minimum spanning trees, T and T'. Then $w(T) = w(T') = s$. As shown in Problem 4.8, there are two edges e and f such that $T_1 = T - e + f$ and $T_2 = T' - f + e$ are both spanning trees in G. Now $w(T_1) = s - w(e) + w(f)$, and $w(T_2) = s - w(f) + w(f)$. If $w(e) > w(f)$, $w(T_1) < s$, which is a contradiction since T_1 is an M.S.T. If $w(e) < w(f)$, $w(T_2) < s$, which is also a contradiction. So there is only one M.S.T.

4.10 Show that if a connected weighted graph G contains a unique edge e of minimum weight, e is an edge of every M.S.T. of G.

Solution. Suppose T is an M.S.T. of G and e is not an edge of T. Let f be any edge of the fundamental cycle $C_T(e)$ other than e. Then $T' = T - f + e$ is spanning tree of G, and $w(T') = w(T) - w(f) + w(e)$. Since $w(e) < w(f)$, $w(T') < w(T)$, and this contradicts the assumption that T is an M.S.T.

4.11 Show that a spanning tree T in a weighted graph is a minimum spanning tree if and only if every edge of T is a minimum weight edge in the fundamental cut set relative to that edge.

Solution. Let e be an edge of an M.S.T. in a weighted graph G. Suppose $w(f) < w(e)$ for some chord of T belonging to the fundamental cut set $D_T(e)$. Then $T' = T - e + f$ with $w(T') < w(T)$, which is a contradiction. So the condition is necessary. On the other hand, suppose T is a spanning tree that satisfies the given condition. Let T' be an M.S.T. of G. If T and T' are not the same, let $e_1 = \{i, j\}$ be an edge of T that is not an edge of T'. Since e_1 is a chord of T', we have the fundamental cycle $C_{T'}(e_1)$ that contains edge f_1 (other than e_1), an edge belonging to $D_T(e_1)$. Let $T'' = T' - f_1 + e_1$. By assumption, $w(e_1) \le w(f_1)$. So $w(T'') \le w(T)$. If $w(e_1) < w(f_1)$, $w(T'') < w(T')$, which is a contradiction. So $w(e_1) = w(f_1)$. Thus $T_1 = T - e_1 + f_1$ is a spanning tree of the graph such that $w(T) = w(T_1)$. If $T_1 = T'$, T is an M.S.T. Otherwise, we consider edge e_2 of T_1, which is not an edge of T'. This edge e_2 is an edge of T and a chord of T'. We proceed as before. Obtain edge f_2 in $D_T(e_2)$, and construct the spanning tree $T_2 = T_1 - e_2 + f_2$ such that $w(T_2) = w(T_1) = w(T)$. If $T_2 = T'$, T is an M.S.T. Otherwise, we continue this process until we get a spanning tree T_k such that $T_k = T'$.

4.12 Show that if in a weighted connected graph G there is an edge e that is a maximum weight edge in any cycle that contains e, there is an M.S.T. in G that does not contain e. In particular, if the weight of e exceeds the weight of every edge in any cycle that contains e, then no M.S.T. in G contains e as an edge.

Solution. Let e be an edge in a weighted connected graph such that $w(e) \ge w(f)$ for any cycle in the graph in which both e and f are edges. Let T be an M.S.T. in G. If e is not an edge of T, we are done. Suppose e is an edge of T. Then $w(e) \le w(e')$ for any e' in the fundamental cut set $D_T(e)$, as proved in Problem 4.11. Since e' is a chord, the fundamental cycle $C_T(e')$ contains e as an edge, as proved in Problem 4.6. So by hypothesis, $w(e) \ge w(e')$. Thus $w(e) = w(e')$. Consequently, $T' = T - e + e'$ is an M.S.T. that does not contain e. Next we show that if $w(e) > w(f)$ for any cycle in which both e and f are edges, no M.S.T. contains e. Suppose this is not the case. Then there is a minimum spanning tree T that contains e. Proceeding as before, we get a spanning tree $T' = T - e + e'$ such that $w(T') < w(T)$, which is a contradiction.

4.13 A spanning tree T in a graph G is a minimum spanning tree if and only if every edge not in T is a maximum weight edge in the fundamental cycle defined by that edge.

Solution. Suppose there is a chord of the minimum spanning tree T such that in the fundamental cycle $C_T(e)$ there exists an edge f violating the maximality condition. In other words, $w(e) < w(f)$. Then $T' = T - f + e$ is an M.S.T. whose weight is less than $w(T)$, which is a contradiction. So the condition is necessary. To prove that the condition is sufficient, suppose T is a spanning tree satisfying the given condition. If f is an edge in T. any chord e of T belonging to the fundamental cut set $D_T(f)$ will define a fundamental cycle $C_T(e)$ in which f is an edge. By assumption, $w(e) \ge w(f)$. In other words, whenever f is an edge of T, we have the inequality $w(e) \le w(f)$, where e is any edge in $D_T(f)$. So T is an M.S.T. because of the condition established in Problem 4.11.

4.14 Prove Theorem 4.1: Kruskal's algorithm solves the M.S.T. problem in a network.

Solution. In this algorithm, an edge e is discarded in favor of an edge of larger weight only because the inclusion of e would have created a cycle C. At the same time, the weight of the discarded edge is greater than or equal to the weight of any other edge in C. So by the optimality condition established in Problem 4.13, if a spanning tree is obtained by this method, it has to be a minimum spanning tree.

4.15 If e is an edge incident at vertex x of a connected weighted graph G and if $w(e) \le w(f)$ for every edge f incident at x, there exists an M.S.T. of G that contains e as an edge. In particular, if $w(e) < w(f)$, every M.S.T. of G contains e.

Solution. Let T be an M.S.T. If $e = \{x, y\}$ is an edge of T, we are done. Let e be a chord of T. According to the optimality condition in Problem 4.11, $w(e) \ge w(f)$, where f is any edge in the fundamental cycle $C_T(e)$. Suppose f is an edge in $C_T(e)$ that is adjacent to x or y. Then, by hypothesis, $w(e) \le w(f)$. Hence, $T' = T - f + e$ is also an M.S.T. In particular, if $w(e) < w(f)$, every M.S.T. of G contains e.

4.16 If W is the set of vertices of any subgraph H of a minimal spanning tree T of a graph $G = (V, E)$ and if e is any edge of minimum weight in the disconnecting set $D = (W, V - W)$, there exists an M.S.T. that contains e as an edge and that has H as a subgraph.

Solution. If e is an edge of T, there is nothing to prove. So assume that T is a chord of T defining the fundamental cycle $C_T(e)$. Then $e \in C_T(e) \cap D$. So there should be an edge f of T (other than e) in $C_T(e) \cap D$. By hypothesis, $w(e) \leq w(f)$. The edge f of T defines the fundamental cut set $D_T(f)$. Thus $f \in C_T(e) \cap D_T(f)$. There should be at least one more edge in $C_T(e) \cap D_T(f)$. The only choice is for e. Since e is in $D_T(f)$ and since T is an M.S.T., $w(f) \geq w(e)$. Thus $w(f) = w(e)$. Let $T' = T - f + e$. Then T' is an M.S.T. that contains e as an edge and that has H as a subgraph.

4.17 Prove Theorem 4.2: Prim's algorithm solves the M.S.T. problem.

Solution. In this algorithm, we start from an arbitrary vertex and add edges one at a time. We maintain a spanning tree T on a set W of vertices of a graph $G = (V, E)$ such that the edge adjoining to T is a minimum weight edge in the disconnecting set $(W, V - W)$. The correctness of the procedure follows immediately from the result established in Problem 4.16.

4.18 Obtain a minimum spanning tree in the graph shown in Fig. 4-18.

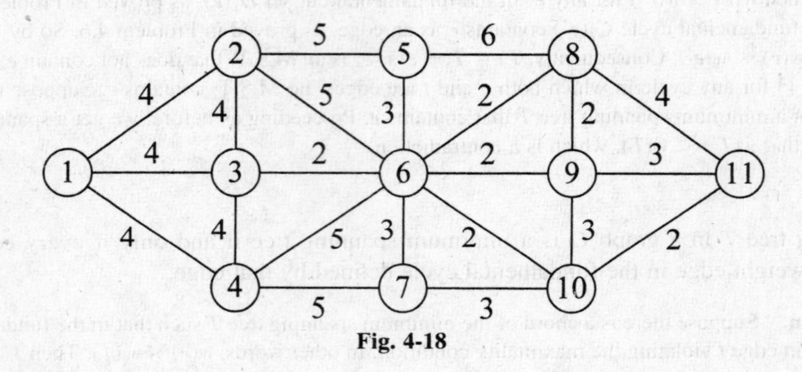

Fig. 4-18

Solution. A minimum spanning tree of weight 28 is displayed in Fig. 4-19.

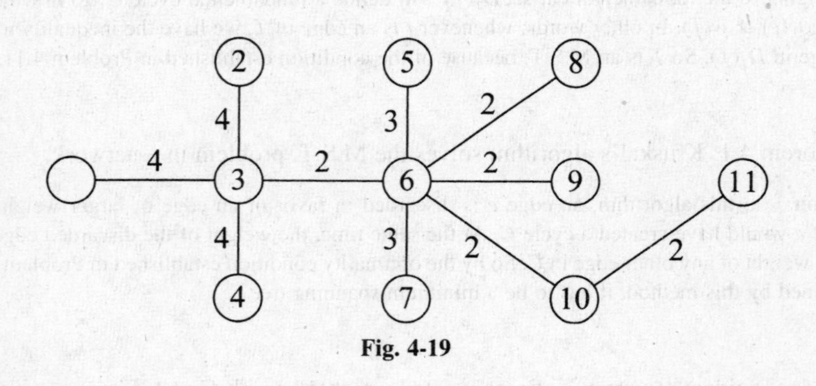

Fig. 4-19

4.19 Suppose we are using Prim's algorithm to obtain an M.S.T. in the graph of Fig. 4-18. At the current stage, $W = \{3, 6, 8, 9\}$ is the set of vertices of the tree H, which is a subtree of a minimum spanning tree T to be obtained by this method. Select the next edge for inclusion, and list the edges after this selection is made.

Solution. The edges of H are $\{3, 6\}$, $\{6, 8\}$, and either $\{6, 9\}$ or $\{8, 9\}$. The edge to be selected is $\{6, 10\}$ in either case.

4.20 Obtain a maximum weight spanning tree in the graph of Fig. 4-18.

Solution. A maximum weight spanning tree of weight 44 is the tree shown in Fig. 4-20.

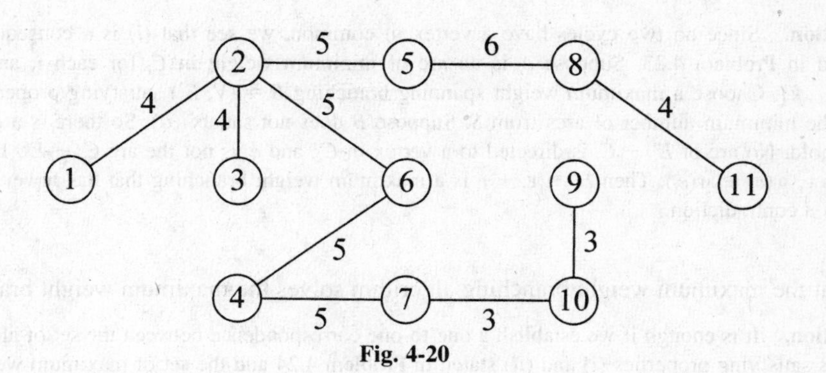

Fig. 4-20

MAXIMUM WEIGHT BRANCHINGS

4.21 Let $G = (V, E)$ be a weighted digraph, and let $B = (V', E')$ be a branching in G. An arc $e \in (E - E')$ is called a B-eligible arc if the arcs in $E'' = E' + e - \{f \in E': t(e) = t(f)\}$ constitute a branching. Show that e is B-eligible if and only if there is no directed path in B from $t(e)$ to $s(e)$.

Solution. The set E'' will not constitute a branching if and only if it contains a cycle. Any such cycle C contains arc e, and the arcs in $(C - e)$ form a directed path from $t(e)$ to $s(e)$. Thus e is B-eligible if and only if there is no directed path in B from $t(e)$ to $s(e)$. (Notice that eligibility is on an individual basis. If e is B-eligible and if f is B-eligible, it is not necessary that they both be B-eligible at the same time.)

4.22 Let C be a directed cycle in $G = (V, E)$, and let $B = (V', E')$ be a branching in G. Show that no arc in $(C - E')$ is B-eligible if and only if $(C - E')$ contains exactly one arc.

Solution. The set $(C - E')$ is obviously nonempty. If it contains the unique arc e, there is a directed path in B from $t(e)$ and $s(e)$. So, by Problem 4.21, the arc e is not B-eligible. To prove the converse, suppose $(C - E') = \{e_i: i = 1, 2, \ldots, k\}$. Assume that these arcs appear clockwise in the cycle such that e_{i+1} follows e_i with no other arc from the set $(C - E')$ in this cycle between these two arcs. Hence, $t(e_{i-1}) = s(e_i)$, or there is a path in $C \cap E'$ from $t(e_{i-1})$ to $s(e_i)$. Also, $t(e_k) = s(e_1)$, or there is a path in $C \cap E'$ from $t(e_k)$ to $s(e_1)$. Suppose no arc in $(C - E')$ is B-eligible. Then there is a directed path in E' from $t(e_i)$ to $s(e_i)$ for each i. Since there is a directed path in $C \cap E'$ from $t(e_{i-1})$, to $s(e_i)$ and since E' is a branching, either there is a directed path P: $t(e_{i-1}) \rightarrow t(e_i) \rightarrow s(e_i)$ in E' or the directed path Q: $t(e_i) \rightarrow t(e_{i-1}) \rightarrow s(e_i)$ in E'. If path P exists, it has to be the same as the unique path in the branching from $t(e_{i-1})$ to $s(e_i)$, which is completely in C. So the subpath $t(e_{i-1}) \rightarrow t(e_i)$ should also be in cycle C, implying that there should be an arc in $C \cap E'$ other than arc e_i directed to vertex $t(e_i)$. But there cannot be two arcs directed to a vertex in a cycle, so path P does not exist. Suppose path Q exists. This implies that there is a path $t(e_i) \rightarrow t(e_{i-1})$ in the branching for $i = 1, 2, \ldots, k$. In that case, we can obtain the cycle $t(e_k) \rightarrow t(e_{k-1}) \rightarrow \cdots \rightarrow t(e_1) \rightarrow t(e_k)$ in the branching, which is a contradiction. So Q does not exist, implying that $(C - E')$ cannot have more than one B-eligible arc.

4.23 Show that if H is a critical subgraph of a weighted digraph $G = (V, E)$, there is a maximum weight branching $B = (V, E')$ in G such that for every cycle C in H, the set $(C - E')$ has exactly one arc.

Solution. Let $H = (V, E_1)$ be a critical graph. From the collection of all maximum weight branchings, select the branching $B = (V, E')$ that has the maximum number of arcs in common with E_1. Let e be an arc in $(E_1 - E')$. If e is B-eligible, the arcs in the set $E'' = E' + e - \{f: f \in E' \text{ and } t(e) = t(f)\}$ will form a maximum weight branching that has more arcs in common with H than B. So no arc in H is B-eligible. In particular, no arc in cycle C in H is B-eligible. So the cardinality of $(C - E')$ is one as established in Problem 4.22.

4.24 Show that the cycles in a critical graph H of a weighted digraph $G = (V, E)$ are C_i ($i = 1, 2, \ldots, k$), there exists a maximum weight branching $B = (V, E')$ of G such that (*i*) $C_i - E'$ has exactly one arc for each i; and (*ii*) if no arc in $(E' - C_i)$ is directed to a vertex in the cycle C_i, the unique arc in $(C_i - E')$ is an arc of minimum weight in C_i for each i.

Solution. Since no two cycles have a vertex in common, we see that (*i*) is a consequence of the result established in Problem 4.23. Suppose e_i is an arc of minimum weight in C_i for each i, and let $S = \{e_i: i = 1, 2, \ldots, k\}$. Choose a maximum weight spanning branching $B = (V, E')$ satisfying property (*i*) such that E' contains the minimum number of arcs from S. Suppose B does not satisfy (*ii*). So there is a cycle C_j where (*ii*) does not hold. No arc of $E' - C_j$ is directed to a vertex in C_j, and e_j is not the arc $C_j - E'$. If the unique arc in $C_j - E'$ is e, $w(e) \geq w(e_j)$. Then $E' - e_j + e$ is a maximum weight branching that has fewer edges from S than E'. This is a contradiction.

4.25 Prove that the maximum weight branching algorithm solves the maximum weight branching problem.

Solution. It is enough if we establish a one-to-one correspondence between the set of all maximum weight branchings satisfying properties (*i*) and (*ii*) stated in Problem 4.24 and the set of maximum weight branchings in a condensed graph. Let $G = (V, E)$ be a digraph with a weight function w defined on its arcs, H be a critical graph in G with cycles C_i ($i = 1, 2, \ldots, k$), and $G_1 = (V_1, E_1)$ be the corresponding condensed graph. A new weight function w_1 is defined on the E_1 as follows.

If e is an arc that is not directed to any of the cycles in H, $w_1(e) = w(e)$. If e is directed to a vertex in C_i, $w_1(e) = w(e) - w(f) + w(e_i)$, where f is the unique arc in C_i that is directed to $t(e)$ and e_i is an arc of minimum weight in the cycle C_i. Let $B = (V, E')$ be any branching in G that satisfies the two properties listed in Problem 4.24 using these k cycles. Arc e in E' such that both $s(e)$ and $t(e)$ are not in the same cycle in H defines a unique arc in A_1. Let D_1 be the set of arcs thus defined. Then $B_1 = (V_1, D_1)$ is a branching in G_1. Thus once a critical graph in G is defined, a branching in G defines a branching in the condensed graph G_1. Now consider the condensed graph defined by a critical graph H. Let the cycles in H be C_i ($i = 1, 2, \ldots, k$), and let $B_1 = (V_1, A_1)$ be a branching in G_1. If the indegree in B_1 of the condensed vertex corresponding to C_i is 0, let $C_i' = C_i - e_i$. If the indegree is 1, there is a unique arc f (belonging to E) in C_i directed to that condensed vertex. In this case, let $C_i' = C_i - f$. Thus in either case, from each cycle, we take all arcs except one. Let X be the set of arcs thus obtained from these k cycles. Now consider arcs in B_1 that are not directed to condensed vertices. Each such arc corresponds to a unique arc in E. Let Y be the set of arcs thus obtained. The union of X and Y forms a branching in G. Next, we turn our attention to optimality. If P is the sum of the weights of the k cycles and Q is the sum of the weights of the minimum arc weights in these cycles, $w(B) - w_1(B_1) = P - Q$. Thus the one-to-one correspondence is established.

4.26 Show that a maximum weight branching in the network shown in Fig. 4-21 cannot be obtained by the greedy algorithm.

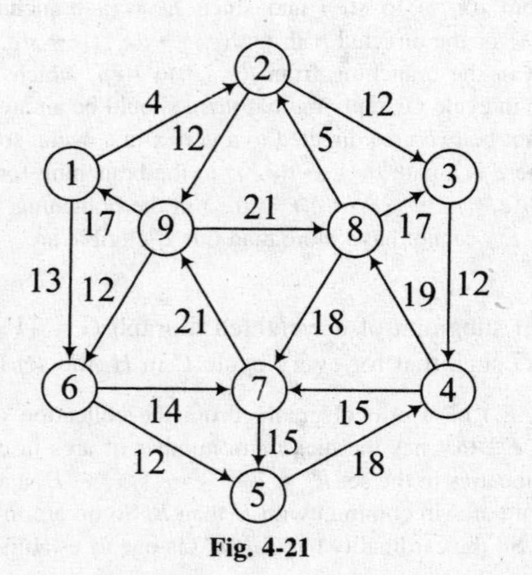

Fig. 4-21

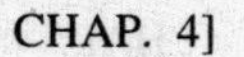

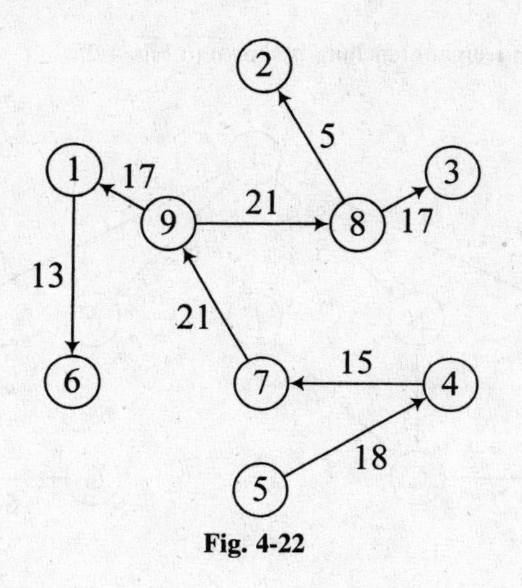

Fig. 4-22

Solution. The branching (Fig. 4-22) obtained by the greedy method has a weight of 127, whereas the maximum weight branching (Fig. 4-23) has a weight of 128.

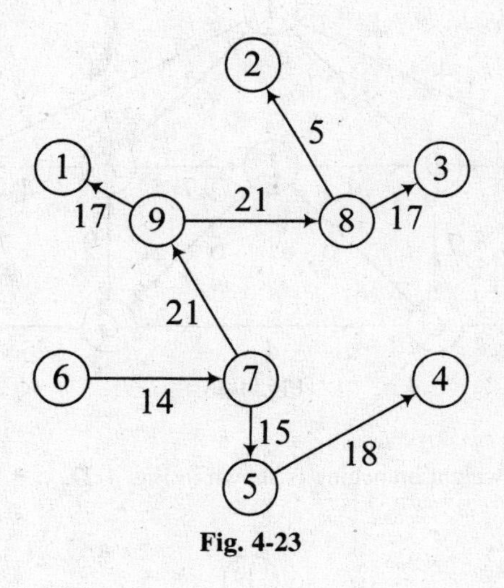

Fig. 4-23

4.27 Find a maximum weight branching in the network shown in Fig. 4-24.

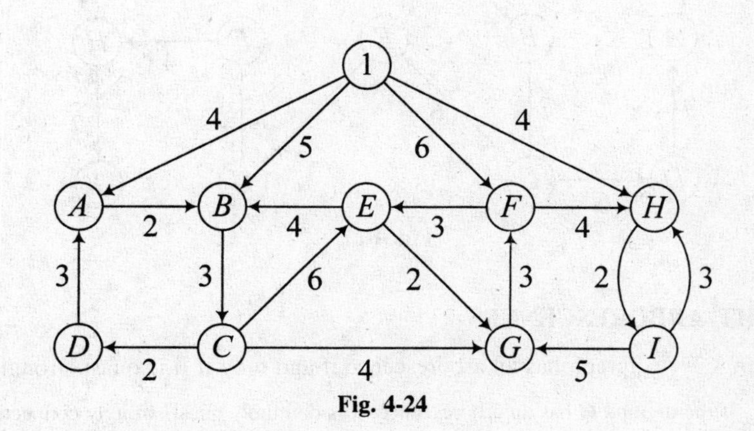

Fig. 4-24

Solution. A maximum weight branching is shown in Fig. 4-25.

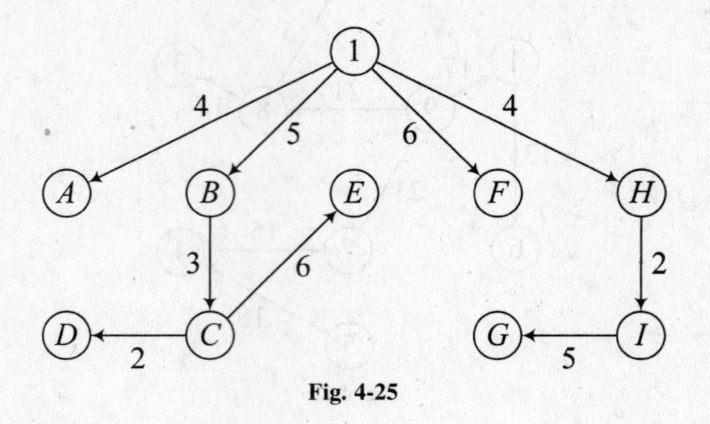

Fig. 4-25

4.28 Find a maximum weight branching on the network shown in Fig. 4-26.

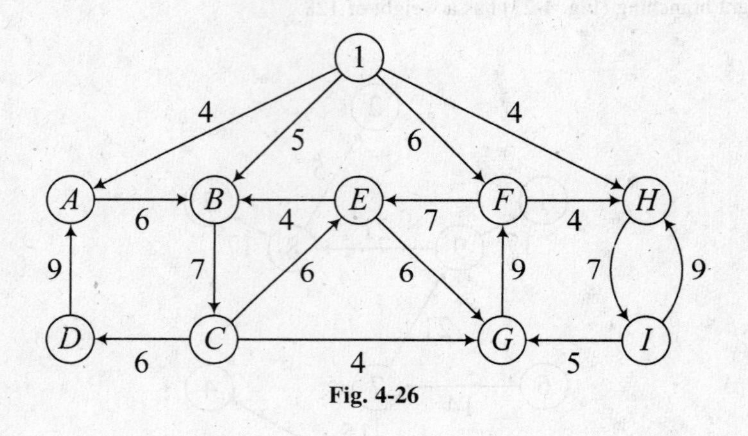

Fig. 4-26

Solution. A maximum weight branching is shown in Fig. 4-27.

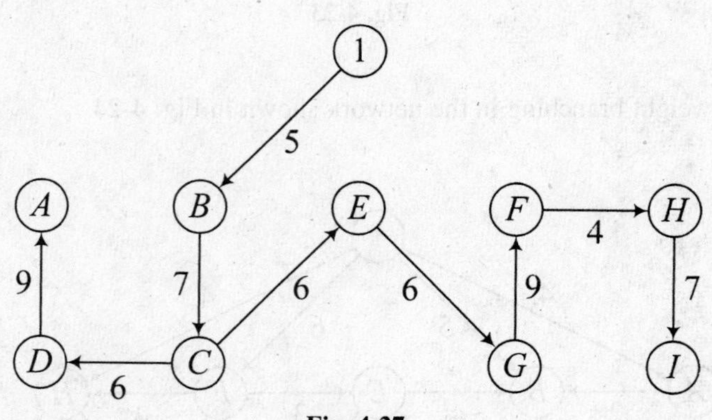

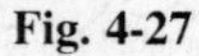

Fig. 4-27

MINIMUM WEIGHT ARBORESCENCES

4.29 Prove Theorem 4.3: A digraph has an arborescence if and only if it is quasi-strongly connected.

Solution. If the digraph G has an arborescence, it is definitely quasi-strongly connected. To prove the converse, consider a quasi-strongly connected digraph $G = (V, E)$, where $V = \{1, 2, \ldots, n\}$. Let $H = (V, E')$ be a maximal quasi-strongly connected subgraph such that the deletion of one more arc from E' will destroy the quasi-strong connectivity of H. There exists a vertex x_1 such that there are paths (in H) from x_1 to vertices 1 and 2. There exists a vertex x_2 such that there are paths (in H) from x_2 to vertices 3 and x_1. Proceeding like this, one can establish that there exists a vertex v from which there is a path (in H) to every other vertex. Let w be any vertex other than v. The indegree (in H) of w is at least 1. Suppose the indegree is more than 1. So there are (at least) two vertices p and q such that (p, w) and (q, w) are arcs in H, which implies that H will continue to be quasi-strongly connected if one of these two arcs is deleted. This violates the maximality assumption. So the indegree in H of every vertex is at most 1. Finally, if there is an arc in H directed to v, the deletion of that arc will not affect the quasi-strong connectivity of H. So the indegree of v is 0. So there is an arborescence with v as a root.

4.30 Prove Theorem 4.4: Let $T^* = (V, E^*)$ be a minimum weight arborescence rooted at vertex r in a weighted digraph $G = (V, E)$ with distinct arc weights, and let $H = (V, E')$ be the subgraph obtained from G by choosing the arc of minimum weight directed to each vertex other than r. Then the set $(C - E^*)$ has exactly one arc for every cycle C in H.

Solution. The set $(C - E^*)$ should have at least one arc. Suppose e' and f' are two arcs in this set. Since T^* is an arborescence, every vertex is reachable from the root using arcs from E^*. So there should be arcs e and f in E^* such that $t(e) = t(e')$ and $t(f) = t(f')$ with $w(e') < w(e)$ and $w(f') < w(f)$. In that case, the arcs in the set $E^* - e + e'$ (or in the set $E^* - f + f'$) will form an arborescence T', where $w(T') < w(T)$, violating the minimality of T.

4.31 Obtain a minimum weight arborescence in the digraph shown in Fig. 4-24.

Solution. A minimum weight arborescence is shown in Fig. 4-28.

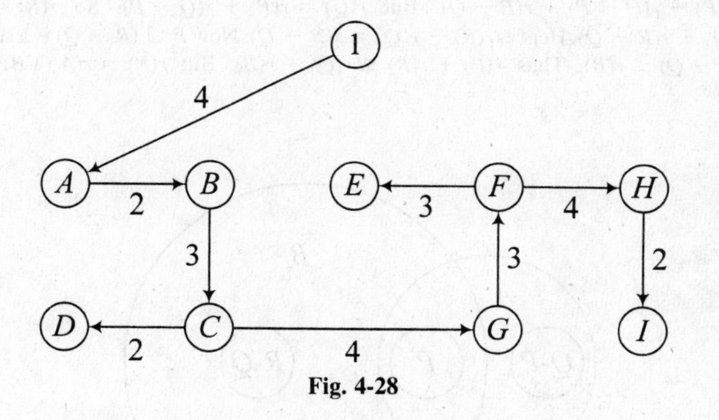

Fig. 4-28

4.32 Obtain a minimum weight arborescence in the digraph shown in Fig. 4-26.

Solution. A minimum weight arborescence is shown in Fig. 4-29.

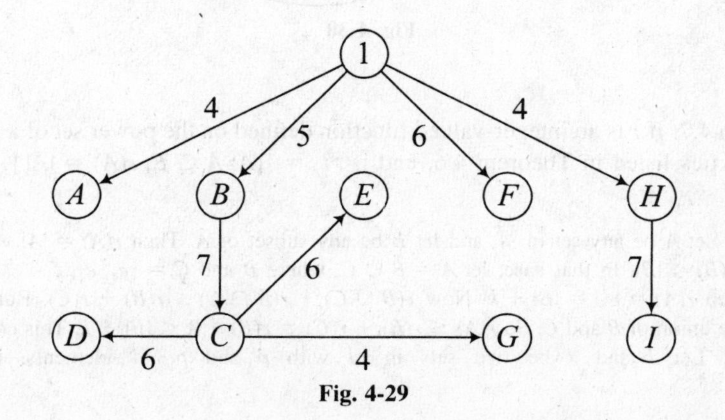

Fig. 4-29

MATROIDS AND THE GREEDY ALGORITHM

4.33 Prove Theorem 4.5: An independent system is a matroid if and only whenever I and J are two independent sets with the property that J has more elements than I, there exists an element $e \in (J - I)$ such that $I \cup \{e\}$ is an independent set.

Solution. Suppose I and J are two independent sets (in a matroid), and suppose the latter has more elements than the former. Obviously, I is not maximal in $I \cup J$. So there exists an independent set $K \subseteq (I \cup J)$ such that I is a proper subset of K. In other words, there exists e in $(J - I)$ such that $I + e$ is an independent set. On the other hand, suppose A and B are two sets that are maximal in a set D of an independent system that satisfies the given hypothesis. If the two sets do not have the same number of elements, suppose there are more elements in B. In that case, there exists e in $(B - A)$ such that $A' = A + e$ is an independent set. This implies that there exists an independent set A' that properly contains A and that is contained in D, violating the maximality of A in D. Thus both A and B have the same cardinality. Hence, the independent system is a matroid.

4.34 Prove Theorem 4.6: The rank function r of a matroid satisfies the following four properties: (*i*) the rank of the empty set is 0, (*ii*) $r(A) \le r(B)$ if $A \subset B$, (*iii*) the function r is submodular: $r(A \cup B) + r(A \cap B) \le r(A) + r(B)$, and (*iv*) the rank of a singleton set is either 0 or 1.

Solution. Properties (*i*), (*ii*), and (*iv*) are obvious. Suppose P is a maximal independent subset of $(A \cap B)$, where A and B are arbitrary subsets of the ground set E of a matroid. Since P is an independent subset of A, we can enlarge P by adjoining elements from $(A - B)$ to an independent set Q that is maximal in A. See Fig. 4-30. Thus $Q = (Q - P) \cup P$ and $(Q - P) \subset (A - B)$. Since A is a subset of $(A \cup B)$, set Q can be enlarged by adjoining elements from $(B - A)$ to an independent set R that is maximal in $(A \cup B)$. Hence, $R = Q \cup (R - Q)$ and $(R - Q) \subset (B - A)$. Thus R is the so disjoint union of three independent sets P, $(Q - P)$ and $(R - Q)$. Thus $r(R) = r(P) + r(Q - P) + r(R - Q)$. But $r(Q) = r(P) + r(Q - P)$. So $r(R) = r(P) + r(Q) - r(P) + r(R - Q) = r(Q) + r(R - Q)$. Hence, $r(R) - r(Q) = r(R - Q)$. Now $P \cup (R - Q)$ is a subset of B, which implies that $r(P) + r(R - Q) \le r(B)$. Thus $r(P) + r(R) \le r(Q) + r(B)$. But $r(P) = r(A \cap B)$, $r(R) = r(A \cup B)$, and $r(Q) = r(A)$.

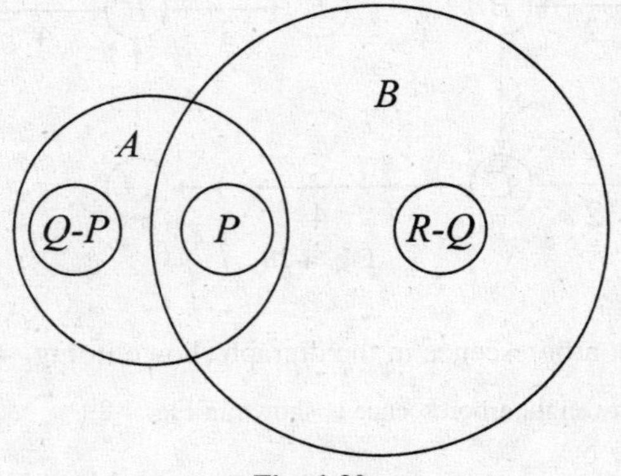

Fig. 4-30

4.35 Prove Theorem 4.7: If r is an integer-valued function defined on the power set of a finite set E satisfying the four properties listed in Theorem 4.6, and if $\mathcal{A} = \{A: A \subseteq E, r(A) = |A|\}$, the pair $(E, \mathcal{A})$ is a matroid.

Solution. Let A be any set in $\mathcal{A}$, and let B be any subset of A. Then $r(A) = |A|$, $r(B) \le |B|$, and $r(B) \le r(A)$. Suppose $r(B) < |B|$. In that case, let $A = B \cup C$, where B and $C = \{e_1, e_2, \ldots, e_k\}$ have no elements in common. Then $r(A) = |A| = |B| + k$. Now $r(B \cup C) + r(B \cap C) \le r(B) + r(C)$. But B and C are disjoint sets, and A is the union of B and C. So $r(A) \le r(B) + r(C) = r(B) + k < |B| + k$. This contradiction establishes that $r(B) = |B|$. Let I and J be two sets in $\mathcal{A}$ with p and $p + 1$ elements, respectively. Let $I =$

$\{e_1, e_2, \ldots, e_q, e_{q+1}, e_{q+2}, \ldots, e_p\}$ and $J = \{e_1, e_2, \ldots, e_q, f_{q+1}, f_{q+2}, \ldots, f_p, f_{p+1}\}$ (where e_i and f_j are unequal for any i and j) be two sets in $\mathcal{A}$. Suppose $r(I + f_i) = r(I)$ for every f_i.

Now $r(I \cup \{f_i, f_j\}) + r(I) \leq r(I + f_i) + r(I + f_j)$. So $r(I \cup \{f_i, f_j\}) + p \leq p + p$ and hence $r(I \cup \{f_i, f_j\}) \leq p$. But $r(I + f_i) = p$. So, $r(I \cup \{f_i, f_j\}) = p$. Proceeding like this, we establish that $r(I \cup \{f_{q+1}, f_{q+2}, \ldots, f_{p+1}\}) = p$. In other words, $r(J) = p$, which is a contradiction. Thus the assumption that $r(I + f_i) = p$ for every i is false. So there exists an element f in J such that $r(I + f) = p + 1$.

4.36 Prove Theorem 4.8: A solution of the problem of finding a maximum weight independent set in an independent system can be obtained by using the greedy algorithm for every nonnegative weight function defined on its ground set if and only if the independent system is a matroid.

Solution.

(*a*) If I and J are two independent sets in an independent system with p and $(p + 1)$ elements, respectively, let $w(e) = (p + 2)$ for all e in I, $w(e) = (p + 1)$ for all e in $(J - I)$, and $w(e) = 0$ for all other e in the ground set. Then $w(J) \geq (p + 1)(p + 1) > p(p + 2) = w(I)$; hence, I is not a solution. By the greedy procedure, we take I and then take an element from $(J - I)$. In other words, there exists an element e in the set $(J - I)$ such that $I + e$ is an independent set. So the independent system is a matroid.

(*b*) Suppose that by applying the greedy algorithm, an independent set $I = \{e_1, e_2, \ldots, e_r\}$ is obtained (in a matroid) in which the elements are arranged in nondecreasing order by weight. If $J = \{f_1, f_2, \ldots, f_r\}$ is any independent set in the matroid, it is enough if we prove by induction that $w(f_i) \leq w(e_i)$ for every i. It is true when $i = 1$. Suppose it holds for $i = 1, 2, \ldots, (m - 1)$. We have to prove it holds when $i = m$. Suppose $w(f_m) > w(e_m)$. Let $D = \{e_1, e_2, \ldots, e_{m-1}\}$ and $A = \{e\colon w(e) \geq w(f_m)\}$. Then D is an independent set and, by the induction hypothesis, is a subset of A. If D is not maximal in A, there exists e in A such that $D + e$ is independent. But if e is in A, $w(e) \geq w(f_m) > w(e_m)$, which implies that after picking e_{m-1}, the greedy algorithm would have selected e and not e_m. So D is maximal in A. Since D has $(m - 1)$ elements, any independent subset of A cannot have more than $(m - 1)$ elements. But $\{f_1, f_2, \ldots, f_m\}$ is an independent subset of A. This contradiction shows that $w(f_m) \leq w(e_m)$.

4.37 If $\{E_1, E_2, \ldots, E_k\}$ is a partition of a finite set E and if $\mathcal{A} = \{I \subset E\colon |I \cap E_i| \leq 1 \text{ for each } i\}$, show that the pair $(E, \mathcal{A})$ is a matroid (known as a **partition matroid** on E), and find its rank function.

Solution. If J is a subset of I that belongs to class $\mathcal{A}$, the intersection of J with every set in the class is at most 1. In other words, the pair is an independent system. Suppose J has p elements and I has $(p + 1)$ elements such that they have r elements in common. Specifically, let $J \cap E_i = e_i$ for $i = 1, 2, \ldots, p$; $I \cap E_i = e_i$ for $i = 1, 2, \ldots, r$, and $I \cap E_i = f_i$ for $i = r + 1, \ldots, p, p + 1$. Then the set J' obtained by adjoining f_{p+1} to J is in class $\mathcal{A}$, and the pair is a matroid, as proved in Problem 4.33. If A is a subset of E, define $r(A)$ to be the number of sets in the partition that have nonempty intersection with the set A. Then it is easily verified that r is a rank function on the ground set E.

4.38 If $G = (X, Y, E)$ is a bipartite graph, show that we can obtain two partition matroids on E: one using the set X and the other using Y.

Solution. Let $X = \{v_1, v_2, \ldots, v_x\}$ and $Y = \{w_1, w_2, \ldots, w_y\}$ be such that every edge in G is between some vertex in X and some vertex in Y. Let $A_i = \{e \in E\colon e \text{ and } v_i \text{ are adjacent}\}$ and $B_j = \{e \in E\colon e \text{ and } w_j \text{ are adjacent}\}$. Then the partition matroid on E defined by the partition $\mathcal{A}_1 = \{A_1, A_2, \ldots, A_x\}$ is the **left partition matroid,** and the partition matroid on E defined by the partition $\mathcal{A}_2 = \{B_1, B_2, \ldots, B_y\}$ is the **right partition matroid.**

4.39 If $G = (V, E)$ is a directed graph, show that it is possible to obtain two partition matroids on the set E of the arcs on the digraph.

Solution. For each vertex i, let H_i be the set of arcs that are directed to that vertex, and let T_i be the set of arcs that are directed from that vertex. Then the matroid on E using the partition $\{H_i\}$ is called the **head partition matroid,** and the matroid using the partition $\{T_i\}$ is called the **tail partition matroid.**

4.40 Let E be a finite set with a weight function w defined on E, and let $(E, \mathscr{A}_i)$ be a collection of k matroids defined on the same ground set E. The ***k*-matroid cardinality intersection problem** is the problem of finding a subset of maximum cardinality that is independent in each of the k matroids. The ***k*-matroid weighted intersection problem** is the problem of finding a subset of maximum weight that is independent in each of the k matroids. Show that *(a)* the former is a special case of the latter and *(b)* the maximum weight branching problem and the minimum weight arborescence problem are 2-matroid weighted intersection problems.

Solution.

(a) If the weight of each element is 1, a set of maximum cardinality is a set of maximum weight and vice versa.

(b) A set I of arcs will constitute a branching in the digraph if and only if I is an independent set both in the head partition matroid and in the circuit matroid defined on E by treating each arc as an edge. If G is a quasi-strongly connected digraph of order n, a set I of $(n - 1)$ arcs is an arborescence if and only if I is independent both in the head partition matroid and the circuit matroid.

4.41 The **span** of any subset A of the ground set of a matroid is a maximal superset of A having the same rank as A. Show that the span of a set is unique.

Solution. Suppose A_1 and A_2 are two spans of A. Let $r(A) = r(A_1) = r(A_2) = p$. By definition, the rank of any set that properly contains A_1 (or, for that matter, A_2) is more than p. Hence, if e_2 is in $(A_2 - A_1)$, $r(A_1 + e_2) > p$. Let I be a subset of A with p elements, and let J be a subset of $(A_1 + e_2)$ with $(p + 1)$ elements. So there is an element in $(J - I)$ that can be adjoined to I to obtain a larger independent set. The only such element is e_2. So $(I + e_2)$ is an independent set with $(p + 1)$ elements. But $(I + e_2)$ is a subset of A_2, and the rank of A_2 is only p. This is a contradiction.

Supplementary Problems

4.42 Find the weight of a minimum spanning tree in the network whose weight matrix is the following matrix:

$$\begin{bmatrix} 0 & 4 & - & 5 & 2 & - & - & - & - \\ 4 & 0 & 4 & - & 7 & - & - & - & - \\ - & 4 & 0 & - & 2 & 5 & - & - & - \\ 5 & - & - & 0 & 1 & - & 3 & - & - \\ 2 & 7 & 2 & 1 & 0 & 1 & 6 & 7 & 6 \\ - & - & 5 & - & 1 & 0 & - & - & 3 \\ - & - & - & 3 & 6 & - & 0 & 4 & - \\ - & - & - & - & 7 & - & 4 & 0 & 4 \\ - & - & - & - & 6 & 3 & - & 4 & 0 \end{bmatrix}$$

Ans. 20

4.43 In Problem 4.42, suppose we have to find a spanning tree T of minimum weight that contains the two edges of maximum weight in the network. Find the weight of T. *Ans.* 26

4.44 Find the weight of a maximum weight spanning tree in the network of Problem 4.42. *Ans.* 44

4.45 Find the weight of a maximum weight branching in the directed network $G = (V, E)$, where $V = \{1, 2, 3, 4, 5, 6, 7, 8\}$ and $E = \{(1, 2), (1, 7), (2, 3), (2, 7), (3, 4), (3, 5), (4, 5), (6, 3), (6, 5), (6, 7), (7, 8), (8, 1)\}$ with weights 5, 8, 8, 8, 4, 5, 1, 6, 8, 6, 8, and 2, respectively. *Ans.* 34

4.46 Find the weight of a maximum weight branching in the directed network $G = (V, E)$, where $V = \{1, 2, 3, 4, 5, 6\}$ and $E = \{(1, 2), (1, 3), (1, 4), (2, 3), (3, 5), (5, 6), (6, 4), (4, 3)\}$ with weights 1, 0, 1, 2, 3, 2, 2, and 3, respectively. *Ans.* 10

4.47 Find the weight of a maximum weight branching in the directed network $G = (V, E)$, where $V = \{1, 2, 3, 4, 5, 6\}$ and $E = \{(1, 4), (1, 5), (2, 1), (2, 3), (3, 1), (4, 2), (4, 3), (4, 5), (4, 6), (5, 4), (6, 5)\}$ with weights 7, 2, 4, 1, 3, 4, 1, −5, −1, 6, and 1, respectively. *Ans.* 16

4.48 Find the weight of a minimum weight arborescence rooted at vertex 1 in the directed network $G = (V, E)$, where $V = \{1, 2, 3, 4, 5, 6\}$ and $E = \{(1, 2), (1, 4), (1, 6), (2, 3), (2, 6), (3, 2), (4, 2), (4, 5), (5, 6), (6, 3), (6, 4)\}$ with weights 16, 15, 19, 1, 10, 4, 11, 5, 3, 8, and 2, respectively. *Ans.* 33

4.49 Find the weight of a minimum weight arborescence rooted at vertex 1 in the directed network $G = (V, E)$, where $V = \{1, 2, 3, 4, 5, 6\}$ and $E = \{(1, 2), (1, 4), (1, 6), (2, 3), (3, 5), (4, 3), (4, 5), (5, 2), (5, 6), (6, 2), (6, 4)\}$ with weights 1, 4, 6, 12, 9, 9, 5, 11, 7, 10, and 6, respectively. *Ans.* 25

4.50 Give an example of a matroid and two subsets of the ground set such that the inequality in the submodularity relation involving these two subsets is a strict inequality. [*Hint:* In the ground set $E = \{a, b, c\}$ of the three edges in the complete graph with three vertices, take $A = \{a, b\}$ and $B = \{b, c\}$.]

4.51 Find the circuits and the bases of a k-uniform matroid.
Ans. Any subset of the ground set with $k + 1$ elements is a circuit. Any subset with k elements is a base.

Chapter 5

Shortest Path Problems

5.1 TWO SHORTEST PATH ALGORITHMS

If there is a path from vertex u to vertex v in a network G, any path of minimum length from u to v is a **shortest path (SP)** from u to v, and its weight is the **shortest distance (SD)** from u to v. The problem of finding shortest paths in networks is called the **shortest path problem.**

In this section, we investigate two well-known SP algorithms. In the first case, we have a network in which no arc weight is negative. Using the algorithm, we can obtain the SP and SD from a fixed vertex v (the source vertex) to every other vertex u provided that there is a path in the network from v to u. In the second case, the weight function need not be nonnegative. Furthermore, using the algorithm, we can obtain the SD and an SP from any vertex u to any vertex v if there is a path from u to v. The existence of negative cycles in the network can also be detected. The algorithm terminates as soon as a negative cycle is located because the presence of a negative cycle in a network makes the SP problem "unbounded" in the sense that if we allow repeated vertices in a path, the shortest distance between two vertices may become unbounded.

Dijkstra's Algorithm

Let $G = (V, E)$, where $V = \{1, 2, \ldots, n\}$ is a directed network in which the weight of every arc is nonnegative. This algorithm can be used to find the SP and SD from any fixed vertex (say vertex 1) to any vertex i if there is a directed path from 1 to i. Let $a(i, j)$ be the weight of the arc from i to j. If there is no arc from i to j, $a(i, j)$ is $+\infty$. Each vertex i is assigned a label that is either permanent or tentative. The permanent label $L(i)$ is the SD from 1 to i. The tentative label $L'(i)$ is an upper bound of $L(i)$. At each stage of the algorithm, P is the set of vertices with permanent labels and T is the set of vertices with tentative labels. Initially, P is the set $\{1\}$ with $L(1) = 0$ and $L'(j) = a(1, j)$ for all j. The procedure terminates when $P = V$. Each iteration consists of two steps;

> **Step 1.** Find a vertex k in T for which $L'(k)$ is finite and minimum. If there is no k, stop; there is no path from 1 to any unlabeled vertex. Otherwise, declare k to be permanently labeled, and adjoin k to P. Stop if $P = V$. Label the arc (i, k), where i is a labeled vertex that determines the minimum value of $L'(k)$.
>
> **Step 2.** Replace $L'(j)$ by the smaller value of $L'(j)$ and $L(k) + a(k, j)$ for every j in T. Go to step 1.

Notice that if G has n vertices and it is possible to obtain the SD from the starting vertex v to every other vertex, the set of $(n - 1)$ arcs obtained by this method will form an arborescence rooted at v that will give both the SD and SP from v to every other vertex.

Theorem 5.1. Dijkstra's algorithm finds the SD from a fixed vertex v to any vertex i in the network if there is a path from v to i. (See Solved Problem 5.1)

Example 1. Obtain the SD and SP from vertex 1 to every other vertex in the network shown in Fig. 5-1.

Iteration 1:

Step 1. $P = \{1\}$, and $L(1) = 0$. $L'(2) = 4$, $L'(3) = 6$, and $L'(4) = 8$. Adjoin vertex 2 to P. The arc (1, 2) is labeled.

Step 2. $P = \{1, 2\}$, and $L(2) = 4$. $L'(3) = \min\{6, L(2) + a(2, 3)\}$, $L'(4) = \min\{8, L(2) + a(2, 4)\}$, $L'(5) = \min\{\infty, L(2) + a(2, 5)\}$, $L'(6) = \min\{\infty, L(2) + a(2, 6)\}$, and $L'(7) = \min\{\infty, L(2) + a(2, 7)\}$.

Iteration 2:

Step 1. $P = \{1, 2\}$, $L(1) = 0$, and $L(2) = 4$. $L'(3) = 5$, $L'(4) = 8$, and $L'(5) = 11$. Adjoin vertex 3 to P. The arc (2, 3) is labeled.

Step 2. $P = \{1, 2, 3\}$, and $L(3) = 5$. $L'(4) = \min\{8, L(3) + a(3, 4)\}$, $L'(5) = \min\{11, L(3) + a(3, 5)\}$, $L'(6) = \min\{\infty, L(3) + a(3, 6)\}$, and $L'(7) = \min\{\infty, L(3) + a(3, 7)\}$.

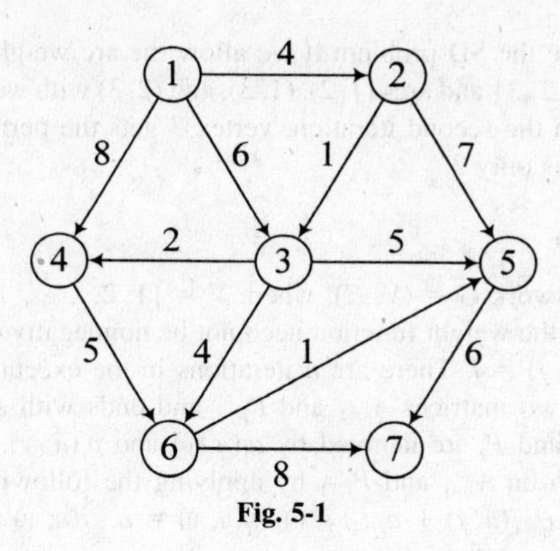

Fig. 5-1

Iteration 3:

Step 1. $P = \{1, 2, 3\}$, $L(1) = 0$, $L(2) = 4$, and $L(3) = 5$. $L'(4) = 7$, $L'(5) = 10$, and $L'(6) = 9$. Adjoin vertex 4 to P. The arc (3, 4) is labeled.

Step 2. $P = \{1, 2, 3, 4\}$, and $L(4) = 7$. $L'(5) = \min\{10, L(4) + a(4, 5)\}$, $L'(6) = \min\{9, L(4) + a(4, 6)\}$, and $L'(7) = \min\{\infty, L(4) + a(4, 7)\}$.

Iteration 4:

Step 1. $P = \{1, 2, 3, 4\}$, $L(1) = 0$, $L(2) = 4$, $L(3) = 5$, and $L(4) = 7$. $L'(5) = 10$, and $L'(6) = 9$. Adjoin vertex 6 to P. The arc (3, 6) is labeled.

Step 2. $P = \{1, 2, 3, 4, 6\}$, and $L(6) = 9$. $L'(5) = \min\{10, L(6) + a(6, 5)\}$, and $L'(7) = \min\{\infty, L(6) + a(6, 7)\}$.

Iteration 5:

Step 1. $P = \{1, 2, 3, 4, 6\}$, $L(1) = 0$, $L(2) = 4$, $L(3) = 5$, $L(4) = 7$, and $L(6) = 9$. $L'(5) = 10$, and $L'(7) = 17$. Adjoin vertex 5 to P. The arc (3, 5) is labeled.

Step 2. $P = \{1, 2, 3, 4, 6, 5\}$, and $L(5) = 10$. $L'(7) = \min\{17, L(5) + a(5, 7)\}$.

Iteration 6:

Step 1. $P = \{1, 2, 3, 4, 6, 5\}$, $L(1) = 0$, $L(2) = 4$, $L(3) = 5$, $L(4) = 7$, $L(6) = 9$, and $L(5) = 10$. $L'(7) = 16$. Adjoin vertex 7 to P. The arc (5, 7) is labeled.

Step 2. $P = \{1, 2, 3, 4, 6, 5, 7\}$, and $L(7) = 16$. At this stage, $P = V$.

The labeled arcs (1, 2), (2, 3), (3, 4), (3, 5), (3, 6), and (5, 7) constitute a shortest path arborescence rooted at vertex 1, as shown in Fig. 5-2.

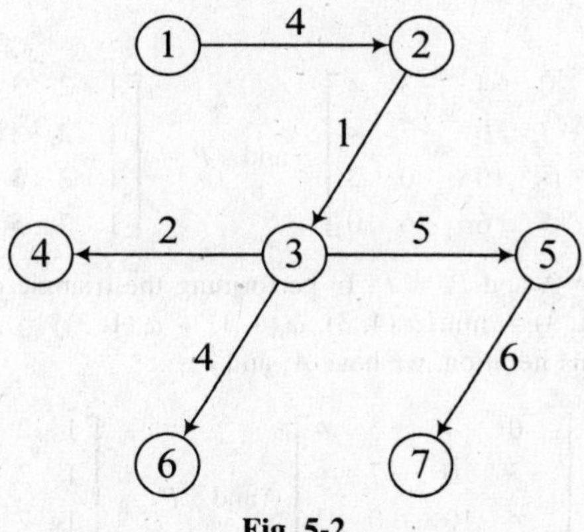

Fig. 5-2

This algorithm need not solve the SD problem if we allow the arc weights to be negative. For example, consider the digraph with $V = \{1, 2, 3\}$ and arcs (1, 2), (1, 3), and (2, 3) with weights 5, 4, and -3, respectively. The starting vertex is vertex 1. In the second iteration, vertex 3 gets the permanent label with $L(3) = 4$, but the SD from vertex 1 to vertex 3 is only 2.

The Floyd–Warshall Algorithm

The weight matrix A of a network $G = (V, E)$, where $V = \{1, 2, \ldots, n\}$, is defined as in the case of Dijkstra's algorithm. In this case, the weight function need not be nonnegative. The initial path matrix P is the $n \times n$ matrix $[p(i, j)]$, where $p(i, j) = j$. There are n iterations in the execution of the algorithm. Iteration j (based at vertex 1) begins with two matrices A_{j-1} and P_{j-1} and ends with A_j and P_j. Initially, $A_0 = A$ and $P_0 = P$. The (u, v) entries in A_j and P_j are denoted by $a_j(u, v)$ and $p_j(u, v)$, respectively. For a fixed j, the matrices A_j and P_j are obtained from A_{j-1} and P_{j-1} by applying the following rules, known as the triangle (triple) operation: If $a_{j-1}(u, v) \le a_{j-1}(u, j) + a_{j-1}(j, v)$, $a_j(u, v) = a_{j-1}(u, v)$ and $p_j(u, v) = p_{j-1}(u, v)$. Otherwise, $a_j(u, v) = a_{j-1}(u, j) + a_{j-1}(j, v)$ and $p_j(u, v) = p_{j-1}(u, j)$.

When the algorithm terminates, we are left with the **SD matrix A_n and the SP matrix P_n**. The (u, v) entry in the first matrix is the shortest distance between these two vertices, and the (u, v) entry in the second matrix denotes the first vertex (after u) in a shortest path from u to v.

Theorem 5.2. The Floyd–Warshall algorithm using the triangle operation correctly solves the SD and SP problem. (See Solved Problem 5.4.)

Example 2. Obtain the SD matrix and the SP matrix of the network shown in Fig. 5-3.

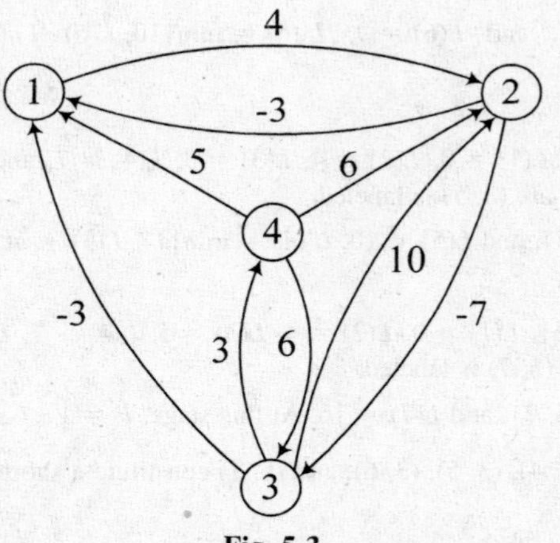

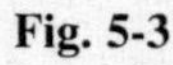

Fig. 5-3

The matrices are

$$A = \begin{bmatrix} 0 & 4 & -3 & \infty \\ -3 & 0 & -7 & \infty \\ \infty & 10 & 0 & 3 \\ 5 & 6 & 6 & 0 \end{bmatrix} \quad \text{and} \quad P = \begin{bmatrix} 1 & 2 & 3 & 4 \\ 1 & 2 & 3 & 4 \\ 1 & 2 & 3 & 4 \\ 1 & 2 & 3 & 4 \end{bmatrix}$$

Iteration 1: We begin with $A_1 = A$ and $P_0 = P$. In performing the triangle operation based at vertex 1, the only change is at the (4, 3) entry: $a_1(4, 3) = \min\{a_0(4, 3), a_0(4, 1) + a_0(1, 3)\} = \min\{6, 5 - 3\} = 2$. Then $p_1(4, 3) = p_0(4, 1) = 1$. Thus at the end of this iteration, we have A_1 and P_1:

$$A_1 = \begin{bmatrix} 0 & 4 & -3 & \infty \\ -3 & 0 & -7 & \infty \\ \infty & 10 & 0 & 3 \\ 5 & 6 & 2 & 0 \end{bmatrix} \quad \text{and} \quad P_1 = \begin{bmatrix} 1 & 2 & 3 & 4 \\ 1 & 2 & 3 & 4 \\ 1 & 2 & 3 & 4 \\ 1 & 2 & 1 & 4 \end{bmatrix}$$

Iteration 2: We begin with A_1 and P_1 and obtain the matrices A_2 and P_2 after applying the triangle operation:

$$A_2 = \begin{bmatrix} 0 & 4 & -3 & \infty \\ -3 & 0 & -7 & \infty \\ 7 & 10 & 0 & 3 \\ 3 & 6 & -1 & 0 \end{bmatrix} \quad \text{and} \quad P_2 = \begin{bmatrix} 1 & 2 & 3 & 4 \\ 1 & 2 & 3 & 4 \\ 2 & 2 & 3 & 4 \\ 2 & 2 & 2 & 4 \end{bmatrix}$$

Iteration 3: At the end of this iteration,

$$A_3 = \begin{bmatrix} 0 & 4 & -3 & 0 \\ -3 & 0 & -7 & -4 \\ 7 & 10 & 0 & 3 \\ 3 & 6 & -1 & 0 \end{bmatrix} \quad \text{and} \quad P_3 = \begin{bmatrix} 1 & 2 & 3 & 3 \\ 1 & 2 & 3 & 3 \\ 2 & 2 & 3 & 4 \\ 2 & 2 & 2 & 4 \end{bmatrix}$$

Iteration 4: At the end of this iteration,

$$A_4 = \begin{bmatrix} 0 & 4 & -3 & 0 \\ -3 & 0 & -7 & -4 \\ 6 & 9 & 0 & 3 \\ 3 & 6 & -1 & 0 \end{bmatrix} \quad \text{and} \quad P_4 = \begin{bmatrix} 1 & 2 & 3 & 3 \\ 1 & 2 & 3 & 3 \\ 4 & 4 & 3 & 4 \\ 2 & 2 & 2 & 4 \end{bmatrix}$$

At this stage, we have the SD between every pair of vertices in the network as the appropriate entries in SD matrix A_4. An SP from one vertex to another can be obtained from the entries in SP matrix P_4. For example, the SD from 1 to 2 is 4, as can be seen from the (1, 2) entry in the SD matrix. In the SP matrix, the (1, 2) entry is 2. So an SP from 1 to 2 is the arc from 1 to 2. The SD from 1 to 4 is 0. The (1, 4) entry in the SP matrix is 3. So in an SP from 1 to 4, the first vertex after 1 is 3. The (3, 4) entry in the SP matrix is 4. So an SP from 1 to 4 is $1 \to 3 \to 4$.

Locating Negative Cycles

Consider a network with the weight matrix

$$A = \begin{bmatrix} 0 & \infty & \infty & 1 \\ 2 & 0 & 1 & 3 \\ \infty & \infty & 0 & \infty \\ 2 & -4 & 3 & 0 \end{bmatrix}$$

At the end of iteration 2,

$$A_2 = \begin{bmatrix} 0 & \infty & \infty & 1 \\ 2 & 0 & 1 & 3 \\ \infty & \infty & 0 & \infty \\ 2 & -4 & 3 & -1 \end{bmatrix} \quad \text{and} \quad P_2 = \begin{bmatrix} 1 & 2 & 3 & 4 \\ 1 & 2 & 3 & 1 \\ 1 & 2 & 3 & 4 \\ 2 & 2 & 2 & 2 \end{bmatrix}$$

The (4, 4) diagonal entry is negative (instead of 0), indicating the presence of a negative cycle. From the second matrix, we see that this cycle (seen as a path from 4 to 4) is $4 \to 2 \to 1 \to 4$ with weight $-4 + 2 + 1 = -1$.

5.2 THE STEINER NETWORK PROBLEM

The Steiner network problem is the optimization problem of finding a tree $T = (W', F)$ in G such that $W \subset W'$ and the weight $w(F)$ is as small as possible, given a set W of vertices in an undirected graph $G = (V, E)$ with a nonnegative weight function w defined on E. The tree T is a **Steiner tree** of the set W. The vertices in $(W' - W)$ are the **Steiner points** of W with respect to T. A Steiner tree with respect to a pair of vertices in an undirected network is any shortest path P between them, and the Steiner points of this pair with respect to P are the intermediate vertices of this path. At the other extreme, any minimum spanning tree in the network is a Steiner tree of the set V. So both the SP problem and the M.S.T. problem can be considered special cases of the Steiner network problem.

Notice that if W is a proper subset of V, a minimum spanning tree of the subgraph H induced by the set W need not be a Steiner tree with respect to W. This can be seen in the following example.

Example 3. In the network shown in Fig. 5-4, the weight of a minimum spanning tree of the subgraph induced by the set $W = \{1, 2, 3, 4\}$ is easily seen to be 9. But it is possible to obtain a tree T that spans these four vertices and vertex 5 such that the weight of the edge set of T is only 8.

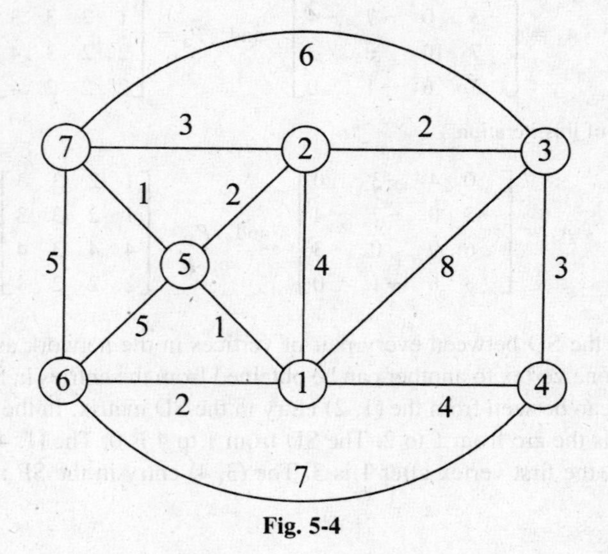

Fig. 5-4

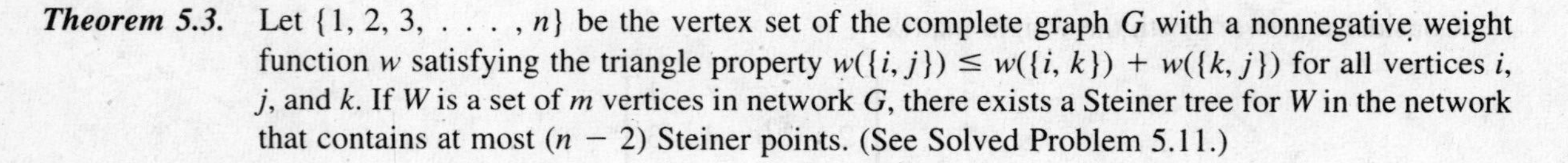

Theorem 5.3. Let $\{1, 2, 3, \ldots, n\}$ be the vertex set of the complete graph G with a nonnegative weight function w satisfying the triangle property $w(\{i, j\}) \leq w(\{i, k\}) + w(\{k, j\})$ for all vertices i, j, and k. If W is a set of m vertices in network G, there exists a Steiner tree for W in the network that contains at most $(n - 2)$ Steiner points. (See Solved Problem 5.11.)

Spanning Tree Enumeration Algorithm

The input is a set W of m vertices in a connected network $G = (V, E)$ of order n in which a nonnegative weight is defined on each edge. Construct the complete network $G' = (V, F)$ in which the weight of an edge joining two vertices is the SD between them. Let $\mathcal{S} = \{S \subset (V - W) : |S| \leq (m - 2)\}$. For each subset S in $\mathcal{S}$, find an M.S.T. of the subgraph of G' induced by $W \cup S$ on G'. Among the trees thus obtained, select a tree T' of minimum weight. Construct a tree T from T' by replacing each edge joining two vertices by the set of edges in a shortest path between them. Those vertices of T that are not in W will form a set of Steiner points for W.

Example 4. Find the Steiner trees with respect to $W = \{3, 6, 7\}$ in the network shown in Fig. 5-4.

The SD matrix for this network is

$$D = \begin{bmatrix} 0 & 3 & 5 & 4 & 1 & 2 & 2 \\ 3 & 0 & 2 & 5 & 2 & 5 & 3 \\ 5 & 2 & 0 & 3 & 4 & 7 & 5 \\ 4 & 5 & 3 & 0 & 5 & 6 & 6 \\ 1 & 2 & 4 & 5 & 0 & 3 & 1 \\ 2 & 5 & 7 & 6 & 3 & 0 & 4 \\ 2 & 3 & 5 & 6 & 1 & 4 & 0 \end{bmatrix}$$

Since $m = 3$, S is any subset of $\{1, 2, 4, 5\}$ with at most one element. There are five choices for $(W \cup S)$: $W_1 = \{3, 6, 7\}$, $W_2 = \{3, 6, 7, 1\}$, $W_3 = \{3, 6, 7, 2\}$, $W_4 = \{3, 6, 7, 4\}$, and $W_5 = \{3, 6, 7, 5\}$. The minimum weight spanning trees of the subgraphs of G' induced by these sets have weights 9, 9, 9, 12, and 8, respectively. So we take the subgraph induced by W_5 on the complete network G'. In this subgraph, a minimum spanning tree T' is as shown in Fig. 5-5.

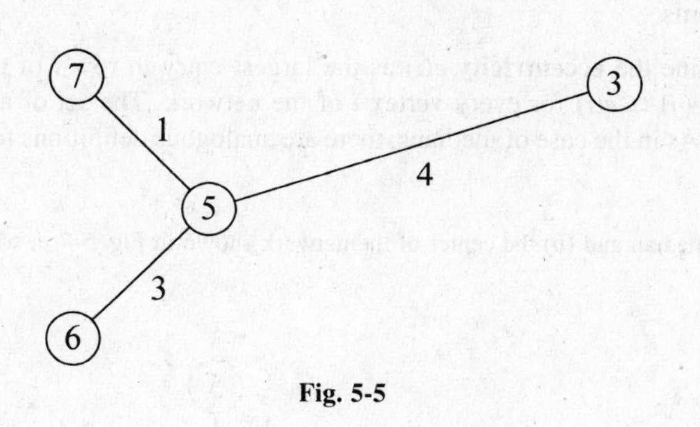

Fig. 5-5

The weight of the edge in T' between 3 and 5 is 4. We replace this edge by a shortest path between 3 and 5 that is also necessarily of weight 4. This SP is 5 — 2 — 3. Likewise the edge between 6 and 5 is replaced by the path 5 — 1 — 6. Thus the Steiner tree T with respect to {3, 6, 7} is as shown in Fig. 5-6. The Steiner points are 1, 2, and 5.

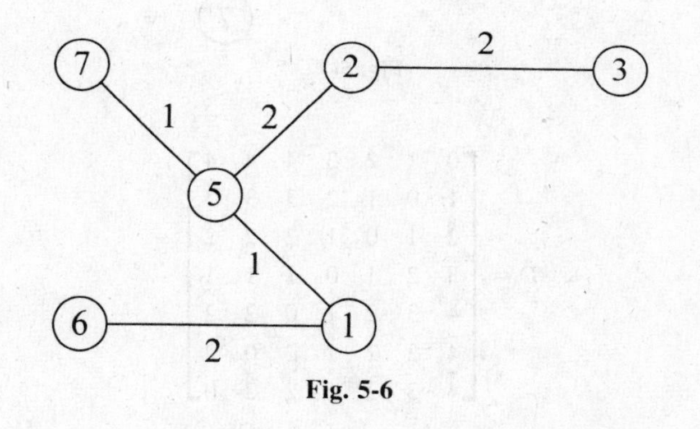

Fig. 5-6

5.3 FACILITY LOCATION PROBLEMS

The problem of deciding the exact place in a community where a facility (such as a school, a post office, or a fire station) should be located to serve the needs of the community as economically and efficiently as possible is known as the facility location problem. Such location problems can often be modeled as networks in which the facility could be located at one or more vertices.

If the facility is an institution like a post office or a school, it is desirable to locate it such that the sum of the distances from the facility to several parts of the community is as small as possible. This category, in which the aim is to minimize the sum of several weights, is aptly called a **minsum problem.** On the other hand, if the facility is an institution like a fire station, it is always desirable to locate it such that the distance from the fire station to the farthest point in the community is minimized. This category, in which the aim is to minimize the maximum shortest distance, is known as a **minmax problem.**

Median (Minsum) Problems

Let $G = (V, E)$, where $V = \{1, 2, \ldots, n\}$ is a weighted network with a nonnegative weight function defined on E, and let $D = [d(i, j)]$ be its SD matrix, where $d(i, j)$ is the SD from i to j in the network. For each i, let $s(i)$ be the sum of all the elements in row i of D. Vertex j is called a **median vertex** if $s(j) \leq s(i)$ for every vertex i of the network. The **median** is the set of all median vertices. In a more general setting, suppose that a nonnegative weight function is defined on the set V in addition to the weight function w on the set E. In that case, we define $s'(i) = w(1)d(i, 1) + w(2)d(i, 2) + \cdots + w(n)d(i, n)$. The vertex j is a **weighted median**

vertex if $s'(j) \leq s'(i)$ for every vertex i of the network, and the **weighted median** is the set of all weighted median vertices.

Center (Minmax) Problems

For each vertex i, define the **eccentricity** $e(i)$ as the largest entry in row i of the SD matrix. Vertex j is called a **center vertex** if $e(j) \leq e(i)$ for every vertex i of the network. The set of all center vertices is called the **center** of the network. As in the case of medians, there are analogous definitions for weighted center vertices and weighted centers.

Example 5. Find (*a*) the median and (*b*) the center of the network shown in Fig. 5-7 in which the weight of each edge is 1.

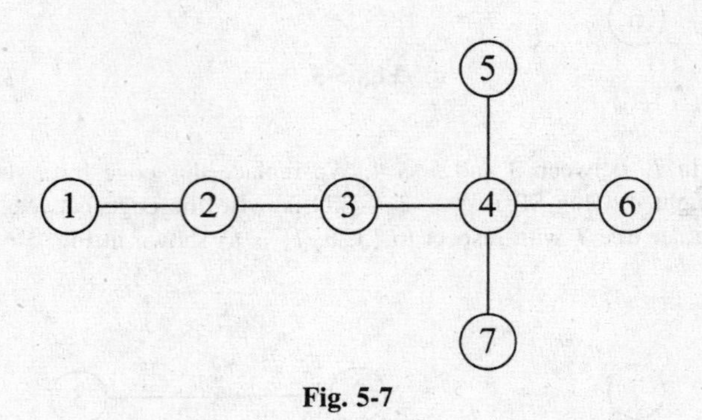

Fig. 5-7

The SD matrix is

$$D = \begin{bmatrix} 0 & 1 & 2 & 3 & 4 & 4 & 4 \\ 1 & 0 & 1 & 2 & 3 & 3 & 3 \\ 2 & 1 & 0 & 1 & 2 & 2 & 2 \\ 3 & 2 & 1 & 0 & 1 & 1 & 1 \\ 4 & 3 & 2 & 1 & 0 & 2 & 2 \\ 4 & 3 & 2 & 1 & 2 & 0 & 2 \\ 4 & 3 & 2 & 1 & 2 & 2 & 0 \end{bmatrix}$$

(*a*) The row sums are 18, 13, 10, 9, 14, 14, and 14. Row 4 has the lowest sum, so 4 is the only median vertex. The median is the set {4}.

(*b*) The eccentricities of the seven vertices are 4, 3, 2, 3, 4, 4, and 4, respectively, so 3 is the only center vertex. The center is the set {3}.

Solved Problems

TWO SHORTEST PATH ALGORITHMS

5.1 Prove Theorem 1: Dijkstra's algorithm finds the SD from a fixed vertex (v) to any vertex i in the network if there is a path from v to i.

Solution. The set of vertices is $V = \{1, 2, \ldots, n\}$. Vertex 1 is taken as the root. At each stage, vertex i has either the permanent label $L(i)$, which is the SD from the root to i, or a tentative label $L'(i)$, which is an upper bound of the SD from the root to i. The set of vertices with permanent labels is denoted by P, and the set of tentative labels is denoted by T. The proof is by induction on the cardinality of P. The induction hypothesis is that (*a*) $L(i)$ is the SD from 1 to each i in P and (*b*) for any vertex j in T, $L'(j)$ is the length of a shortest path from 1 to j, of which every vertex other than j is in P. Statement (*a*) is true when P is the set {1}. Suppose it is true when

P has $(k - 1)$ elements. Specifically, let $P = \{1, 2, \ldots, k - 1\}$. Suppose the next vertex to be assigned to P is k. By the induction hypothesis, $L'(k)$ is the length of a SP from 1 to k in which every vertex other than k is in P. By our definition, $L(k) = L'(k)$. We claim that $L(k)$ is the SD from 1 to k. If this is not the case, $L(k) > d$, where d is the SD. Hence, any shortest path from 1 to k has at least one intermediate vertex from the set T. Let v be the first vertex (from T) in this path. If the SD from 1 to v is d', $d' \leq d < L(k)$. But because k is in P and v is in T, it is implied that $L(k) \leq d'$. So (a) is true when P has k vertices. We now prove (b), as follows. Vertex k is used to revise the labels of the remaining vertices in T. The label $L'(j)$ remains the same or gets the new label $L(k) + a(k, j)$. Thus (b) holds when P has k vertices.

5.2 Using Dijksra's algorithm, find a shortest distance arborescence rooted at vertex 1 of the directed network shown in Fig. 5-8.

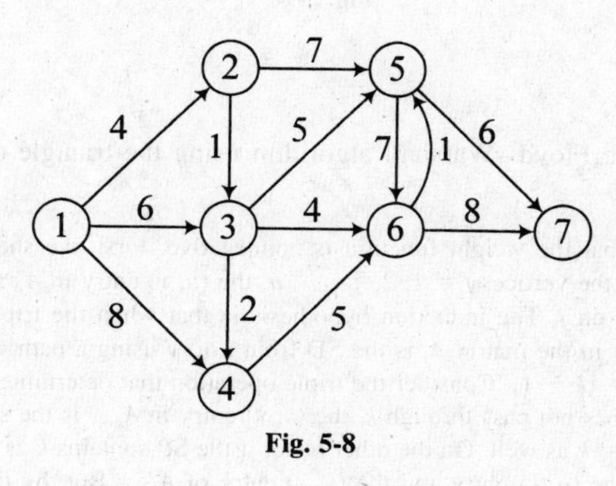

Fig. 5-8

Solution. The shortest distance arborescence is the network shown in Fig. 5-9.

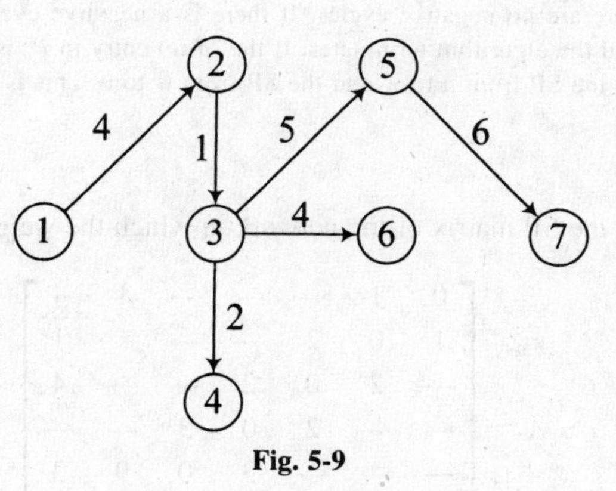

Fig. 5-9

5.3 Using Dijkstra's algorithm, find a shortest distance arborescence for the network whose weight matrix is

$$A = \begin{bmatrix} 0 & — & 4 & 10 & 3 & — & — \\ — & 0 & 1 & 1 & 2 & 11 & 0 \\ — & 9 & 0 & 8 & 3 & 2 & 1 \\ — & 4 & 0 & 0 & 8 & 6 & 3 \\ — & 0 & 1 & 2 & 0 & 3 & 1 \\ — & 1 & 1 & 3 & 2 & 0 & 0 \\ — & 4 & 3 & — & — & 2 & 0 \end{bmatrix}$$

Solution. The shortest distance arborescence rooted at vertex 1 is the network shown in Fig. 5-10.

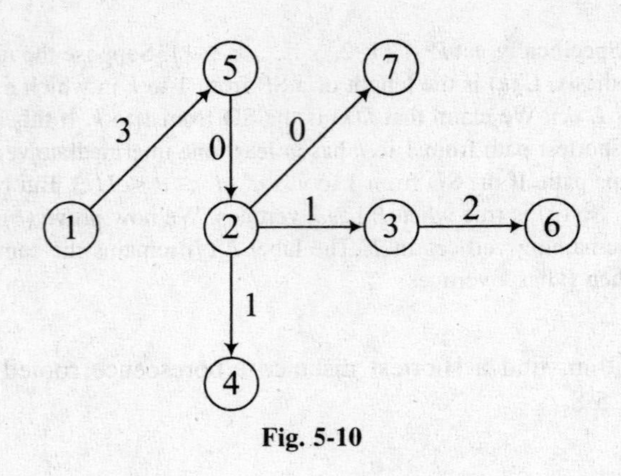

Fig. 5-10

5.4 Prove Theorem 5.2: The Floyd–Warshall algorithm using the triangle operation correctly solves the SD and SP problem.

Solution. Assume that the weight function is nonnegative. First, we show that if we perform the triple operation successively for the vertices $j = 1, 2, \ldots, n$, the (u, v) entry in A_n at the end is the SD from u to v. The proof is by induction on j. The induction hypothesis is that when the triple operation based at vertex j is completed, the (u, v) entry in the matrix A_j is the SD from u to v using a path with intermediate vertices $w \leq j$. Suppose this is true for $j = (k - 1)$. Consider the triple operation that determines the (u, v) entry in A_k. If the SP from u to v at this stage does not pass through k, the (u, v) entry in A_{k-1} is the same as the (u, v) entry in A_k. So the hypothesis holds for $j = k$ as well. On the other hand, if the SP contains k as an intermediate vertex, the (u, v) entry will be the sum of the (u, k) entry and the (k, v) entry of A_{k-1}. But, by the induction hypothesis, each of these entries corresponds to an optimal path with intermediate vertices $\leq k$. In this case, the hypothesis also holds for k. This completes the inductive argument when the weight function is nonnegative. This procedure works well as long as there are no negative cycles. If there is a negative cycle, a diagonal entry will, at some stage, become negative and the algorithm terminates. If the (u, v) entry in P_n is w, an SP from u to v is obtained by joining (concatenating) the SP from u to w and the SP from w to v. This is also the consequence of the triple operation.

5.5 Find the SD matrix and the SP matrix of the network in which the weight matrix A is

$$A = \begin{bmatrix} 0 & 1 & - & - & - & 1 & 4 \\ 1 & 0 & 2 & - & - & - & 1 \\ - & 2 & 0 & 2 & - & - & 4 \\ - & - & 2 & 0 & 3 & - & - \\ - & - & - & 3 & 0 & 9 & 3 \\ 1 & - & - & - & 9 & 0 & - \\ 4 & 1 & 4 & - & 3 & - & 0 \end{bmatrix}$$

Solution. The SD matrix A_7 and the SP matrix P_7 are

$$A_7 = \begin{bmatrix} 0 & 1 & 3 & 5 & 5 & 1 & 2 \\ 1 & 0 & 2 & 4 & 4 & 2 & 1 \\ 3 & 2 & 0 & 2 & 5 & 4 & 3 \\ 5 & 4 & 2 & 0 & 3 & 6 & 5 \\ 5 & 4 & 5 & 3 & 0 & 6 & 3 \\ 1 & 2 & 4 & 6 & 6 & 0 & 3 \\ 2 & 1 & 3 & 5 & 3 & 3 & 0 \end{bmatrix} \quad \text{and} \quad P_7 = \begin{bmatrix} - & 2 & 2 & 2 & 2 & 6 & 2 \\ 1 & - & 3 & 3 & 7 & 1 & 7 \\ 2 & 2 & - & 4 & 4 & 2 & 2 \\ 3 & 3 & 3 & - & 5 & 3 & 3 \\ 7 & 7 & 4 & 4 & - & 7 & 7 \\ 1 & 1 & 1 & 1 & 1 & - & 1 \\ 2 & 2 & 2 & 2 & 5 & 2 & - \end{bmatrix}$$

5.6 Find the SD matrix and the SP matrix of the network for which the weight matrix is

$$A = \begin{bmatrix} 0 & - & 4 & 10 & 3 & - & - \\ - & 0 & -1 & -1 & 2 & 11 & 0 \\ - & 9 & 0 & 8 & 3 & 2 & 1 \\ - & 4 & 0 & 0 & 8 & 6 & 3 \\ - & 0 & 1 & 2 & 0 & 3 & -1 \\ - & -1 & -1 & 3 & 2 & 0 & 0 \\ - & 4 & 3 & - & - & 2 & 0 \end{bmatrix}$$

Solution. The SD matrix A_7 and the SP matrix P_7 are

$$A_7 = \begin{bmatrix} 0 & 3 & 2 & 2 & 3 & 4 & 2 \\ - & 0 & -1 & -1 & 2 & 1 & 0 \\ - & 1 & 0 & 0 & 3 & 2 & 1 \\ - & 1 & 0 & 0 & 3 & 2 & 1 \\ - & 0 & -1 & -1 & 0 & 1 & -1 \\ - & -1 & -2 & -2 & 1 & 0 & -1 \\ - & 1 & 0 & 0 & 3 & 2 & 0 \end{bmatrix} \quad \text{and} \quad P_7 = \begin{bmatrix} - & 5 & 5 & 5 & 5 & 5 & 5 \\ 1 & - & 3 & 4 & 5 & 3 & 7 \\ 1 & 6 & - & 6 & 5 & 6 & 7 \\ 1 & 3 & 3 & - & 3 & 3 & 3 \\ 1 & 2 & 2 & 2 & - & 2 & 7 \\ 1 & 2 & 2 & 2 & 2 & - & 2 \\ 1 & 6 & 6 & 6 & 6 & 6 & - \end{bmatrix}$$

5.7 In Problem 5.6, find an SP from vertex 4 to vertex 2.

Solution. The (4, 2)-entry in the path matrix is 3. So in an SP from 4 to 2, the first vertex after 4 is 3. The (3, 2) entry in the path matrix is 6, so the next vertex in the SP is 6. The (6, 2) entry in the path matrix is 2, so the last vertex in the SP is 2. Thus an SP from 4 to 2 is $4 \to 3 \to 6 \to 2$ with a weight of $0 + 2 + (-1) = 1$, which is the (4, 2) entry in the SD matrix.

5.8 Construct a shortest distance arborescence rooted at vertex 1 in the network of Problem 5.6, the path matrix.

Solution. One such arborescence is the following shortest path from the root to vertex 3: $1 \to 5 \to 7 \to 6 \to 2 \to 4 \to 3$.

5.9 Suppose the negative entry (-1) that appears in the fourth column of the weight matrix in problem 5.6 is replaced by a smaller negative entry (-3). Find a negative cycle in the modified network.

Solution. There are two negative cycles: $2 \to 4 \to 3 \to 6 \to 2$ and $2 \to 4 \to 3 \to 7 \to 6 \to 2$.

5.10 In Problem 5.6, obtain the distance matrix and the path matrix that will give the shortest distance between every pair of vertices using paths that will not use vertices 5, 6, and 7 as intermediate vertices. Use these two matrices to obtain a path of minimum length from vertex 7 to vertex 4 that does not use vertices 5 and 6 as intermediate vertices.

Solution. The matrices A_4 and P_4 will ignore the paths that contain 5, 6, and 7 as intermediate vertices. These matrices are

$$A_4 = \begin{bmatrix} 0 & 13 & 4 & 10 & 3 & 6 & 5 \\ - & 0 & -1 & -1 & 2 & 1 & 0 \\ - & 9 & 0 & 8 & 3 & 2 & 1 \\ - & 4 & 0 & 0 & 3 & 2 & 1 \\ - & 0 & -1 & -1 & 0 & 1 & -1 \\ - & -1 & -2 & -2 & 1 & 0 & -1 \\ - & 4 & 3 & 3 & 6 & 2 & 0 \end{bmatrix} \quad \text{and} \quad P_4 = \begin{bmatrix} - & 3 & 3 & 4 & 5 & 3 & 3 \\ 1 & - & 3 & 4 & 5 & 3 & 7 \\ 1 & 2 & - & 4 & 5 & 6 & 7 \\ 1 & 2 & 3 & - & 3 & 3 & 3 \\ 1 & 2 & 2 & 2 & - & 3 & 7 \\ 1 & 2 & 2 & 2 & 2 & - & 2 \\ 1 & 2 & 3 & 2 & 2 & 6 & - \end{bmatrix}$$

A path of minimum distance from 7 to 4 (which will not pass through 5 and 6) will have a weight of 3, which is the (7, 4) entry in the weight matrix. Using the path matrix, we see that this path is $7 \to 2 \to 1$ with weight $4 + (-1) = 3$. As we saw in Problem 5.6, the SP from 7 to 4 that allows any vertex to be an intermediate vertex has weight 0. Thus the path is $7 \to 6 \to 2 \to 4$ with total weight $2 + (-1) + (-1)$.

THE STEINER NETWORK PROBLEM

5.11 Prove Theorem 3: Let $\{1, 2, 3, \ldots, n\}$ be the vertex set of the complete graph G with a nonnegative weight function w satisfying the triangle property $w(\{i, j\}) \leq w(\{i, k\}) + w(\{k, j\})$ for any three vertices i, j, and k. If W is a set of m vertices in network G, there exists a Steiner tree for W in the network that contains at most $(n - 2)$ Steiner points.

Solution. Let the number of Steiner points of W with respect to a tree T be p. Then T has $(m + p)$ vertices and $(m + p - 1)$ edges. If x is the average degree (in T) of a Steiner point and y is the average degree (in T) of a non-Steiner point, the degree sum $px + my$ is equal to $2(m + p - 1)$. At the same time, $x \geq 3$ (because of the triangle property), and y is at least 1. Hence, $p \leq (m - 2)$.

5.12 Find a Steiner tree for the set $W = \{1, 2, 3, 4\}$ in the network shown in Fig. 5-4.

Solution. Here $m = 4$. The subsets of vertices to be considered are $\{1, 2, 3, 4\}$, $\{1, 2, 3, 5\}$, $\{1, 2, 3, 4, 6\}$, $\{1, 2, 3, 4, 7\}$, $\{1, 2, 3, 4, 5, 6\}$, $\{1, 2, 3, 4, 5, 7\}$, and $\{1, 2, 3, 4, 6, 7\}$. The weights of the minimum spanning trees induced by these sets on a complete network with the weight matrix as the SD matrix of the given network are 8, 8, 10, 10, 9, 9, and 11, respectively. So the weight of a Steiner tree is 8. Vertex 5 is the Steiner point, and the corresponding tree has $\{1, 5\}$, $\{2, 5\}$, $\{2, 3\}$, and $\{3, 4\}$ as edges with weights 1, 2, 2, and 3, giving a total of 8.

5.13 Find the Steiner points of the set $\{1, 3, 5\}$ in the network of Problem 5.5.

Solution. From the SD matrix, we see that the weights of the minimum spanning trees in G' induced by $\{1, 3, 5\}$, $\{1, 3, 5, 2\}$, $\{1, 3, 5, 4\}$, $\{1, 3, 5, 6\}$, and $\{1, 3, 5, 7\}$ are 8, 7, 8, and 9, respectively. So in G' the Steiner point is at vertex 2. The M.S.T. in G' consists of edges $\{1, 2\}$, $\{2, 3\}$, and $\{2, 5\}$. Edge $\{2, 5\}$ is the shortest path, $2 — 7 — 5$, in G. So the Steiner points are 2 and 7. The edges of the Steiner tree are $\{1, 2\}$, $\{2, 3\}$, $\{2, 7\}$, and $\{5, 7\}$.

FACILITY LOCATION PROBLEMS

5.14 Find the median of the network of Problem 5.5.

Solution. The row sums in the SD matrix are 17, 14, 19, 25, 26, 22, and 17. The minimum is for vertex 2. So the median is the set $\{2\}$.

5.15 Find the weighted median of the network of Problem 5.5 if the weights of the vertices are 1, 0, 3, 3, 3, 1, and 2, respectively.

Solution. The row sums in this case are 44, 34, 34, 36, 41, 55, and 38. The median is the set $\{2, 3\}$.

5.16 Find the center of the network of Problem 5.5.

Solution. The maximum entries in the rows of the SD matrix are 5, 4, 5, 6, 6, 6, and 5. The minimum is for row 2. So the center is the set $\{2\}$.

5.17 In a network, the minimum value of the eccentricity is the **radius $r(G)$**, and the maximum value is the **diameter $d(G)$**. Show that $r(G) \leq d(G) \leq 2r(G)$.

Solution. Of course, $r(G) \leq d(G)$. Let the SD between two vertices u and v be denoted by $d(u, v)$. There is a vertex x such that $d(x, v) \leq r(G)$ for every vertex v. There are also vertices p and q such that $d(p, q) = d(G)$. Now $d(p, q) \leq d(p, w) + d(w, q) \leq r(G) + r(G)$.

5.18 Show that if each edge of a tree has a nonnegative weight, the cardinality of its center (median) is at most 2.

Solution. If w is a vertex of degree 1 and if w is adjacent to v, the degree of w is at least 2. Moreover, $d(w, u) \geq d(v, u)$, where u is any other vertex. Thus $e(w) \geq e(v)$. Consequently, if T' is the tree obtained by deleting all vertices of degree 1 from T, both trees will have the same center. Then we delete all vertices of degree 1 from T'. Continue this process. Eventually, we get a tree with either one vertex or two vertices. The proof is analogous for median.

Supplementary Problems

5.19 List the arcs of the shortest distance arborescence rooted at vertex 1 of the network with $V = \{1, 2, 3, 4, 5, 6\}$ and $E = \{(1, 2), (1, 3), (1, 4), (2, 3), (2, 5), (3, 6), (4, 5), (5, 6)\}$ with weights 4, 7, 3, 3, 2, 2, 3, and 2, respectively. *Ans.* (1, 2), (1, 3), (1, 4), (4, 5), and (5, 6)

5.20 List the arcs of the shortest distance arborescence rooted at vertex 1 of the network with $V = \{1, 2, 3, 4, 5, 6\}$ and $E = \{(1, 2), (1, 5), (2, 3), (2, 4), (2, 5), (3, 4), (4, 5), (4, 6), (5, 3)\}$ with weights 3, 7, 7, 2, 5, 3, 2, 5, and 1, respectively. *Ans.* (1, 2), (2, 4), (4, 5), (5, 3), and (4, 6)

5.21 Find the SD matrix A_5 and the SP matrix P_5 of the undirected network with $V = \{1, 2, 3, 4, 5\}$ and $E = \{\{1, 2\}, \{1, 5\}, \{2, 3\}, \{2, 4\}, \{3, 4\}, \{4, 5\}\}$ with weights 1, 4, 5, 1, 2, and 1, respectively.

Ans.
$$A_5 = \begin{bmatrix} 0 & 1 & 4 & 2 & 3 \\ 1 & 0 & 3 & 1 & 2 \\ 4 & 3 & 0 & 2 & 3 \\ 2 & 1 & 2 & 0 & 1 \\ 3 & 2 & 3 & 1 & 0 \end{bmatrix}; \quad P_5 = \begin{bmatrix} 1 & 2 & 2 & 2 & 2 \\ 1 & 2 & 4 & 4 & 4 \\ 4 & 4 & 3 & 4 & 4 \\ 2 & 2 & 3 & 4 & 5 \\ 4 & 4 & 4 & 4 & 5 \end{bmatrix}$$

5.22 Find the SD matrix and the SP matrix of the network with the following weight matrix:

$$\begin{bmatrix} — & 15 & — & — & 9 & — \\ — & — & 35 & 3 & — & — \\ — & — & — & 6 & — & 21 \\ — & — & — & — & 2 & 7 \\ — & 4 & — & 2 & — & — \\ — & — & 5 & — & — & — \end{bmatrix}$$

Ans.
$$A_6 = \begin{bmatrix} — & 13 & 23 & 11 & 9 & 18 \\ — & — & 15 & 3 & 5 & 10 \\ — & 12 & — & 6 & 8 & 13 \\ — & 6 & 12 & — & 2 & 7 \\ — & 4 & 14 & 2 & — & 9 \\ — & 17 & 5 & 11 & 13 & — \end{bmatrix}; \quad P_6 = \begin{bmatrix} — & 5 & 5 & 5 & 5 & 5 \\ — & — & 4 & 4 & 4 & 4 \\ — & 4 & — & 4 & 4 & 4 \\ — & 5 & 6 & — & 5 & 6 \\ — & 2 & 4 & 4 & — & 4 \\ — & 3 & 3 & 3 & 3 & — \end{bmatrix}$$

5.23 Find the weight of a Steiner tree of the set {1, 2, 3} in the network whose weight matrix is

$$\begin{bmatrix} - & 6 & 5 & 1 & 3 & - & - \\ 6 & - & - & - & 2 & 3 & - \\ 5 & - & - & 3 & - & 7 & 2 \\ 1 & - & 3 & - & 2 & - & 1 \\ 3 & 2 & - & 2 & - & - & 4 \\ - & 3 & 7 & - & - & - & 4 \\ - & - & 2 & 1 & 4 & 4 & - \end{bmatrix}$$

Ans. 8

Chapter 6

Flows, Connectivity, and Combinatorics

6.1 FLOWS IN NETWORKS AND MENGER'S THEOREM

The Maximum Flow Problem

A digraph with an integer-valued function c (known as the **capacity function**) defined on its set of arcs is called a **capacitated network.** Two vertices in the network are specially designated: the **source,** with indegree zero, and the **sink,** with outdegree zero. Every other vertex is an **intermediate vertex.** We assume that the set of vertices is $\{1, 2, \ldots, n\}$ in which vertex 1 is the source and vertex n is the sink. If $e = (i, j)$ is an arc in G, the integer $c(e) = c(i, j)$ is the **capacity** of the arc. It is assumed that the capacity of each arc is nonnegative. A **flow** f in the network is an integer-valued function defined on its set of arcs such that $0 \leq f(e) \leq c(e)$ for each arc e. The integer $f(e)$ is the **flow along arc e.** The sum of the flows along all the arcs directed to vertex i is the **inflow into i,** and the sum of the flows along all the arcs directed from vertex i is the **outflow from i.** A flow is a **feasible flow** if it satisfies the **conservation condition:** the inflow into i is equal to the outflow from i for every vertex i other than the source and the sink. If f is a feasible flow in a capacitated network G, the **value $f(G)$ of the flow** is the outflow from the source. A feasible flow in a capacitated network such that the value of the flow is as large as possible is called a **maximum flow** in the network. The problem of finding a feasible flow in a network such that its flow value is maximum is known as the maximum flow problem.

[If there are arcs in either direction between a pair of vertices in a network, a new vertex can be inserted on one of the two arcs, replacing that arc by two arcs in the same direction with the same capacity. So without loss of generality, it can be assumed that the digraph is asymmetric; that is, for distinct vertices i and j, not both (i, j) and (j, i) are arcs.]

Example 1. Figure 6-1 shows a digraph representing a capacitated network (with 1 as the source and 4 as the sink) and a feasible flow defined on its set of arcs. Along each arc are two integers separated by a comma. The first number is the flow along the arc, and the second is its capacity. The flow value is 3, and it is easy to see that it can be increased by four more units using arc (1, 4). So the feasible flow as depicted is not a maximum flow.

Cuts in a Capacitated Network

Consider any partition of the vertex set of a capacitated network $G = (V, E)$ into two sets S and T such that the source is in S and the sink is in T. The set $(S, T) = \{(i, j) : (i, j) \in E, i \in S, j \in T\}$ is called a **cut** (more appropriately, a **source-sink cut**) in the network since no flow can be sent from the source to the sink if all the arcs in the cut are deleted. The sum of the capacities of all the arcs in cut (S, T) is the **capacity $c(S, T)$** of the cut. A cut is called a **minimum cut** if its capacity does not exceed the capacity of any other cut. If f is a feasible flow in the network, the sum of the flows along all the arcs in cut (S, T) is the **flow $f(S, T)$** along the cut. Obviously, $0 \leq f(S, T) \leq c(S, T)$.

[Observe that every capacitated network (V, E) with source s and sink t always has a feasible flow and a source-sink cut: the cut $(\{s\}, V - \{s\})$ is a source-sink cut, and the trivial flow $f = 0$ is a feasible flow.]

Theorem 6.1. If f is any feasible flow in a capacitated network G and if (S, T) is any cut in the network, $f(G) = f(S, T) - f(T, S)$.

Corollary 1: The value $f(G)$ of any feasible flow f in a network is also equal to the inflow into the sink.

Corollary 2: If f is any feasible flow and if (S, T) is any cut, $f(G) \leq c(S, T)$. (See Solved Problems 6.2 and 6.3.)

Example 2. In the network shown in Fig. 6-2, both the inflow into the sink and the outflow from the source for the current flow f are 8, which is the current flow value. If $S = \{1, 2, 3\}$ and $T = \{4, 5, 6\}$, the flow value is also equal to $f(S, T) - f(T, S) = (3 + 4 + 7) - (0 + 6) = 8$. Furthermore, the flow value is less than or equal to the capacity

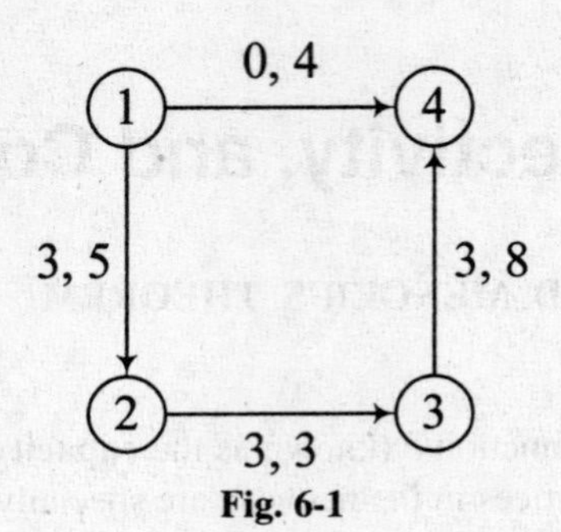

Fig. 6-1

$(7 + 4 + 8) = 19$ of this cut. Suppose $S = \{1, 4, 5\}$ and suppose T is its complement. The flow value again is $f(S, T) - f(T, S) = (5 + 0 + 1 + 6) - 4 = 8$, and the capacity of this cut is $(8 + 4 + 5 + 9) = 26$, which also exceeds the flow value.

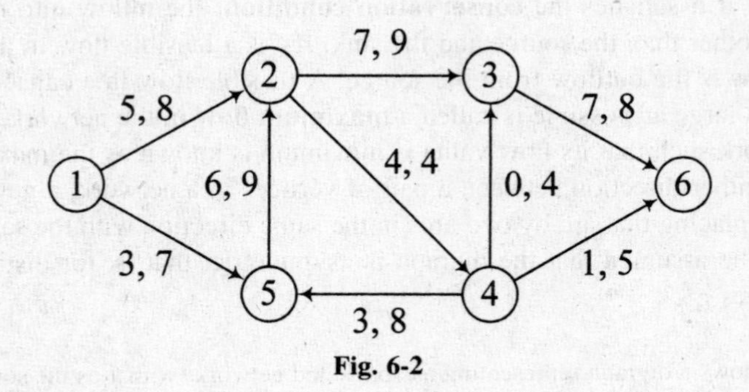

Fig. 6-2

If f is a feasible flow in a capacitated network, the arc (i, j) is ***f*-saturated** if $f(i, j) = c(i, j)$, is ***f*-zero** if $f(i, j) = 0$, and is ***f*-positive** if $f(i, j)$ is positive and less than $c(i, j)$.

Theorem 6.2. (*a*) If f is a feasible flow and if (S, T) is any cut, $f(G) = c(S, T)$ if and only if every arc in (S, T) is f-saturated and every arc in cut (T, S) is f-zero. (*b*) If f is a feasible flow and if (S, T) is any cut such that $f(G) = c(S, T)$, f is a maximum flow and (S, T) is a minimum cut. (See Solved Problem 6.4.)

Example 3. In the digraph shown in Fig. 6-3, the flow value is 12. The capacity of cut (S, T), where $S = \{1, 2, 3, 5\}$ and $T = \{4, 6\}$, is also 12. The arcs in (S, T) are (2, 4) and (3, 6). Both these arcs are f-saturated. The arcs in (T, S) are (4, 5) and (4, 3). Both these arcs are f-zero. The current flow is a maximum flow, and the cut is a minimum cut.

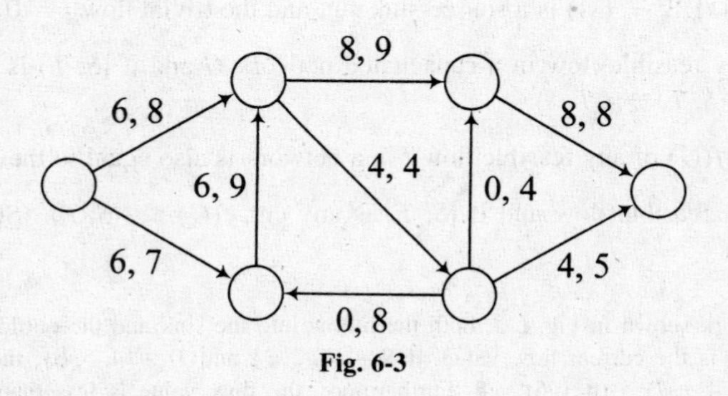

Fig. 6-3

Flow-Augmenting Paths

An alternating sequence P of vertices and arcs of the form $v_0, e_1, v_1, e_2, \ldots, e_k, v_k$ in which no vertex is repeated is called a **semipath** from v_0 to v_k. Arc e_i in this path is a **forward arc** in P if it is directed to v_i. Otherwise, it is a **backward arc** in P. A semipath is ***f*-unsaturated** if no forward arc is f-saturated and no backward arc is f-free. An f-unsaturated path from the source to the sink is called an ***f*-augmenting path.** For each arc e_i in an f-augmenting path P, define $\delta_i(P)$ to be $c(e_i) - f(e_i)$ if e_i is a forward arc and $f(e_i)$ if e_i is a backward arc. The **excess flow capacity of semipath P is $\delta(P)$** $= \min\{\delta_i(P) : e_i \in P\}$, which is a positive integer.

Example 4. In the semipath displayed in Fig. 6-4, the source is 1 and the sink is 9. The only backward arc is the arc from 3 to 4. The excess flow capacity of this path is the minimum in the set $\{8 - 3, 7 - 6, 2, 9 - 4\}$, which is 1.

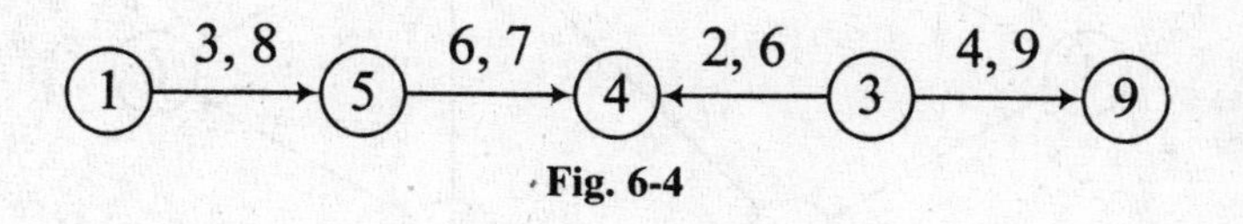

Fig. 6-4

If P is an f-augmenting path with excess flow capacity $\delta(P)$, a new feasible flow f' can be obtained by increasing the flow along each forward arc by $\delta(P)$ and at the same time decreasing the flow along each backward arc by $\delta(P)$ such that $f'(G) = f(G) + \delta(P)$. Once this is accomplished, the semipath is no longer an augmenting path. In the semipath of Example 4, we can increase the flow value by increasing the flow along each forward arc by one unit and at the same time decreasing the flow from the backward arc by one unit. In other words, if there is an f-augmenting path in a network, flow f is not a maximum flow since its flow value can be increased using this augmenting path. It turns out that the converse of this assertion is also true.

Theorem 6.3. A flow f in a capacitated network is a maximum flow if and only if there is no f-augmenting path in the network. (See Solved Problem 6.5.)

Theorem 6.4 **(Ford–Fulkerson Theorem).** In a capacitated network, the value of a maximum flow is equal to the capacity of a minimum cut. This theorem is also known as the **max-flow min-cut theorem.** (See Solved Problem 6.6.)

The Edmonds–Karp Algorithm to Solve the Maximum Flow Problem

Input: $G = (V, E)$ is a capacitated network with a capacity function c and an initial feasible flow f (which could be the trivial flow). The set V is $\{1, 2, \ldots, n\}$ in which the source is 1 and the sink is n. The capacity of each arc is assumed to be a positive integer.

Step 1. Construct a digraph $D(f) = (V, E')$ as follows: (*a*) If (i, j) is an f-saturated arc in E, (j, i) is an arc in E'. (*b*) If (i, j) is an f-zero arc in E, (i, j) is also an arc in E'. (*c*) If (i, j) is an f-positive arc in E, both (i, j) and (j, i) are arcs in E'.

Step 2. Starting from the source, apply a breadth first search (BFS) in $D(f)$ to obtain a directed path (with a minimum number of arcs) from the source to the sink. If there is no such path, go to step 4.

Step 3. The directed path in the digraph $D(f)$ from the source to the sink n obtained in step 2 is a semipath in G with positive excess flow capacity, so it is an f-augmenting path. Increase the flow value in the network by using this path. Go to step 1.

Step 4. The maximum flow value is the outflow from the source. Let S be the set of vertices that can be reached from vertex 1 in $D(f)$, and let T be its complement. (S, T) is a minimum cut. (See Solved Problem 6.7.)

Vertex-Capacitated Networks

In a capacitated network, each arc is associated with a positive integer known as its capacity. If, in addition, each vertex also is associated with a positive integer, the network is called a vertex-capacitated network. A feasible flow in the network is known as a **generalized feasible flow** if the outflow at each intermediate vertex does not exceed the capacity of that vertex. As before, there is a unique vertex known as the source and another unique vertex known as the sink. A set X of arcs and vertices is called a **generalized source-sink cut** if every directed path from the source to the sink will contain at least one element from X. In the network shown in Fig. 6-5, the set $\{(1, 2), (2, 3), (3, 6), 5\}$ is a generalized source-sink cut.

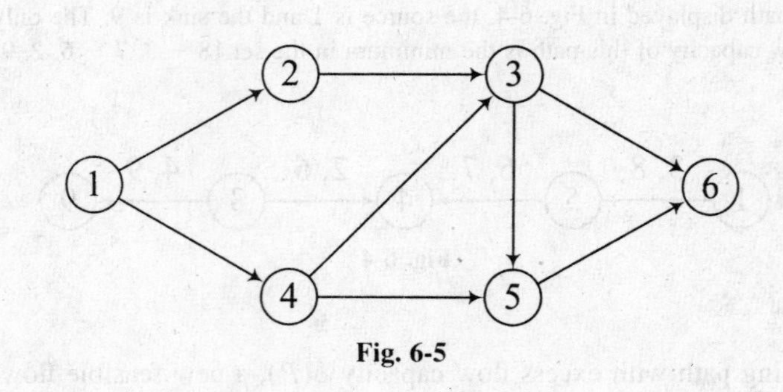

Fig. 6-5

The capacity of a generalized cut is the sum of the capacities of its elements. If the capacity of each vertex is infinite, no vertex can be an element of a minimum generalized cut. Suppose X is a minimum generalized cut and S is the set of all vertices that can be reached from the source without using an arc from X. Then $(S, V - S)$ is a cut such that every arc in it is an arc in X. Thus a minimum cut in the usual sense is a minimum cut in the generalized sense when the capacity of each vertex is infinite.

Theorem 6.5 **(Generalized Max-Flow Min-Cut Theorem).** The maximum value of a generalized source-sink flow in a vertex-capacitated network G is equal to the capacity of a minimum generalized source-sink cut. (See Solved Problem 6.11.)

Example 5. A generalized maximum flow and a generalized minimum cut in the network represented by Fig. 6-6 are as follows: the generalized cut is $\{(2, 4), 3\}$, and the flow is as indicated on the arcs.

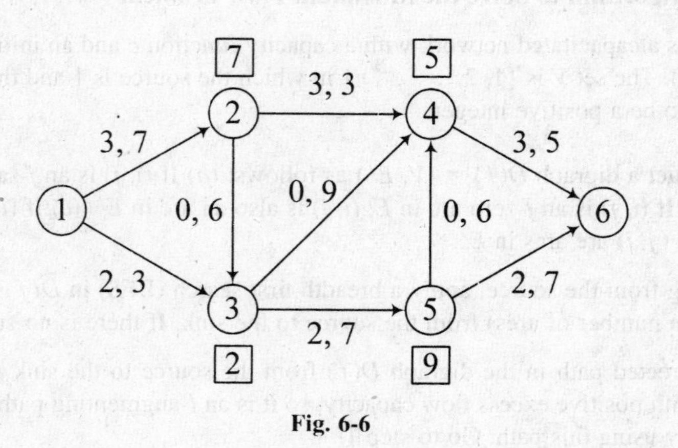

Fig. 6-6

Two paths from source s to sink t in a network are called **internally disjoint paths** from s to t if they have no vertices in common other than s and t. Any two paths are called **arc-disjoint paths (edge-disjoint paths** in the case of undirected graphs) if they have no arcs (edges) in common.

Theorem 6.6 **(Menger's Theorem).**

(*a*) *Vertex form for undirected graphs:* The maximum number of internally disjoint paths between any two nonadjacent vertices in a graph is equal to the minimum number of vertices whose deletion results in a graph in which there are no paths between those two vertices.

(*b*) *Edge form for undirected graphs:* The maximum number of edge-disjoint paths between two vertices s and t in a graph is equal to the minimum number of edges whose deletion results in a graph in which there are no paths between the two vertices.

(*c*) *Vertex form for directed graphs:* The maximum number of internally disjoint paths from vertex s to vertex t in a digraph in which there are no arcs from s to t is equal to the minimum number of vertices whose deletion results in a digraph in which there are no paths from s to t.

(*d*) *Arc form for directed graphs:* The maximum number of arc-disjoint paths from vertex s to vertex t in a digraph is equal to the minimum number of arcs whose deletion results in a digraph in which there are no paths from s to t.

(See Solved Problems 6.13 through 6.16.)

Theorem 6.7. The vertex form of Menger's theorem, the arc (edge) form of Menger's theorem, and the Ford–Fulkerson theorem are equivalent. (See Solved Problem 6.19.)

6.2 MORE ON CONNECTIVITY

Vertex Connectivity

Recall that a graph is connected if and only if there is a path between every pair of vertices in it. Since loops do not have any significance in flow problems as well as in problems related to connectivity, we assume that the graphs and digraphs we investigate in this chapter have no loops. A set W of vertices in a graph $G = (V, E)$ is a **separating set** (also known as **a vertex cut**) **of** $\boldsymbol{G}$ if $G - W$ has more than one component. If a separating set consists of a single vertex w, w is known as a **cut vertex** (or **articulation vertex**). The **connectivity number** $\boldsymbol{\kappa(G)}$ of a graph G is the minimum size of a separating set in it. Since a complete graph has no separating set, we adopt the convention that the connectivity number of a complete graph of order n is $(n - 1)$ for all n. A graph G is said to be ***k*-connected** if $\kappa(G) \geq k$. Thus K_n is $(n - 1)$-connected for all n, and a graph that is not complete is k-connected if and only if every separating set in it has at least k vertices. The connectivity number of a graph is zero if and only if it is either a disconnected graph or the trivial graph.

Theorem 6.8. A graph of order n is k-connected (where $1 \leq k \leq n - 1$) if the degree of each vertex is at least $(n + k - 2)/2$. (See Solved Problem 6.37.)

Example 6. In the graph of Fig. 6-7, there are six vertices. If we take $k = 2$, we see that the degree of each vertex is at least $(6 + 2 - 2)/2 = 3$. So the graph is 2-connected. Notice that the deletion of the vertices 2 and 5 results in a disconnected graph. Thus the connectivity number of the graph is 2. Since $(6 + 3 - 2)/2$ is more than 3, the graph is not 3-connected because there is a vertex of degree 3 in the graph.

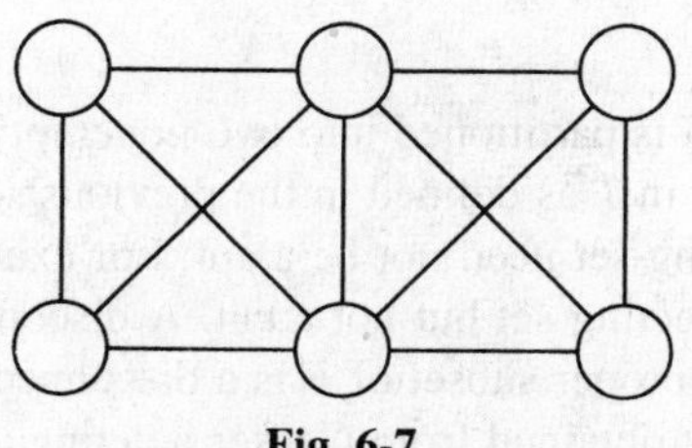

Fig. 6-7

A **nonseparable graph** is a connected nontrivial graph with no cut vertex. A **block** of a graph G is a maximal nonseparable subgraph H; that is, if H' is a nonseparable subgraph of G such that H is a subgraph of H', $H = H'$. Observe that a graph is 2-connected if and only if it is a block with at least two edges. K_2 is the only block that is not 2-connected. Any cycle in a graph is a block of that graph. If H is a subgraph of G, it is not necessary that $\kappa(H) \leq \kappa(G)$; consider the case that H is a block of a 1-connected graph G that is not 2-connected. Obviously, if H is a spanning subgraph of G, $\kappa(H) = \kappa(G)$. (See Problem 6.32 for a complete characterization of blocks due to Harary.)

As mentioned before, a graph is connected (that is, 1-connected) if and only if there is at least one path between every pair of vertices in it. More generally, a graph is k-connected if and only if there are at least k paths between every pair of vertices in it. The pleasant fact is that no two of these paths have any intermediate vertex in common. This remarkable characterization of k-connected graphs, which can be easily established using (the vertex form) of Menger's theorem, is due to Whitney.

Theorem 6.9 **(Whitney's Theorem).** A graph with at least $(k + 1)$ vertices is k-connected if and only if any two distinct vertices in the graph are connected by at least k internally disjoint paths. In particular, a graph with at least three vertices is a block if and only if every two vertices lie on a common cycle. (See Solved Problem 6.30.)

Example 7. In the graph of Fig. 6-7, the number of internally disjoint paths between every pair of vertices is 2. So, by Theorem 6.9, the graph is 2-connected but not 3-connected.

Edge Connectivity

A set F of edges in a graph G is a **disconnecting set** if $G - F$ has more than one component. If a disconnecting set consists of a single edge, that edge is called a **bridge** (also known as a **cut edge** or an **isthmus**). A graph is said to be ***k* edge connected** if every disconnecting set has at least k edges. The **edge-connectivity number $\lambda(G)$** of a graph G is the minimum size of a disconnecting set in it and, *by definition,* is zero when G is the trivial graph. Thus $\lambda(G)$ is zero if and only if G is disconnected or trivial, and it is k edge connected if and only if $\lambda(G) \geq k$.

The following result is known as **Whitney's inequality.** For any graph G, $\kappa(G) \leq \lambda(G) \leq \delta(G)$, where $\delta(G)$ is the minimum vertex degree of the graph. (See Solved Problems 2.11 and 2.12. So any k-connected graph is k edge connected.

Example 8. In the graph G of Fig. 6-7, the minimum degree $\delta(G)$ is 3, and $\kappa(G) = 2$. So $\lambda(G) = 2$ or 3. The graph does not become disconnected by removing any set of two edges, but it becomes disconnected if the three edges adjacent to a vertex of degree 3 are deleted. So $\lambda(G) = 3$.

The following characterization of k-edge-connected graphs is the "edge version" of Theorem 6.9.

Theorem 6.10. A graph is k edge connected if and only if any two distinct vertices in it are connected by at least k edge-disjoint paths. (See Solved Problem 6.39.)

Example 9. In the graph of Fig. 6-7, joining any two vertices are three edge-disjoint paths. This confirms that its edge-connectivity number is 3.

Cuts and Cut Sets

If the set V of vertices in a graph G is partitioned into two nonempty subsets, cut (S, T) of all edges of the graph joining vertices in S and vertices in T as defined in the previous section is indeed a disconnecting set of the graph. But an arbitrary disconnecting set need not be a cut. For example, the set of the three edges in the complete graph of order 3 is a disconnecting set but not a cut. A disconnecting set F of the graph G is called a **cut set** (also known as a **bond**) if no proper subset of F is a disconnecting set. Suppose F is such a minimal disconnecting set and H is a component obtained from G after deleting all the edges of F from the graph. If W

is the set of vertices of H, the minimality of F implies that F is cut $(W, V - W)$. Thus every cut set is a cut. But the converse is not true: the set of two edges in $K_{1,2}$ forms a cut, but it is not a cut set. Since every cut set is a cut as well as a minimal disconnecting set, the edge-connectivity number of a graph is the minimum size of a cut. So the following characterization of the class of k-edge-connected graphs is obvious.

Theorem 6.11. A graph is k edge connected if and only if the number of edges in any cut is at least k.

6.3 SOME APPLICATIONS TO COMBINATORICS

Matchings and Coverings

The number of vertices in a maximum independent set in graph G is denoted by $\alpha(G)$, and the number of vertices in a minimum vertex cover in G is denoted by $\beta(G)$. Likewise, the number of edges in a maximum matching in a graph G is denoted by $\alpha_1(G)$, and the number of edges in a minimum edge cover is denoted by $\beta_1(G)$. It has been proved that $\alpha(G) + \beta(G) = \alpha_1(G) + \beta_1(G) = n$, where n is the order of G. See Solved Problems 2.19, 2.20, 2.26, 2.27, and 2.28.

Theorem 6.12 **(Konig's Theorem).** In a bipartite graph G, $\alpha_1(G) = \beta(G)$. [Consequently, $\alpha(G) = \beta_1(G)$ if G has no vertex of degree 0. This equality is known as Konig's other theorem.] (See Solved Problems 6.44 and 6.46.)

Example 10. In the bipartite graph of Fig. 6-8, a maximum matching consists of three edges $\{1, 6\}$, $\{2, 8\}$, and $\{5, 9\}$. A minimum vertex cover consists of vertices 1, 5, and 8. The six vertices $\{2, 3, 4, 6, 7, 9\}$ form a maximum independent set, and the set of all edges in the graph except edge $\{5, 8\}$ forms a minimum edge cover.

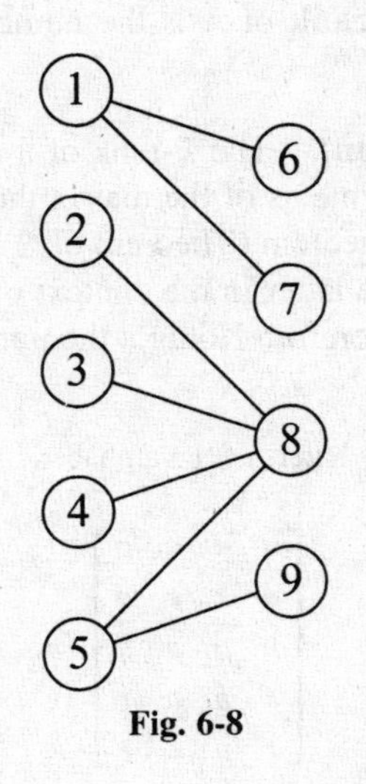

Fig. 6-8

In a bipartite graph (X, Y, E), a **complete matching from X to Y** is a matching M such that every vertex in X is incident to an edge in M, and a **perfect matching** is a matching that is complete from X to Y as well as from Y to X. If both X and Y have the same number of elements, a complete matching from one to the other is a perfect matching.

Systems of Distinct Representatives

Suppose we have a family of N sets. The possibility that two sets in the family are the same is not to be ruled out. In other words, the family could be just a collection, not a set. If it is possible to select one element from each set in the family such that the N selected elements constitute a set X, X is called a **system of distinct representatives (SDR)** of the family. For example, if the family consists of the sets $\{1, 2, 3\}$, $\{3, 4\}$, $\{1, 2, 3\}$, and $\{5, 6, 7\}$, one can select 1, 3, 2, and 5, respectively, from these four sets to form an SDR of the family.

Theorem 6.13 **(Hall's Marriage Theorem).** A family of N sets (not necessarily distinct) will have a system of distinct representatives if and only if the following condition, known as the **marriage condition (MC),** is satisfied: The union of any subfamily of k sets from the family should have at least k elements for every k from the set $\{1, 2, \ldots, N\}$. (Konig's theorem implies this theorem. See Solved Problem 6.47.)

This theorem, which was first proved by Philip Hall in 1935 settling the following issue known as the marriage problem, has several proofs. Suppose there are m women and n men in a party, where $m \leq n$. A necessary and sufficient requirement that we can form a set of m couples (each couple consisting of a woman and a man known to each other) is that every set of k women in the party collectively know at least k men in the party for every choice of k. The following theorem is a graph-theoretic version of Hall's marriage theorem.

Theorem 6.14. In the bipartite graph (X, Y, E), a complete matching from X to Y exists if and only if $|f(A)| \geq |A|$ for every subset A of X, where $f(A)$ is the set of those vertices in Y that are adjacent to at least one vertex in A.

The P-Rank of a Matrix

A **line** in a matrix A is either a row or a column. Suppose P is a property that an element in A may not have. A collection of elements in the matrix satisfying the property P is ***P*-independent** if no two elements in the collection lie on the same line. The ***P*-rank** of A is the number of elements in a largest P-independent collection in A.

Theorem 6.15 **(Konig–Egervary Theorem).** The P-rank of a matrix is equal to the minimum number of lines that contain all the elements of the matrix that possess the property P. [Notice that both this theorem and Konig's theorem (Theorem 6.12) make the same assertion: the former in the context of matrices, and the latter in the context of bipartite graphs. Hall's marriage theorem implies this theorem and therefore Konig's theorem. See Solved Problem 6.48.]

Example 11. In the matrix consisting of a few letters of the alphabet,

$$\begin{bmatrix} a & e & a & p \\ q & r & e & q \\ p & p & u & q \\ s & p & a & r \\ t & p & e & e \end{bmatrix}$$

an element is supposed to have the property P if it is a vowel. All the vowels in the matrix can be covered by three lines: row 1, row 5, and column 3. A largest P-independent set consists of the first vowel a in row 1, the vowel e from row 2, and the last vowel e from row 5. Thus the P-rank is 3, which is equal to the minimum number of lines needed to cover all the vowels in the matrix.

Theorem 6.16. Konig's theorem implies Menger's theorem. (See Solved Problem 6.49.)

Theorem 6.17 **(Konig's Marriage Theorem).** If a bipartite graph $G = (X, Y, E)$ is k-regular (where k is positive), there is a perfect matching in the graph. (Hall's marriage theorem implies this theorem. See Solved Problem 6.50.)

Partially Ordered Sets and Dilworth's Theorem

A **partially ordered set** (or **poset**) consists of a set X and an order relation $\leq$ among its elements satisfying the following three properties: (*a*) reflexivity: $x \leq x$ for every x in X, (*b*) antisymmetry: $x \leq y$ and $y \leq x$ imply that $x = y$, and (*c*) transitivity: $x \leq y$ and $y \leq z$ imply that $x \leq z$. Given a poset with a finite number of elements, one can construct a digraph G such that there is a one-to-one correspondence between the set of elements in the poset and the set of vertices of the digraph and such that an arc is drawn from the vertex that corresponds to element x to the vertex that corresponds to element y if and only if $x \leq y$. If all the arcs that are present in G due to transitivity are deleted, we get a (unique) subgraph H known as the **Hasse diagram** of the poset.

Example 12. The digraph of Fig. 6-9 represents a poset with 13 elements. There is an arrow from 1 to 6 and an arrow from 6 to 9. In addition to the relations $1 \leq 6$ and $6 \leq 9$, we also have, due to transitivity, the relation $1 \leq 9$, which is not explicitly shown by an arrow from 1 to 9.

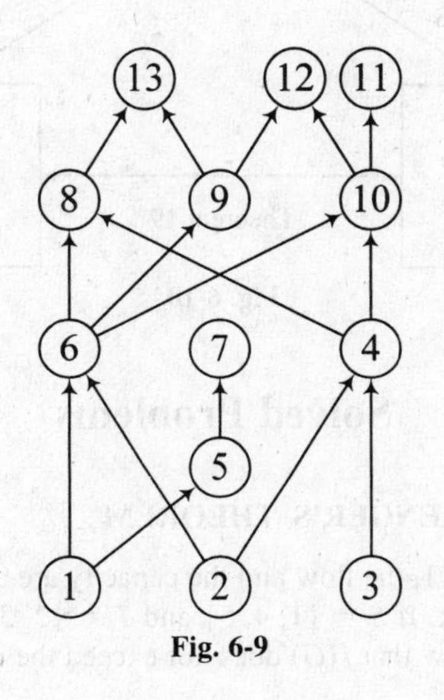

Fig. 6-9

Two elements x and y in a poset are **comparable** if either $x \leq y$ or $y \leq x$. Otherwise, they are **incomparable.** A set of elements belonging to a poset is called a **chain** (or **linear order**) in the poset if every pair in the set is a comparable pair. A set C of n elements in a poset is a chain if and only if the elements in C can be labeled as y_i $(i = 1, 2, \ldots, n)$ such that $y_1 \leq y_2 \leq \cdots \leq y_n$. An **antichain** in a poset, on the other hand, is a set in which no two elements are comparable.

Theorem 6.18 **(Dilworth's Theorem).** In a finite poset, the maximum size of an antichain is equal to the minimum number of chains into which the set of elements of the poset can be partitioned. (The Konig–Egervary theorem implies this theorem. See Solved Problem 6.51.)

Example 13. In the poset represented in Fig. 6-9, the set of the 13 elements can be partitioned into four chains $\{1, 5, 7\}$, $\{2, 6, 8, 13\}$, $\{3, 4, 10, 11\}$, and $\{9, 12\}$. The set $\{7, 8, 9, 10\}$ is an antichain with four elements in the poset.

Theorem 6.19. Dilworth's theorem implies Hall's marriage theorem. (See Solved Problem 6.52.)

Theorem 6.20 **(The Equivalence Theorem).** Menger's theorem, the Ford–Fulkerson theorem, Konig's theorem, the Konig–Egervary theorem, Dilworth's theorem, and Hall's marriage theorem are equivalent.

Some of the implications connecting these six famous theorems and resulting in their equivalence are indicated in Fig. 6-10.

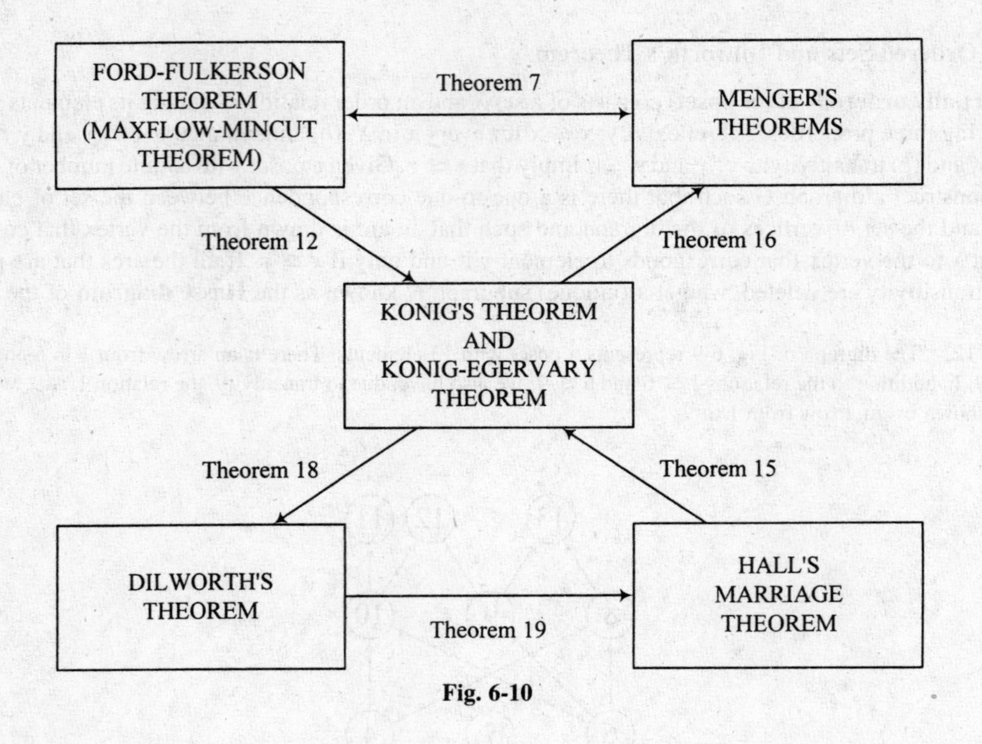

Fig. 6-10

Solved Problems

FLOWS IN NETWORKS AND MENGER'S THEOREM

6.1 In network G shown in Fig. 6-11, the flow and the capacity are as indicated on the arcs. Vertex 1 is the source, and vertex 6 is the sink. If $S = \{1, 4, 5\}$ and $T = \{2, 3, 6\}$, show that the flow value $f(G)$ is equal to $f(S, T) - f(T, S)$. Show that $f(G)$ does not exceed the capacity $c(S, T)$ of cut (S, T).

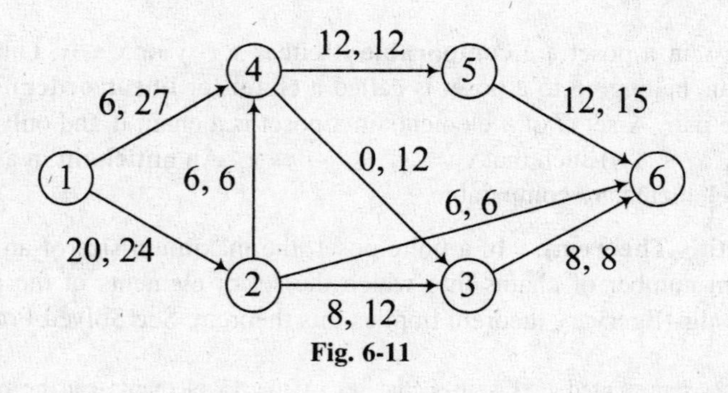

Fig. 6-11

Solution. Here the flow value $f(G)$ is $6 + 20 = 12 + 6 + 8 = 26$, $f(S, T) = 20 + 0 + 12 = 32$, and $f(T, S) = 6$. Thus $f(G) = f(S, T) - f(T, S)$. The capacity of cut (S, T) is $27 + 12 + 15 = 54$.

6.2 Prove Theorem 6.1: If f is any feasible flow in a capacitated network and if (S, T) is any cut in the network, $f(G) = f(S, T) - f(T, S)$.

Solution. The vertex set is $V = \{1, 2, \ldots, n\}$. S is any set of vertices that contains vertex 1 (the source), and T is its complement that contains vertex n (the sink). Notice that $\Sigma_j f(i, j) - \Sigma_j f(j, i)$ is $f(G)$ when $i = 1$. So $\Sigma_{i \in S} \Sigma_j f(i, j) - \Sigma_{i \in S} \Sigma_j f(j, i) = f(G)$. If i and j are both in S, the term $f(i, j)$ appears in the first summation

$\Sigma_{i \in S} \Sigma_j f(i, j)$ as well as in the second summation $\Sigma_{i \in S} \Sigma_j f(j, i)$. So it is enough if we let the subscript j vary for all j in T. Hence, $\Sigma_{i \in S} \Sigma_{j \in T} f(i, j) - \Sigma_{i \in S} \Sigma_{j \in T} f(j, i) = f(G)$.

6.3 Prove the two corollaries of Theorem 6.1. (*a*) Corollary 1: The value $f(G)$ of any feasible flow f in a network is also equal to the inflow into the sink. (*b*) Corollary 2: If f is any feasible flow and if (S, T) is any cut, $f(G) \leq c(S, T)$.

Solution. (*a*) In Theorem 6.1, let $S = V - \{n\}$. In that case $f(T, S)$ is zero, and $f(S, T)$ is the inflow into the sink. (*b*) $f(G) = f(S, T) - f(T, S) \leq f(S, T) \leq c(S, T)$.

6.4 Prove Theorem 6.2: (*a*) If f is a feasible flow and if (S, T) is any cut, $f(G) = c(S, T)$ if and only if every arc in (S, T) is f-saturated and every arc in cut (T, S) is f-zero. (*b*) If f is a feasible flow and if (S, T) is any cut such that $f(G) = c(S, T)$, f is a maximum flow and (S, T) is a minimum cut.

Solution.

(*a*) $f(G) = c(S, T)$ if and only if $f(S, T) - f(T, S) = c(S, T)$. This is possible if and only if every arc in cut (T, S) is f-zero and every arc in cut (S, T) is f-saturated.

(*b*) Let f' be a maximum flow and (S', T') be a minimum cut in the network. Then $f(G) \leq f'(G) \leq c(S', T') \leq c(S, T)$. If $f(G)$ and $c(S, T)$ are equal, we have the chain $f(G) = f'(G) = c(S', T') = c(S, T)$, which implies that f is indeed a maximum flow and (S, T) is a minimum cut.

6.5 Prove Theorem 6.3: A flow f in a capacitated network is a maximum flow if and only if there is no f-augmenting path in the network.

Solution. If there is an f-augmenting path in the network, the current flow value can be increased using this path; therefore, the current flow is not a maximum flow. So if the current flow f is a maximum flow, there are no f-augmenting paths. We now show that if there are no f-augmenting paths with respect to flow f, f is indeed a maximum flow. Let S be the set of all vertices i such that there is an f-unsaturated path from the source (vertex 1) to i. Obviously, 1 is a vertex in S and the sink (vertex n) is not a vertex in S. Thus there is a cut (S, T) in the network. Let (i, j) be any arc in this cut. Since there is no f-unsaturated path from the source to j, arc (i, j) is necessarily f-saturated. Likewise, any arc in cut (T, S) is f-zero. So the current flow value is equal to the capacity of cut (S, T). Hence, the current flow is a maximum flow.

6.6 Prove Theorem 6.4. (Ford–Fulkerson Theorem): In a capacitated network, the value of a maximum flow is equal to the capacity of a minimum cut.

Solution. Let f be a maximum flow, and let S be the set of vertices i such that there is an f-unsaturated path from the source to i. Then the complement T of S is nonempty; thus there is a cut (S, T). Each arc in this cut is f-saturated. Moreover, each arc in cut (T, S) is f-zero. So the flow value $f(G)$ is equal to $c(S, T)$, which is a minimum cut.

6.7 Let $G = (V, E)$ be a capacitated network with capacity function c and initial feasible flow f (which could be the trivial flow). The set V is $\{1, 2, \ldots, n\}$ in which the source is 1 and the sink is n. The capacity of each arc is assumed to be a positive integer. Construct a digraph $D(f) = (V, E')$ as follows. (1) If (i, j) is an f-saturated arc in E, (j, i) is an arc in E'. If (i, j) is an f-zero arc in E, (i, j) is also an arc in E'. (3) If (i, j) is an f-positive arc in E, both (i, j) and (j, i) are arcs in E'. Show that there is an f-augmenting path in the network if and only if there is a directed path from the source to the sink in $D(f)$, and show that a shortest source-sink path in $D(f)$ has the same length as a shortest f-augmenting path.

Solution. Let Q be a directed path in $D(f)$ of the form $v_0, e_1, v_1, e_2, \ldots, e_k, v_k$ in which no vertex is repeated. This alternating sequence of vertices and arcs forms a semipath Q' in the network. Arc e_i in Q from v_{i-1} to v_i corresponds to an arc in Q' that is either from v_{i-1} to v_i or in the opposite direction. In the former case, it is an f-unsaturated arc. In the latter case, it is an arc that is not f-zero. Thus no forward arc in Q' is f-saturated, and no backward arc is f-zero. So Q' is an f-unsaturated path. Similarly, it can be shown that any f-unsaturated path Q' in the network corresponds to a directed path Q in $D(f)$. Furthermore, both Q and Q' have the same number of arcs. So a shortest source-sink path in the digraph has the same number of arcs as a shortest f-augmenting path.

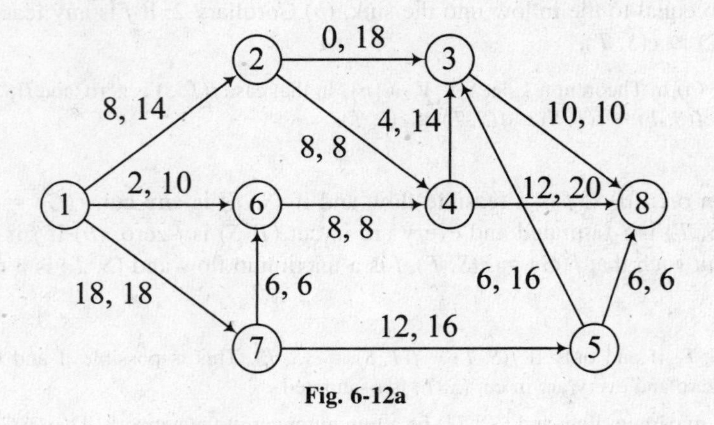

Fig. 6-12a

6.8 Obtain a maximum flow and a minimum cut in the network shown in Fig. 6.12(a).

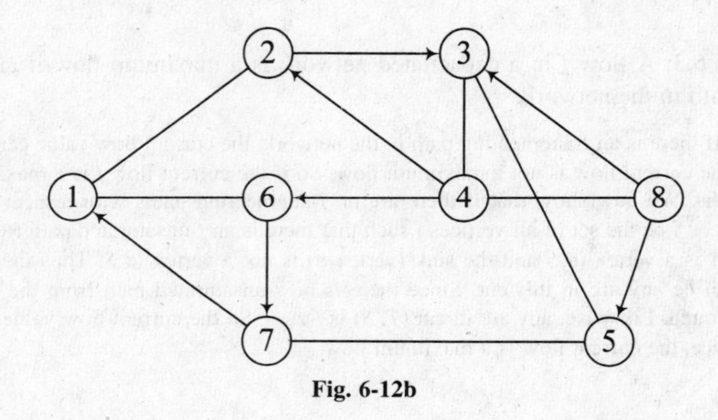

Fig. 6-12b

Solution. Here 1 is the source and 8 is the sink. The digraph $D(f)$ corresponding to the current flow in which the undirected edges are considered arcs in either direction is shown (for the sake of convenience) as a mixed graph in Fig. 6.12(b). In this digraph, $1 \rightarrow 2 \rightarrow 3 \rightarrow 4 \rightarrow 8$ is a directed path from the source to the sink, defining the flow augmenting (semi)path $1 \rightarrow 2 \rightarrow 3 \leftarrow 4 \rightarrow 8$ along which a flow of four units can be sent from 1 to 8. The updated network is shown in Fig. 6-12(c).

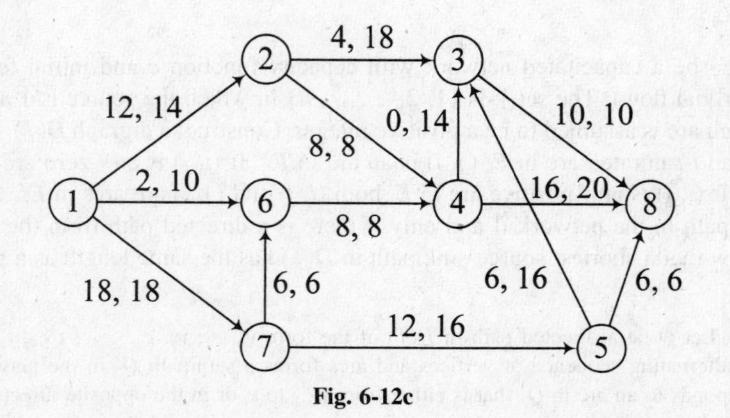

Fig. 6-12c

The digraph corresponding to this revised flow in which there is no directed path from 1 to 8 is shown in Fig. 6-12(d). So the updated flow is a maximum flow with flow value 32. In this digraph, the vertices not reachable from 1 are 4 and 8. Thus $T = \{4, 8\}$ and $S = V - T$. Cut $(S, T) = \{(2, 4), (6, 4), (3, 8), (5, 8)\}$ with a capacity of $8 + 8 + 10 + 6 = 32$ is a minimum cut.

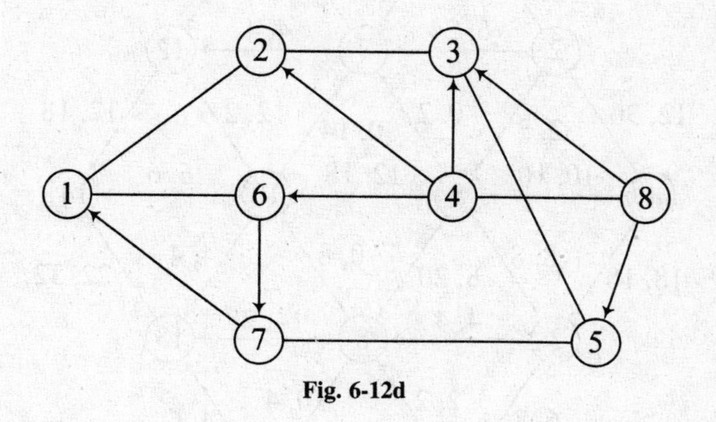

Fig. 6-12d

6.9 Obtain a maximum flow and a minimum cut in the network shown in Fig. 6-13(a).

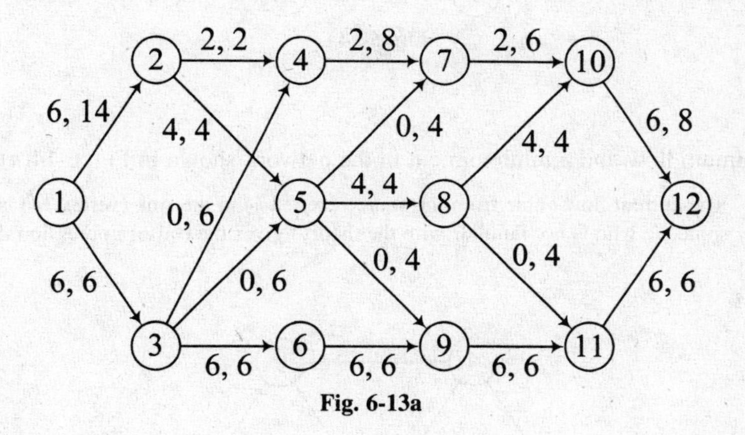

Fig. 6-13a

Solution. The current flow value from the source (vertex 1) to the sink (vertex 12) is 12. The mixed graph corresponding to this flow is shown in Fig. 6-13(b). In this mixed graph, there is no directed path from the source to the sink. So the current flow with flow value 12 is a maximum flow. The only vertex reachable from the source is vertex 2. Thus $S = \{1, 2\}$, and T is the remaining set of vertices. Cut (S, T), consisting of arcs (1, 3), (2, 4) and (2, 5) with capacity $6 + 2 + 4 = 12$, is a minimum cut.

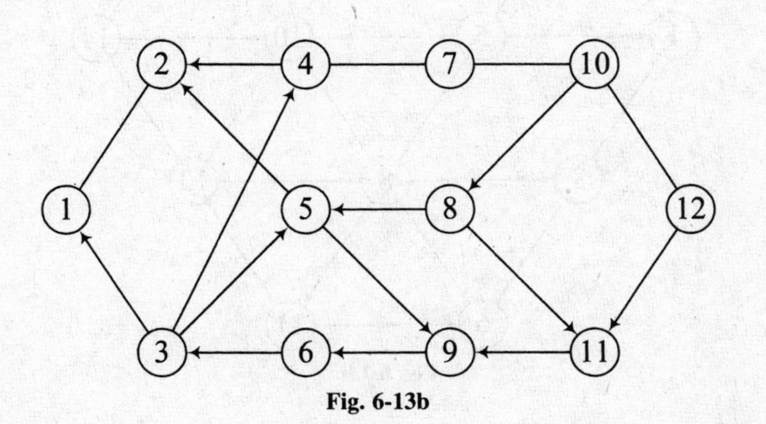

Fig. 6-13b

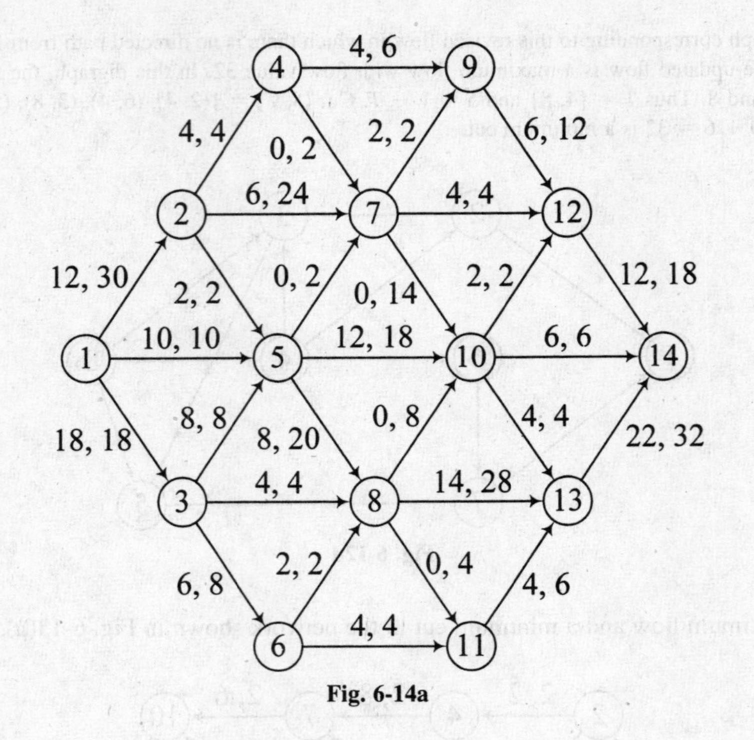

Fig. 6-14a

6.10 Obtain a maximum flow and a minimum cut in the network shown in Fig. 6-14(*a*).

Solution. The current flow value from the source (vertex 1) to the sink (vertex 14) is 40. A cursory look at the network (by someone who is not familiar with the theory) gives the (false) impression that it is not possible to

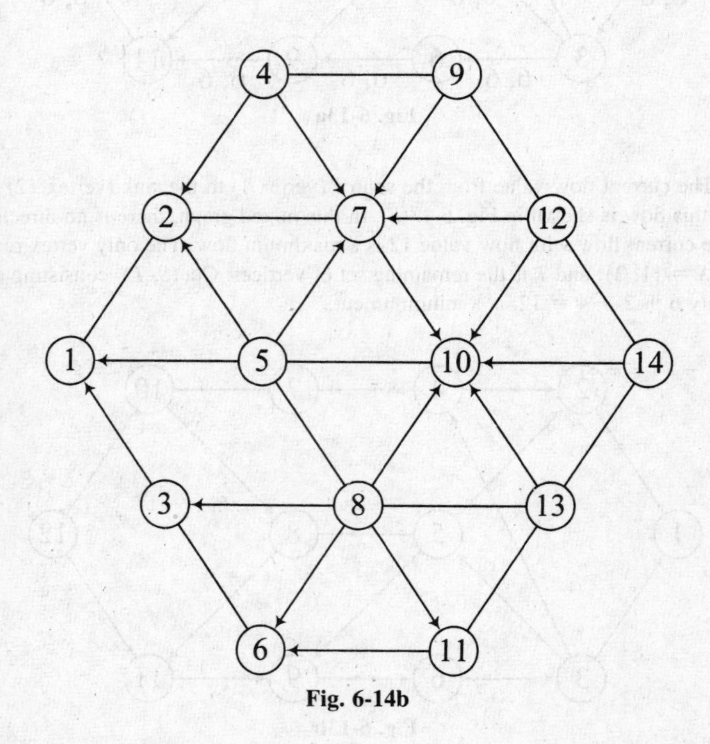

Fig. 6-14b

increase the flow value. That there is a directed path in the mixed graph shown in Fig. 6-14(*b*) from the source to the sink indicates that the current flow is not a maximum flow.

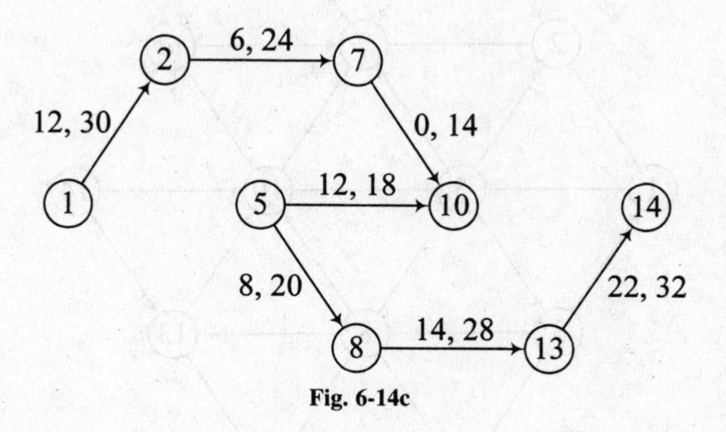

Fig. 6-14c

A directed path in the mixed graph from the source to the sink defines the semipath as shown in Fig. 6-14(*c*), which is a flow augmenting path. Using this semipath, the flow value can be increased by 10 more units. The updated network with a flow value of 50 units is shown in Fig. 6-14(*d*). The mixed graph corresponding to the updated flow in which there is no directed path from the source to the sink is shown in Fig. 6-14(*e*). So the updated flow is a maximum flow.

In the mixed graph, the set T contains vertices that are not reachable from the source are 4, 9, 12, and 14. The remaining vertices constitute set S. Cut (S, T), consisting of arcs (2, 4), (7, 9), (7, 12), (10, 12), (10, 14), and

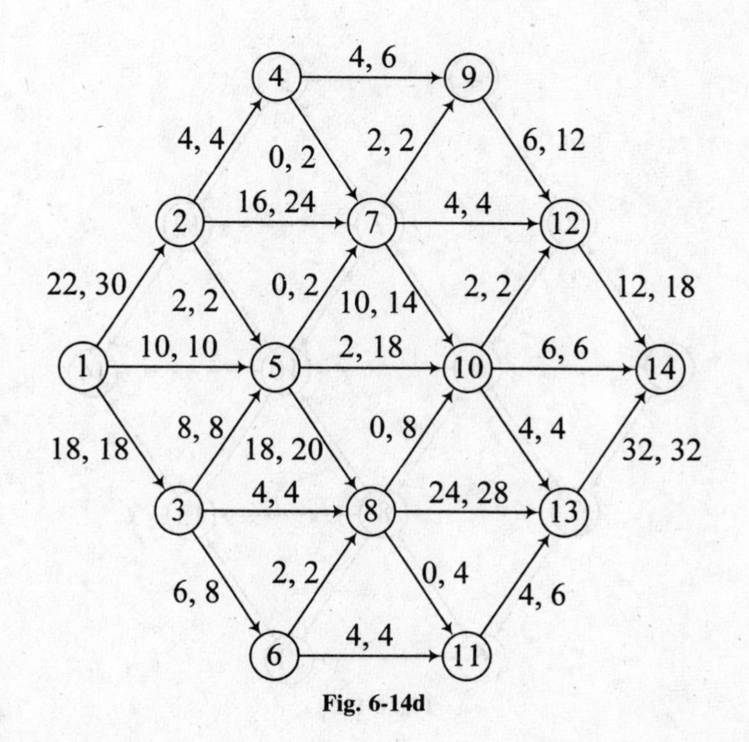

Fig. 6-14d

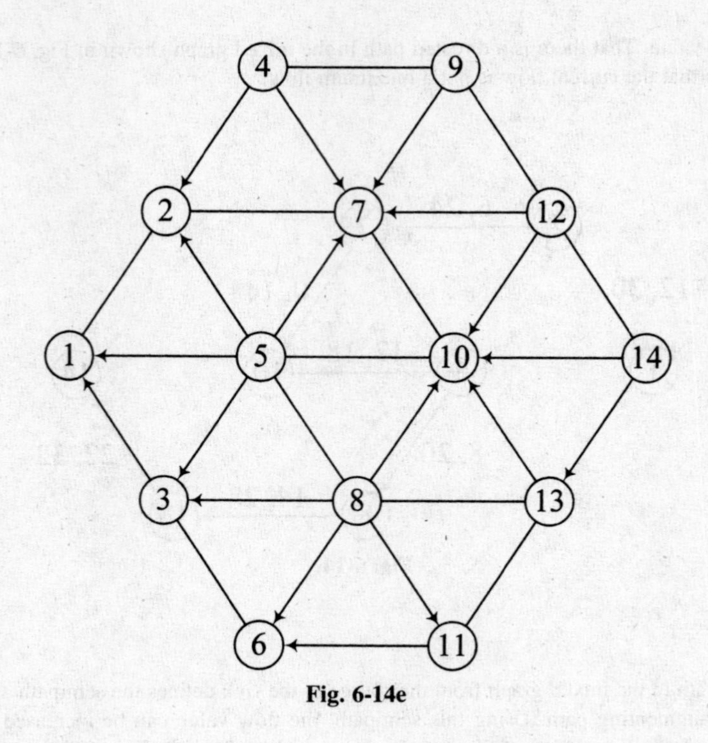

Fig. 6-14e

(13, 14), has a capacity of 4 + 2 + 4 + 2 + 6 + 32, which is equal to the current flow value. If these six arcs are deleted from the network, we get a digraph as shown in Fig. 6-14(f) in which the dotted lines are the deleted arcs. It is easy to see that there is no directed path from the source to the sink in the digraph without the dotted lines.

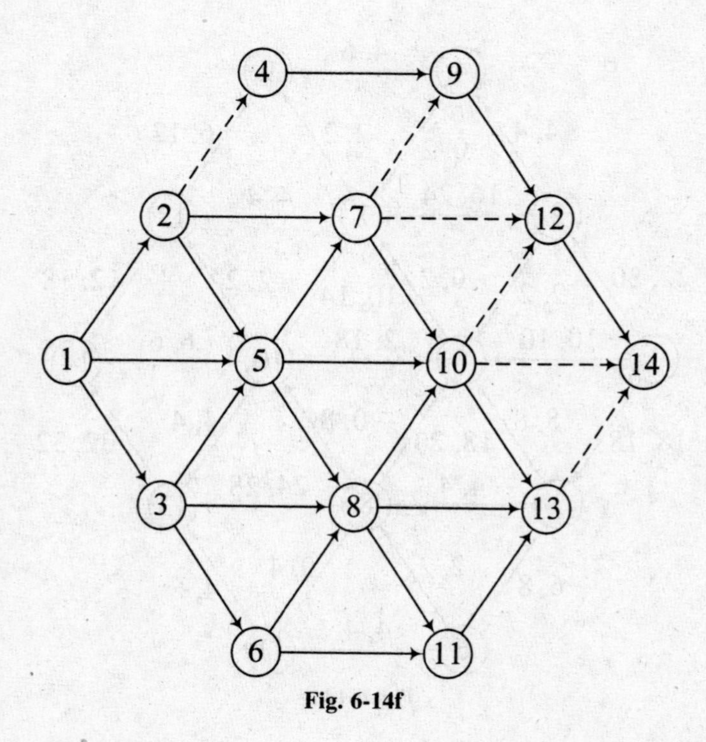

Fig. 6-14f

6.11 Prove Theorem 6.5: The maximum value of a generalized source-sink flow in a vertex-capacitated network G is equal to the capacity of a minimum generalized source-sink cut.

Solution. Let s be the source, and let t be the sink of G. Then we can construct an s-t network G' (which is not vertex-capacitated) as follows. Each intermediate vertex i of weight w_i is replaced by two intermediate vertices i' and i'' along with an arc (i', i'') of capacity w_i. At the same time, every arc (u, v) of the network is replaced by arc (u'', v'). (See Problem 6.12, which illustrates this construction.) Let $s = s' = s''$ and $t = t' = t''$. Any flow entering i' must pass through i'', and all flow leaving f'' must come from f. So there is a one-to-one correspondence between generalized flows in G and feasible flows in G'. Thus the maximum-flow minimum-cut theorem in G' implies the generalized theorem in the vertex-capacitated network G.

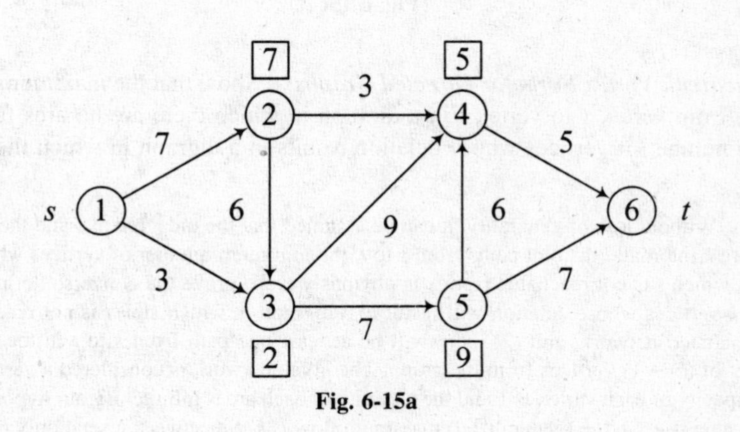

Fig. 6-15a

6.12 Obtain a generalized maximum flow and a generalized minimum cut in the vertex-capacitated network in which vertex 1 is the source and vertex 6 is the sink shown in Fig. 6-15(a).

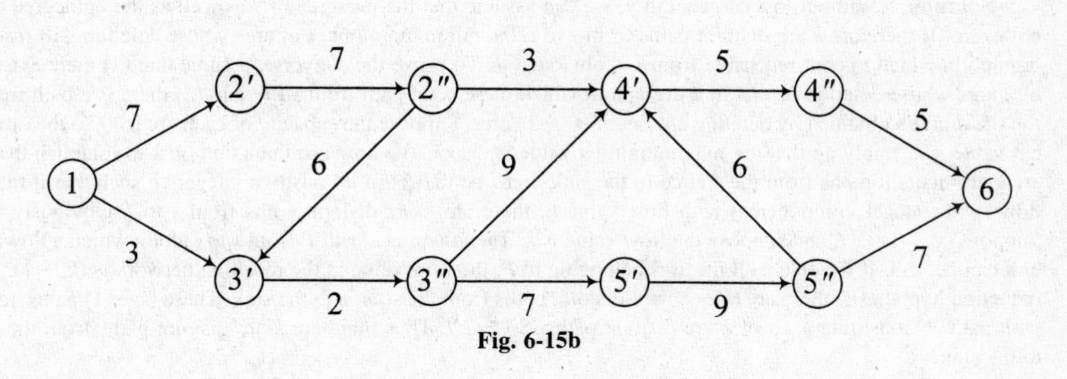

Fig. 6-15b

Solution. The expanded network G' is shown in Fig. 6-15(b). The maximum flow value in G' is 5. A minimum cut is $\{(2'', 4'), (3', 3'')\}$. Arc $(2', 4'')$ corresponds to arc (2, 4) in G. Arc $(3', 3'')$ corresponds to vertex 3 in G. Thus a generalized minimum cut consists of arc (2, 4) and vertex 3. The generalized maximum flow in the given network is shown in Fig. 6-15(c).

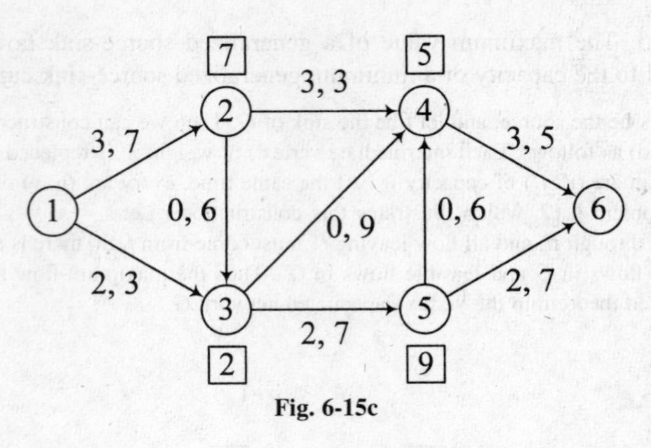

Fig. 6-15c

6.13 (*Menger's Theorem: Vertex Form for Directed Graphs*) Show that the maximum number of internally disjoint paths from vertex s to vertex t in a digraph in which there are no arcs from s to t is equal to the minimum number of vertices whose deletion results in a digraph in which there are no paths from s to t.

Solution. Without loss of generality, it can be assumed that the indegree of s and the outdegree of t are both zero. If there are k internally disjoint paths from s to t, the minimum number of vertices whose deletion will result in a digraph in which t is not reachable from s is obviously k. To prove the converse, let us assume that (1) there exists a set of k vertices whose deletion will result in a digraph in which sink t is not reachable from source s in the vertex-capacitated network, and (2) there will be at least one path from s to t in the network obtained after deleting any set of $(k - 1)$ vertices from the graph. The given network is considered a vertex-capacitated network in which the capacity of each vertex is 1 and the capacity of each arc is infinite. By our hypothesis, the (generalized) minimum cut value is k. So the (generalized) maximum flow is k. Since we can send only one unit of flow through a vertex from the source to the sink and since there is no arc from s to t, there should be exactly k internally disjoint paths from the source to the sink.

6.14 (*Menger's Theorem: Arc Form for Directed Graphs*) Show that the maximum number of arc-disjoint paths from vertex s to vertex t in a digraph is equal to the minimum number of arcs whose deletion results in a digraph in which there are no paths from s to t.

Solution. Without loss of generality, we can assume that the indegree of s as well as the outdegree of t are both zero. If there are k arc-disjoint paths from s to t, the minimum number of arcs whose deletion will result in a digraph in which t is not reachable from s is obviously k. To prove the converse, assume that (1) there exists a set of k arcs whose deletion results in a digraph in which there is no path from s to t and (2) there is a path from s to t in the digraph obtained by deleting any set of $(k - 1)$ arcs. Suppose the capacity of each arc is 1. So the minimum cut value is k, implying that the maximum flow value is also k. We now use induction on k to establish that there are k arc-disjoint paths from the source to the sink. Let I be the set of all positive integers n such that if there is a flow (with integer components) with flow value n, there are n arc-disjoint paths from s to t. Obviously, $1 \in I$. Suppose $(k - 1) \in I$, and suppose the flow value is k. Then there is a path P from s to t along which a flow of one unit can be sent. If we delete all the arcs belonging to P, the flow value in the resulting network is $(k - 1)$. By the induction hypothesis, there are $(k - 1)$ arc-disjoint paths from the source to the sink. These $(k - 1)$ paths, together with path P, constitute a set of k arc-disjoint paths. So $k \in I$. Thus there are k arc-disjoint paths from the source to the sink.

6.15 (*Menger's Theorem: Vertex Form for Undirected Graphs*) Show that the maximum number of internally disjoint paths between any two nonadjacent vertices s and t in a graph is equal to the minimum number of vertices whose deletion results in a graph in which there are no paths between those two vertices.

Solution. If G is the given graph, construct the associated digraph $D(G)$ by replacing each edge of the graph joining two vertices u and v by two arcs (u, v) and (v, u). Delete all arcs directed to s and all arcs directed from t.

Let the resulting digraph be G'. There is a one-to-one correspondence between the set of directed paths from s to t in G' and the set of paths between s and t in G. Hence, this theorem follows as a consequence of the result established in Problem 6.13.

6.16 (*Menger's Theorem: Edge Form for Undirected Graphs*) **Show that the maximum number of edge-disjoint paths between two vertices s and t in a graph is equal to the minimum number of edges whose deletion results in a graph in which there are no paths between the two vertices.**

Solution. If G is the given graph, construct the associated digraph $D(G)$ by replacing each edge of the graph joining two vertices u and v by two arcs (u, v) and (v, u). Delete all arcs directed to s and all arcs directed from t. Let the resulting digraph be G'. There is a one-to-one correspondence between the set of directed paths from s to t in G' and the set of paths between s and t in G. Hence, this theorem follows as a consequence of the result established in Problem 6.14.

6.17 Show that the vertex form of Menger's theorem implies the edge (arc) form.

Solution. Let G be any undirected graph in which s and t are two vertices. Introduce two new vertices x and y. Join x and s by constructing a new edge. Join t and y by constructing another new edge. The enlarged graph is G', and its line graph is $L(G')$. Let s' be the vertex in the line graph that corresponds to the new edge joining x and s in G'. Likewise, let t' be the vertex in the line graph that corresponds to the new edge joining t and y. Notice that s' and t' are not adjacent since s and t are distinct. (See Problem 6.18.) Any cut in G corresponds to a set of vertices in $L(G)$, the deletion of which results in a graph in which there is no path between s' and t'. Furthermore, any pair of edge-disjoint paths between s and t in G become two internally disjoint paths between s' and t' in $L(G)$. So the vertex form of Menger's theorem in the line graph implies its edge form in G. The proof for digraphs is the same.

6.18 Show that there are three edge-disjoint paths between s and t in the graph G shown in Fig. 6.16(a) by showing that there are three internally disjoint paths between s' and t' in the line graph $L(G')$ of G', where G' is obtained by enlarging G, as explained in Problem 6.17.

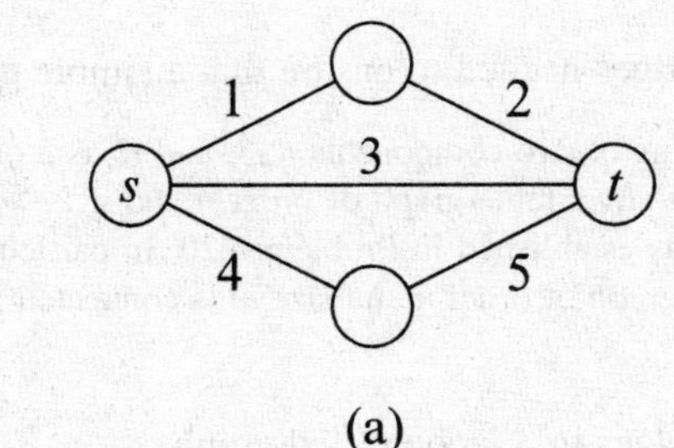

(a)

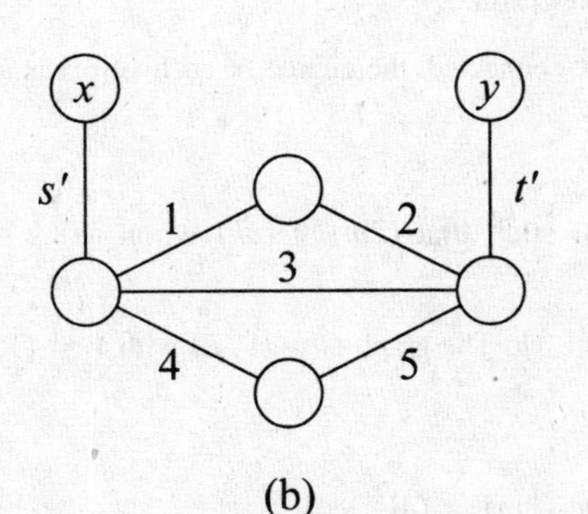

(b)

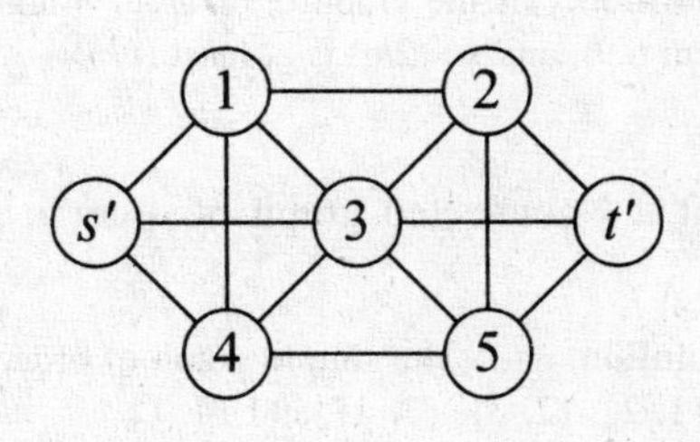

(c)

Fig. 6-16

Solution. The five edges of G are marked 1, 2, 3, 4, and 5. The three edge-disjoint paths between s and t are (*i*) the path consisting of the edges 1 and 2, (*ii*) the path consisting of the edge 3, and (*iii*) the path consisting of the edges 4 and 5. The enlarged graph G' is shown in Fig. 6-16(*b*). The edge joining the new vertex x and s becomes the source s' in $L(G')$, as shown in Fig. 6-16(*c*). Likewise, the edge joining the new vertex y and the sink t becomes the new sink t' in the line graph. There are three internally disjoint paths between s' and t' in the line graph: (*i*) s'—1—2—t' (*ii*) s'—3—t', and (*iii*) s'—4—5—t'.

6.19 Prove Theorem 6.7: The vertex form of Menger's theorem, the arc (edge) form of Menger's theorem, and the Ford–Fulkerson theorem are equivalent.

Solution. See Problem 6.13 for a proof that the Ford–Fulkerson theorem implies the vertex form of Menger's theorem. In Problem 6.17, it was shown that the vertex form implies the edge (arc) form. So it is enough if we prove that the edge (arc) form of Menger's theorem implies the Ford–Fulkerson theorem. Let G be a capacitated network with source s and sink t as specified in this chapter. The capacity of each arc joining vertex i and vertex j is a positive integer c_{ij}. Replace each such arc by c_{ij} arcs, resulting in a capacitated multigraph G' with unit capacity on each arc. The flow value p of a maximum flow in G is equal to the number of arc-disjoint paths from s to t in G'. The capacity q of a minimum cut in G is equal to the number of arcs in a minimum cut in G'. The arc form of Menger's theorem implies that $p = q$. In other words, the maximum flow value in G is equal to the capacity of a minimum cut in it. So the arc form of Menger's theorem implies the Ford–Fulkerson theorem.

MORE ON CONNECTIVITY

6.20 Show that if a simple graph of order n and size m has k components, $m \le \frac{1}{2}(n-k)(n-k+1)$.

Solution. The conclusion remains valid even if we assume that each component is a complete graph. Suppose H_i and H_j are two such components with n_i and n_j vertices, where $n_i \ge n_j \ge 1$. If we replace these two components by two complete graphs of order $(n_i + 1)$ and $(n_j - 1)$, respectively, the total number of vertices will remain unchanged but the number of edges will *increase* by $n_i - n_j + 1$. So the number of edges of a simple graph of order n with k components will be a *maximum* if there are $(k - 1)$ isolated vertices and one component that is a complete graph with $(n - k + 1)$ vertices with $\frac{1}{2}(n-k)(n-k+1)$ edges.

6.21 Find the minimum number of edges needed to ensure that a simple graph is connected.

Solution. The graph consisting of two components K_{n-1} and K_1 is a disconnected graph of order n and size $\frac{1}{2}(n-1)(n-2)$. If m is the size of any simple graph of order n and if $m > \frac{1}{2}(n-k)(n-k+1)$, the number of its components is $(k - 1)$ or less, as established in Problem 6.20. In particular, if $m > \frac{1}{2}(n-2)(n-2+1)$, the graph is connected. Thus a simple graph of order n and size m is connected if $m > \frac{1}{2}(n-1)(n-2)$.

6.22 Find the minimum number of edges in a k-connected graph.

Solution. If the graph G of order n and size m is k-connected, the degree of each vertex is at least k (see Theorem 6.9), and so $(2m)$ is at least (nk).

6.23 Exhibit a k-connected graph of order n and size m such that $(2m) = (nk)$ when (*a*) $k = 1$ and (*b*) $k = 2$.

Solution. (*a*) The simple graph of order 2 and size 1. (*b*) The graph $G = (V, E)$ with $V = \{1, 2, 3, 4\}$ and edges $\{1, 2\}$, $\{2, 3\}$, $\{3, 4\}$, and $\{4, 1\}$.

6.24 Exhibit a k-connected of order n and size m such that $(2m) = (nk) + 1$.

Solution. If G is a k-connected graph with n vertices and m edges and if (nk) is odd, $(2m)$ is at least $(nk) + 1$. It can be easily verified that the graph in Fig. 6-17 with five vertices and eight edges is a 3-connected graph.

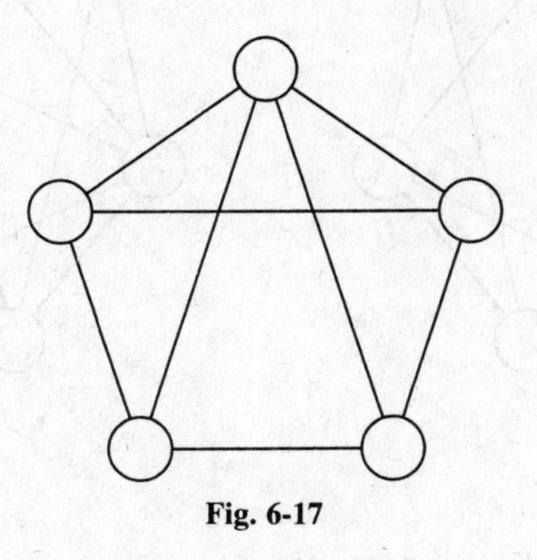

Fig. 6-17

6.25 If $1 \leq k < n$, the **Harary graph $H_{k,n}$** of order n is constructed as follows. The n vertices are placed on the circumference of a circle. (*a*) If $k = 2r$, join each vertex to the nearest r vertices in each direction around the circle. (*b*) If $k = 2r + 1$ and n is even, join each vertex to the nearest r vertices in each direction on the circle and also to the vertex exactly opposite to it. (*c*) Suppose $k = 2r + 1$ and n is odd. First the graph $H_{2r,n}$ is constructed as in part (*b*). Define $tn + i = i$ for any positive integer t, and using this (modulo) addition rule, construct additional edges by joining vertex i and vertex $\frac{1}{2}(n + 3)$ for $1 \leq i \leq (\frac{1}{2})(n + 1)$. Find the size of the Harary graph $H_{k,n}$.

Solution. In parts (*a*) and (*b*), the degree of each vertex is k, so the sum of the degrees of the n vertices is (nk). In part (*c*), there are $(n - 1)$ vertices of degree k and one vertex of degree $(k + 1)$. The sum of the degrees of the n vertices in this case is $(nk) + 1$. Thus $(2m) = (nk)$ or $(nk) + 1$, where m is the size of the graph.

6.26 Construct the Harary graphs $H_{k,n}$ for (*a*) $n = 6$, $k = 4$; (*b*) $n = 6$, $k = 5$; (*c*) $n = 7$, $k = 4$; and (*d*) $n = 7$, $k = 5$.

Solution. These graphs are shown in Figs. 6-18 through 6-21.

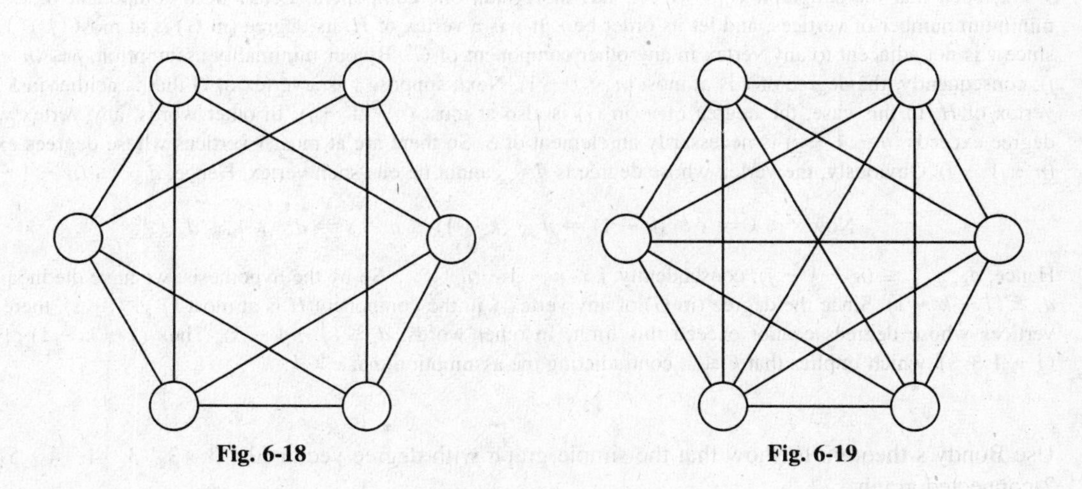

Fig. 6-18 **Fig. 6-19**

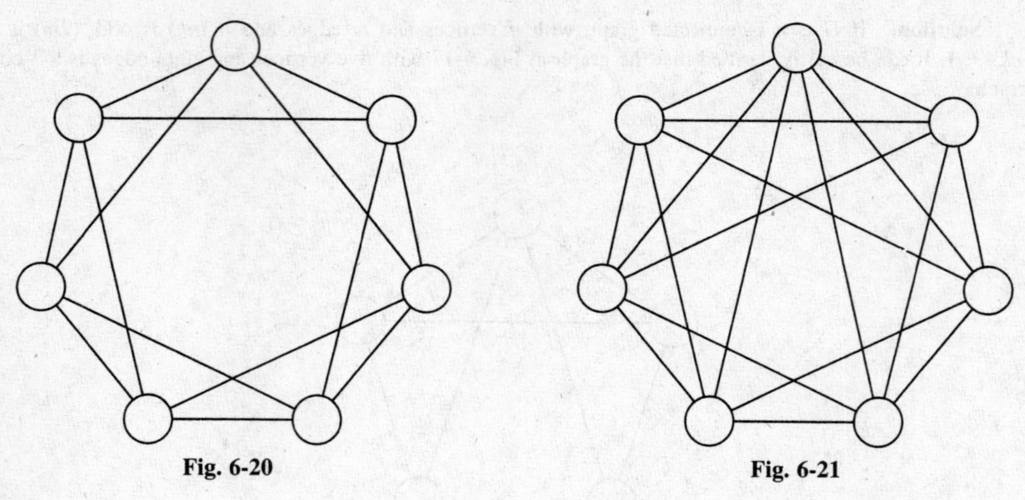

Fig. 6-20

Fig. 6-21

6.27 Show that the Harary graph $K_{k,n}$ is k-connected.

Solution. Let $k = 2r$ and $H_{k,n}$ be denoted by G. Let V be the set of n vertices of G, and let W be any subset of V such that $|W| < 2r$. We shall establish that there is a path between every pair of vertices in the subgraph G' obtained after deleting W from G. Let p and q be any nonadjacent vertices in G'. Now the n vertices of G can be placed on the circumference of a circle. Let X be the set of vertices (of G) situated on the circle between p and q (excluding p and q) as we move clockwise from p to q. Likewise, let Y be vertices (of G) situated on the circle between p and q (excluding p and q) as we move counterclockwise from p to q. Since $|W \cap X| + |W \cap Y| < 2r$, at least one of the sets $W \cap X$ or $W \cap Y$ should have fewer than r elements. If $|W \cap X| < r$, there is a path between p and q in G' consisting of vertices exclusively from the set $(X - W)$ since every vertex v is adjacent (in G) to r vertices on either side of v. Thus G' is a connected graph. In other words, the graph G remains connected after deleting a set of $2r$ vertices from it. So $\kappa(G) \geq 2r$. But $\kappa(G) \leq 2r$ since the minimum degree of G is $2r$. Thus the Harary graph $H_{k,n}$ is k-connected when k is even. The proof is similar when k is odd.

6.28 Bondy's theorem shows that if k is a fixed positive integer less than n and if the degree vector $[d_1 \quad d_2 \quad \cdots \quad d_n]$ (in nondecreasing order) of a simple graph G satisfies the inequality $d_j \geq (j + k - 1)$ whenever $1 \leq j \leq (n - 1 - d_{n-k+1})$, the graph G is k-connected.

Solution. Suppose G is not k-connected. Then $\kappa(G) < k$. So there exists a set S of s vertices of G (where $s < k$) such that the subgraph $G' = G - S$ has more than one component. Let H be a component of G' with minimum number of vertices, and let its order be j. If v is a vertex of H, its degree (in G) is at most $(j - 1 + s)$ since v is not adjacent to any vertex in any other component of G'. By our minimality assumption, $j \leq (n - s - j)$; consequently, the degree of v is at most $(n - j - 1)$. Next, suppose v is a vertex of G that is neither in S nor a vertex of H. In this case, the degree of v (in G) is also at most $(n - 1 - j)$. In other words, any vertex whose degree exceeds $(n - 1 - j)$ is necessarily an element of S. So there are at most s vertices whose degrees exceed $(n - 1 - j)$. Obviously, the vertex whose degree is d_{n-s} cannot be one such vertex. Hence, $d_{n-s} \leq (n - 1 - j)$.

$$\text{Now } s < k \Rightarrow s \leq (k - 1) \Rightarrow n - (k - 1) \leq n - s \Rightarrow d_{n-(k-1)} \leq d_{n-s}.$$

Hence, $d_{n-(k-1)} \leq (n - 1 - j)$; consequently, $j \leq n - 1 - d_{n-(k-1)}$. So by the hypothesis, we have the inequality $d_j \geq (j + k - 1)$. Since the degree (in G) of any vertex v in the component H is at most $(j - 1 + s)$, there are j vertices whose degrees cannot exceed this limit. In other words, $d_j \leq (j - 1 - s)$. Thus $(j + k - 1) \leq d_j \leq (j - 1 + s)$, which implies that $k \leq s$, contradicting the assumption that $s > k$.

6.29 Use Bondy's theorem to show that the simple graph with degree vector $[2 \quad 3 \quad 3 \quad 3 \quad 4 \quad 4 \quad 5]$ is a 2-connected graph.

Solution. (There is a simple graph G for which this vector is the degree vector. See Example 9 in Chapter 1.) Here $n = 7$, $k = 2$, and $n - 1 + d_{n-k+1} = 6 - d_6 = 2$. So the choices for j (to apply Bondy's theorem) are 1 and 2. For $j = 1$, $d_1 = 2$ and $1 + k - 1 = 2$. For $j = 2$, $d_2 = 3$ and $2 + k - 1 = 3$. Thus the required inequalities are satisfied. Hence, G is 2-connected.

6.30 Prove Theorem 6.9 (Whitney's theorem): A graph with at least $(k + 1)$ vertices is k-connected if and only if any two distinct vertices in the graph are connected by at least k internally disjoint paths. In particular, a graph with at least three vertices is a block if and only if every two vertices lie on a common cycle.

Solution. Let G be a k-connected graph. So $\kappa(G) \geq k$. Let p be the maximum number of internally disjoint paths between u and v. Suppose $p < k$. If u and v are not adjacent, $\kappa(G) = p < k$ by Menger's theorem (vertex form), which is a contradiction. Suppose u and v are adjacent. Let G' be the graph obtained after deleting the edge joining the vertices u and v. Then $\kappa(G') = p - 1 < k - 1$. So there exists a set S of s vertices ($s < k - 1$) such that $G' - S$ is a disconnected graph. So either $G - \{S \cup u\}$ or $G - \{S \cup v\}$ is disconnected, which implies that $\kappa(G) \leq s + 1 < (k - 1) + 1 = k$, arriving at the same contradiction as before. Conversely, assume that the number of internally disjoint paths between any pair of vertices is at least k. Suppose G is not k-connected. In that case, $\kappa(G) < k$. Obviously, the graph is not a complete graph. Let S be the set of vertices such that $G - S$ becomes disconnected. If u and v are vertices belonging to two different components, these two vertices cannot be adjacent vertices of the graph. So there are at least k internally disjoint paths between these two vertices. So by Menger's theorem, S should have at least k vertices. In other words, $\kappa(G) \geq k$.

6.31 Give an example of a k-connected graph such that the number of internally disjoint paths between any pair of vertices is equal to k.

Solution. In the complete bipartite graph $K_{n,n+1} = (X, Y, E)$ with n vertices in X, the maximum number of internally disjoint paths between a pair of two vertices in Y is n.

6.32 (*Harary's Characterization of Blocks of a Graph*) In a connected graph G with three or more vertices, the following statements are equivalent:
(1) G is a block. (2) If u and v are two distinct vertices of G, there is a cycle that passes contains those two vertices. (3) If u is a vertex and e is an edge, there is a cycle that contains u and e. (4) If e and f are two distinct edges, there is a cycle that contains those two edges. (5) If u and v are two vertices and e is an edge, there is a path between those two vertices that contains edge e. (6) For every three vertices, there is a path between two of them that contains the third. (7) For every three vertices, there is a path between two of them that does not contain the third.

Solution.

(1) implies (2): Suppose G is a block. Since G has more than two vertices, it is 2-connected. So by Whitney's theorem, there are at least internally disjoint paths between these two vertices, forming a cycle.

(2) implies (3): Let e be the edge joining v and w. By hypothesis, there is a cycle C that contains u and v. If C passes through w, we are done. Otherwise, let P be a path between w and u that does not pass through v. If P does not pass through any vertex of C, there are two cycles that contain v and edge e. Otherwise, let u' be the first vertex of P in cycle C. Then the path P, the path in C from u to v that does not contain u', and edge e together form the desired cycle.

(3) implies (4): Let e be the edge joining u and v. Then there is a cycle that passes through u and v that also contains edge f.

(4) implies (5): Let u and v be two vertices, and let e be an edge. Let e' be an edge adjacent to u, and let f' be an edge adjacent to v. Since, by assumption, there is a cycle that contains both these edges, there is a cycle that passes through u and v; therefore, the hypothesis of (2) is satisfied. So there exists a cycle C that contains u and edge e. Likewise, there is a cycle C' that contains v and x. We are done if u is in C' or when v is in C. Otherwise, we construct a path P starting from u using the vertices of C until we reach vertex w that belongs to C'. Then we continue the construction of the path from w to u using the vertices of C' and edge e.

(5) implies (6): Let u, v, and w be three distinct vertices. Suppose e is an edge adjacent to w. There is a path between u and v that contains edge e; therefore, the vertex w.

(6) implies (7): Let u, v, and w be three distinct vertices as before. There is a path P between u and v that passes through w by hypothesis. The subpath P' of P between u and w does not pass through v.

(7) implies (1): Let u and v be any two vertices. Suppose w is another vertex. Then there is a path P between u and v that does not pass through w. Consequently, there are at least two internally disjoint paths between u and v. So G is a block.

6.33 Let G be a k-connected graph and G' be the graph obtained from G by constructing a new vertex w and joining it to k or more vertices in G. Then G' is k-connected.

Solution. Let W be a separating set of the enlarged graph G'. If w is a vertex in W, $W' = W - v$ is a separating set of G, implying that W has at least $k + 1$ elements. If w is not a vertex in W, the set X of vertices adjacent to w is a subset of W. In this case, W has at least k vertices. Thus G' is k-connected.

6.34 (*Fan Lemma of Dirac*) A set of k paths from vertex v of a graph to each vertex in a set X of k vertices is called a **(v, X) fan of size k** if no two paths in the set have a vertex in common other than v. Show that a graph is k-connected if and only if it has at least $(k + 1)$ vertices and for any choice of a vertex v and any choice Y of vertices (where $v \notin Y$) with k or more vertices, there is a (v, X) fan of size k, where $X \subset Y$.

Solution. Let G be a k-connected graph. Enlarge the graph as in Problem 6.33 by constructing a new vertex w and joining w to each vertex in Y. The enlarged graph G' is also k-connected. So there are k internally disjoint paths between v and w. If we delete the new edges from these paths, we get a (v, X)-fan of size k. To prove the converse, assume that the graph G is not k-connected. So there is a separating set S (with less than k vertices) such that $G - S$ has more than one component. Let v and w be vertices belonging to different components of $G - S$. Let X be a set of k vertices that contain w and S but not v. In that case, every path between v and w will pass through a vertex belonging to S, showing that there is no (v, X) fan of size k, which is a contradiction.

6.35 (*Dirac's Theorem on* k-*Connectivity*) In a k-connected graph with three or more vertices, for every set of k vertices there is a cycle that will pass through these k vertices. (The converse is not true. A cycle with k vertices is not k-connected when $k > 3$.)

Solution. If $k = 2$, the result follows as an immediate consequence of Whitney's theorem. So let $k > 2$. Let W be any set of k vertices in the graph. Of all the cycles in the graph that have vertices in common with W, choose a cycle C that has the maximum number of vertices in common with W. Let m be the number of vertices common to C and W, and let $W' = \{w_1, w_2, \ldots, w_m\}$ be the set of common vertices. Suppose $m < k$. So there is a vertex w in $W - W'$. Since the graph is k-connected, there are k paths joining w to each vertex in W' such that no two paths have a vertex in common other than w, as established in Problem 6.34. Suppose w_i and w_j are adjacent vertices on C. Let Q_i be a path joining w and w_i, and let Q_j be a path joining w and w_j. If the only vertices in C are the vertices of W', it is possible to construct a cycle C' using these paths and the path between w_i and w_j obtained from C after deleting the edge joining these two vertices. Cycle C' has more vertices in common with W' than W, violating the maximality assumption. So there should be at least one vertex in C that should not belong to set W. Let w_{m+1} be a vertex in C that is not in W. The vertices w and w_{m+1} are distinct. The assumption that $m < k$ implies that $m + 1 \le k$. Since G is k-connected, there exist paths P_i joining w and w_i $(i = 1, 2, \ldots, m + 1)$ such that no two of these $(m + 1)$ paths have a vertex in common other than vertex w. Let v_i be the first vertex in path P_i belonging to cycle C as we move from w to w_i.

Denote the subpath of P_i between w and v_i by P_i'. The $(m + 1)$ vertices v_i are distinct vertices on C, and the possibility that $v_i = w_i$ is not ruled out. Cycle C defines two paths between every pair of vertices v_j and v_k. Since C and W have exactly m vertices in common, there are integers j and k such that one of the paths defined by C between v_j and v_k has no vertex belonging to W. Let P be the other path between these two vertices. Then the three paths P, P_j', and P_k' together constitute a cycle that has more than m vertices belonging to the set W, which is a contradiction. So $m = k$, as we wished to prove.

6.36 Give an example of a k-connected graph such that an arbitrary set of $(k + 1)$ vertices need not lie on a cycle in the graph.

Solution. Consider the complete bipartite graph $K_{k,k+1} = (X, Y, E)$ with k vertices in X that is k-connected. There is no cycle that passes through all the vertices in the set Y.

6.37 Show that a graph of order n is k-connected (where $1 \leq k \leq n - 1$) if the degree of every vertex is at least $(n + k - 2)/2$.

Solution. If the graph is complete, it is k-connected for $k \leq (n - 1)$. Assume that the graph is not complete and not k-connected. So there is a disconnecting set S of s vertices such that $G - S$ is a disconnected graph. Let H be a component of $G - S$ with as few vertices as possible. If the order of H is r, $r \leq (n - s - r)$, which gives the upper bound $(n - s)/2$ for the order r. If v is any vertex of H, the degree of v in the graph G cannot exceed $(r - 1) - s$. Thus $\deg v \leq (n - s)/2 - 1 + s = (n + s - 2)/2 < (n + k - 2)/2$, violating the given inequality.

6.38 Show that the sufficient condition in Problem 6.37 for k-connectivity is not a necessary condition.

Solution. Here is a counterexample. For $n \geq 6$, let W_n be the graph obtained from the cyclic graph C_{n-1} with $(n - 1)$ vertices by introducing a new vertex and joining it to each vertex of the cycle. This graph is known as the **wheel** since the shape of the graph looks like a wheel if the vertices of the cycle are placed symmetrically on the circumference of a circle and the new vertex is placed at the center. The wheel thus defined is 3-connected; the minimum degree of a vertex is 3, which is less than $(n + 3 - 2)/2$ whenever $n > 5$.

6.39 Prove Theorem 6.10: A graph is k edge connected if and only if any two distinct vertices in it are connected by at least k edge-disjoint paths.

Solution. If there are k or more pairwise edge-disjoint paths between every pair of vertices, we need at least k edges to disconnect the graph, so the edge-connectivity number $\lambda(G)$ is at least k. Hence, the graph is k edge connected. Conversely, if the graph is k edge connected, it is not possible to disconnect it by deleting any set of $(k - 1)$ edges. Let p the maximum number of edge-disjoint paths between two vertices x and y in the graph. Now by Menger's theorem (edge form), there are p edges, the deletion of which will result in a disconnected graph. So it is possible to disconnect the graph by deleting p edges. Hence, $p > (k - 1)$. So the number of edge-disjoint paths between the two vertices is at least k.

6.40 (*Chvátal and Erdös Theorem*) Show that the graph G is Hamiltonian if $\kappa(G) \geq \alpha(G)$, where $\alpha(G)$ is its internal stability (independence) number.

Solution. The internal stability number $\alpha(G)$ is the maximum cardinality of a set of vertices in a graph such that no two vertices in it are adjacent. If $\alpha(G) = 1$, the graph is complete; therefore, it is Hamiltonian. Otherwise, $\kappa(G) = k \geq 2$, implying that G has at least one cycle. Let C be a cycle in G with the maximum number of vertices. Then C has at least k vertices since every set of k vertices belongs to some cycle, as proved in Problem 6.35. It can be shown that C is a Hamiltonian cycle. Suppose C is not a Hamiltonian cycle. If X is the set of all vertices of C, there exists a vertex $w \notin X$ and a (w, X) fan of size k as proved in Problem 6.34. If w is adjacent to any of the vertices in X, there exists a larger circle passing through all the vertices in X and vertex w, violating the maximality assumption. So w is not adjacent to any vertex in X. In that case, there is an independent set consisting of $(k + 1)$ vertices, contradicting the hypothesis.

6.41 Let $G = (V, E)$ be a connected graph, and let S be a proper subset of V. The subgraph induced by S is denoted by $G(S)$, and the subgraph induced by $T = V - S$ is denoted by $G(T)$. Show that the disconnecting set $D = (S, T)$ is a cut set if and only if both $G(S)$ and $G(T)$ are connected.

Solution. Suppose the two subgraphs are connected, and suppose D is not a cut set. Then there exists a proper subset D' of D that is a cut set. Let e be an edge in $D - D'$ joining vertex x in S and vertex y in T. If v is any vertex in $G(S)$, there is a path between v and x that passes through vertices from set S. Likewise, if w is any vertex in $G(T)$, there is a path between y and w consisting of vertices exclusively from T. In other words, when all the edges belonging to D' are deleted, the graph G still remains connected, implying that D' is not a disconnecting set. So D is a cut set. Conversely, suppose D is a cut set, and let e be an edge in D joining x in S and y in T. Then

$D' = D - e$ is not a cut set; therefore, $G - D'$ is a connected graph. Since e is the only edge between S and T in $G - D'$, there should be a path between every vertex in $G(S)$ and vertex x. Thus $G(S)$ and, similarly, $G(T)$ are connected graphs.

6.42 Let $G = (V, E)$, where $V = \{v_1, v_2, \ldots, v_n\}$, and for each i, let G_i be the subgraph obtained by deleting vertex v_i from G. Show that G is connected if and only if at least two of these subgraphs are connected.

Solution. Let G be a connected graph, and let T be any spanning tree in the graph. Any vertex v_i of degree 1 (in T) cannot be a cut vertex of the graph; therefore, G_i is a connected subgraph. But T has at least two vertices of degree 1. So two of these subgraphs are connected subgraphs. Conversely, suppose G_i and G_j are two connected subgraphs among these n subgraphs. Any vertex v_k other than these two is a vertex of both G_i and G_j. Hence, there is a path between v_j and v_k in the connected graph G_i, and there is a path between v_i and v_k in the connected graph G_j. So G is a connected graph.

6.43 Let $G = (V, E)$, where $V = \{v_1, v_2, \ldots, v_n\}$, and for each i, let G_i be the subgraph obtained by deleting vertex v_i from G. If each G_i is a connected graph with exactly one cycle, what can we say about G?

Solution. Notice that each G_i has $(n - 1)$ vertices and $(n - 1)$ edges. So the degree of each vertex in G is $m - (n - 1)$, where m is the size of G. Hence, $(2m) = (n)(m - n + 1)$. So $m(n - 2) = n(n - 1)$, implying that $n = 4$ and $m = 6$. Thus G is the complete graph with four vertices. So the only graph for which each subgraph G_i contains (is) a unique cycle is K_4.

SOME APPLICATIONS TO COMBINATORICS

6.44 Show that the max-flow min-cut theorem implies Konig's theorem.

Solution. Construct a digraph G' from the given bipartite graph $G = (X, Y, E)$ by converting each edge between vertex x in X and vertex y in Y to an arc (directed edge) from x to y and by introducing two vertices s and t such that there is an arc from s to every vertex in X and there is an arc from every vertex in Y to t. The capacity of any arc directed from s is 1. Likewise, the capacity of any arc directed to t is also 1. The capacity of every other arc is assumed to be infinite. Thus G' can be considered a capacitated network with s as the source and t as the sink. (See Problem 6.45.) Any feasible flow with flow value k obviously corresponds to a matching of cardinality k in the bipartite graph. So a maximum flow in G' defines a maximum cardinality matching in the bipartite graph G. Now if W is any covering in G, there is no arc in G' from a vertex in $(X - W)$ to a vertex in $(Y - W)$. So if $S = \{s\} \cup (X - W) \cup (Y \cap W)$ and $T = \{t\} \cup (Y - W) \cup (X \cap W)$, we have cut (S, T) in the network with capacity k. In other words, any covering of cardinality k defines a cut with cut value equal to k. On the other hand, any cut (S, T) with finite cut value k consists of k arcs, each of capacity 1. Let W_1 be the set of vertices in X that are adjacent to s, and let W_2 be the set of vertices in Y that are adjacent to t in this cut. Then the union of W_1 and W_2 is a covering. Thus any minimum cut in the digraph G' defines a minimum covering in the bipartite graph G. Thus the cardinality of a maximum matching in G is equal to the cardinality of a minimum covering in G since the former is equal to the maximum flow value in G' and the latter is equal to the minimum cut value in G'.

6.45 Illustrate Konig's theorem for the bipartite graph in Fig. 6-22(a) by converting it into a capacitated network as described in Problem 6.44.

Solution. It is easy to see that $M = \{\{x_1, y_2\}, \{x_3, y_3\}, \{x_4, y_4\}\}$ is a matching and $W = \{x_3, x_4, y_2\}$ is a covering. Both M and W have the same number of elements, so by Konig's theorem, the former is a maximum matching and the latter is a minimum covering. The network corresponding to the bipartite graph is shown in Fig. 6-22(b). Then $S = \{s, x_1, x_2, y_2\}$ and $T = \{t, x_3, x_4, y_1, y_3, y_4, y_5\}$. So cut (S, T) consists of arcs $\{s, x_3\}$, $\{s, x_4\}$ and $\{y_2, t\}$. The capacity of this cut is 3, which is also equal to the maximum flow value.

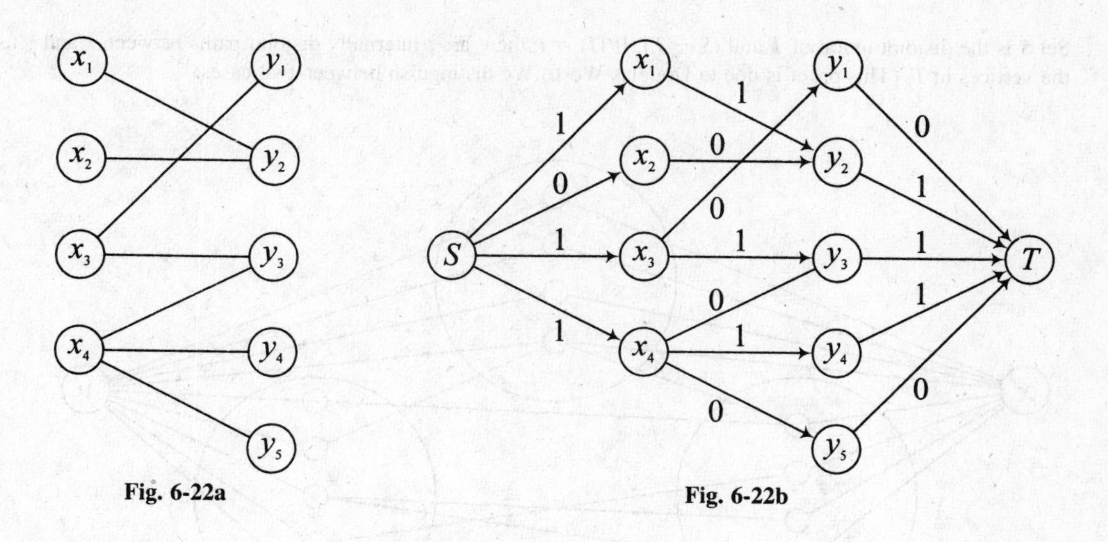

Fig. 6-22a

Fig. 6-22b

6.46 Show that Menger's theorem implies Konig's theorem.

Solution. For the given bipartite graph, we construct the directed graph G' as in Problem 6.44. By Menger's theorem, the maximum number of internally disjoint paths from s to t is equal to the minimum number of vertices whose deletion destroys all paths from s to t. But the former is the cardinality of a (maximum) matching, and the latter is the cardinality of a (minimum) covering.

6.47 Show that Konig's theorem implies Hall's marriage theorem.

Solution. Let $F = \{X_1, X_2, \ldots, X_m\}$ be a family of sets, and let $X = \{a_1, a_2, \ldots, a_n\}$ be the union of all the sets in the family. In the bipartite graph $G = (F, X, E)$, there is an edge between X_i and a_j if and only if a_j is an element of X_i. The marriage condition (M.C.) states that for any choice of a set F' of k vertices from F, the union of these k sets has at least k elements. Hall's theorem asserts that the M.C. is a necessary and sufficient condition for the existence of an SDR in the family. Obviously, the M.C. is a necessary condition. To prove the sufficiency part, let us assume that the M.C. holds but that the family has no SDR. In that case, let the largest subfamily that has an SDR consist of r vertices from F, where $r < m$. In other words, the size of a maximum matching is r; therefore, by Konig's theorem, there exists a covering W consisting of r vertices out of which at least one vertex is necessarily from X. Suppose $F - W = \{X_1, X_2, \ldots, X_k\}$. Then $|X \cap W|$ is $r - (m - \mathrm{k})$. Now the M.C. implies that the cardinality of the union of the k sets in the subfamily $(F - W)$ is at least k. So $|X \cap W| \geq k$; hence, $r - (m - \mathrm{k}) \geq k$, which contradicts the assumption that $r < m$.

6.48 Show that Hall's marriage theorem implies the Konig–Egervary theorem.

Solution. If the matrix is $m \times n$, construct a bipartite graph $G = (X, Y, E)$ with m vertices in $X = \{1, 2, \ldots, m\}$ corresponding to the rows and with n vertices in $Y = \{1, 2, \ldots, n\}$ corresponding to the n columns. Join vertex i in X to vertex j in Y if and only if the (i, j) element in the matrix has property P. Thus the Konig–Egervary theorem and Konig's theorem are equivalent. And Hall's marriage theorem implies the latter.

6.49 Show that Konig's theorem implies Menger's theorem.

Solution. Let x and y be two nonadjacent vertices in a graph $G = (V, E)$ with n vertices. The neighborhood $N(x)$ of x is the set of all vertices adjacent to x, and $N(y)$ is the set of all vertices adjacent to y. The intersection $N(x) \cap N(y)$ is denoted by T. Let $X = N(x) - T$ and $Y = N(y) - T$. We thus have a partition of V into four sets: T, X, Y, and U, where $U = V - T - X - Y$. Set S of vertices in the graph is called an x, y separating set if the deletion of S from the graph will result in a subgraph in which there is no path between x and y. Obviously, T is a subset of any such separating set. Suppose k is the size of a minimum x, y separating set. So there is a path between x and y when any set of $(k - 1)$ vertices is deleted, and there exists an x, y separating set of cardinality k.

Set S is the disjoint union of T and $(S - T)$. If $|T| = t$, there are t internally disjoint paths between x and y using the vertices in T. (This proof is due to Douglas West.) We distinguish between two cases:

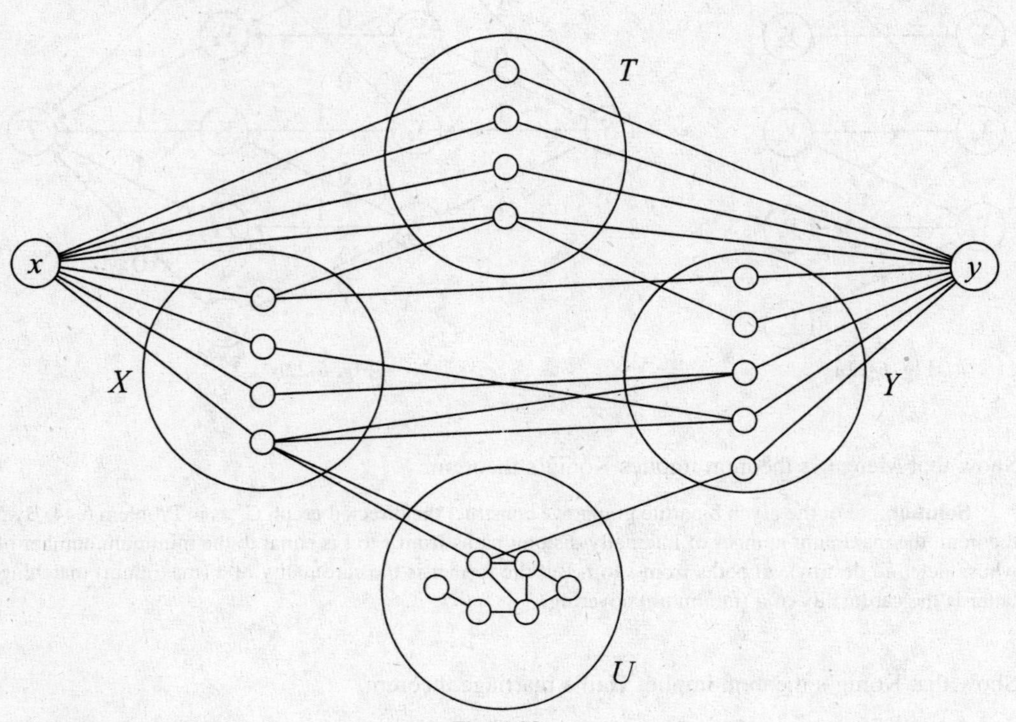

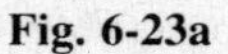

Fig. 6-23a

Case (i): If S is a minimum $x - y$ separating set, $S \cap U = \emptyset$. See Fig. 6-23(a). Let $G' = (X, Y, E')$ be the bipartite graph in which the edges are precisely those edges in G joining vertices in X and Y. See Fig. 6-23(b). Since $(S - T)$ is a set of vertices of minimum cardinality in the bipartite graph and since the exclusion of one of its vertices will no longer make S a separating set, $(S - T)$ is necessarily a minimum vertex cover in the bipartite graph. So, by Konig's theorem, there exists a (maximum) matching of size $(k - t)$ in the bipartite graph, giving $(k - t)$ internally disjoint paths between x and y using the vertices in $(S - T)$. Thus there are k internally disjoint paths between x and y, out of which t paths have two edges and the rest have three edges.

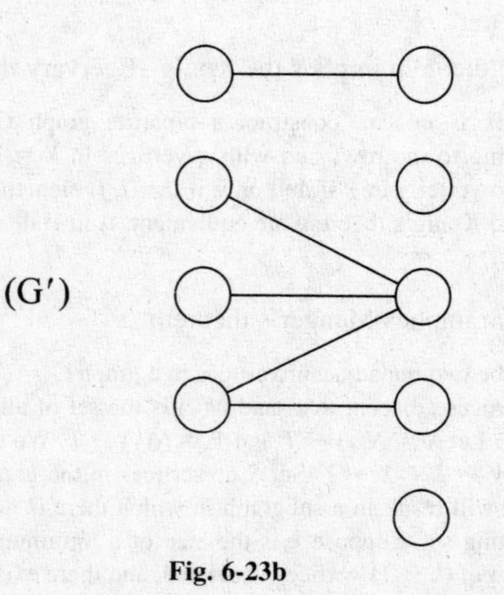

Fig. 6-23b

Case (ii): There is a minimum x, y separating set S of size k such that $S \cap U \neq \emptyset$. See Fig. 6-24. It can be proved by induction on n (the order of G) that there are k internally disjoint paths between x and y. Assume that Menger's theorem is true for all graphs of order less than n. Let $G(x)$ be the subgraph of G consisting of all paths between x and the vertices in S such that no two such paths have a vertex in common other than x. Construct the graph $G'(x)$ by introducing an artificial vertex x' and joining x' to each vertex in S. The size of a minimum x, x' separating set in $G'(x)$ cannot be more than k; if it is less than k, the minimality requirement regarding S will be violated. Since the paths chosen to construct $G(x)$ are pairwise disjoint (except at x), there is at least one vertex in $N(x)$ that is not a vertex of $G'(x)$. So its order is less than n. So by the induction hypothesis, there are k internally disjoint paths between x and x'. Hence, there are k pairwise disjoint paths between x and the vertices in S. Similarly, it is established that there are k such paths between y and the vertices in S. If we splice the k paths from x in $G(x)$ and the k paths from y in $G(y)$, we get k internally disjoint paths between x and y.

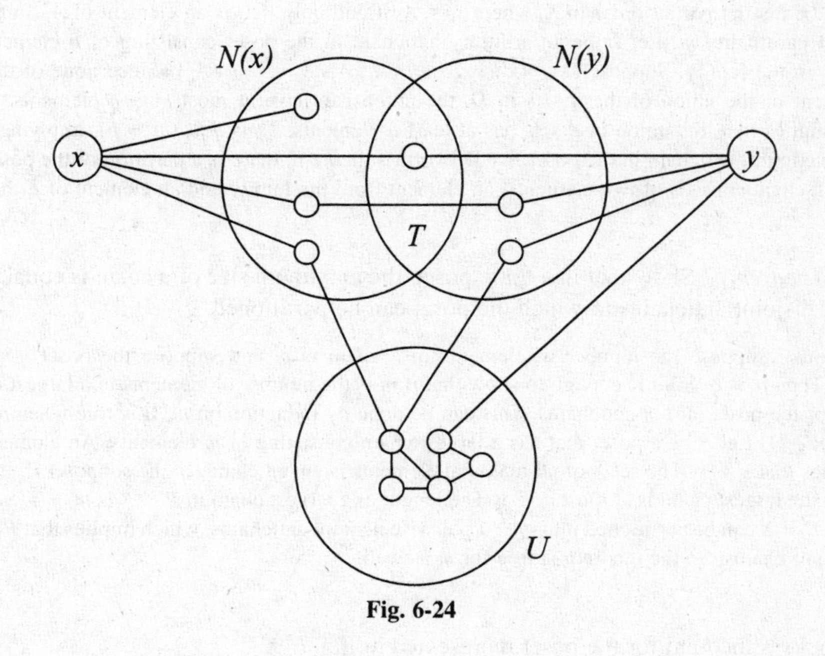

Fig. 6-24

6.50 Prove Theorem 6.17 (Konig's marriage theorem): If a bipartite graph $G = (X, Y, E)$ is k-regular (where k is positive), there is a perfect matching in that graph.

Solution. Let A be any subset of X, and let $f(A)$ be the set of vertices in Y that are adjacent to at least one vertex in A. Let E_1 be the set of edges adjacent to the vertices in A, and let E_2 be the set of edges adjacent to the vertices in $f(A)$.Then $|E_1| \leq |E_2|$. But $|E_1| = (k)|A|$ and $|E_2| = (k)|f(A)|$. Thus $|f(A)| \geq |A|$ for any set subset A of X, satisfying the hypothesis of Hall's theorem. So there is a complete matching from X and Y. But both X and Y have the same number of vertices since k is positive. So there is a perfect matching between X and Y.

6.51 Prove Theorem 6.18, (Dilworth's theorem): In a finite poset, the maximum size of an antichain is equal to the minimum number of chains into which the set of elements of the poset can be partitioned.

Solution. The finite poset $P = (\{x_1, x_2, \ldots, x_n\}, \leq)$ can be represented by an $n \times n$ matrix $\mathbf{A} = [a_{ij}]$, where $a_{ij} = 1$ if and only if $x_i < x_j$. So $a_{ij} = 1$ implies that $a_{ji} = 0$. Assume that a set of 1's in the matrix is an independent set if no two 1's belonging to the set are on the same row or column. Every chain in the poset consisting of two or more elements defines an independent set. Suppose a chain decomposition of P consists of p_i nonsingleton chains of length k_i $(i = 1, 2, \ldots, r)$ and q singleton chains. Then $n = \Sigma p_i q_i + q$. Now this decomposition will define an independent set in which the number of 1's will be m, where $m = \Sigma p_i(q_i - 1) = \Sigma p_i q_i - \Sigma p_i = n - q - p$, where $p = \Sigma p_i$. Since $p + q$ is the total number of chains, the cardinality of the poset is the sum of the number of chains in the decomposition and the number of 1's in the corresponding independent set. Therefore, if there exists a maximum independent set of size t (the term rank of the matrix), there is a minimum chain decomposition of q singleton chains and $n - t - q$ chains of length more than 1. Then by the Konig–Egervary theorem, the matrix can be covered by t lines and cannot be covered by less than t lines. This minimal cover corresponds to a set D of $n - t - q$ elements of P (one from each of the $n - t - q$ chains). The set D' of elements that constitute the singleton chains is of cardinality q. Then the union of D and D' is a set of incomparable elements of cardinality $n - t$. So there is a chain decomposition of the poset P consisting of $n - t$ chains and an antichain of cardinality $n - t$ in P.

6.52 Prove Theorem 6.19: Dilworth's theorem implies Hall's marriage theorem.

Solution. Suppose $\{A_i : i = 1, 2, \ldots, n\}$ is a family of subsets (not necessarily distinct) of a finite set $E = \{x_1, x_2, \ldots, x_m\}$ satisfying Hall's marriage condition. Consider the set $X = \{x_1, x_2, \ldots, x_m, A_1, A_2, \ldots, A_m\}$. Let $<$ be a strict partial order in X, where $x_i < A_j$ if and only if x_i is an element of A_j. In this poset, E is an antichain of cardinality m. Let D be an arbitrary antichain in the poset consisting of p elements from E and q elements from the family. Suppose $D = \{x_1, x_2, \ldots, x_p, A_1, A_2, \ldots, A_q\}$. Since none of the p elements can be an element of the union of the q sets in D, the union can have at most $m - p$ elements. But the marriage condition implies that this union of q sets has at least q elements. Thus $q \leq (m - p)$, implying that $p + q \leq m$. So E is a maximum antichain in the poset. By Dilworth's theorem, there is a partition of the poset consisting of m chains. Each chain consists of two elements: an element from the family and an element of E that belongs to it.

6.53 (*Mirsky's Theorem*) Show that in a finite poset, the maximum size of a chain is equal to the minimum number of disjoint antichains to which the poset can be partitioned.

Solution. Suppose the number of elements in a chain is p, and suppose the poset is partitioned into q antichains. Then $p \leq q$. So it is enough to show that if m is the number of elements in a largest chain, there exists a partition of the poset into m antichains. This can be done by induction on m. It is true when $m = 1$. Suppose it is true for $m - 1$. Let P be a poset that has a largest chain consisting of m elements. An element x is maximal if $x \leq y$ implies that $x = y$. The set X of all maximal elements is an antichain. In the subposet $P - X$, the number of elements in the largest chain is at most $n - 1$. The length of a largest chain in $P - X$ is $m - 1$. So by the induction hypothesis, $P - X$ can be partitioned into $m - 1$ pairwise disjoint antichains, which implies that P can be partitioned into m disjoint chains. So the theorem is true for m as well.

6.54 Verify Mirsky's theorem for the poset represented in Fig. 6-9.

Solution. The set $\{1, 6, 8, 13\}$ is a chain with maximum number of elements. The four antichains $\{1, 2, 3\}$, $\{4, 5, 6\}$, $\{11, 12, 13\}$, and $\{7, 8, 9, 10\}$ constitute a partition of the poset.

Supplementary Problems

6.55 Find the maximum flow value and a minimum cut in the network with six vertices with vertex 1 as the source and vertex 6 as the sink and with the following weight matrix:

$$\begin{bmatrix} - & 24 & - & 27 & - & - \\ - & - & 15 & 6 & - & 6 \\ - & - & - & - & - & 8 \\ - & - & 12 & - & 12 & - \\ - & - & - & - & - & 15 \\ - & - & - & - & - & - \end{bmatrix}$$

Ans. The maximum flow value is 26, and a minimum cut is $C(S, T)$, where S consists of vertices 1, 2, 3, and 4.

6.56 Find the maximum flow value and a minimum cut in the network with eight vertices with vertex 1 as the source and vertex 8 as the sink and with the following weight matrix:

$$\begin{bmatrix} - & 16 & 24 & 12 & - & - & - & - \\ - & - & - & - & 30 & - & - & - \\ - & - & - & - & 9 & 6 & 12 & - \\ - & - & - & - & - & - & 21 & - \\ - & - & - & - & - & 9 & - & 15 \\ - & - & - & - & - & - & - & 9 \\ - & - & - & - & - & - & - & 18 \\ - & - & - & - & - & - & - & - \end{bmatrix}$$

Ans. The maximum flow value is 42, and a minimum cut is $C(S, T)$, where T consists of only the sink.

6.57 Find the maximum flow value and a minimum cut in the network with 12 vertices with vertex 1 as the source and vertex 12 as the sink in which the arcs are (1, 2), (1, 3), (2, 4), (2, 5), (3, 4), (3, 5), (3, 6), (4, 7), (5, 7), (5, 8), (5, 9), (6, 9), (7, 10), (8, 10), (8, 11), (9, 11), (10, 12), and (11, 12) with weights 14, 6, 2, 4, 6, 6, 6, 8, 4, 8, 4, 6, 6, 4, 4, 6, 8, and 6, respectively.
Ans. The maximum flow value is 12, and a minimum cut is $C(S, T)$, where S consists of the source and vertex 2.

6.58 Find the maximum flow value and a minimum cut in the network with seven vertices with vertex 1 as the source and vertex 7 as the sink and with the following weight matrix:

$$\begin{bmatrix} - & 3 & 18 & - & 12 & - & - \\ - & - & - & 12 & 6 & - & - \\ - & - & - & - & - & 12 & - \\ - & - & - & - & - & - & 27 \\ - & 9 & - & 9 & - & 3 & - \\ - & - & - & - & 6 & - & 12 \\ - & - & - & - & - & - & - \end{bmatrix}$$

Ans. The maximum flow value is 27, and a minimum cut is $C(S, T)$, where S consists of the source and vertex 3.

6.59 Find the maximum flow value and a minimum cut in the network with eight vertices with vertex 1 as the source and vertex 8 as the sink and with the following weight matrix:

$$\begin{bmatrix} - & 14 & - & - & - & 10 & 18 & - \\ - & - & 18 & 8 & - & - & - & - \\ - & - & - & - & - & - & - & 10 \\ - & - & 14 & - & - & - & - & 20 \\ - & - & 16 & - & - & - & - & 6 \\ - & - & - & 8 & - & - & - & - \\ - & - & - & - & 16 & 6 & - & - \\ - & - & - & - & - & - & - & - \end{bmatrix}$$

Ans. The maximum flow value is 32, and a minimum cut is $C(S, T)$, where T consists of the sink and vertex 4.

6.60 Find the maximum flow value and a minimum cut in the network with 10 vertices with vertex 1 as the source and vertex 10 as the sink in which the arcs are (1, 2), (1, 3), (1, 4), (1, 5), (2, 6), (3, 2), (3, 7), (4, 7), (4, 8), (5, 4), (6, 7), (6, 10), (7, 10), (8, 9), (8, 10), (9, 5), and (9, 10) with weights 12, 12, 8, 8, 6, 10, 12, 10, 16, 10, 8, 8, 16, 10, 8, 8, and 6, respectively.
Ans. The maximum flow value is 38, and a minimum cut is $C(S, T)$, where S consists of the source and vertex 2.

6.61 Find a generalized minimum cut in the vertex capacitated network with vertices 1 (source), 2, 3, 4, 5, 6 (sink) with weights 0, 1, 1, 3, 2, 0 and with arcs (1, 2), (1, 3), (2, 4), (3, 2), (3, 5), (4, 3), (4, 6), (5, 4), and (5, 6) and with weights 5, 2, 3, 2, 5, 1, 7, 1, and 4, respectively.
Ans. The generalized minimum cut consists of vertices 2 and 3 and no arcs.

6.62 Verify Konig's theorem and Konig's "other" theorem in the case of the bipartite graph (V, W, E), where $V = \{1, 2, 3, 4, 5\}$, $W = \{6, 7, 8, 9\}$, and $E = \{1, 6\}, \{2, 6\}, \{3, 7\}, \{3, 9\}, \{4, 6\}, \{5, 7\}, \{5, 8\}$. [*Hint:* A maximum matching consists of three edges. A maximum independent set consists of six vertices.]

6.63 Construct the binary matrix corresponding to the bipartite graph in Problem 6.62 with five rows corresponding to the vertices in V and four columns corresponding to the vertices in W such that (i, j) entry in the matrix is positive if and only if there is an edge between i and j. Verify the Konig–Egervary theorem for this binary matrix. [*Hint:* The number of independent elements is 3.]

6.64 Show that the family $\{A_1, A_2, A_3, A_4, A_5, A_6\}$ of sets, where $A_1 = \{a, b, c\}$, $A_2 = \{b, c\}$, $A_3 = \{c, e, f\}$, $A_4 = \{a, b\}$, $A_5 = \{a, c\}$, and $A_6 = \{d, e, f\}$, does not have an SDR. [*Hint:* Consider the union of sets A_1, A_2, A_4, and A_5.]

Chapter 7

Matchings and Factors

7.1 MORE ON MATCHINGS

A set M of edges in a graph G, as defined earlier, is a matching (or an independent edge set) in G if no two edges in M have a vertex in common. An edge in a matching in a graph is a **matched edge,** and an edge of the graph that is not in the matching is a **free edge.** A vertex that is incident to an edge of the matching M is called a **matched vertex** (with respect to M), and any other vertex is an **exposed vertex** with respect to the matching. A path between two vertices is an ***M*-alternating path** if its edges are alternately free and matched with respect to M. An ***M*-augmenting path** between two vertices u and v is an M-alternating path in which both u and v are exposed. A **maximum matching** (also known as a **maximum cardinality matching**) is a matching of maximum cardinality, whereas a **maximal matching** is one that is not a proper subset of another matching. Of course, a maximum matching is necessarily a maximal matching. The following characterization of a maximum matching in a graph is due to C. Berge.

Theorem 7.1 **(Berge's Theorem).** A matching M in a graph is a maximum matching if and only if there is no M-augmenting path in the graph. (see Solved Problem 7.3.)

In a bipartite graph $G = (X, Y, E)$, a matching is a complete matching from X to Y if every vertex in X is incident to an edge of the matching. It was shown in Chapter 6 that a necessary and sufficient condition for the existence of such a complete matching is that $|f(A)| \geq |A|$ for every subset of A of X, where $f(A)$ is the set of vertices that are adjacent to at least one vertex in A. Recall that this statement is the graph-theoretic formulation of Hall's marriage theorem; see Theorems 6.13 and 6.14. See also Solved Problem 7.4 for a proof that Berge's theorem implies Hall's marriage theorem.

A matching in a graph is a **perfect matching** if every vertex of the graph is incident to an edge in the matching. Given an arbitrary graph $G = (V, E)$, it is quite natural to ask whether it has a perfect matching so that the vertices in V can be grouped in pairs using this matching. Hall's marriage theorem implies that every k-regular bipartite graph has a perfect matching; see Theorem 6.17. A necessary and sufficient condition for an arbitrary graph to have a perfect matching was obtained by W. T. Tutte in 1947.

Theorem 7.2 **(Tutte's Theorem).** The graph $G = (V, E)$ has a perfect matching if and only if the number of odd components of $(G - S)$ does not exceed $|S|$ for every $S \subset V$. (See Solved Problem 7.8.)

7.2 THE OPTIMAL ASSIGNMENT PROBLEM

If each edge e of a bipartite graph is assigned a nonnegative weight $w(e)$, the problem of finding a matching M in G such that the sum $w(M)$ of the weights of the edges in M is as small as possible is known as the optimal assignment problem. It is assumed that $w(e)$ is a nonnegative integer. Without loss of generality, we may assume that the bipartite graph under consideration is $K_{n,n} = (X, Y, E)$ by introducing artificial vertices and artificial edges and by assigning the weight $+\infty$ to artificial edges. The **weight matrix** of the graph is $A = [a_{ij}]$, where a_{ij} is the weight of the edge joining vertex x_i in X to vertex y_j in Y. Thus a solution of the optimal assignment problem consists of a choice of n elements from the matrix such that (i) no two selected elements lie in the same row or same column and (ii) the sum of the n selected entries is as small as possible. A choice of n such elements then defines an optimum matching M, also known as an **optimal assignment.** If there exists a permutation P of the $n \times n$ identity matrix I such that the nonzero elements in P lie in the same position as n of the zeros in the weight matrix, the selection of these n elements from A will produce an optimal assignment M such that $w(M) = P.A$ (the "dot product" of the two matrices) is the sum of the n^2 pairwise products of the entries of the two matrices. In this case, we say A is matched with P.

Suppose we add an integer t to each entry in a row (or a column) of the weight matrix A, thereby modifying the weight matrix to a new matrix A'. Then $P.A' = P.A + t$. In other words, the choice of an optimal assignment is unaffected if we modify the weight matrix in this manner.

Thus given an arbitrary weight matrix A, we first try to find out whether we can obtain a modified matrix A' (with nonnegative integers) by systematically subtracting positive numbers from columns and rows such that A' can be matched with a permutation matrix. By subtracting the smallest number of a row from each entry in that row and by continuing this process for each row, we get a modified matrix in which each row has at least one zero. Then we can carry out the same procedure for each column. If we are able to obtain a modified matrix that can be matched with a permutation matrix, we are done.

Example 1. Matrix A given below is modified by subtracting 2 from row 1, 3 from row 2, 2 from row 3, and 2 from row 4. Then we subtract 3 from column 1. The modified matrix A' can be matched with the permutation matrix P. The three matrices are as follows:

$$A = \begin{bmatrix} 5 & 4 & 2 & 4 \\ 6 & 3 & 3 & 5 \\ 6 & 2 & 5 & 2 \\ 6 & 3 & 2 & 7 \end{bmatrix}, \quad A' = \begin{bmatrix} 0 & 2 & 0 & 2 \\ 0 & 0 & 0 & 2 \\ 1 & 0 & 3 & 0 \\ 1 & 1 & 0 & 5 \end{bmatrix}, \quad \text{and} \quad P = \begin{bmatrix} 1 & 0 & 0 & 0 \\ 0 & 1 & 0 & 0 \\ 0 & 0 & 0 & 1 \\ 0 & 0 & 1 & 0 \end{bmatrix}$$

If $X = \{x_1, x_2, x_3, x_4\}$ and $Y = \{y_1, y_2, y_3, y_4\}$ are the sets of vertices in the complete bipartite graph (X, Y, E) then a minimum weight matching M (corresponding to the nonzero entries in P) consisting of edges $\{x_1, y_1\}$, $\{x_2, y_2\}$, $\{x_3, y_4\}$, and $\{x_4, y_3\}$ with a total weight of $5 + 3 + 2 + 2 = A'.P = 12$.

If an association like this exists between the modified matrix and a permutation matrix, an easy procedure to obtain the permutation matrix is as follows. Locate a row or a column with the smallest number of zeros. In a row, identify a zero entry in that row, draw a vertical line through that entry, and ignore that line in any future consideration. If a column is chosen, identify a zero entry in that column, draw a horizontal line through that entry, and ignore that line in any future consideration. Continue this process. If we are able to draw n lines like this, the n identified elements in the matrix will correspond to the nonzero entries of P.

Example 2. We can apply this procedure to obtain a minimum assignment for the weight matrix A and its modified matrix A':

$$A = \begin{bmatrix} 4 & 1 & 3 & 2 & 4 \\ 6 & 2 & 2 & 4 & 5 \\ 1 & 3 & 4 & 1 & 1 \\ 5 & 2 & 3 & 4 & 1 \\ 7 & 6 & 5 & 3 & 3 \end{bmatrix} \quad \text{and} \quad A' = \begin{bmatrix} 3 & 0 & 2 & 1 & 3 \\ 4 & 0 & 0 & 2 & 3 \\ 0 & 2 & 3 & 0 & 0 \\ 4 & 1 & 2 & 3 & 0 \\ 4 & 3 & 2 & 0 & 0 \end{bmatrix}$$

In row 1 of A', there is only one zero. The (1, 2) entry is identified, and a line is drawn along column 2 that is to be ignored in future considerations. At this stage, there is only one zero entry in column 3. So the (2, 3) entry is identified, and a line is drawn along row 2. Then we go to row 4, with a single zero. The (4, 5) entry is identified, and a line is drawn along column 5. Then the (5, 4) entry is identified in row 5, and a line is drawn along column 4. Then the (3, 1) entry is identified, and a line is drawn either along column 1 or along row 3. Thus we are able to draw five lines. The five identified entries of the matrix correspond to the nonzero entries of a permutation matrix. The weight of a minimum matching is therefore $1 + 2 + 1 + 3 + 1 = 8$.

Obviously, such an association between a matrix modified by this method and a permutation matrix need not always exist. For example, consider a modified matrix in which the only zero of row i and the only zero of row j both lie on column k. In cases like this, we have to redistribute the zeros of the modified matrix so that it can be associated with a permutation matrix. According to the Konig–Egervary theorem, the number of edges in maximum matching (using the zeros of the reduced matrix) is equal to the minimum number of lines that can be drawn to cover all the zeros of the matrix. Suppose the number of lines needed to cover the zeros is k, which is less than n. If t is the smallest uncovered entry, we subtract t from all the entries in each of the uncovered rows. This will convert the zero entries (in the covered columns) into negative entries. At this stage, we add t to all the entries in each covered column. Then we have an updated matrix with a redistribution of zeros. We continue this process until we get n lines and n entries.

Example 3. The matrix A and its modified matrix A' are

$$A = \begin{bmatrix} 4 & 9 & 3 & 11 & 4 \\ 9 & 8 & 3 & 10 & 8 \\ 7 & 5 & 3 & 8 & 6 \\ 9 & 5 & 3 & 4 & 6 \\ 10 & 11 & 7 & 10 & 11 \end{bmatrix}, \quad A' = \begin{bmatrix} 0 & 4 & 0 & 7 & 0 \\ 5 & 3 & 0 & 6 & 4 \\ 3 & 0 & 0 & 4 & 2 \\ 5 & 0 & 0 & 0 & 2 \\ 2 & 2 & 0 & 2 & 3 \end{bmatrix}, \quad \text{and} \quad P = \begin{bmatrix} 0 & 0 & 0 & 0 & 1 \\ 0 & 0 & 1 & 0 & 0 \\ 0 & 1 & 0 & 0 & 0 \\ 0 & 0 & 0 & 1 & 0 \\ 1 & 0 & 0 & 0 & 0 \end{bmatrix}$$

The zeros of A' can be covered by drawing four lines: row 1, row 4, column 2, and column 3. The smallest uncovered entry is 2. We subtract 2 from each entry in row 1 and from each entry in row 2. Then we add 2 to each entry in column 3 and to each entry in column 4. The corresponding permutation matrix P is shown above. The weight of an optimal assignment is $P.A' = 4 + 3 + 5 + 4 + 10 = 26$. (This method of solving the assignment problem is known as the **Hungarian method.**)

Example 4. We apply the Hungarian method to matrix A:

$$A = \begin{bmatrix} 2 & 2 & 2 & 5 & 2 \\ 1 & 3 & 3 & 1 & 4 \\ 6 & 4 & 7 & 4 & 2 \\ 5 & 6 & 5 & 5 & 2 \\ 9 & 8 & 5 & 5 & 2 \end{bmatrix}$$

Here $n = 5$. By subtracting the smallest element from each row we get modified matrix B, which has a zero in each line:

$$B = \begin{bmatrix} 0 & 0 & 0 & 3 & 0 \\ 0 & 2 & 2 & 0 & 3 \\ 4 & 2 & 5 & 2 & 0 \\ 3 & 4 & 3 & 3 & 0 \\ 7 & 6 & 3 & 3 & 0 \end{bmatrix}$$

Iteration 1: At least three lines are needed to cover the zeros of matrix B. Thus $k = 3$. The zeros of B are covered with three lines: row 1, row 2, and column 5. The smallest uncovered entry is $t = 2$. Subtract 2 from each entry of the uncovered rows and add 2 to each entry of the covered columns. We get matrix C:

$$C = \begin{bmatrix} 0 & 0 & 0 & 3 & 2 \\ 0 & 2 & 2 & 0 & 5 \\ 2 & 0 & 3 & 0 & 0 \\ 1 & 2 & 1 & 1 & 0 \\ 5 & 4 & 1 & 1 & 0 \end{bmatrix}$$

Iteration 2: At least four lines are needed to cover the zeros of the matrix C. Thus $k = 4$. The zeros of the matrix C can be covered with four lines: row 1, row 2, row 3, and column 5. The smallest uncovered entry is $t = 1$. Subtract 1 from each entry of the uncovered rows and add 1 to each entry of the covered columns. We get matrix D:

$$D = \begin{bmatrix} 0 & 0 & 0 & 3 & 3 \\ 0 & 2 & 2 & 0 & 6 \\ 2 & 0 & 3 & 0 & 0 \\ 0 & 1 & 0 & 0 & 0 \\ 4 & 3 & 0 & 0 & 0 \end{bmatrix}$$

Iteration 3: We need five lines to cover the zeros of D. Thus there is a permutation matrix P that can be matched with D. An optimal matching has weight $2 + 1 + 4 + 5 + 2 = 14$.

7.3 THE TRAVELING SALESPERSON PROBLEM (TSP)

A Hamiltonian cycle in a (directed) graph G is a (directed) cycle that passes through every vertex of G. The problem of finding a Hamiltonian cycle (if it exists) of minimum weight if each (arc) edge has a nonnegative weight is known as the optimal Hamiltonian problem (OHP). In other words, we are looking for a closed path of minimum weight that passes through each vertex *exactly* once. In many practical situations, a more meaningful question is related to the problem of finding a closed path of minimum weight that passes through each vertex *at least* once. This problem is known as the optimal salesperson problem (OSP). The OHP is usually known as the traveling salesperson problem (TSP).

In the weight matrix of a weighted graph (digraph) $G = (V, E)$ and in its shortest distance matrix, each entry is nonnegative and each diagonal entry is zero. Suppose $V = \{1, 2, \ldots, n\}$. Let $G' = (V, E')$ be the graph (digraph) in which there is an edge (arc) from i to j if and only if there is a path from i to j in G; in that case, the weight of the edge (arc) is the shortest distance from i to j. The (i, j) entry in the weight matrix will be undefined if there is no edge (arc) from i to j. Likewise, the (i, j) entry in the shortest distance matrix will be undefined if there is no path (directed path) from i to j. By replacing each diagonal entry and each undefined entry by ∞, let A be the $n \times n$ matrix obtained from the weight matrix, and by replacing each diagonal entry and each undefined entry by ∞, let D be the $n \times n$ matrix obtained from the shortest distance matrix. Matrix A will be used when we consider the OHP, whereas we use matrix D when the OSP is being considered.

Example 5. Suppose weight matrix A and shortest distance D (after replacing the diagonal entries by ∞) of the digraph $G = (V, E)$, where $V = \{1, 2, 3, 4\}$, are

$$A = \begin{bmatrix} \infty & 5 & 19 & 11 \\ \infty & \infty & 4 & 7 \\ \infty & 5 & \infty & 14 \\ 9 & \infty & 6 & \infty \end{bmatrix}, \quad D = \begin{bmatrix} \infty & 5 & 9 & 11 \\ 16 & \infty & 4 & 7 \\ 21 & 5 & \infty & 12 \\ 9 & 11 & 6 & \infty \end{bmatrix}$$

A solution to the OHP (using matrix A) is $1 \to 2 \to 3 \to 4 \to 1$ with weight $5 + 4 + 14 + 9 = 32$. A solution to the OHP (using matrix D) is $1 \to 3 \to 2 \to 4 \to 1$ with weight $9 + 5 + 7 + 9 = 30$. A shortest path from 3 to 4 is $3 \to 2 \to 4$. Thus a solution to the OSP is $1 \to 2 \to 3 \to 2 \to 4 \to 1$, which is a closed path visiting vertex 2 twice.

A Branch and Bound Method to Solve TSP

Given a graph (or digraph) with $V = \{1, 2, \ldots, n\}$, we construct the complete bipartite graph (X, Y, E), where $X = \{x_i : i = 1, 2, \ldots, n\}$ and $Y = \{y_j : j = 1, 2, \ldots, n\}$ with the weight of the edge joining x_i and y_j the same as the (i, j) entry in A (or in D). Then we solve the optimal assignment problem for this matrix. If this optimal assignment (matching) in A defines a Hamiltonian cycle in G, the OHP is readily solved. Similarly, if the optimal assignment in D defines a Hamiltonian cycle in G', the OSP is readily solved. In many cases, however, this optimal assignment need not define a Hamiltonian cycle. Consider the simple problem where $n = 6$. Suppose an optimal matching in the bipartite graph is $\{(x_1, y_2), (x_2, y_3), (x_3, y_1), (x_4, y_5), (x_5, y_6), (x_6, y_4)\}$. Here we get two disjoint cycles (subtours) in the graph and not a Hamiltonian cycle. Such subtours have to be eliminated to get an optimal solution.

Suppose A is the $n \times n$ weight matrix of G with $w(A)$ as the weight of an optimal assignment in A. If this assignment gives a Hamiltonian cycle, we are done. Otherwise, $w(A)$ is a lower bound of weight Z of an optimal Hamiltonian cycle. If the graph is not Hamiltonian, Z is ∞. Suppose the optimal assignment for A gives a subtour passing through set $S = \{v_1, v_2, \ldots, v_k\}$ of k vertices. Any Hamiltonian cycle in G should contain at least one arc from a vertex in S to a vertex in $S' = V - S$. So we create k subproblems corresponding to the k vertices in S as follows. For the ith subproblem, replace the weight of each arc from v_i to every other vertex in S by ∞ so that these arcs will not be used, compelling us to use an arc (if it exists) from v_i to a vertex in $V - S$. The matrix thus obtained is the **branching matrix** for vertex v_i. We thus have k branching matrices and k optimal assignment problems. Each matrix will give an optimal assignment, the weight of which cannot be less than $w(A)$. Since any Hamiltonian cycle in G should contain an arc from a vertex in S to a vertex in S', it is enough if we examine only these k optimal solutions in our future computations. The subtour caused by set S is eliminated, and at the same time, no Hamiltonian cycles are lost in this branching process. If any of these subproblems gives a Hamiltonian cycle and if its weight does not exceed the minimum weight of the other subproblems, that cycle is optimal. Otherwise, we take the subproblem with the **smallest lower bound** (among the existing subproblems) and branch out again from its matrix using one of its subtours. This branch and bound method eventually leads us to an optimal solution if it exists or it leads us to the conclusion that there is no Hamiltonian cycle. See Example 6.

Example 6. Consider a problem in which the initial optimal assignment is non-Hamiltonian with weight 50. We write $w = 50$(NH) as the lower bound for the first iteration in the algorithm. NH stands for non-Hamiltonian assignment, whereas H stands for Hamiltonian assignment. For this optimal assignment, suppose there is a subtour consisting of five vertices that gives the optimal assignments with weights 53(NH), 54(NH), 56(NH), 58(H), and ∞. We can ignore the Hamiltonian cycle with weight 58 as well as the assignment with ∞. We can also conclude that there is no Hamiltonian cycle with weight less than 53. It is not known whether there is a Hamiltonian cycle with weight less than 56 at this stage. So the current lower bound for the second iteration is 53. Suppose a subtour for the non-Hamiltonian subproblem (with weight 53) consists of three vertices, giving optimal assignments 55(NH), 57(H), and ∞. The assignments 57(H) and ∞ can be ignored. Then the current lower bound for the third iteration becomes 54. We now start with the matrix that gives the non-Hamiltonian assignment with weight 54, choose one of its subtours, and proceed as before. Suppose we have the optimal assignments 55(H), 55(NH), 59(H), and 61(NH) at this stage. The branch and bound procedure ends with the report that the optimal solution is the Hamiltonian cycle with weight 55 obtained in this branch. We thus have a **tree enumeration scheme** depicted by an arborescence with the root at the initial weight matrix and vertices at subsequent branching matrices. See Fig. 7-1.

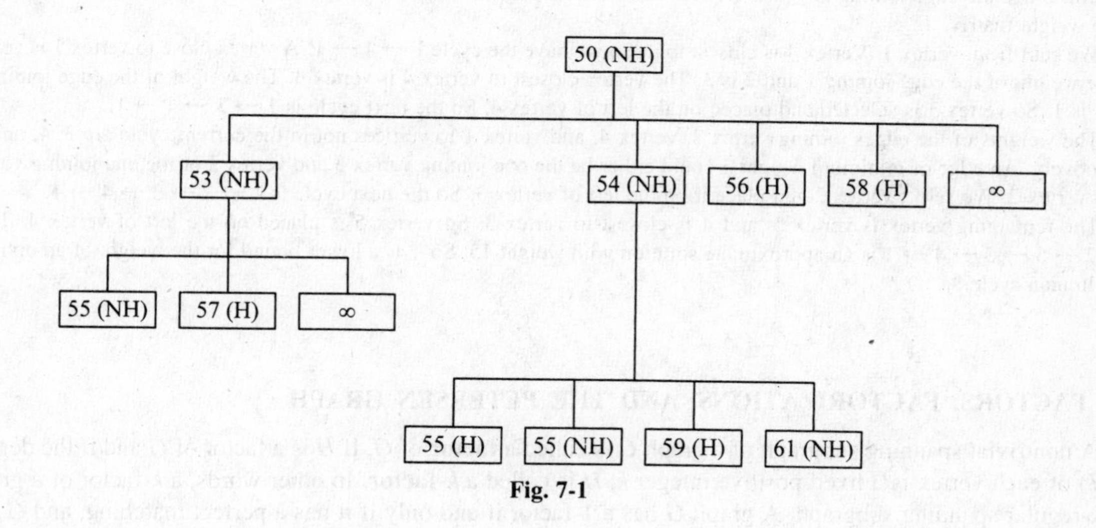

Fig. 7-1

Obtaining an Approximate Solution

Eliminating all the subtours is not an easy problem. In a typical problem, the number of such subtours could be very large, even when the number of vertices is not. So in many cases, it will be helpful to find a Hamiltonian cycle (assuming it exists) whose weight is as close as possible to the weight of an optimal Hamiltonian cycle. We can find an "approximate solution" of this kind in certain types of graphs. If we consider a complete undirected graph (with a nonnegative weight function defined on its set of edges) in which the edge joining every pair of vertices is a shortest path between them, it is possible to obtain a Hamiltonian cycle whose weight does not exceed twice the weight of an optimal Hamiltonian cycle. This quick method involves the following steps:

Step 1. Choose any vertex v_1 as the initial cycle C_1 with one vertex

Step 2. Let C_k be a cycle with k vertices. The vertices in this cycle are arranged on a line, with v_1 as

the first vertex as well as the last vertex. From each vertex (except the first vertex) in this line, select a vertex not in C_k that is nearest to that vertex. Among those selected vertices, choose a vertex u that is closest to the vertices in C_k. Let v be the vertex that is adjacent to u.

Step 3. **Let C_{k+1} be the cycle obtained by inserting u adjacent to v on its left on the line.**

Step 4. **Repeat steps 2 and 3 until all the vertices are included in the cycle.**

(The weight of the Hamiltonian cycle thus obtained does not exceed twice the weight of an optimal Hamiltonian cycle. See Solved Problem 7.35 for a proof.)

Example 7. In the complete graph with five vertices whose weight matrix is

$$A = \begin{bmatrix} \infty & 3 & 3 & 2 & 7 \\ 3 & \infty & 3 & 4 & 5 \\ 3 & 3 & \infty & 1 & 4 \\ 2 & 4 & 1 & \infty & 5 \\ 7 & 5 & 4 & 5 & \infty \end{bmatrix}$$

the method described above can be used to obtain a Hamiltonian cycle as an approximate solution of the optimal Hamiltonian problem. Since the edge joining any two vertices is a shortest path between them, the shortest distance matrix is the same as the weight matrix.

We start from vertex 1. Vertex 4 is closest to it. So we have the cycle $1 \to 4 \to 1$. A vertex close to vertex 1 is vertex 2. The weight of the edge joining 1 and 2 is 3. The vertex closest to vertex 4 is vertex 3. The weight of the edge joining 3 and 4 is 1. So vertex 3 is selected and placed on the left of vertex 4. So the next cycle is $1 \to 3 \to 4 \to 1$.

The weights of the edges joining vertex 3, vertex 4, and vertex 1 to vertices not in the current cycle are 3, 4, and 3, respectively. An edge of minimum weight 3 could either be the one joining vertex 3 and vertex 2 or the one joining vertex 1 and vertex 2. We select vertex 2 and place it on the left of vertex 3. So the next cycle is $1 \to 2 \to 3 \to 4 \to 1$.

The remaining vertex is vertex 5, and it is closest to vertex 3. So vertex 5 is placed on the left of vertex 4. Thus $1 \to 2 \to 5 \to 3 \to 4 \to 1$ is an approximate solution with weight 15. So $\frac{15}{2}$ is a lower bound for the weight of an optimal Hamiltonian cycle.

7.4 FACTORS, FACTORIZATIONS, AND THE PETERSEN GRAPH

A nontrivial spanning subgraph of a graph G is called a **factor** of G. If H is a factor of G and if the degree (in H) of each vertex is a fixed positive integer k, H is called a ***k*-factor.** In other words, a k-factor of a graph is a k-regular spanning subgraph. A graph G has a 1-factor if and only if it has a perfect matching, and G has a connected 2-factor if and only if it is Hamiltonian. A necessary and sufficient condition (Tutte's theorem) for the existence of a 1-factor in G is that the number of odd components of $(G - S)$ does not exceed the cardinality of S, whereas a necessary condition (Theorem 3.4) for the existence of a connected 2-factor is that the number of components of $(G - S)$ does not exceed the cardinality of S, where S is any set of vertices of G. It is an immediate consequence of Hall's marriage theorem that any nontrivial regular bipartite graph has a 1-factor.

A set $\{H_i = (V, E_i) : i = 1, 2, \ldots, p\}$ of factors of a graph $G = (V, E)$ is a **factorization** of G if the set $\{E_i : i = 1, 2, \ldots, p\}$ forms a partition of E. If G has such a factorization, it can be **factored into** p factors. A factorization of G consisting of k-factors is called a ***k*-factorization.** If $\{H_1, H_2, \ldots, H_p\}$ is a k-factorization of G, we write $G = H_1 \oplus H_2 \oplus \cdots \oplus H_p$ and say that G is ***k*-factorable.** If a graph is k-factorable, it is a regular graph in which the degree of each vertex is a multiple of k. A subgraph H of a graph G is an **isofactor** of G if G has a factorization (consisting of at least two factors) such that each factor in the factorization is isomorphic to H. If G has an isofactor H, we say that G is ***H*-factorable** with an **isomorphic factorization** into the factor H.

The problem involving factors and factorizations of graphs that has received the most attention is the characterization of graphs that have a 1-factor and graphs that are 1-factorable. In the former case, even though we have Tutte's characterization of such graphs which besides being considered as one of the basic theorems in graph theory serves as an important tool in the investigation of factors and factorizations of graphs, no easily applicable criterion has been found to determine whether an arbitrary graph has a 1-factor. In the latter case, graphs that are 1-factorable have not yet been classified. Only certain classes of regular graphs are known to be 1-factorable.

Theorem 7.3. (*a*) A complete graph of even order is 1-factorable. (*b*) Any r-regular nontrivial bipartite graph is 1-factorable. (See Solved Problems 7.45 and 7.46.)

Theorem 7.4. A simple graph is 2-factorable if and only if it is r-regular, where r is even. (See Solved Problem 7.48.)

Theorem 7.5. (*a*) The complete graph of order $(2n + 1)$ can be factored into n Hamiltonian cycles. (*b*) The complete graph of order $2n$ can be factored into $(n - 1)$ Hamiltonian cycles and a 1-factor. (See Solved Problems 7.49 and 7.53.)

Trivially, every 1-regular graph has a 1-factor and is 1-factorable. On the other hand, a 2-regular graph G has a 1-factor if and only if every component of G is an even cycle; in that case, it is 1-factorable. Figure 7-2 shows that an arbitrary 3-regular (known as a **cubic graph**) graph need not have 1-factor. By deleting vertex v, we get three odd components, so by Tutte's theorem, this cubic graph has no perfect matching.

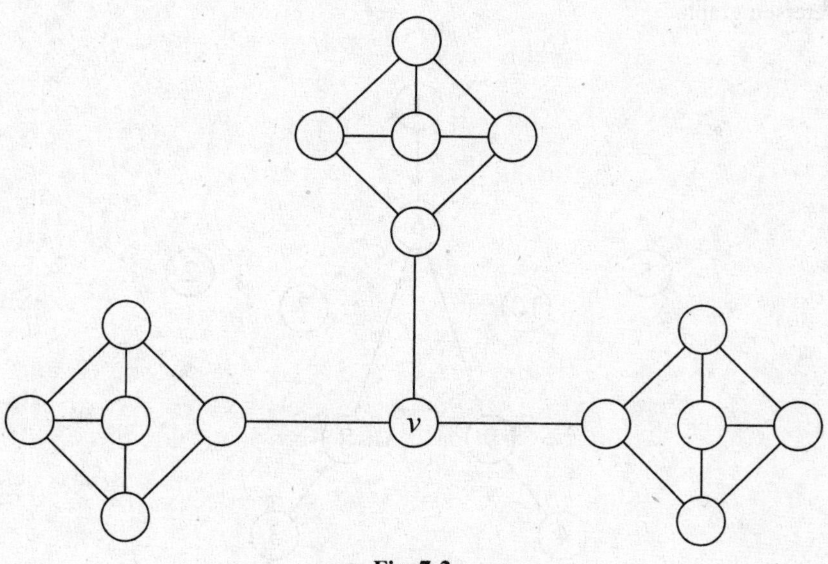

Fig. 7-2

When will a cubic graph have a 1-factor? The following theorem gives an easily verifiable sufficient condition.

Theorem 7.6 **(Petersen's Theorem).** A cubic graph in which no edge is a bridge can be factored into a 2-factor and a 1-factor. (See Solved Problem 7.59.)

Even though a bridgeless cubic graph has a 1-factor, it need not be a 1-factorable graph in general. The **Petersen graph** shown in Fig. 7-3 is a counterexample.

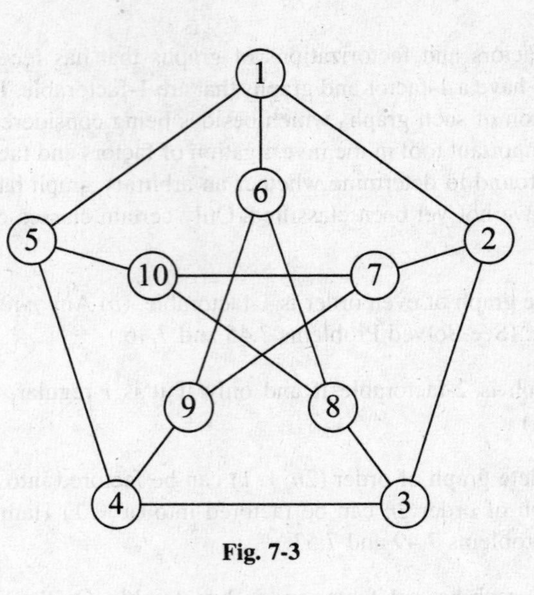

Fig. 7-3

Theorem 7.7. The Petersen graph is not 1-factorable. (See Solved Problem 7.60.)

Observe that the size of an isofactor of a graph G is a divisor of the size of G. So if the Petersen graph has an isofactor, it should have five or three edges. Now $5K_2$ is not an isofactor of the Petersen graph since it is not 1-factorable. It can be easily verified, however, that the graph shown in Fig. 7-4 with five edges is an isofactor of the Petersen graph.

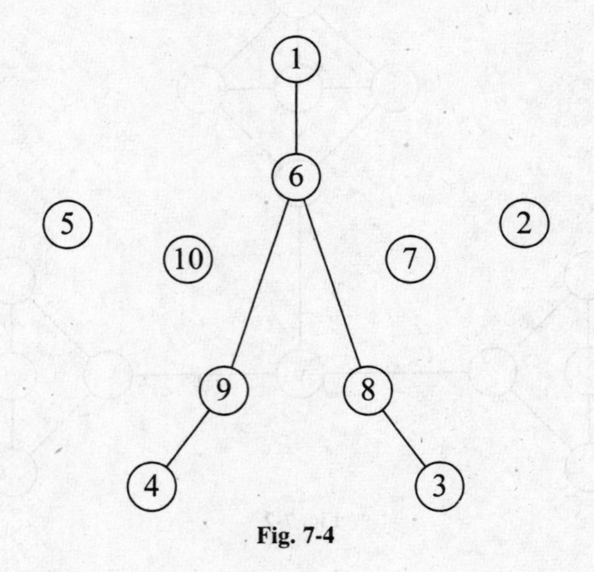

Fig. 7-4

Solved Problems

MORE ON MATCHINGS

7.1 Find the number of perfect matchings in $K_{n,n}$ and in K_{2n}.

Solution. In the case of bipartite graphs, a vertex can be arbitrarily chosen from one of its partite sets and can be matched with a vertex in the other set in n ways. Delete these two vertices, and then continue the process.

So in this case there are $n!$ perfect matchings. In the case of K_{2n}, suppose there are f_n perfect matchings. There are $(2n - 1)$ ways of matching an arbitrary vertex with any one of the remaining vertices. So we have the recursion formula $f_n = (2n - 1)f_{n-1}$ with $f_1 = 1$. Thus $f_n = 1 \cdot 3 \cdot 5 \cdot \ldots \cdot (2n - 1)$.

7.2 If M is a matching in a graph G and if P is an M-augmenting path in G, show that the symmetric difference $(M \Delta P)$ is also a matching in G with one more edge than M.

Solution. Suppose e and f are any two edges in $(M \Delta P)$, the disjoint union of $(P - M)$ and $(M - P)$. If both e and f are in $P - M$ (or, for that matter, in $M - P$), they cannot have a vertex in common since P is an M-alternating path. Suppose e is in $(P - M)$ and f is in $(M - P)$. If e and f have vertex v in common, v is necessarily a terminal vertex of the M-augmenting path P, implying that v is not an exposed vertex, which is a contradiction. Thus no two edges in $(M \Delta P)$ have a vertex in common, so it is indeed a matching. Notice that the augmenting path P has k matched edges and $(k + 1)$ free edges. If M has m edges, $(M - P)$ has $m - k$ edges and $(P - M)$ has $(2k + 1) - k = (k + 1)$ edges. Thus $(M \Delta P)$ has $m + 1$ edges.

7.3 Prove Theorem 7.2 (Berge's Theorem): A matching M in a graph $G = (V, E)$ is a maximum matching if and only if there is no M-augmenting path in G.

Solution. Suppose there exists an M-augmenting path P with respect to a maximum matching M. Then (as shown in Problem 7.2), the symmetric difference $(M \Delta P)$ is a matching that has more edges than the maximum matching, which is a contradiction. So if M is a maximum matching, there is no M-augmenting path. To prove the converse, it is enough to show that there is an M-augmenting path if the given matching M is not a maximum matching. Suppose M' is a maximum matching. Let $G' = (V, M \Delta M')$. In G', the degree of each vertex is either 1 or 2. So each component G is either an even cycle (with an equal number of edges belonging to both the matchings) or a path with edges alternately belonging to the two matchings. There is at least one component P (which is not a cycle) with an odd number of edges; otherwise, both M and M' will have the same cardinality. Since M' has more edges than M, the first edge and the last edge of P belong to M'. In that case, P is an M-augmenting path. In other words, if a matching M is not a maximum matching, there is always an M-augmenting path.

7.4 Show that Berge's theorem implies Hall's marriage theorem.

Solution. It is enough if we establish the following graph-theoretical analog (Theorem 6.14) of Hall's marriage theorem. In a bipartite graph $G = (X, Y, E)$, there is a complete matching from X to Y (matching each vertex in X to some vertex in Y) if and only if $f(A) \geq |A|$ for every $A \subset X$, where $f(A)$ is the set of vertices in Y that are adjacent to at least one vertex in A. Obviously, if there is a complete matching from X to Y, this inequality has to be satisfied. To prove the converse, assume that this inequality is satisfied for all A but that there is no complete matching from X to Y. Suppose M is a maximum matching from X to Y. Since there is no complete matching from X to Y, there is a vertex u in X that is exposed under M. Let S be the set of all vertices in X such that there is an M-alternating path from u, and let T be the set of all vertices in Y such that there is an M-alternating path from u. So every vertex in $S - u$ is matched with a vertex in T and vice versa; hence, $|T| = |S| - 1$, implying that $|T| < |S|$. If vertex v is in $f(S) - T$, there will be an M-augmenting path between u and v, violating the maximality of M. So $T = f(S)$, which implies that $|f(S)| = |T| < |S|$. But $|f(S)| \geq |S|$, according to the hypothesis.

7.5 If the order of a graph G is even and if S is any set of vertices of the graph, the number of odd components of the graph $(G - S)$ is odd if and only if $|S|$ is odd.

Solution. The total number of vertices in the even components is even. If the number of odd components is odd, the total number of vertices in the odd components is odd. Thus the total number of vertices in all the components together is odd. Hence, the number of vertices in S is odd since the order of the graph is even.

7.6 Suppose W is the set of all vertices of degree $(n - 1)$ in a graph G of order n, where n is even. Show that G has a perfect matching if the number of odd components of $(G - W)$ does not exceed $|W|$ and if every component of $(G - W)$ is complete.

Solution. Each even component has a perfect matching. If one vertex is removed from each odd component, the remaining vertices in that component can be matched in pairs. Suppose the vertices removed from the set of odd components constitute a set of k vertices. By assumption, S has $k + q$ vertices, where $q \geq 0$. By hypothesis, each vertex in S is adjacent to every other vertex in the graph. Choose k vertices at random from S and match them arbitrarily in pairs with the k vertices chosen from the odd components. At this stage, there are q unmatched vertices forming set Q. Now q is always even since k is odd if and only if $k + q$ is odd. Moreover, every vertex in Q is adjacent to every other vertex in Q. Thus the set of vertices in Q can also be matched in pairs. So there is a perfect matching in G.

7.7 Show that if G' is the graph obtained from the graph G by joining two of its nonadjacent vertices so that G becomes a spanning subgraph of G', the number of odd components of $(G' - S)$ cannot exceed the number of odd components of $(G - S)$.

Solution. Let $o(G)$ denote the number of odd components of any graph. We have to prove that $o(G' - S) \leq o(G - S)$. Let u and v be two nonadjacent vertices in G, and let G' be the graph obtained from G by joining these two vertices. If either u or v belongs to S or if both u and v belong to the same component, both $(G - S)$ and $(G' - S)$ have the same number of odd components and the same number of even components. If these two vertices belong to two odd components in $(G - S)$, these two odd components join together to become an even component of $(G' - S)$; in this case, $o(G - S) = o(G' - S) + 2$. If u belongs to an odd component of $(G - S)$ and if v belongs to an even component of $(G - S)$, these two components join together as an odd component of $(G' - S)$; then $o(G - S) = o(G' - S)$. If both u and v belong to two different even components, $o(G - S) = o(G' - S)$ also. In other words, $o(G' - S) \leq o(G - S)$.

7.8 Prove Theorem 7.2 (Tutte's theorem): The graph $G = (V, E)$ has a perfect matching if and only if the number of odd components of $(G - S)$ does not exceed $|S|$ for every $S \subset V$.

Solution. Let $o(G)$ denote the number of odd components of G. Suppose M is a perfect matching in the graph. Then the order of G is even. Let S be any set of vertices of G. If S is the empty set, $o(G - S) = o(G) = 0 = |S|$. Suppose S is not empty. If $(G - S)$ has no odd components, $o(G - S) = 0 < |S|$. Consider any odd component H of $(G - S)$. Then there exists at least one vertex v in H that is not matched with any vertex in H. Since there is a perfect matching in G, there is a matched edge e joining v and vertex w that has to be in S. So for each odd component, there is a corresponding vertex in S. Thus if $o(G - S) = k$, $|S| \geq k$.

To prove the sufficiency part of the theorem, it has to be shown that there is a perfect matching in G if $o(G - S) \leq |S|$ for every S. This inequality implies that G has no odd component (by taking S as the empty set); hence, the order n of the graph is even. If the graph is complete, it has a perfect matching. So we assume that G is not complete. Suppose G' is the graph obtained from G by joining any two of its nonadjacent vertices so that G becomes a spanning subgraph of G'. Then, as established in Problem 7.7, $o(G' - S) \leq o(G - S)$. Hence, the inequality $o(G - S) \leq |S|$ implies that $o(G' - S) \leq o(G - S) \leq |S|$. Thus, without loss of generality, we assume that the graph G under consideration is maximal in the sense that there will be a perfect matching in the enlarged graph when any two nonadjacent vertices in G are joined by an edge. It is enough to show that there is a perfect matching in such as maximal graph G satisfying the inequality if $o(G - S) \leq |S|$ for every set S of vertices of G.

Let W be the set of vertices of degree $n - 1$ in G. If each component of $(G - W)$ is complete, there is a perfect matching in G, as proved in Problem 7.6. Suppose $(G - W)$ has a component H that is not complete. So there are three vertices x, y, and z in H such that x and z are not adjacent and y is adjacent to them both. Since y is not in W, its degree is less than $(n - 1)$, which implies that there is a vertex w that is not adjacent to y. This vertex w cannot be in W since the degree of each vertex in W is $(n - 1)$. So w is a vertex in one of the components of $(G - W)$. By the maximality assumption on G, we get a perfect matching M_1 in the enlarged graph G_1 obtained from G by joining x and z by edge e_1. Similarly, we get a perfect matching M_2 in the enlarged graph G_2 obtained from G by joining y and w by edge e_2. Obviously, e_1 is in $M_1 - M_2$ and e_2 is in $M_2 - M_1$. Let F be the spanning subgraph of G whose edges are either in M_1 or in M_2 but not in both. If the edge joining p and q is in both the matchings, the degree of p (and q) in F is 0. If the edge joining p and q is in M_1 but not in M_2, there exists a vertex r adjacent to p such that the edge joining p and r is in M_2. So the degree of p (and similarly of q) in F is 2. Thus the degree of each vertex in F is either 0 or 2. In other words, F is the disjoint union of cycles. Furthermore, in each cycle, the edges belonging to these two matchings occur alternately. Thus each cycle has an even number of edges. If edge e_1 belongs to cycle C_1 and edge e_2 belongs to another cycle C_2, a matching M in G, is readily obtained. M consists of all edges in C_1 that are matched under the M_2 matching and all other edges (from the other cycles) that are matched under the M_1 matching.

Finally, suppose both e_1 and e_2 belong to the same cycle C shown in Fig. 7-5. Every edge in C other than e_1 and e_2 is an edge of G. Edge a (in G) joining x and y cannot be a matched edge under either of these matchings. So a is not an edge of cycle C. Similarly, edge b joining y and z cannot be an edge of C. One of these two edges (say edge a) will divide C into two even cycles, with a as a common edge. Then we can get a matching of the vertices in C consisting of edge a, the edges under the matching M_1 on one of the cycles, and the edges under the matching M_2 in the other cycle. These matchings from C, together with the matchings under either M_1 or M_2, constitute a perfect matching in G. This completes the proof. [The proof of Tutte's theorem presented here is due to L. Lovasz (1973).]

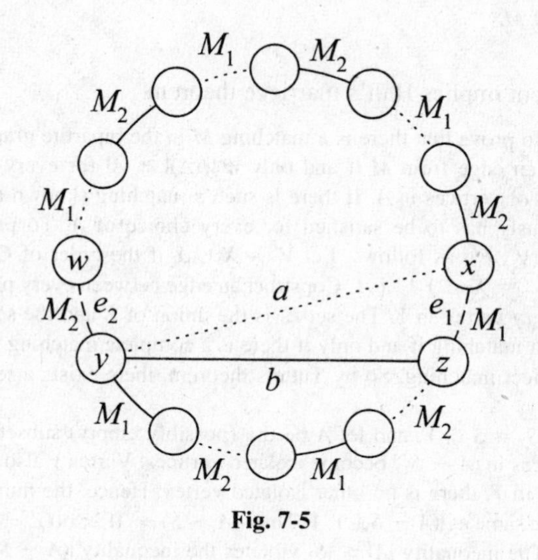

Fig. 7-5

7.9 Show that G has no perfect matching if and only if there exists a set S of vertices of the graph such that the number of odd components of $(G - S)$ is at least $|S| + 2$.

Solution. By Tutte's theorem, G has no perfect matching if and only if there exists a set S of vertices such that $o(G - S) > |S|$. But both $o(G - S)$ and $|S|$ have the same parity: either both are odd or both are even. Hence, $o(G - S) \geq |S| + 2$.

7.10 Show that the minimum number of vertices that cannot be matched in a graph of order n is t if and only if $o(G - S) \leq |S| + t$ for every set S of vertices of the graph.

Solution. Since $n - t$ is even, $n + t$ is also even. Let $G' = (V', E')$ be the graph of order $n + t$ obtained by introducing a set W of t new vertices and by joining each new vertex to the other $n + t - 1$ vertices so that G is a subgraph of G'. Then there will be a matching M in G such that all but t vertices in it are matched if and only if G' has a perfect matching. This is equivalent to the assertion that $o(G' - S') \leq |S'|$, where S' is any subset of V' by Tutte's theorem. There are two cases to be considered: (1) V' is a subset of S'. Let $S = S' - V'$. Then $o(G' - S') = o(G - (S' - V')) = o(G - S) \leq |S| + t = |S'|$. (2) V' is not a subset of S'. In that case, $G' - S'$ is a connected graph. So $o(G' - S') \leq 1$, which implies that $o(G' - S') \leq |S'|$. Thus in either case G' has a perfect matching whenever $o(G - S) \leq |S| + t$ for every set S of vertices of the graph.

7.11 Show that a tree cannot have more than one perfect matching.

Solution. Suppose M and M' are two perfect matchings in a tree such that there is an edge e_1 joining vertices v_1 and v_2 in $M - M'$. Edge e_1 cannot be a terminal edge since a terminal edge of a tree has to be an edge in every perfect matching. So there is a nonterminal vertex v_3 and edge e_2 joining v_2 and v_3 such that e_2 is in $M' - M$. If this process continues, we get a cycle that will terminate at v_1. But the graph is acyclic.

7.12 A tree has a perfect matching if and only if it has exactly one odd component when an arbitrary vertex is deleted from it.

Solution. If v is a vertex in a tree T, each component of the disconnected graph is a subtree. If the tree has a perfect matching, the number of odd subtrees is at most 1 by Tutte's theorem. Suppose all subtrees are even. Now one edge incident to v in T is a matched edge. Suppose this matched edge is the edge joining v and w, where w is a vertex in one of the components. If we exclude these two matched vertices, the total number of remaining vertices in the tree is odd. This is a contradiction since T has a perfect matching. Conversely, suppose $o(T - v) = 1$ for every vertex v of tree T. Suppose there is no perfect matching. Let M be a maximum matching. So there is an unmatched vertex v such that no edge incident to v is a matched edge under M. By hypothesis, $T - v$ has exactly one odd component. In that odd component, the number of unmatched vertices is odd. In particular, there is an unmatched vertex w in that component such that there is an M-augmenting path from v to w, violating the maximality of M.

7.13 Show that Tutte's theorem implies Hall's marriage theorem.

Solution. We have to prove that there is a matching M in the bipartite graph $G = (X, Y, E)$ such that every vertex in X is incident to an edge from M if and only if $|f(A)| \geq |A|$ for every $A \subset X$, where $f(A)$ is the set of vertices adjacent to the set of vertices in A. If there is such a matching (known as a **complete matching from X to Y**), this equality obviously has to be satisfied for every choice of A. To prove the reverse implication, we construct the graph $G' = (V', E')$ as follows. Let $V' = X \cup Y$ if the order of G is even. Otherwise, construct a new vertex v; in that case, $V' = X \cup Y \cup \{v\}$. Construct an edge between every pair of vertices in Y. Also construct an edge between v and every vertex in Y. The set E' is the union of E and the set of new edges thus constructed. Obviously, G' has a perfect matching if and only if there is a complete matching from X to Y in the bipartite graph G. Suppose G' has no perfect matching. So by Tutte's theorem, there exists a set $S \subset V'$ such that $o(G' - S) \geq |S| + 2$.

Let $S_1 = S \cap X$ and $S_2 = S \cap Y$, and let A be the (possibly empty) subset of X such that $f(A) \subset S_2$. If we delete S from G', the vertices in $(A - S_1)$ become isolated vertices. Vertex v also becomes isolated. Because of the construction of new edges in Y, there is no other isolated vertex. Hence, the number of components of $G' - S$ is $|(A - S_1)| + 1$, which is the same as $|(A - S)| + 1$. Thus $|(A - S) + 1| \geq o(G' - S) \geq |S| + 2$. But by hypothesis, $|A| \leq |f(A)| \leq |S_2| \leq |S|$. The inequality $|A| \leq |S|$ violates the inequality $|(A - S)| + 1 \geq |S| + 2$. Hence, G' has a perfect matching.

7.14 The problem of finding a closed walk in an undirected connected weighted network that has at least two odd vertices such that the walk contains each edge at least once and such that its weight is a minimum is known as the undirected **Chinese postman problem (CPP).** Solve this problem if the number of odd vertices is exactly two.

Solution. If there are no odd vertices, the graph is Eulerian; therefore, any Eulerian circuit is a solution to the problem. If the graph is not Eulerian, it can be converted into an Eulerian graph by constructing additional edges so that the degree of each vertex is even. Construct an artificial edge between the two odd vertices so that the enlarged graph G' is Eulerian. Let the weight of this new edge be the shortest distance between these two vertices. Now locate an Eulerian circuit C in G'. Replace the new edge in C' by a shortest path between x and y. The closed trail thus constructed is a solution to the problem.

7.15 Find a solution to the CPP in the network shown in Fig. 7-6.

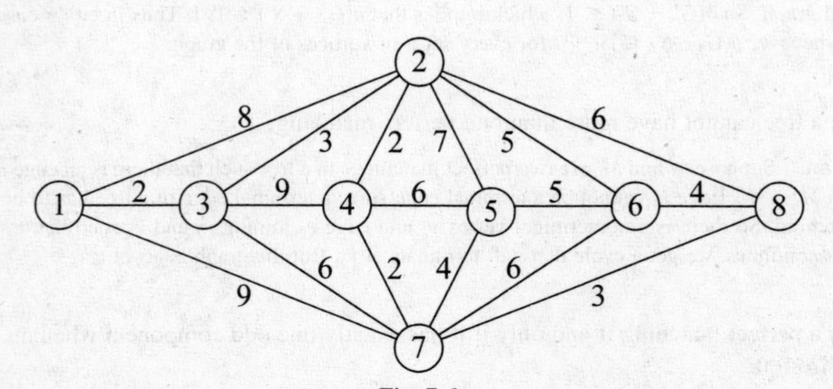

Fig. 7-6

Solution. Vertices 1 and 8 are odd. The shortest path connecting these two vertices is $1 \to 3 \to 2 \to 4 \to 7 \to 8$ with a total weight of 12. These five edges will have to be repeated. The total weight of all the edges in G is 87. Thus the weight of an optimal Chinese postman route is $87 + 12 = 99$. The enlarged Eulerian graph is shown in Fig. 7-7 in which the dotted lines indicate repeated edges. Starting from vertex 1, the postman goes to 3 and returns to 1. At this stage, both the edges joining 1 and 3 are deleted. After that, the vertices visited consecutively are 2, 3, 2, 4, 2, 5, 4, 3, 7, 4, 7, 5, 6, 7, 8, 2, 6, 8, 7, and finally 1. This closed walk contains 22 edges. The network has 17 edges, and five of them are being repeated.

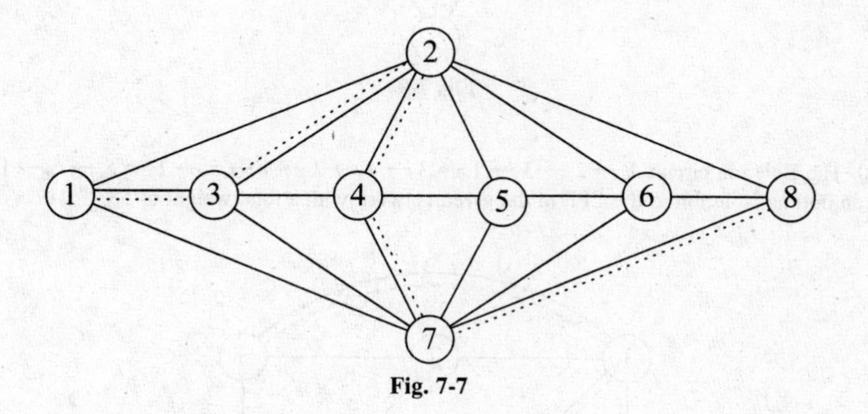

Fig. 7-7

7.16 Discuss the method for solving the CPP if the number of odd vertices is more than two.

Solution. Suppose the number of odd vertices is $2k$. Find the shortest distance between every pair of these odd vertices, and construct a complete graph with $2k$ vertices in which the weight of the edge between two vertices is the shortest distance between them. Suppose M is a minimum weight perfect matching in this complete graph. The edges belonging to M are edges in the shortest paths between pairs of odd vertices. All these edges become repeated edges in the given network. As a result, we have an enlarged network that is Eulerian. Any Eulerian circuit in the enlarged network defines an optimal route in the given network.

7.17 Obtain an optimal solution of the CPP in the network shown in Fig. 7-8.

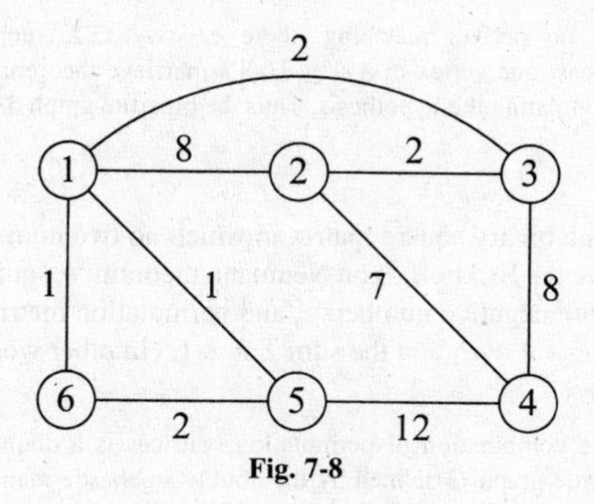

Fig. 7-8

Solution. The odd vertices form the set {2, 3, 4, 5}. The complete graph with these four vertices is displayed in Fig. 7-9, in which the weight of an edge between two vertices is the shortest distance between them in the given network. Of the three perfect matchings in this complete graph, the matching consisting of edges {2, 4} and {3, 5} with weight $7 + 3 = 10$ is the minimum weight perfect matching. Edge {2, 4} in the complete graph represents the path $2 \to 4$, and this edge is to be repeated. Edge {3, 5} in the complete graph represents the path $3 \to 1 \to 5$, and the two edges in this path are also repeated. The enlarged graph with repeated edges is shown in

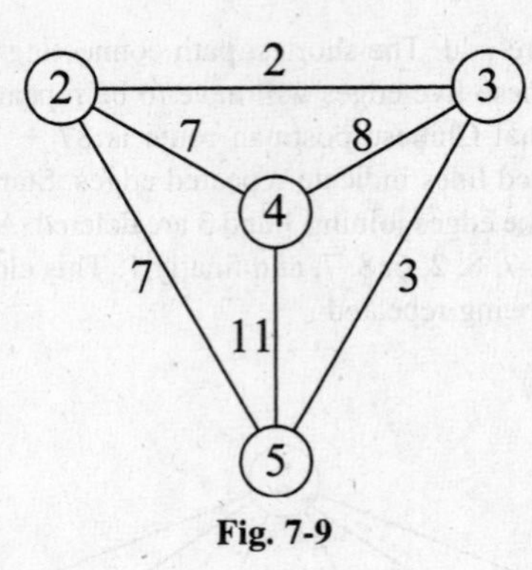

Fig. 7-9

Fig. 7-10. The Eulerian circuit 1 → 2 → 3 → 1 → 3 → 4 → 2 → 4 → 5 → 1 → 5 → 6 → 1 in the enlarged graph is an optimal solution to the CPP in the given network with a total weight of 53.

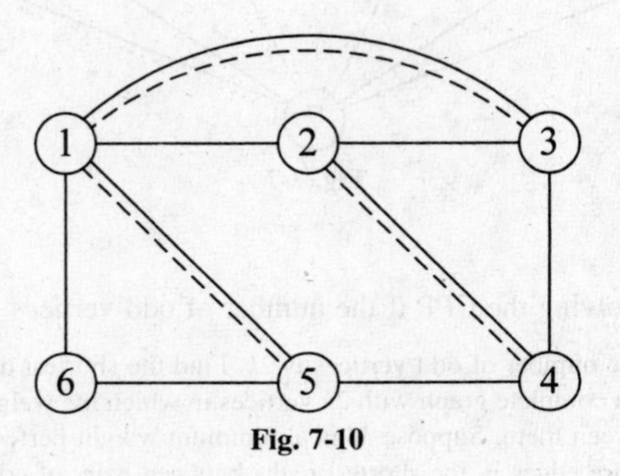

Fig. 7-10

7.18 A matrix $D = [d_{ij}]$ is **doubly stochastic** if each entry is nonnegative and if the sum of the entries in any row or column is equal to 1. Suppose $G = (X, Y, E)$ is a bipartite graph in which each vertex in X corresponds to a row of D and each vertex in Y corresponds to a column of D. Furthermore, there is an edge between vertex x_i in X and vertex y_j in Y if and only if the entry d_{ij} is positive. Show that there is a perfect matching in G.

Solution. If there is no perfect matching, there exists $A \subset X$ such that $|f(A)| < |A|$, where $f(A) = \{y \in Y : y \text{ is adjacent to at least one vertex in } A\}$, by Hall's marriage theorem. This inequality implies that some column sum is more than 1, violating the hypothesis. Thus the bipartite graph defined by a doubly stochastic matrix has a perfect matching.

7.19 A **permutation matrix** is a binary square matrix in which no two nonzero elements appear in the same row or same column. Prove the Birkhoff–von Neumann theorem: A square matrix D is doubly stochastic if and only if there exist nonnegative numbers r_i and permutation matrices P_i $(i = 1, 2, \ldots, k)$ such that $D = r_1P_1 + r_2P_2 + \cdots + r_kP_k$ and the sum $\Sigma\, r_i$ is 1. (In other words, D is a **convex combination** of permutation matrices.)

Solution. Any convex combination of permutation matrices is a doubly stochastic matrix. To prove the converse, consider the bipartite graph G defined by the doubly stochastic matrix D. If D is a permutation matrix, we are done. Assume that this is not the case. Let M_1 be a perfect matching in G, and let P_1 be the permutation matrix that corresponds to this matching. Now each edge has a nonnegative weight d_{ij}, and since D is not a permutation matrix, there is a weight that is less than 1. Choose edge e_1 of minimum weight a_1 in the matching M. Then we can write $D = a_1P_1 + (1 - a_1)D_1$, where D_1 is doubly stochastic with more zero entries than D. At the next stage, the doubly stochastic matrix D_1 is expressed as $a_2P_2 + (1 - a_2)D_2$, where D_2 has more zeros than D_1. We continue this process until we reach a stage where the current matrix D_k is a permutation matrix, which is P_k. Thus we get a convex combination of permutation matrices.

7.20 Represent the following doubly stochastic matrix as a convex combination of permutation matrices:

$$\begin{bmatrix} \frac{1}{2} & \frac{1}{2} & 0 \\ \frac{1}{4} & 0 & \frac{3}{4} \\ \frac{1}{4} & \frac{1}{2} & \frac{1}{4} \end{bmatrix}$$

Solution. The bipartite graph is $G = (X, Y, E)$, where both X and Y have three vertices. There is no edge between x_1 and y_3. A perfect matching M_1 consists of edges $\{x_1, y_2\}$, $\{x_2, y_1\}$, and $\{x_3, y_3\}$. Edge $a_1 = \{x_2, y_1\}$ is an edge of minimum weight in M_1. The permutation matrix P_1 corresponding to M_1 has rows [0 1 0], [1 0 0], and [0 0 1]. The weight of a_1 is $\frac{1}{4}$. We write $D = (\frac{1}{4})P_1 + (\frac{3}{4})D_1$, where the rows of the matrix D_1 are $[\frac{2}{3}\ \ \frac{1}{3}\ \ 0]$, [0 0 1], and $[\frac{1}{3}\ \ \frac{2}{3}\ \ 0]$. Notice that D_1 has more zeros than D. A perfect matching M_2 in the bipartite graph defined by D_1 contains $\{x_1, y_2\}$, $\{x_2, y_3\}$, and $\{x_3, y_1\}$ with weights $\frac{1}{3}$, 1, and $\frac{1}{3}$, respectively. Thus the permutation matrix P_2 has rows [0 1 0], [0 0 1], and [1 0 0], and $a_2 = \frac{1}{3}$. So we write $D_1 = (\frac{1}{3})P_2 + (\frac{2}{3})D_2$, where the rows of D_2 are [1 0 0], [0 0 1], and [0 1 0]. Since we have obtained a doubly stochastic matrix D_2 that is a permutation matrix, we write $D_2 = P_3$ and stop. Thus $(\frac{1}{4})P_1 + (\frac{3}{4})D_1 = (\frac{1}{4})P_1 + (\frac{3}{4})[(\frac{1}{3})P_2 + (\frac{2}{3})D_2] = (\frac{1}{4})P_1 + (\frac{1}{4})P_2 + (\frac{1}{2})P_3$ is a convex combination of the doubly stochastic matrix D.

7.21 Express the following double stochastic matrix D as a convex combination of permutation matrices:

$$D = \begin{bmatrix} 0.5 & 0 & 0 & 0.5 \\ 0 & 0.25 & 0.25 & 0.5 \\ 0.25 & 0.5 & 0.25 & 0 \\ 0.25 & 0.25 & 0.5 & 0 \end{bmatrix}$$

Solution. By using the procedure outlined in Problem 7.20, we have the representation $D = \frac{1}{2}P_1 + \frac{1}{4}P_2 + \frac{1}{4}P_3$, where the three permutation matrices are

$$P_1 = \begin{bmatrix} 1 & 0 & 0 & 0 \\ 0 & 0 & 0 & 1 \\ 0 & 1 & 0 & 0 \\ 0 & 0 & 1 & 0 \end{bmatrix}, \quad P_2 = \begin{bmatrix} 0 & 0 & 0 & 1 \\ 0 & 1 & 0 & 0 \\ 0 & 0 & 1 & 0 \\ 1 & 0 & 0 & 0 \end{bmatrix}, \quad \text{and} \quad P_3 = \begin{bmatrix} 0 & 0 & 0 & 1 \\ 0 & 0 & 1 & 0 \\ 1 & 0 & 0 & 0 \\ 0 & 1 & 0 & 0 \end{bmatrix}$$

THE OPTIMAL ASSIGNMENT PROBLEM

7.22 The weights of the edges of the complete bipartite graph $K_{5,5} = (X, Y, E)$ are

$$A = \begin{bmatrix} 8 & 3 & 2 & 10 & 5 \\ 10 & 7 & 10 & 6 & 6 \\ 4 & 9 & 4 & 2 & 9 \\ 8 & 10 & 5 & 3 & 3 \\ 9 & 5 & 8 & 5 & 9 \end{bmatrix}$$

Find a perfect matching of minimum weight in the bipartite graph. Here $X = \{x_i : i = 1, 2, 3, 4, 5\}$ and $Y = \{y_i : i = 1, 2, 3, 4, 5\}$.

Solution. By subtracting the smallest element in each row from all the elements of that row and doing likewise for all the columns as well, we get the following modified matrix B, which can be matched with the permutation matrix P (or P') such that the nonzero entries in P (or P') are matched with the zeros of B:

$$B = \begin{bmatrix} 4 & 1 & 0 & 8 & 8 \\ 2 & 1 & 4 & 0 & 0 \\ 0 & 7 & 2 & 0 & 7 \\ 3 & 7 & 2 & 0 & 0 \\ 2 & 0 & 3 & 0 & 4 \end{bmatrix}, \quad P = \begin{bmatrix} 0 & 0 & 1 & 0 & 0 \\ 0 & 0 & 0 & 0 & 1 \\ 1 & 0 & 0 & 0 & 0 \\ 0 & 0 & 0 & 1 & 0 \\ 0 & 1 & 0 & 0 & 0 \end{bmatrix}, \quad P' = \begin{bmatrix} 0 & 0 & 1 & 0 & 0 \\ 0 & 0 & 0 & 1 & 0 \\ 1 & 0 & 0 & 0 & 0 \\ 0 & 0 & 0 & 0 & 1 \\ 0 & 1 & 0 & 0 & 0 \end{bmatrix}$$

Notice that the dot products $A.P$ and $A.P'$ are both equal to 20, which is the weight of a minimum weight perfect matching. We may choose 4, 5, 2, 3, 6 (or 4, 5, 2, 6, 3) from columns 1, 2, 3, 4, and 5. An optimal matching consists of edges $\{x_1, y_3\}$, $\{x_2, y_5\}$, $\{x_3, y_1\}$, $\{x_4, y_4\}$, and $\{x_5, y_2\}$. Another optimal matching consists of edges $\{x_1, y_3\}$, $\{x_2, y_4\}$, $\{x_3, y_1\}$, $\{x_4, y_5\}$, and $\{x_5, y_2\}$.

7.23 The weights of the edges of the complete bipartite graph $K_{5,5} = (X, Y, E)$ are

$$A = \begin{bmatrix} 8 & 3 & 2 & 10 & 5 \\ 10 & 7 & 10 & 6 & 6 \\ 4 & 9 & 4 & 2 & 9 \\ 8 & 10 & 5 & 3 & 3 \\ 9 & 5 & 8 & 5 & 9 \end{bmatrix}$$

Find a perfect matching of maximum weight in the bipartite graph. Here $X = \{x_i : i = 1, 2, 3, 4, 5\}$ and $Y = \{y_i : i = 1, 2, 3, 4, 5\}$.

Solution. By subtracting each entry of the matrix from an integer that is greater than or equal to the largest entry in the matrix, this problem can be converted to a minimization problem. After subtracting each element from 10, we obtain the matrix C. Then C is modified by subtracting the smallest element from each row and each column. The modified matrix D and a matching permutation matrix P are

$$D = \begin{bmatrix} 2 & 7 & 8 & 0 & 5 \\ 0 & 3 & 0 & 4 & 4 \\ 5 & 0 & 5 & 7 & 0 \\ 2 & 0 & 5 & 7 & 7 \\ 0 & 4 & 1 & 4 & 0 \end{bmatrix}, \quad P = \begin{bmatrix} 0 & 0 & 0 & 1 & 0 \\ 0 & 0 & 1 & 0 & 0 \\ 0 & 0 & 0 & 0 & 1 \\ 0 & 1 & 0 & 0 & 0 \\ 1 & 0 & 0 & 0 & 0 \end{bmatrix}$$

The weight of an optimal (maximum) weight matching is $A.P = 49$. The entries in D that correspond to the nonzero entries in P give the weights of the edges in a maximum weight matching. An optimal matching consists of edges $\{x_1, y_4\}$, $\{x_2, y_3\}$, $\{x_3, y_5\}$, $\{x_4, y_2\}$, and $\{x_5, y_1\}$.

7.24 The weights of the edges of the complete bipartite graph $K_{5,5} = (X, Y, E)$ are

$$A = \begin{bmatrix} 4 & 7 & 8 & 9 & 2 \\ 6 & 4 & 5 & 6 & 2 \\ 5 & 3 & 4 & 5 & 2 \\ 1 & 1 & 1 & 5 & 3 \\ 2 & 3 & 1 & 0 & 3 \end{bmatrix}$$

Find a perfect matching of minimum weight in the bipartite graph. Here $X = \{x_i : i = 1, 2, 3, 4, 5\}$ and $Y = \{y_i : i = 1, 2, 3, 4, 5\}$.

Solution. The zeros of the modified matrix B can be covered by drawing lines along the last two rows and a line along the last column. The smallest uncovered entry is 1, which is subtracted from the entries of each uncovered row. Then we add 1 to each element in the covered column. The zeros of the resulting matrix C can be covered by drawing lines along the last three rows and drawing a line along the last column. The smallest uncovered entry is 1, which is subtracted from the elements of each row. We add 1 to each element in the covered column. The resulting matrix is D.

$$B = \begin{bmatrix} 2 & 5 & 6 & 7 & 0 \\ 4 & 2 & 3 & 4 & 0 \\ 3 & 1 & 2 & 3 & 0 \\ 0 & 0 & 0 & 4 & 2 \\ 2 & 3 & 1 & 0 & 3 \end{bmatrix}, \quad C = \begin{bmatrix} 1 & 4 & 5 & 6 & 0 \\ 3 & 1 & 2 & 3 & 0 \\ 2 & 0 & 1 & 2 & 0 \\ 0 & 0 & 0 & 4 & 3 \\ 2 & 3 & 1 & 0 & 4 \end{bmatrix}, \quad D = \begin{bmatrix} 0 & 3 & 4 & 5 & 0 \\ 2 & 0 & 1 & 2 & 0 \\ 2 & 0 & 1 & 2 & 1 \\ 0 & 0 & 0 & 4 & 4 \\ 2 & 3 & 1 & 0 & 5 \end{bmatrix}$$

We need five lines to cover the zeros of this matrix. So at this stage, by identifying five of the zeros in D with the nonzero entries of a corresponding permutation matrix, we have an optimal solution that selects 4 from row 1, 2 from row 2, 3 from row 3, the third 1 from row 4, and the 0 from row 5. The weight of this assignment is $4 + 2 + 3 + 1 + 0 = 10$. An optimal matching consists of $\{x_1, y_1\}$, $\{x_2, y_5\}$, $\{x_3, y_2\}$, $\{x_4, y_3\}$, and $\{x_5, y_4\}$.

7.25 Four candidates A, B, C, and D were tested by a firm to fill three different types of jobs classified as P, Q, and R. A scored 10, 9, and 4 points, respectively, for these jobs. B scored 10, 6, and 8 points; C scored 9, 10, and 10 points; and D scored 8, 9, and 8 points. Based on these scores, the firm has to hire three of them such that the sum of the scores of those who were selected is a maximum. Show that the firm cannot come to an equitable decision about hiring these four candidates at this stage.

Solution. To obtain a square matrix, we introduce a nonexistent job S and assume that each candidate scored zero points when tested for this job. Then we have a candidate versus job matrix with four rows and four columns in which each entry in the last column is zero. Since we have a maximization problem, we subtract the largest entry from each entry and obtain matrix M, which can be modified into M', where M and M' are

$$M = \begin{bmatrix} 0 & 1 & 6 & 10 \\ 0 & 4 & 2 & 10 \\ 1 & 0 & 0 & 10 \\ 2 & 1 & 2 & 10 \end{bmatrix}, \qquad M' = \begin{bmatrix} 0 & 0 & 5 & 1 \\ 0 & 3 & 1 & 0 \\ 2 & 0 & 0 & 1 \\ 2 & 0 & 1 & 0 \end{bmatrix}$$

By identifying M' with an appropriate permutation matrix, we find two optimal solutions: (1) A gets P, B is not hired, C gets R, and D gets Q, with a total score of $10 + 0 + 10 + 9 = 29$. (2) A gets Q, B gets P, C gets R, and D is not hired, with a total score of $9 + 10 + 10 + 0 = 29$. Since there is no unique optimal solution, the firm cannot come to a definite and fair conclusion regarding the selection of candidates for these three jobs based on their test scores.

7.26 Show that the optimal assignment problem can be interpreted as a two-matroid intersection problem.

Solution. For every bipartite graph $G = (X, Y, E)$, we can define two partition matroids: one for X and the other for Y. These two are known as the left partition matroid and the right partition matroid. See Solved Problem 4.38. Any set I of edges of G is a matching if and only if I is an independent set in both these matroids. If both X and Y have n elements, the problem of finding a minimum weight matching in G is equivalent to the problem of finding a set I of cardinality n that is independent in both the matroids such that the weight of the sum of the edges in I is a minimum.

THE TRAVELING SALESPERSON PROBLEM (TSP)

7.27 Let C be a Hamiltonian cycle in an undirected network G. (*a*) If T is any minimum weight spanning tree in G, show that $w(T) \le w(C)$. (*b*) If in the set of edges of the graph incident at vertex v the edges p and q are of minimum weight, show that $w(T) + w(p) + w(q) \le w(C)$, where T is a minimum weight spanning tree in the $G - v$.

Solution.

(*a*) Suppose T' is the spanning tree obtained from C by deleting one of its edges. Then $w(T) \le w(T') \le w(C)$.

(*b*) Let e and f be the two edges of C adjacent to v in C. Then $C - \{e, f\}$ is a spanning tree in $G - v$. So $w(T) \le w(C) - w(e) - w(f)$. Hence, $w(T) + w(p) + w(q) \le w(T) + w(e) + w(f) \le w(C)$.

7.28 Obtain lower bounds for the weight of an optimal Hamiltonian cycle in the undirected network shown in Fig. 7-11.

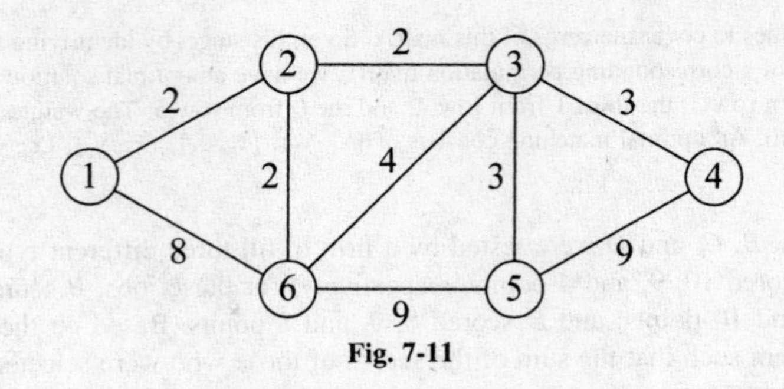

Fig. 7-11

Solution. The weight of a minimum weight spanning tree T in the network is 12, which is a lower bound for any Hamiltonian cycle in the network. If we delete vertex 1, the lower bound is 10 + 2 + 8 = 20, as outlined in Problem 7.27. If we delete vertex 2, the lower bound is 18 + 2 + 2 = 22. If we delete vertex 3, the lower bound is 22 + 2 + 3 = 27. If we delete vertex 5, the lower bound is 9 + 3 + 9 = 21. If we delete vertex 6, the lower bound is 9 + 3 + 9 = 21. The best lower bound among these six is 22.

7.29 Using the optimal assignment method, obtain a lower bound for an optimal Hamiltonian cycle in the network shown in Fig. 7-11.

Solution. Since the undirected network has six vertices, we construct the symmetric complete bipartite graph $K_{6,6} = (X, Y, E)$ in which the weight of the edge between x_i and y_j is the weight of the edge in G between i and j. Then we have the symmetric 6×6 weight matrix A for this bipartite graph in which the diagonal entries as well as the entries corresponding to pairs of nonadjacent vertices are assigned the arbitrarily large (in this case) weight 10:

$$A = \begin{bmatrix} 10 & 2 & 10 & 10 & 10 & 8 \\ 2 & 10 & 2 & 10 & 10 & 2 \\ 10 & 2 & 10 & 3 & 3 & 4 \\ 10 & 10 & 3 & 10 & 9 & 10 \\ 10 & 10 & 3 & 9 & 10 & 9 \\ 8 & 2 & 4 & 10 & 9 & 10 \end{bmatrix}$$

The modified (after a few iterations) matrix B is

$$B = \begin{bmatrix} 2 & 0 & 7 & 1 & 1 & 0 \\ 0 & 14 & 5 & 7 & 7 & 0 \\ 8 & 6 & 13 & 0 & 0 & 2 \\ 2 & 8 & 0 & 1 & 0 & 2 \\ 2 & 8 & 0 & 0 & 1 & 1 \\ 0 & 0 & 1 & 1 & 0 & 2 \end{bmatrix}$$

An optimal assignment corresponds to the entries (1, 2), (2, 6), (6, 1), (3, 4), (4, 5), (5, 3) in these matrices with a total weight of 27. Thus 27 is a lower bound for any optimal Hamiltonian cycle in the network.

7.30 Obtain an optimal Hamiltonian cycle in the digraph whose weight matrix is

$$A = \begin{bmatrix} - & 5 & 19 & 11 \\ - & - & 4 & 7 \\ - & 5 & - & 14 \\ 9 & - & 6 & - \end{bmatrix}$$

Solution. The vertices are labeled 1, 2, 3, and 4. By inspection, we see that there are only two directed Hamiltonian cycles: $1 \to 2 \to 3 \to 4 \to 1$ with weight 32 and $1 \to 3 \to 2 \to 4 \to 1$ with weight 47. The former is the unique optimal Hamiltonian cycle.

7.31 Obtain a closed walk in the network of Problem 7.30 passing through each vertex as least once such that the sum of the weights of the edges in this walk is a minimum.

Solution. The shortest distance matrix D for this network is

$$D = \begin{bmatrix} - & 5 & 9 & 11 \\ 16 & - & 4 & 7 \\ 21 & 5 & - & 12 \\ 9 & 11 & 6 & - \end{bmatrix}$$

There are six Hamiltonian cycles (not listed here) with weights 30, 34, 38, 39, 47, and 48. The cycle $1 \to 2 \to 3 \to 4 \to 1$ with weight $5 + 4 + 12 + 9 = 30$ is the optimal cycle. The arc from 3 to 4 here is the shortest path $3 \to 2 \to 4$ in the original network. Thus an optimal closed walk is $1 \to 2 \to 3 \to 2 \to 4 \to 1$, which goes through vertex 2 twice and every other vertex once.

7.32 Find an optimal Hamiltonian cycle in a directed network whose weight matrix A is

$$A = \begin{bmatrix} - & 17 & 10 & 15 & 17 \\ 18 & - & 6 & 10 & 20 \\ 12 & 5 & - & 14 & 19 \\ 12 & 11 & 15 & - & 7 \\ 16 & 21 & 18 & 6 & - \end{bmatrix}$$

Solution. The vertices are labeled 1, 2, 3, 4, and 5. The optimal assignment gives the minimum as 46 with two subtours $1 \to 3 \to 2 \to 1$ and $4 \to 5 \to 4$. So we do not get a Hamiltonian cycle at this stage. If there is a Hamiltonian cycle in this graph, its weight should be at least 46. Thus we write $w(A) = 46$(NH) at the root of the tree of enumeration. Set S is the set of vertices of one of these subtours. We take S as the set consisting of vertices 4 and 5. Corresponding to vertex 4, we construct the matrix $A(4)$ by deleting the (4, 5) entry from A. Likewise, corresponding to vertex 5, we construct the matrix $A(5)$ by deleting the (5, 4) entry from A. The optimal assignment in $A(4)$ with weight 46 gives two subtours: $2 \to 3 \to 2$ and $1 \to 5 \to 4 \to 1$. So we write $w(A(4)) = 46$(NH). The optimal assignment in $A(5)$ with weight 48 gives a Hamiltonian cycle $1 \to 3 \to 2 \to 4 \to 5 \to 1$. So 48 is an upperbound for the weight of an optimal Hamiltonian cycle. At this stage, we branch from matrix $A(4)$ with $S = \{2, 3\}$, resulting in two matrices: $A(4{:}2)$ and $A(4{:}3)$. The optimal assignment in the former gives a Hamiltonian cycle $1 \to 3 \to 2 \to 5 \to 4 \to 1$ with weight 53, and the optimal assignment in the latter gives the Hamiltonian cycle $1 \to 5 \to 4 \to 2 \to 3 \to 1$ with weight 52. At this stage, the branch and bound search for an optimal Hamiltonian cycle ends with the conclusion that $1 \to 3 \to 2 \to 4 \to 5 \to 1$ is an optimal solution.

7.33 Obtain an optimal Hamiltonian cycle in the digraph whose weight matrix A is

$$A = \begin{bmatrix} - & 1 & - & - & 2 & - \\ 2 & - & 1 & - & 1 & 6 \\ - & 2 & - & 1 & - & - \\ - & - & 2 & - & - & 2 \\ 1 & - & - & 3 & - & 10 \\ 6 & - & 1 & 3 & - & - \end{bmatrix}$$

Solution.

Iteration 1: The optimal assignment is NH with weight 7 and with $1 \to 2 \to 5 \to 1$ as a subtour. We take $S = \{1, 2, 5\}$.

Iteration 2: We construct $A(1)$, $A(2)$, and $A(5)$. If we delete the appropriate entries from A to construct $A(1)$, there is no solution. So $w(A(1))$ is ∞. If we delete the appropriate entries from A to construct $A(2)$, the optimal assignment is NH with weight 11 and a subtour $1 \to 5 \to 1$. If we delete the appropriate entries from A to construct $A(2)$, the optimal assignment with weight 12 gives a Hamiltonian cycle $1 \to 5 \to 4 \to 6 \to 3 \to 2 \to 1$ with weight 12. So at the end of iteration, the conclusion is that 11 is a lower bound and 12 is an upper bound for the weight of an optimal solution. Now we branch out from matrix $A(2)$.

Iteration 3: We start with $A(2)$ and $S = \{1, 5\}$. When the appropriate deletions are made, there is no solution. The conclusion is that $1 \to 5 \to 4 \to 6 \to 3 \to 2 \to 1$ is an optimal Hamiltonian cycle with weight 12.

7.34 Obtain a closed walk of minimum weight in the digraph of Problem 7.33 that passes through each vertex at least once.

Solution. The shortest distance matrix D is

$$D = \begin{bmatrix} — & 1 & 2 & 3 & 2 & 5 \\ 2 & — & 1 & 2 & 1 & 4 \\ 4 & 2 & — & 1 & 3 & 3 \\ 6 & 4 & 2 & — & 5 & 2 \\ 1 & 2 & 3 & 3 & — & 5 \\ 5 & 3 & 1 & 2 & 4 & — \end{bmatrix}$$

The optimal assignment gives a Hamiltonian cycle $1 \to 2 \to 3 \to 4 \to 6 \to 5 \to 1$ with weight $1 + 1 + 1 + 2 + 4 + 1 = 10$. By replacing the arc from 6 to 5 in this cycle by the shortest path (in the network) $6 \to 3 \to 2 \to 5$, we get the optimal closed walk $1 \to 2 \to 3 \to 4 \to 6 \to 3 \to 2 \to 5 \to 1$ with weight 10.

7.35 Show that the weight of a Hamiltonian cycle obtained by the method of finding an approximate solution described in Section 7.3 does not exceed twice the weight of an optimal Hamiltonian cycle.

Solution. Let C: $1 \to 2 \to 3 \to \cdots (n-1) \cdots 1$ be an optimal Hamiltonian cycle in the complete graph K_n in which the shortest distance between a pair of vertices i and j is the weight $w(i, j)$ of the edge between them, and let P be the optimal Hamiltonian path obtained from C after deleting its last edge. Suppose we start the approximation algorithm from vertex 1. If i is a vertex nearest to i, the algorithm will select the edge joining 1 and i, and we have cycle C_1: $1 \to i \to 1$ with weight $w(C_1) = w(1, i) + w(1, i)$. At the next stage, the algorithm will select a vertex that is nearest to either i or 1. Without loss of generality, assume that vertex j, which is nearest to vertex i, is chosen. At this stage, we have cycle C_2: $1 \to j \to i \to 1$ with weight $w(C_2)$. Then $w(C_2) - w(C_1) = w(1, j) + w(i, j) - w(1, i)$. But $w(1, j) - w(1, i) \le w(i, j)$, by our hypothesis. So $w(C_2) - w(C_1) \le 2w(i, j)$.

Now, while moving from C_1 to C_2, the approximation algorithm selected the edge joining i and j and ignored the edge joining i and $(i + 1)$ that appears in the optimal Hamiltonian path. [The possibility that j and $(i + 1)$ are the same is not ruled out.] So $w(i, j) \le w(i, i + 1)$. Hence, $w(C_2) - w(C_1) \le 2w(i, i + 1)$, where the edge joining i and $(i + 1)$ that appears in P is the "ignored" edge. More generally, $[w(C_{k+1}) - w(C_k)] \le$ (twice the weight of an ignored edge from P). By adding these $(n - 1)$ inequalities, we get the inequality $w(C_n) \le 2w(P) \le 2w(C)$, where C_n is a Hamiltonian cycle obtained by the approximation algorithm.

7.36 Use the approximation algorithm to obtain a Hamiltonian cycle in the complete graph whose weight matrix A is

$$A = \begin{bmatrix} — & 3 & 3 & 2 & 7 \\ 3 & — & 3 & 4 & 5 \\ 3 & 3 & — & 1 & 4 \\ 2 & 4 & 1 & — & 5 \\ 7 & 5 & 4 & 5 & — \end{bmatrix}$$

Solution. Let Z be the weight of an optimal Hamiltonian cycle in this network.

Starting from vertex 1, C_1: $1 \to 4 \to 1$, C_2: $1 \to 3 \to 4 \to 1$, C_3: $1 \to 2 \to 3 \to 4 \to 1$, and C_4: $1 \to 2 \to 5 \to 3 \to 4 \to 1$; $w(C_4) = 16 \le 2Z$.

Starting from vertex 2, C_1: $2 \to 1 \to 2$, C_2: $2 \to 4 \to 1 \to 2$, C_3: $2 \to 3 \to 4 \to 1 \to 2$, and C_4: $2 \to 5 \to 3 \to 4 \to 1 \to 2$; $w(C_4) = 15 \le 2Z$.

Starting from vertex 3, C_1: $3 \to 4 \to 3$, C_2: $3 \to 1 \to 4 \to 3$, C_3: $3 \to 2 \to 1 \to 4 \to 3$, and C_4: $3 \to 2 \to 1 \to 4 \to 5 \to 3$; $w(C_4) = 17 \le 2Z$.

Starting from vertex 4, C_1: $4 \to 3 \to 4$, C_2: $4 \to 3 \to 1 \to 4$, C_3: $4 \to 2 \to 3 \to 1 \to 4$, and C_4: $4 \to 2 \to 3 \to 1 \to 5 \to 4$; $w(C_4) = 22 \le 2Z$.

Starting from vertex 5, C_1: $5 \to 3 \to 5$, C_2: $5 \to 4 \to 3 \to 5$, C_3: $5 \to 1 \to 4 \to 3 \to 5$, and C_4: $5 \to 2 \to 1 \to 4 \to 3 \to 5$; $w(C_4) = 15 \le 2Z$.

7.37 Using the branch and bound method, obtain an optimal Hamiltonian cycle in the network of Problem 7.36.

Solution. The optimal assignment (after redistributing the zeros of the weight matrix a couple of times in the spirit of the Konig–Egervary theorem) gives the Hamiltonian cycle $1 \to 2 \to 5 \to 4 \to 3 \to 1$ with weight 14. (Observe that twice the weight of an optimal Hamiltonian cycle is 30 and that the weight each Hamiltonian cycle obtained by the approximation method in Problem 7.36 does not exceed 30.)

7.38 Show that if T is a minimum weight spanning tree in the complete weighted graph G of order n in which the edge between any pair of vertices is a shortest path between them, it is possible to obtain a Hamiltonian cycle C such that $w(C) \le 2w(T) \le 2w(C')$, where C' is an optimal Hamiltonian cycle in G.

Solution. It has been already proved that $2w(T) \le 2w(C')$. To establish the other inequality, we replace each edge of the tree T by two arcs directed in the opposite direction so that T becomes an Eulerian digraph with $2n$ arcs with a directed Eulerian circuit with weight $2w(T)$. The number of arcs is successively reduced by replacing by an arc (edge) that is not in the circuit two arcs of the circuit two as a time. This will not increase the weight of the path because of our hypothesis. Ultimately, we have a directed circuit with n arcs. See Problem 7.39.

7.39 Obtain a Hamiltonian cycle using the approximation method described in Problem 7.38 for the network in Problem 7.36.

Solution. The edges of a minimum spanning tree T are $\{1, 4\}$, $\{2, 3\}$, $\{3, 4\}$, and $\{3, 5\}$. Each edge is converted into two arcs (in opposite direction), as shown in Fig. 7-12. Arcs (1, 4) and (4, 3) are replaced by arc (1, 3). Arcs (5, 3) and (3, 2) are replaced by arc (5, 2). Arcs (2, 3) and (3, 4) are replaced by arc (2, 4). As a result, we have a directed Hamiltonian cycle C': $1 \to 3 \to 5 \to 2 \to 4 \to 1$ in which we use arcs (3, 5) and (4, 1) from the Eulerian circuit. In this process, we also dropped arcs (1, 4), (4, 3), (5, 6), (3, 2), (2, 3), and (3, 4) from the Eulerian circuit. Now $w(C') = w(1, 3) + w(3, 5) + w(5, 2) + w(2, 4) + w(4, 1) = 3 + 4 + 5 + 4 + 2 = 18 \le [w(1, 4) + w(4, 3)] + w(3, 5) + [w(5, 3) + w(2, 3)] + [w(2, 3) + w(3, 4)] + w(4, 1) = 2w(T) = 20$.

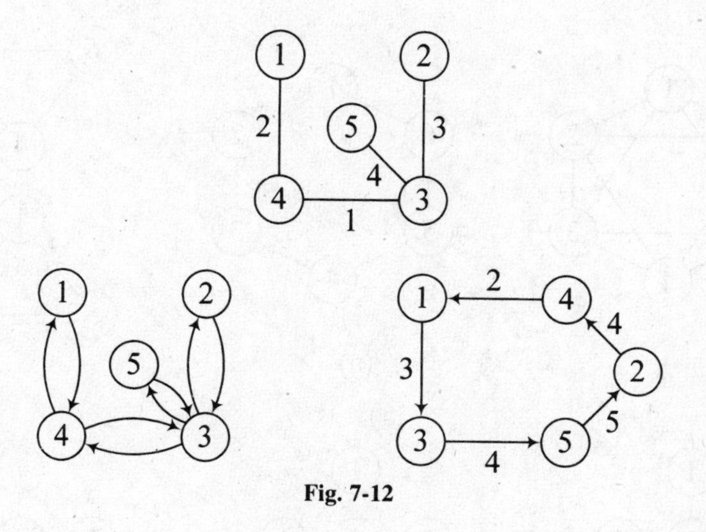

Fig. 7-12

7.40 Show that the problem of finding a directed Hamiltonian cycle in a digraph with n vertices is equivalent to finding a directed Hamiltonian path in a digraph with $(n + 1)$ vertices.

Solution. Suppose $G = (V, E)$ is a digraph with $V = \{1, 2, \ldots, n\}$. Construct a digraph $G' = (V, E')$ with $V = \{1, 2, \ldots, n, (n + 1)\}$ as follows. Both E and E' have the same number of arcs. For each arc in the digraph G directed to vertex 1 from vertex i, construct an arc in G' from i to the new vertex $(n + 1)$. Every other arc in E is an arc in E'. Then a Hamiltonian cycle starting from 1 and ending in 1 in G is equal to a Hamiltonian path from 1 to $(n + 1)$ in G', and vice versa.

7.41 Show that the TSP can be interpreted as a three-matroid intersection problem.

Solution. Given digraph G of n, construct digraph G' with one more vertex as described in Problem 7.40. Then define three matroids corresponding to G': the head-partition matroid, the tail-partition matroid, and the graphic matroid. An optimal Hamiltonian path in G is a set I of arcs of G' such that (1) the cardinality of I is n, (2) the weight of I is a minimum, and (3) I is an independent set in all the three matroids.

FACTORS, FACTORIZATIONS, AND THE PETERSEN GRAPH

7.42 Show that an Eulerian graph cannot have a bridge.

Solution. If a graph is Eulerian, it is a connected graph in which the degree of each vertex is even. Suppose it has a bridge. If this bridge is deleted, there will be two components, and in each component will be exactly one vertex of odd degree. But in any graph, the number of odd vertices is always even.

7.43 If a cubic graph has a bridge, it is not 1-factorable.

Solution. Suppose a cubic graph with a bridge is 1-factorable. The bridge will be in one of the three 3-factors. The removal of the bridge gives two components. Each component will have an even number of odd vertices and a vertex of degree 2. So there cannot be one more 1-factor since the number of vertices in each component is odd.

7.44 (*a*) Give an example of a factorization of a graph consisting of two 2-factors such that the two factors are not isomorphic. (*b*) Give an example of a factorization of a graph consisting of two isomorphic factors that are not regular.

Solution. (*a*) See Fig. 7-13(*a*). (*b*) See Fig. 7-13(*b*).

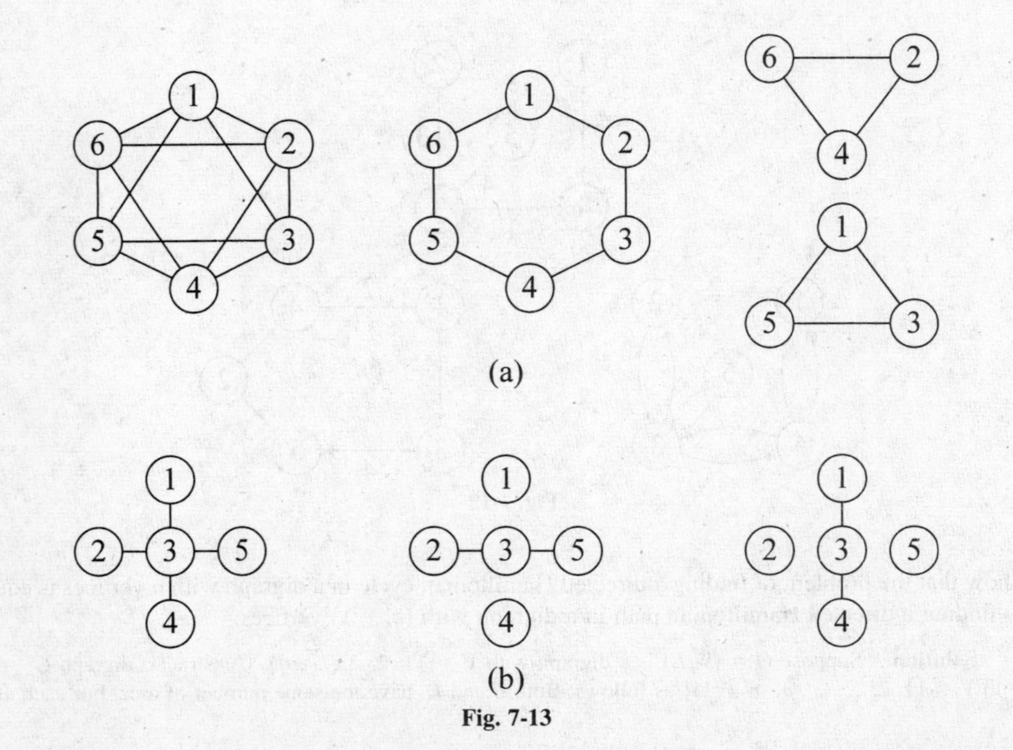

Fig. 7-13

7.45 Show that the complete graph of even order is 1-factorable.

Solution. Suppose the vertices are labeled 1, 2, 3, . . . , $2n$. Construct a regular polygon with $(2n - 1)$ vertices, and label these vertices 2, 3, . . . , $2n$ in the clockwise direction. Assume that the distance between two consecutive vertices in the cycle is one unit. The center of the polygon corresponds to vertex 1 of the complete graph. A 1-factor M_1 consisting of the edge joining 1 and 2 along with $(n - 1)$ more edges is constructed as follows. There are $n - 1$ vertices starting from 2 in the clockwise direction and $n - 1$ vertices in the counterclockwise direction, excluding vertex 2. Join the vertex located i units from vertex 2 in the clockwise direction and the vertex located i units from vertex 2 in the counterclockwise direction, where $i = 1, 2, \ldots, n$. By this method, $(n - 1)$ edges are obtained. The edge joining center vertex 1 and vertex 2 together with these $(n - 1)$ edges form a 1-factor. Then we locate vertices that are equidistant (in either direction) from vertex 3 on the cycle. This will give us a 1-factor consisting of the edge joining vertex 1 and vertex 3 along with $(n - 1)$ new edges. We continue this process. The last 1-factor consists of the edge joining vertex 1 and vertex $2n$ along with $(n - 1)$ new edges. Thus this construction defines $(2n - 1)$ perfect matchings. Since each 1-factor has n edges, the total number of edges in these $(2n - 1)$ matchings is $n(2n - 1)$, which is the total number of edges in the complete graph. Hence, the graph is 1-factorable.

7.46 A regular bipartite graph of degree r (where r is positive) is 1-factorable.

Solution. The proof is by induction on r. If $r = 1$, the result is true. Suppose it is true for $(r - 1)$, where $r \geq 2$. Consider any regular bipartite graph G of degree r. A consequence of Hall's marriage theorem is that there is a 1-factor F in G. Then the graph $G' = G - F$ is a regular bipartite graph of degree $(r - 1)$ that is 1-factorable by the induction hypothesis. Any 1-factorization of G' together with the 1-factor F gives a 1-factorization of G.

7.47 The k-cube Q_k is 1-factorable for every $k \geq 1$.

Solution. The k-cube Qk is a k-regular bipartite graph. (See Solved Problem 1.56.) So, by Problem 7.46, it is 1-factorable. A proof by induction is along the following lines. Q_k can be inductively defined as the Cartesian product $Q_{k-1} \times K_2$ that has 2^k vertices. The vertices of K_2 are labeled 1 and 2. For each vertex u in Q_{k-1}, the ordered pairs $(u, 1)$ and $(u, 2)$ define two adjacent vertices in the product since 1 and 2 are adjacent in K_2. By definition, the ordered pairs (u, i) and (v, i) are adjacent in the product if and only if u and v are adjacent in Q_{k-1}, where $i = 1$ or 2. Suppose the edge joining u and v is edge e in a perfect matching M (with 2^{k-1} edges) in Q_{k-1}. Then this matched edge defines two edges in Q_k: edge e_1 joining $(u, 1)$ and $(v, 1)$ and edge e_2 joining $(u, 2)$ and $(v, 2)$. Thus each matched edge in Q_{k-1} defines two edges in Q_k that have no vertex in common. So for every perfect matching in Q_{k-1}, we obtain a perfect matching in Q_k. If we consider the edge joining $(v, 1)$ and $(v, 2)$ as a matched edge for each vertex v of Q_{k-1}, there is one more perfect matching in Q_k.

7.48 Prove Theorem 7.4: A simple graph is 2-factorable if and only if it is r-regular, where r is even.

Solution. If a graph is 2-factorable, the degree of each vertex is obviously $2r$ for some positive integer r. Suppose the degree of each vertex of a simple graph $G = (V, E)$ of order n is $2r$, where r is positive. So there is an Eulerian circuit C in the graph. Consider this as a directed Eulerian circuit. If $V = \{v_1, v_2, \ldots, v_n\}$, construct a bipartite graph $G' = (X, Y, F)$, where $X = \{x_1, x_2, \ldots, x_n\}$ and $Y = \{y_1, y_2, \ldots, y_n\}$. The edges in F are defined as follows: join x_i and y_j by an edge if and only if there is an arc from v_i to v_j in the directed circuit C. The bipartite graph thus constructed is necessarily r-regular; therefore, it has a 1-factorization $F_1 \oplus F_2 \oplus \cdots \oplus F_r$, as proved in Theorem 7.3. For each factor F_k in this partition, define a bijection f_k on $\{1, 2, \ldots, n\}$ by $f_k(i) = j$ if and only if there is an edge in F_k between v_i and v_j. Since there is no loop in G, $f_k(i) \neq i$ for each i. Since G has no multiple edges, $f_k(i) = j$ implies that $f_k(j) = i$ for every i and j. In other words, each permutation under f_k is of length at least three, so it defines a cycle in the graph. Thus each 1-factor in the factorization of G' defines a collection of pairwise disjoint cycles in G; consequently, G is 2-factorable.

7.49 Show that a complete graph of order $(2n + 1)$ can be factored into n Hamiltonian cycles.

Solution. Let $V = \{v_i : i = 1, 2, \ldots, (2n + 1)\}$ be the set of vertices. Define path P_i between vertices v_i and v_{i+n} passing through all vertices except v_{2n} as follows: $v_i \rightarrow v_{i-1} \rightarrow v_{i+1} \rightarrow v_{i-2} \rightarrow \cdots \rightarrow v_{i+n-1} \rightarrow v_{i+n}$, where all subscripts are taken as integers 1, 2, . . . , $2n$ (mod $2n$). There are n such edge disjoint paths. Construct the Hamiltonian cycle C_i from each path P_i by joining v_{2n} to the endpoints of each path P_i.

7.50 Obtain a 2-factorization of the complete graph with nine vertices.

Solution. The graph has 36 edges. Each Hamiltonian cycle in the graph will have nine edges. So the 2-factorization consists of four edge-disjoint Hamiltonian cycles in the graph. Here $2n + 1 = 9$, so $n = 4$. The subscripts of the vertices are the integers 1, 2, 3, . . . , 8 (mod 8). We construct four edge-disjoint paths:

P_1: $1 \to 8 \to 2 \to 7 \to 3 \to 6 \to 4 \to 5$
P_2: $2 \to 1 \to 3 \to 8 \to 4 \to 7 \to 5 \to 6$
P_3: $3 \to 2 \to 4 \to 1 \to 5 \to 8 \to 6 \to 7$
P_4: $4 \to 3 \to 5 \to 2 \to 6 \to 1 \to 7 \to 8$

Observe that each row has $2n$ entries. The first row starts from 1, and the entries in it alternately increase and decrease. The entries in each column increase (mod $2n$). Change each path into a Hamiltonian cycle by joining vertex 9 to the endpoints of each path.

7.51 Show that the complete graph of order $2n$ can be factored into n Hamiltonian paths; hence, prove it is 1-factorable.

Solution. This is an immediate consequence of the fact that the complete graph of order $(2n + 1)$ can be factored into n Hamiltonian cycles. Each path that leads to the construction of a Hamiltonian cycle in K_{2n+1} (see Problems 7.49 and 7.50) is, in fact, a Hamiltonian path in K_{2n}. Moreover, each such path defines a 1-factor in K_{2n}. Hence, it can be factored into n 1-factors.

7.52 Obtain a 1-factorization of the complete graph with eight vertices.

Solution. In the complete graph with vertices 1, 2, . . . , 9, if vertex 9 is deleted, we get four edge-disjoint Hamiltonian paths that constitute a factorization of K_8 into four 1-factors:

P_1: $1 \to 8 \to 2 \to 7 \to 3 \to 6 \to 4 \to 5$
P_2: $2 \to 1 \to 3 \to 8 \to 4 \to 7 \to 5 \to 6$
P_3: $3 \to 2 \to 4 \to 1 \to 5 \to 8 \to 6 \to 7$
P_4: $4 \to 3 \to 5 \to 2 \to 6 \to 1 \to 7 \to 8$

7.53 Show that a complete graph of order $2n$ can be factored into n Hamiltonian cycles and one 1-factor.

Solution. Suppose the vertices are labeled 1, 2, 3, . . . , $2n$. Construct a regular polygon with $(2n - 1)$ vertices, and label these vertices 2, 3, . . . , $2n$ in the clockwise direction. Assume that the distance between two consecutive vertices in the cycle is one unit. The center of the polygon corresponds to vertex 1 of the complete graph. A 1-factor M_1 consisting of the edge joining 1 and 2 along with $(n - 1)$ more edges is constructed as follows. There are $n - 1$ vertices starting from 2 in the clockwise direction and $n - 1$ vertices in the counterclockwise direction, excluding vertex 2. Join the vertex located i units from vertex 2 in the clockwise direction and the vertex located i units from vertex 2 in the counterclockwise direction, where $i = 1, 2, \ldots, n - 1$. By this method, $(n - 1)$ edges are obtained. The edge joining center vertex 1 and vertex 2 together with these $(n - 1)$ edges forms a 1-factor. Then we locate vertices that are equidistant (in either direction) from vertex 3 on the cycle. This will give us a 1-factor consisting of the edge joining vertex 1 and vertex 3 along with $(n - 1)$ new edges. We continue this process. If M_1 is the first matching, a Hamiltonian cycle C_1 starting from vertex $2n$ can be constructed by alternately using vertices on the clockwise direction and counterclockwise direction in this matching. Use the first $(n - 1)$ 1-factors to construct $(n - 1)$ Hamiltonian cycles. The next 1-factor and these $(n - 1)$ cycles give the desired factorization.

7.54 Obtain a factorization of the complete graph with eight vertices consisting of three Hamiltonian cycles and one 1-factor.

Solution. The vertices are 1, 2, . . . , 8. Keeping vertex 1 at the center of a polygon and the vertices in increasing order as corners in the clockwise direction, the following four 1-factors are obtained:

M_1: (1, 2), (3, 8), (4, 7), (5, 6)
M_2: (1, 3), (4, 2), (5, 8), (6, 7)

M_3: (1, 4), (5, 3), (6, 2), (7, 8)
M_4: (1, 5), (6, 4), (7, 3), (8, 2)

The first three matchings give the following Hamiltonian cycles:

C_1: $1 \to 2 \to 3 \to 8 \to 4 \to 7 \to 5 \to 6 \to 1$

C_2: $1 \to 3 \to 4 \to 2 \to 5 \to 8 \to 6 \to 7 \to 1$

C_3: $1 \to 5 \to 6 \to 4 \to 7 \to 3 \to 8 \to 2 \to 1$

These three cycles and the 1-factor M_4 constitute a partition consisting of three Hamiltonian cycles and one perfect matching.

7.55 If $0 \le r < n$, rn is even if and only if there exists an r-regular graph G of order n.

Solution. If G is r-regular of order n, its size is $(rn)/2$, so rn has to be even. To prove the converse, we examine two cases.

Case (*i*): n is even. In this case, rn is even. Since n is even, $K_n = F_1 \oplus F_2 \oplus \cdots \oplus F_{n-1}$, where each F_i is a 1-factor. Then $G = F_1 \oplus F_2 \oplus \cdots \oplus F_r$ is an r-regular graph of order n.

Case (*ii*): n is odd. So r has to be even. In this case, K_n can be factored into $(n-1)/2$ Hamiltonian cycles. So $K_n = F_1 \oplus F_2 \oplus \cdots \oplus F_{(n-1)/2}$, where each F_i is a 2-factor. Then $G = F_1 \oplus F_2 \oplus \cdots \oplus F_{r/2}$.

7.56 Show that a Hamiltonian cycle in a complete graph of odd order is an isofactor of the graph.

Solution. This is certainly true since any complete graph of order $(2n + 1)$ has a factorization consisting of n Hamiltonian cycles.

7.57 Show that a complete graph of odd order cannot be factored into Hamiltonian paths.

Solution. If the order of the graph is $(2n + 1)$, each Hamiltonian path will have $2n$ edges. The total number of edges in the graph is $n(2n + 1)$. Since $2n$ is not a divisor of $n(2n + 1)$, there is no factorization into Hamiltonian paths.

7.58 (*a*) Find two nonisomorphic-connected 1-factorable cubic graphs of the same order. (*b*) Find two nonisomorphic connected cubic graphs such that each can be factored into a 1-factor and a Hamiltonian cycle.

Solution. The two cubic graphs of order 6 shown in Fig. 7-14 are not isomorphic because one of them has a triangle as a subgraph while the other does not.

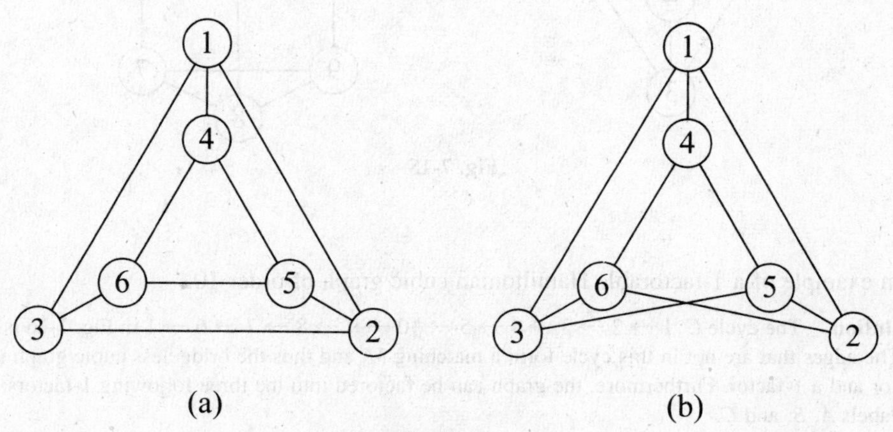

Fig. 7-14

(*a*) The graph in Fig. 7-14(*a*) can be factored into $M_1 = \{(1, 3), (2, 5), (4, 6)\}$, $M_2 = \{(3, 6), (4, 5), (1, 2)\}$, and $M_3 = \{(2, 3), (5, 6), (1, 4)\}$. The graph in Fig. 7-14(*b*) can be factored into $M_1 = \{(1, 3), (2, 5), (4, 6)\}$, $M_2 = \{(3, 6), (4, 5), (1, 2)\}$, and $M_3 = \{(3, 5), (2, 6), (1, 4)\}$.

(*b*) The graph in Fig. 7-14(*a*) can be factored into the Hamiltonian cycle $1 \to 2 \to 3 \to 6 \to 5 \to 4 \to 1$ and the 1-factor $M = \{(1, 3), (2, 5), (4, 6)\}$. The graph in Fig. 7-14(*b*) can be factored into the Hamiltonian cycle $1 \to 2 \to 5 \to 3 \to 6 \to 4 \to 1$ and $M = \{(1, 3), (2, 6), (4, 5)\}$.

7.59 Prove Theorem 7.6 (Petersen's Theorem): A cubic graph G in which no edge is a bridge can be factored into a 2-factor and a 1-factor.

Solution. If G has a 1-factor, the removal of the edges belonging to the 1-factor results in a 2-factor. So it is enough if we show that G has a 1-factor. Let $H_1, H_2, \ldots, H_k$ be the odd components of $G - S$, where S is any set of vertices. Since G has no bridges, the number of edges between S and any odd component H_i is more than 1. Suppose the number m_i of edges between S and H_i is an even number $2q$. If the number of vertices in H_i is $(2q + 1)$, the sum of the degrees of the vertices in H_i will be $3(2q + 1) - 2q$, which will be an odd number. Hence, m_i is at least 3 for each i. Thus $k \le \frac{1}{3}(m)$, where $m = \{m_1 + m_2 + \cdots + m_k\}$. But $m \le 3|S|$. So $k \le |S|$, where k is the number of odd components of $G - S$. Hence, by Tutte's theorem, G has a 1-factor.

7.60 Prove Theorem 7.7: The Petersen graph is not 1-factorable.

Solution. The Petersen graph (see Fig. 7-3) has 15 edges. Vertices 1, 2, 3, 4, and 5 are the outer vertices; the others are the inner vertices. If it is 1-factorable, it will have three 1-factors each with five edges. So if any set F of five edges of the graph is chosen at random, there will be at least one 1-factor that contains at least two edges from F. Let F be the set of five edges joining an outer vertex and the unique inner vertex adjacent to it. Suppose H is the 1-factor that contains edge e that joins vertex 1 and vertex 6 and edge f that joins vertex 2 and vertex 7. But if these four vertices are deleted from the Petersen graph, the remaining six vertices cannot be matched in three pairs.

7.61 Give an example of a cubic bridgeless graph of order 10 that is 1-factorable.

Solution. The bridgeless cubic graph shown in Fig. 7-15 can obviously be factored into a 2-factor (consisting of two disjoint cycles) and a 1-factor. It also can be factored into three 1-factors as follows: $M_1 = \{(1, 2), (3, 4), (5, 6), (7, 8), (9, 10)\}$, $M_2 = \{(1, 3), (2, 4), (6, 7), (8, 9), (5, 10)\}$, and $M_3 = \{(1, 4), (2, 3), (6, 10), (7, 9), (5, 8)\}$.

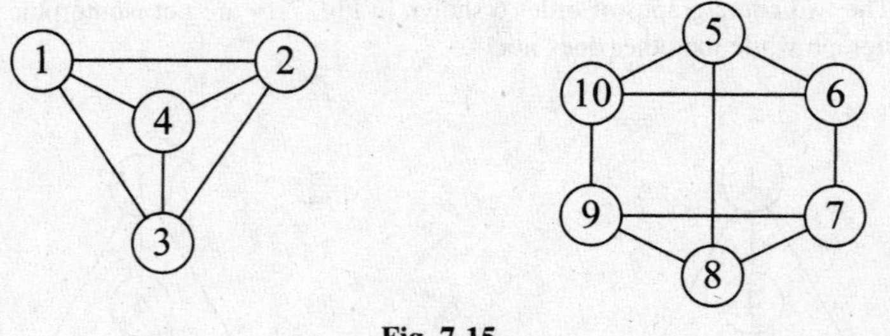

Fig. 7-15

7.62 Give an example of a 1-factorable Hamiltonian cubic graph of order 10.

Solution. The cycle $C: 1 \to 2 \to 3 \to 4 \to 5 \to 10 \to 9 \to 8 \to 7 \to 6 \to 1$ in Fig. 7-16 is a Hamiltonian cycle. The edges that are not in this cycle form a matching M, and thus the bridgeless cubic graph is factored into a 2-factor and a 1-factor. Furthermore, the graph can be factored into the three following 1-factors: arcs indicated by the labels A, B, and C.

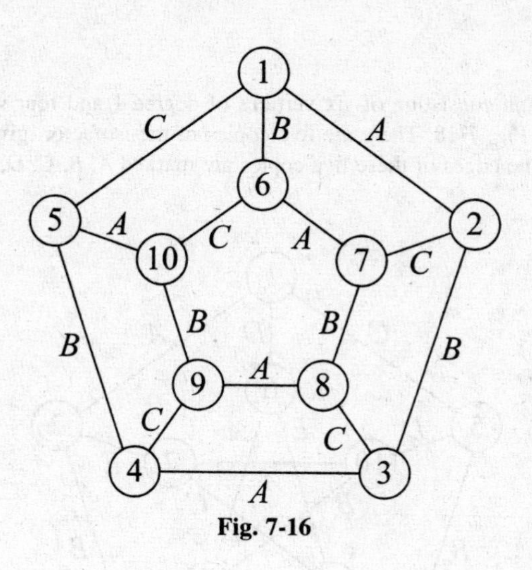

Fig. 7-16

7.63 Show that $K_{2n} = (2n - 1)H$ and $K_{n,n} = nH$, where $H = nK_2$.

Solution. The 1-factorable graph K_{2n} with $(n)(2n - 1)$ edges has an isomorphic factorization consisting of $(2n - 1)$ copies of its isofactor nK_2. Similarly, the 1-factorable bipartite graph $K_{n,n}$ with n^2 edges has an isomorphic factorization consisting of n copies of K_2.

7.64 Find an isofactor of the Petersen graph (with five sides) other than what is shown in Fig. 7-4.

Solution. The graph shown in Fig. 7-17 is another isofactor of the Petersen graph.

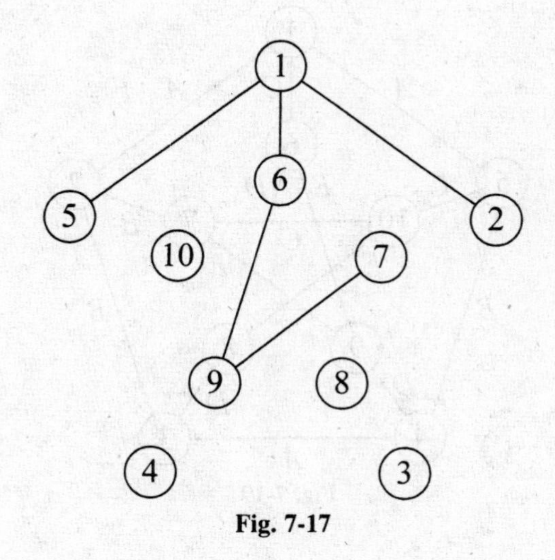

Fig. 7-17

7.65 Show that the Petersen graph is 3-connected.

Solution. Between any two vertices, there are three internally disjoint paths. So, by Whitney's theorem, the graph is 3-connected. Furthermore, the vertex connectivity number and the edge connectivity number are both 3 since the maximum degree is 3.

7.66 Find an isofactor of the Petersen graph with three edges.

Solution.

(*i*) A spanning H subgraph consisting of six vertices of degree 1 and four vertices of degree 0 is an isofactor with three sides. See Fig. 7-18. There are five copies of the isofactor, giving an isomorphic factorization of the Petersen graph. The edges of these five copies are marked A, B, C, D, and E.

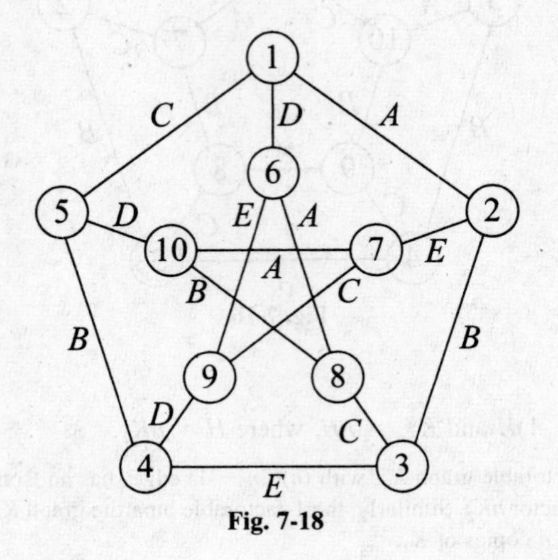

Fig. 7-18

(*ii*) A spanning subgraph with degree vector [2 1 1 1 1 0 0 0 0] is another isofactor with three sides, as shown in Fig. 7-19.

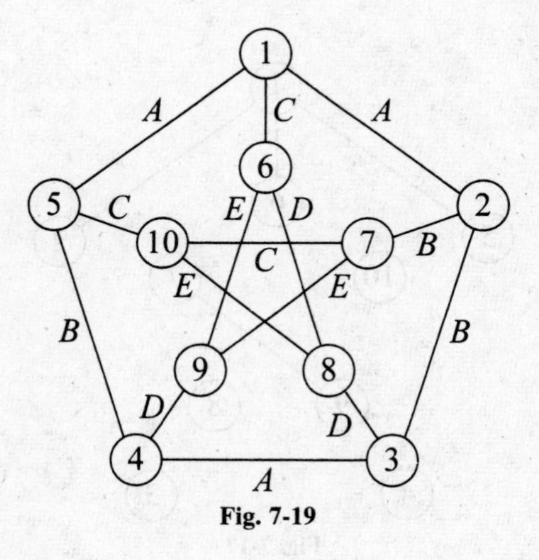

Fig. 7-19

7.67 (*Hartsfield-Ringel Theorem*) Show that the path with three edges is an isofactor of any cubic bridgeless graph.

Solution. Every bridgeless cubic graph can be factored into a 2-factor and a 1-factor. Suppose the edges belonging to a 1-factor are colored blue and the edges belonging to a 2-factor are colored red. So every vertex is incident to one blue edge and two red edges. The blue edges are labeled 1, 2, 3, . . . arbitrarily. Now we traverse the edges belonging to a cycle (a component of the 2-factor) in one direction. While doing so, each red edge from vertex u to vertex v in the cycle is assigned the same number as the label of the blue edge that is incident to vertex u. Then each number appears on three consecutive edges. Thus P_3 is an isofactor.

7.68 Obtain an isomorphic factorization of the Petersen graph such that each factor is a path consisting of three edges.

Solution. In Fig. 7-20, each edge joining an outer vertex to the corresponding inner vertex belongs to a 1-factor, and the five blue edges belonging to this 1-factor are labeled 1, 2, 3, 4, and 5. The 2-factor consists of two components: one is the cycle $A \to B \to C \to E \to A$ and the other is $F \to H \to J \to G \to I \to F$. As we move from vertex A in the counterclockwise direction, edge AB gets label 1. As we move from F in the counterclockwise direction, edge FH gets label 1. Thus the path $H \to F \to A \to B$ is one copy of the isofactor. The remaining four copies correspond to the four paths labeled 2, 3, 4, and 5 respectively.

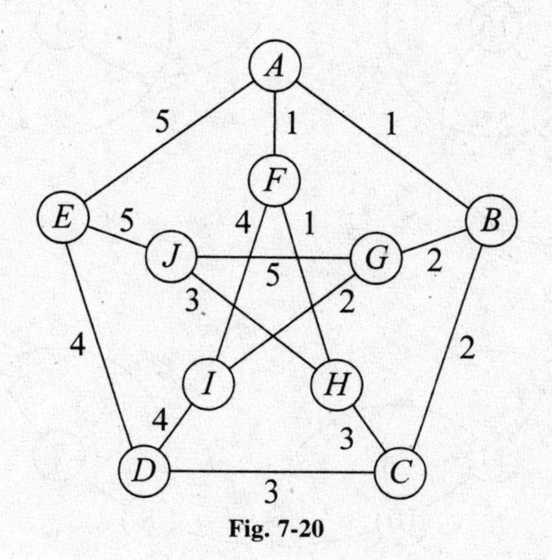

Fig. 7-20

7.69 Give examples of bipartite and nonbipartite cubic bridgeless graphs with a Hamiltonian cycle as a 2-factor.

Solution.

(*i*) The cube graph Q_3 with eight vertices [see Fig. 7-21(*a*)] is a bipartite bridgeless graph that can be factored into a Hamiltonian cycle and a 1-factor consisting of the pairs (1, 4), (2, 7), (3, 6), and (5, 8).

(*ii*) The cubic graph with eight vertices [see Fig. 7-21(*b*)] is a nonbipartite bridgeless graph that can be factored into a Hamiltonian cycle and a 1-factor consistng of the pairs (1, 3), (2, 4), (5, 7), and (6, 8).

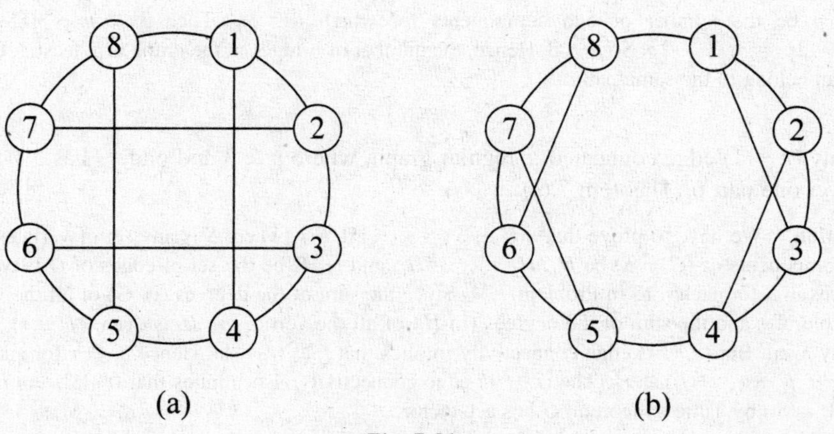

Fig. 7-21

7.70 Construct a nonbipartite cubic bridgeless graph of order 2^k with a Hamiltonian cycle as a 2-factor.

Solution. Label the vertices $1, 2, \ldots, 2^k$, and place them clockwise around a cycle. Join vertex i and vertex $i + 1$ for each i. Thus a 2-factor is constructed. To obtain the 1-factor, join vertex i and $i + 2$. Each $(i, i + 2)$ is a matched pair. See Figure 7-22, when $k = 4$.

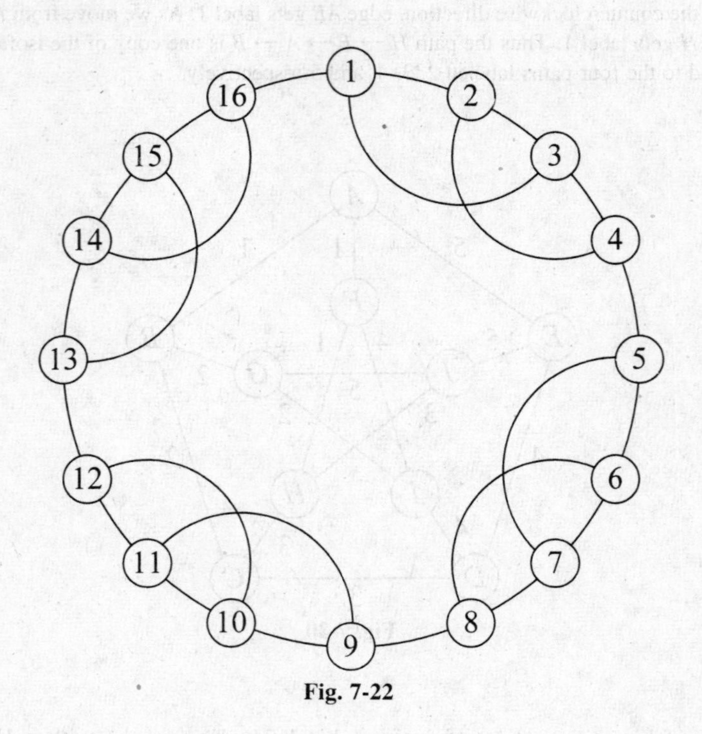

Fig. 7-22

7.71 Show that if a cubic graph G does not have a 1-factor, it will have at least three bridges, not all of them belonging to the same path. (In other words, if the bridges of a cubic graph lie on a single path, it has a 1-factor.)

Solution. By Tutte's theorem, G will have at least one bridge. Furthermore, there exists a set S of vertices such that $o(G - S) = k > |S| = s$. It has been already shown (see Problem 7.5) that k is odd if and only if s is odd. Hence, $k \geq (s + 2)$. Let the odd components of $(G - S)$ be $H_1, H_2, \ldots, H_k$, and let F_i be the set of edges of G between vertices in H_i and vertices in S for each i. Since the sum of the degrees (in G) of all the vertices in H_i is odd and the sum of the degrees (in H_i) of all the vertices in H_i is even, the number $f_i = |F_i|$ is necessarily odd. Let p be the number of odd components for which $f_i = 1$. Then $3s \geq p + 3(k - p) = 3k - 2p \geq 3(s + 2) - 2p = 3s + -2p$. So $p \geq 3$. Hence, the number of bridges in the graph is at least 3. Obviously no three bridges can belong to the same path.

7.72 If G is any $(r - 1)$ edge connected, r-regular graph, where $r \geq 3$ and odd, G has a 1-factor. (If $r = 3$, we recover one part of Theorem 7.6.)

Solution. We have to prove that $o(G - S) = k \leq |S| = s$, where S is any set of vertices of the graph. Let the odd components of $(G - S)$ be $H_1, H_2, \ldots, H_k$, and let F_i be the set of edges of G between vertices in H_i and vertices in S for each i, as in Problem 7.71. Since the sum of the degrees (in G) of all the vertices in H_i is an odd multiple of r and the sum of the degrees (in H_i) of all the vertices in H_i is even, $(f_i - r)$, where $f_i = |F_i|$, is necessarily even. But $(r - 1)$ edge connectivity implies that $f_i \geq (r - 1)$. Hence, $f_i \geq r$ for each odd component H_i. So $(f_1 + f_2 + \cdots + f_k) \geq rk$. The $(r - 1)$ edge connectivity also implies that $(r)(|S|) \geq (f_1 + f_2 + \cdots + f_k)$. Thus $|S| \geq k$. So by Tutte's theorem, G has a 1-factor.

7.73 If G is any $(r - 1)$ connected, r-regular graph, where $r \geq 3$ and odd, G has a 1-factor.

Solution. Any $(r - 1)$ connected graph is $(r - 1)$ edge connected. So the result follows from Problem 7.72.

7.74 Show that the Petersen graph is the complement of the line graph of the complete graph with five vertices.

Solution. The graph K_5 shown in Fig. 7-23 has 10 edges labeled as 1, 2, . . . , 10. Suppose e is the edge joining vertex u and vertex v in this graph. Then u is incident to three edges other than e. Likewise, v is adjacent to three other edges. Edge e is the only edge common to both u and v. In other words, the linegraph is a 6-regular graph with 10 vertices. Its complement is a 3-regular graph with 10 vertices. In the line graph, vertex 1 is adjacent to vertices 3, 4, 10, 7, 8, and 9. So in the complement of the line graph vertex 1 is adjacent to 2, 5, and 6. Likewise, vertex 2 is adjacent to vertices 1, 3 and 7. Thus the complement of the line graph is the graph shown in Fig. 7-3.

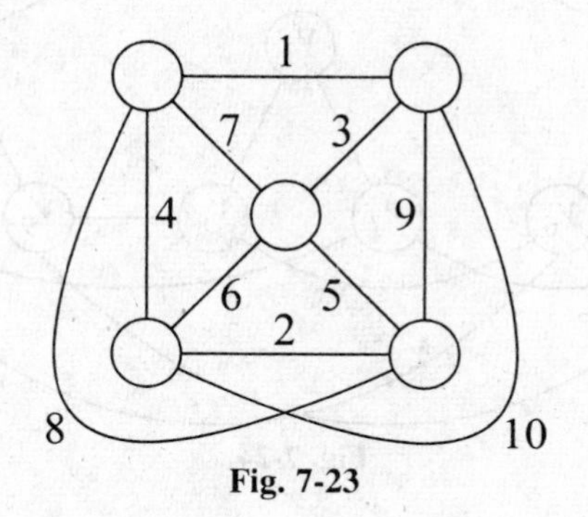

Fig. 7-23

7.75 Show that the Petersen graph has cycles of length 5, 6, 8, and 9.

Solution. In Fig. 7-3, we have cycles $C_5: 1 \to 2 \to 7 \to 10 \to 5 \to 1$, $C_6: 1 \to 2 \to 7 \to 10 \to 8 \to 6 \to 1$, $C_8: 1 \to 6 \to 8 \to 3 \to 4 \to 9 \to 7 \to 2 \to 1$, and $C_9: 1 \to 2 \to 3 \to 4 \to 9 \to 7 \to 10 \to 8 \to 6 \to 1$.

7.76 Show that the Petersen graph is not Hamiltonian. (This proof is by D. West.)

Solution. In Fig. 7-3, we see a 2-factor of the graph consisting of an "outer" cycle with vertices 1, 2, 3, 4, and 5 and an "inner" cycle with vertices 6, 7, 8, 9, and 10. In addition, there is a 1-factor consisting of five "links" connecting an outer vertex and its matching inner vertex. Suppose the graph is Hamiltonian. Then there is a Hamiltonian cycle that should contain either two or four of these connecting links. If the number of such links is 2, their end vertices are not adjacent in at least one of these two cycles. For instance, if the links are (1, 6) and (2, 7), end vertices 6 and 7 are not adjacent in the inner cycle. If the links are (1, 6) and (3, 8), end vertices 1 and 3 are not adjacent in the outer cycle. So if the number of links is 2, there are two end vertices u and v that are adjacent in one cycle and not adjacent in the other cycle. In the cycle in which u and v are not adjacent, there is no path between them that contains the remaining three vertices of that cycle as intermediate vertices. So the number of connecting links in the Hamiltonian cycle is not 2. If the number of links is 4, the number of vertices in a cycle consisting of these four links is 9; hence, there is no Hamiltonian cycle. So PG is not Hamiltonian.

7.77 A ***g*-cage** is a cubic graph with as few vertices as possible such that the number of edges in its smallest cycle is exactly g. Show that the Petersen graph is a 5-cage and that any 5-cage is isomorphic to it.

Solution. The number of edges in the smallest cycle in the Petersen graph PG is 5. To show that PG is a 5-cage, it is enough to prove that the number of vertices in any 5-cage G is at least 10. Let v_1 be a vertex in G that is adjacent to vertices v_2, v_3, and v_4, as in Fig. 7-24. Now v_2 will be adjacent to vertices v_5 and v_6, v_3 will be adjacent to v_7 and v_8, and v_4 will be adjacent to vertices v_9 and v_{10}. These 10 vertices are necessarily distinct since there is no 3-cycle or 4-cycle in the graph. So a 5-cage should have at least 10 vertices. Hence, *PG* (with 10 vertices) is indeed a 5-cage. Since *PG* is a 5-cage and has 10 vertices, any other 5-cage *G* should also be of order 10. Suppose *G* is any 5-cage with vertices as in Fig. 7-24. Since it is a cubic graph, vertex v_5 has to be adjacent to two more vertices. Without any loss of generality whatsoever, we join vertex v_5 to v_7 and v_9. Then we join vertex v_6 to v_8 and v_{10}. We cannot join v_7 to v_9. So v_7 and v_{10} are joined. Finally, we join v_8 and v_9. Thus the cubic graph *G* is constructed. If we relabel the vertices $v_i (i = 1, 2, \ldots, 10)$ as vertices 1, 5, 6, 2, 4, 10, 9, 8, 3, and 7, we have the Petersen graph displayed in Fig. 7-3. Thus the Petersen graph is the unique 5-cage.

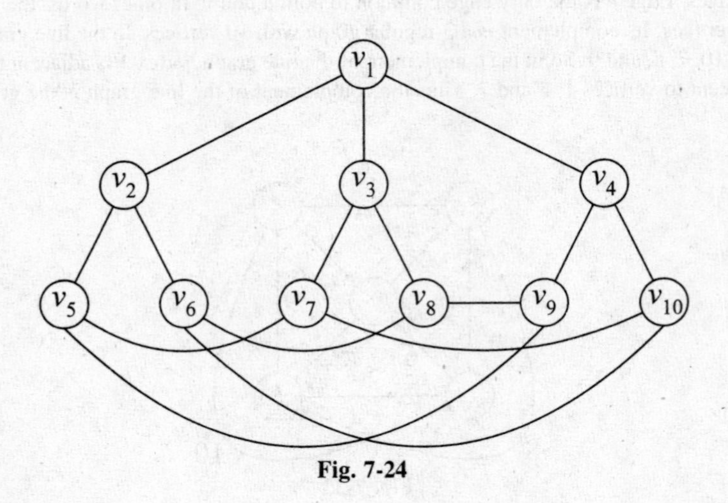

Fig. 7-24

7.78 The **girth** of a graph *G* that is not acyclic is the number of edges in a shortest cycle in *G*. An **(*r*, *g*)-cage** is a *r*-regular graph of girth *g* with as few vertices as possible. (Thus a *g*-cage is a (3, *g*)-cage.) Show that $K_{r,r}$ is the unique $(r, 4)$-cage $(r > 1)$.

Solution. The girth of $K_{r,r}$ is 4, and its order is $2r$. Consider any *r*-regular graph *G* with girth 4. Let *x* be any vertex of *G*, and let *y* be one of the *r* vertices adjacent to *x*. Since the girth of *G* is 4, *y* cannot be adjacent to any of the remaining $(r - 1)$ vertices adjacent to *x*. So there should be another set of $(r - 1)$ vertices adjacent to *y*. Thus *G* should have at least $1 + r + (r - 1) = 2r$ vertices. So any $(r, 4)$-cage should have at least $2r$ vertices. But $K_{r,r}$ is a *r*-regular graph with girth 4 and order $2r$.

7.79 Obtain the unique $(2, g)$-cage, the unique $(r - 1, 3)$-cage, the unique 3-cage, and the unique 4-cage.

Solution. Obviously, the unique $(2, g)$-cage is the cyclic graph of order *g*, the unique $(r - 1, 3)$-cage is the complete graph of order *r*, the unique 3-cage is the complete graph of order 4, and the unique 4-cage is $K_{3,3}$.

7.80 Show that the **Heawood graph *HG*** is the unique 6-cage.

Solution. The 3-regular graph *HG* of order 14 is constructed [see Fig. 7-25(*a*)] as follows. The 14 vertices labeled 1 to 14 are placed clockwise on a circle. Each vertex *i* is adjacent to $(i + 1)$ and $(i + 13)$, where addition is modulo 14. If *i* is odd, it is adjacent to $(i + 9)$. If *i* is even, it is adjacent to $(i + 5)$. Obviously, its girth is 6 and order is 14. It is easy to see that the girth of *HG* is 6. Let *G* be any 6-cage, and let v_1 and v_2 be two adjacent vertices in *G* (see Fig. 7-25(*b*)]. The other vertices adjacent to these two are v_3, v_4, v_5, and v_6, as shown in the figure. These four vertices, in turn, are adjacent to eight more vertices. No two of these 14 vertices can coincide since the girth is 6. So a 6-cage should have at least 14 vertices. Since the order of *HG* is 14, we conclude that *HG* is a 6-cage. To show that it is the unique 6-cage, we first construct a 3-regular graph using the 14 vertices, as shown in Fig.

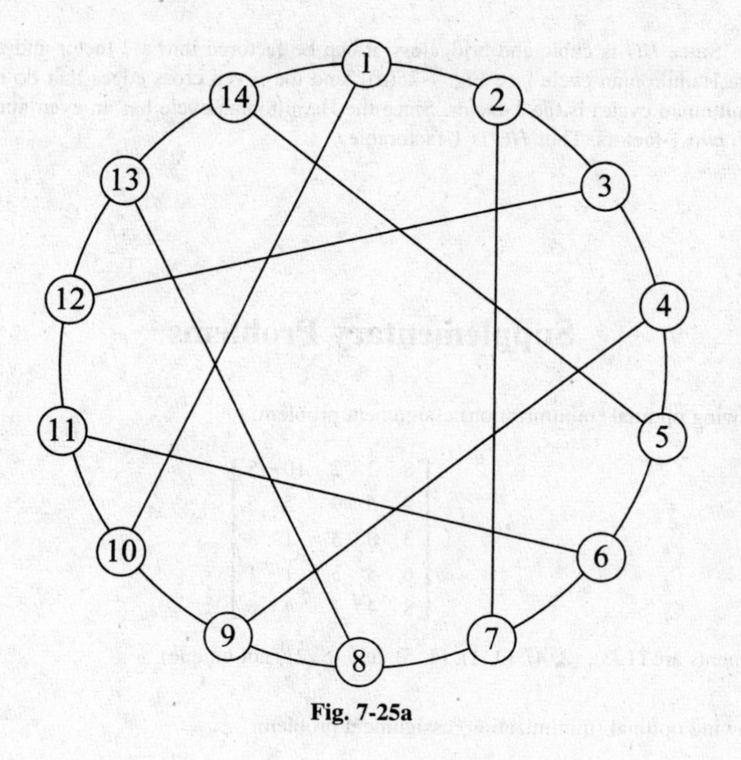

Fig. 7-25a

7-25(*b*), and then prove that the two graphs in Fig. 7-25(*a*) and (*b*) are isomorphic. By relabeling the 14 vertices v_i (i = 1, 2, . . . , 14) as 1, 14, 10, 2, 5, 13, 9, 11, 3, 7, 4, 6, 8, and 12, one can see that the graphs in Fig. 7-25(*a*) and (*b*) are isomorphic.

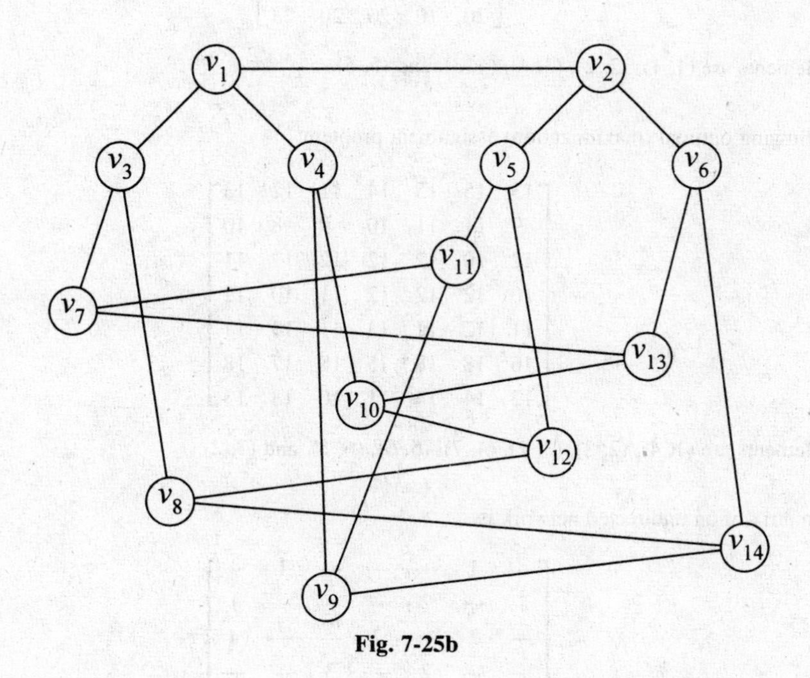

Fig. 7-25b

7.81 Show that the Heawood graph is 1-factorable.

Solution. Since HG is cubic and bridgeless, it can be factored into a 2-factor and a 1-factor. In this case, the 2-factor is a Hamiltonian cycle [see Fig. 7-25(a)], and the seven cross edges that do not belong to the outer circle (the Hamiltonian cycle) is the 1-factor. Since the Hamiltonian cycle has an even number of vertices, it can be factored into two 1-factors. Thus HG is 1-factorable.

Supplementary Problems

7.82 Solve the following optimal (minimization) assignment problem:

$$\begin{bmatrix} 8 & 3 & 2 & 10 & 5 \\ 9 & 6 & 9 & 5 & 5 \\ 3 & 8 & 3 & 1 & 8 \\ 6 & 8 & 3 & 1 & 1 \\ 8 & 4 & 7 & 4 & 8 \end{bmatrix}$$

Ans. The elements are (1, 3), (2, 4), (3, 1), (4, 5), and (5, 2) (not unique).

7.83 Solve the following optimal (minimization) assignment problem:

$$\begin{bmatrix} 21 & 18 & 20 & 14 & 19 \\ 15 & 15 & 12 & 13 & 20 \\ 15 & 17 & 20 & 22 & 20 \\ 15 & 12 & 14 & 12 & 12 \\ 20 & 10 & 20 & 20 & 23 \end{bmatrix}$$

Ans. The elements are (1, 1), (2, 2), (3, 4), (4, 3), and (5, 5).

7.84 Solve the following optimal (maximization) assignment problem:

$$\begin{bmatrix} 13 & 15 & 15 & 14 & 11 & 12 & 13 \\ 9 & 11 & 11 & 10 & 9 & 8 & 10 \\ 12 & 10 & 12 & 12 & 12 & 12 & 11 \\ 10 & 12 & 12 & 12 & 11 & 10 & 12 \\ 11 & 12 & 11 & 13 & 11 & 13 & 11 \\ 16 & 18 & 18 & 15 & 18 & 17 & 18 \\ 12 & 14 & 14 & 11 & 10 & 13 & 13 \end{bmatrix}$$

Ans. The elements are (1, 4), (2, 3), (3, 1), (4, 7), (5, 6), (6, 5), and (7, 2).

7.85 The weight matrix of an undirected network is

$$\begin{bmatrix} - & 1 & - & - & - & 1 & 4 \\ 1 & - & 2 & - & - & - & 1 \\ - & 2 & - & 2 & - & - & 4 \\ - & - & 2 & - & 3 & - & - \\ - & - & - & 3 & - & 9 & 3 \\ 1 & - & - & - & 9 & - & - \\ 4 & 1 & 4 & - & 3 & - & - \end{bmatrix}$$

Construct an extra edge joining vertices 1 and 6 with a weight of 7 units and another extra edge joining vertices 3 and 7 with a weight of 5 units. Obtain an optimal postman route in the enlarged network.

Ans. $1 \to 6 \to 1 \to 7 \to 2 \to 1 \to 6 \to 5 \to 7 \to 5 \to 4 \to 3 \to 7 \to 3 \to 2 \to 1$

7.86 Express the following doubly stochastic matrix as a convex combination of permutation matrices:

$$\begin{bmatrix} 0.21 & 0.13 & 0.38 & 0 & 0.28 \\ 0.31 & 0 & 0.15 & 0.38 & 0.14 \\ 0 & 0.73 & 0.14 & 0.05 & 0.08 \\ 0.38 & 0.14 & 0.05 & 0.36 & 0.07 \\ 0.08 & 0 & 0.28 & 0.21 & 0.43 \end{bmatrix}$$

[*Hint:* This matrix can be written as $0.05P_1 + 0.07P_2 + 0.08P_3 + 0.14P_4 + 0.28P_5 + 0.38P_6$, where the matrices in the convex linear combination are all permutation matrices.]

7.87 Obtain an optimal Hamiltonian cycle in the network with the weight matrix

$$\begin{bmatrix} - & 2 & - & 5 & - & - \\ - & - & 1 & - & 2 & 1 \\ 2 & - & - & - & - & 5 \\ 4 & - & - & - & - & 2 \\ - & 9 & - & 2 & - & - \\ - & - & 2 & - & 2 & - \end{bmatrix}$$

Ans. $1 \to 2 \to 3 \to 6 \to 5 \to 4 \to 1$

7.88 Obtain an optimal Hamiltonian cycle in the network with the weight matrix

$$\begin{bmatrix} - & 12 & 10 & 9 & 10 & 13 & 9 \\ 10 & - & 17 & 10 & 10 & 11 & 10 \\ 9 & 14 & - & 11 & 9 & 12 & 11 \\ 11 & 10 & 11 & - & 12 & 11 & 10 \\ 9 & 11 & 11 & 9 & - & 14 & 12 \\ 10 & 10 & 10 & 10 & 10 & - & 10 \\ 9 & 10 & 9 & 10 & 9 & 10 & - \end{bmatrix}$$

Ans. $1 \to 7 \to 6 \to 3 \to 5 \to 4 \to 2 \to 1$

Chapter 8

Graph Embeddings

8.1 PLANAR GRAPHS AND DUALITY

A graph is a **planar graph** if it is possible to represent it in the plane (that is, to draw it as a diagram on a piece of paper) such that no two edges of the graph intersect except possibly at a vertex to which they are both incident. Any such drawing of a planar graph G in a plane is a **planar embedding** of G. A **plane graph** is a particular representation of a planar graph in the plane drawn in such a way that any pair of edges meet only at their end vertices (if at all they meet). The graph in Fig. 8-1(*a*) is a planar graph since it is isomorphic to the plane graph in Fig. 8-1(*b*).

Theorem 8.1 **(Fary–Stein–Wagner Theorem).** Any simple planar graph has a straight-line representation: it has an embedding in a plane such that each edge in the embedding is a straight line. (See Solved Problem 8.32.)

If x is any point in the plane of a plane graph that is neither a vertex nor a point on an edge, the set of all points in the plane that can be reached from x by traversing along a curve that does not have a vertex of the graph or a point of an edge as an intermediate point is the **region of the graph** that contains x. Thus a plane graph G partitions the plane into the regions of G, and among these regions is exactly one region (the **exterior** or **infinite region**), whose area is not finite. Every other region is an **interior region.** The **boundary** of a region is the subgraph formed by the vertices and edges encompassing that region. If the boundary of the exterior region of a plane graph is a cycle, that cycle is known as the **maximal cycle** of the graph.

If a planar graph G has an embedding on the plane such that the *boundary* of each region (including the unbounded region) is a convex polygon, G is said to have a **convex embedding.** Notice that K_4 has a convex embedding, whereas $K_{2,n}$ does not have a convex embedding whenever $n \geq 4$.

Theorem 8.2. A planar graph G (*i*) can be embedded on the plane such that each region including the exterior region is a polygon if and only if G is 2-connected and (*ii*) has a convex embedding if G is 3-connected. (See Solved Problems 8.5, 8.29, and 8.31.)

The **degree of a region** is the number of edges in a (closed) walk that encloses it. Since a bridge belongs to the boundary of only one region, it contributes to the size of the boundary twice. Thus the sum of the degrees of all the regions in a plane graph is twice the size of the graph. In the plane graph in Fig. 8-1(*b*) are four interior regions of degree 3 and two interior regions of degree 4. The degree of the exterior region is 10. The sum of the degrees (of the regions) is 30, and the graph has 15 edges.

If G_1 and G_2 are two planar graphs, we can take vertex u from the boundary of the exterior face of G_1 and vertex v from the exterior face of G_2 and merge these two vertices to form a new vertex x, creating a new planar graph G with x as a cut vertex. In this case, G is obtained by a **vertex merging** of G_1 and G_2. Likewise, edge e from the boundary of the exterior face of G_1 joining vertices a and b can be merged with edge f from the boundary of the exterior face of G_2 joining the vertices p and q (by identifying a with p and b with q), creating a planar graph G. In this case, G is obtained by an **edge merging** of G_1 and G_2.

If H_1 and H_2 are any two plane graphs isomorphic to a planar graph G, both the plane graphs have the same number of regions. This result is an immediate consequence of the following theorem, due to Euler.

Theorem 8.3 **(Euler's Formula for Plane Graphs).** If a connected plane graph of order n and size m has f regions, $n - m + f = 2$. (See Solved Problem 8.1.)

For example, in Fig. 8-1(*b*), $n = 10$, $m = 15$, and $f = 7$, verifying this formula.

A simple planar graph is called a **maximal planar graph** if the graph becomes nonplanar when any two nonadjacent vertices in it are joined by an edge. A maximal planar graph is necessarily a connected graph. Any planar graph is a spanning subgraph of a maximal planar graph. A planar graph G is a maximal planar graph if and only if the degree of every region (interior as well as exterior) of G is 3. Moreover, if the order of a maximal planar graph is at least 4, the degree of every vertex is at least 3. Since every planar graph has a

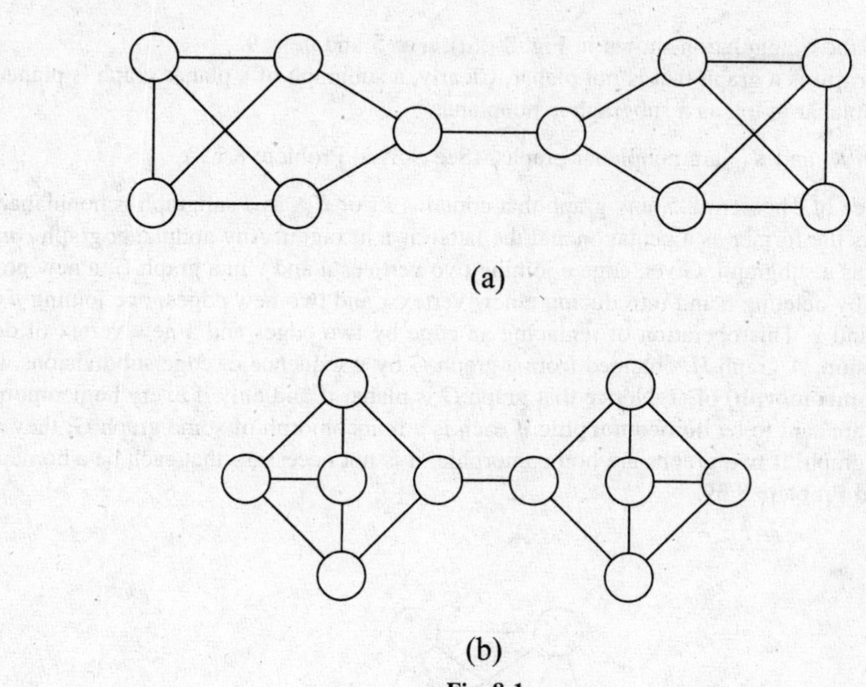

(a)

(b)

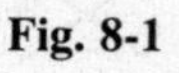

Fig. 8-1

straight-line representation, a maximal planar graph is also known as a **triangulation** or a **triangulated plane graph.** The graph in Fig. 8-2(a) is a triangulation, but the graph in Fig. 8-2(b) is not.

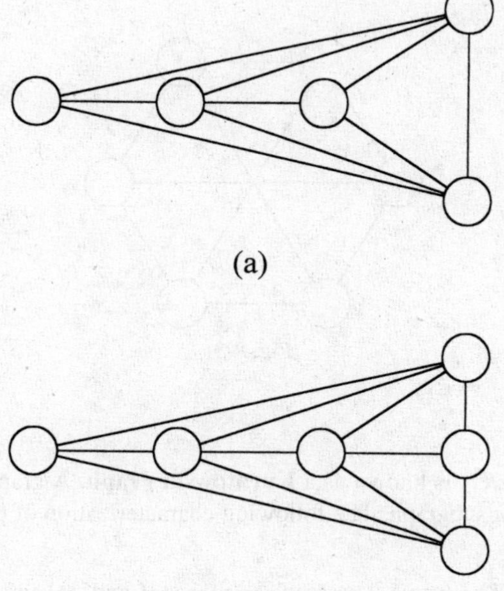

(a)

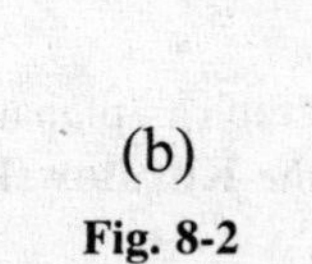

(b)

Fig. 8-2

Theorem 8.4. The number of edges in a triangulation of order n (where $n \geq 3$) is $3n - 6$. (See Solved Problem 8.13.)

For example, in the triangulation shown in Fig. 8-2(*a*), $n = 5$ and $m = 9$.

A **nonplanar graph** is a graph that is not planar. Clearly, a subgraph of a planar graph is planar, and any graph that has a nonplanar graph as a subgraph is nonplanar.

Theorem 8.5. Both K_5 and $K_{3,3}$ are nonplanar graphs. (See Solved Problem 8.15.)

As a consequence of Theorem 8.5, any graph that contains K_5 or $K_{3,3}$ as a subgraph is nonplanar. See Fig. 8-3, which represents the former as a pentagon and the latter as a hexagon. Any nonplanar graph *contains* one of these two graphs as a subgraph. Given edge e joining two vertices u and v in a graph G, a new graph H can be obtained from G by deleting e and introducing a new vertex x and two new edges, one joining u and x and the other joining v and x. This operation of replacing an edge by two edges and a new vertex of degree 2 is called **edge subdivision.** A graph H, obtained from a graph G by a sequence of edge subdivisions, is called a **subdivision** (or a **homeomorph**) of G. Notice that graph G is planar if and only if every homeomorph of G is planar. Two graphs are said to be **homeomorphic** if each is a homeomorph of some graph G; they are subdivisions of the same graph. If two graphs are homeomorphic, it is not necessary that each be a homeomorph of the other; see Solved Problem 8.39.

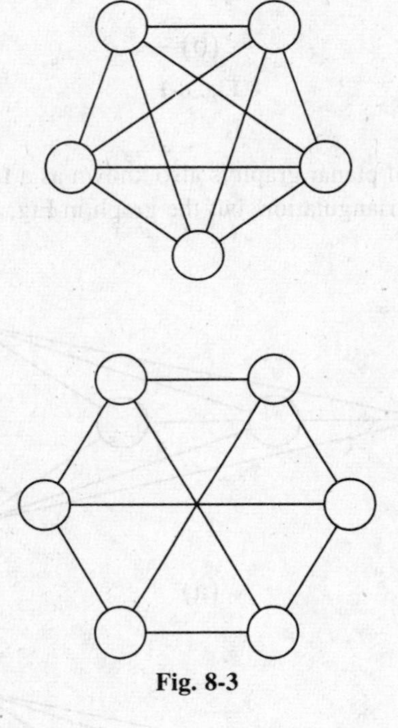

Fig. 8-3

Any homeomorph of K_5 or $K_{3,3}$ is known as a **Kuratowski graph.** A graph is said to have a ***K*-subgraph** if it has a Kuratowski graph as a subgraph. The following characterization of planar graphs is one of the basic theorems in graph theory.

Theorem 8.6 **(Kuratowski's Theorem).** A graph is planar if and only if it does not have a K-subgraph. (This theorem is also known as the **Kuratowski–Pontryagin** theorem.) (See Solved Problem 8.31.)

If e is an edge joining two vertices u and v of a graph G, an **elementary contraction of G by e** is the process of obtaining a simple graph $G.e$ from G by deleting e and by introducing a new vertex (by merging u and v) such that this new vertex is adjacent to those vertices that were adjacent to u or v in G after deleting any multiple edges that may appear in this process. A graph G is **contractible** to graph H if H can be obtained from G by a sequence of elementary contractions; in this case, we say that H is a **contraction** of G. If H is a homeomorph of G, G is a contraction of H. But if H is a contraction of G, it is not necessary that G be a homeomorph of H.

Graph H in Fig. 8-4 is a contraction of G. It can be obtained from G by contracting the edge joining v_3 and v_4 and then contracting the edge joining the merged vertex and v_5. The set of vertices of G can be partitioned into $\{v_1\}$, $\{v_2\}$, $\{v_3, v_4\}$, and $\{v_5, v_6\}$, and the one-to-one correspondence is between these sets and w_1, w_2, w_3, and w_4, respectively. Here H is a contraction of G, but G is not a homeomorph of H.

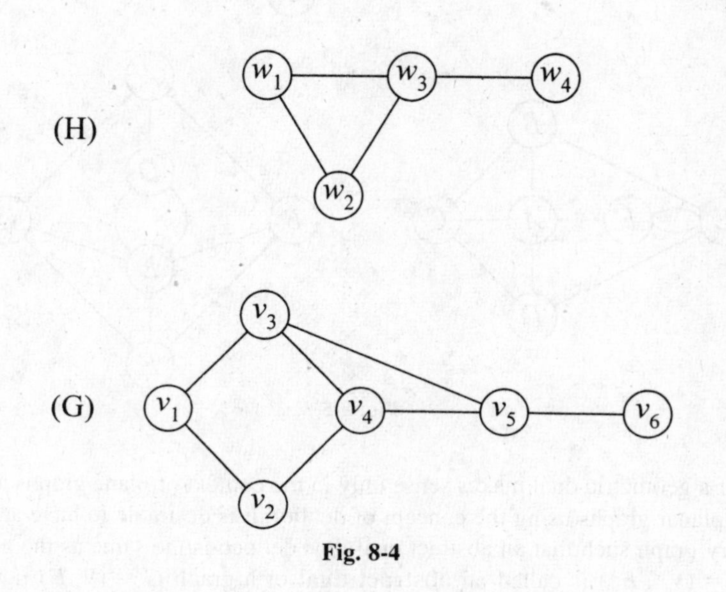

Fig. 8-4

Notice that $H = (W, F)$ is a contraction of $G = (V, E)$ if and only if there exists a partition of V and a one-to-one correspondence between the sets in this partition and the vertices in W such that two elements in W are adjacent if and only if the subgraph of G induced by the union of their corresponding images in the partition is a connected graph.

A **subcontraction** of a graph G is a contraction of a subgraph of G. If G has a subgraph that is a homeomorph of H, H is a subcontraction of G. We now have another important characterization of planar graphs involving K_5 and $K_{3,3}$.

Theorem 8.7 **(Harary–Tutte–Wagner Theorem).** A graph is planar if and only if neither K_5 nor $K_{3,3}$ is a subcontraction of G. (See Solved Problem 8.42.)

Planarity and Duality

Let G be an embedding of a planar graph on the plane. Using this embedding, we can construct a plane graph G', called the **geometric dual of the plane graph *G*,** as follows. Each region of G corresponds to a vertex of G'. If e is an edge of G that has region X on one side and region Y on the other (the two regions could be the same), the corresponding **dual edge *e'*** is an edge joining vertices x and y, which correspond to X and Y, respectively. By the way the dual graph is defined, there is a path between every pair of vertices in the dual graph. In other words, G' is a connected graph. Once the plane graph is embedded, its dual is uniquely defined. More generally, if is G any planar graph, the geometric dual of any plane embedding of G is called **a geometric dual of the planar graph.** It is not at all necessary that two different geometric duals of a planar graph G (corresponding to two different embeddings) are isomorphic. See Fig. 8-5. Both G_1 and G_2 are embeddings on the plane of a planar graph, but their geometric duals G_1' and G_2' are not isomorphic.

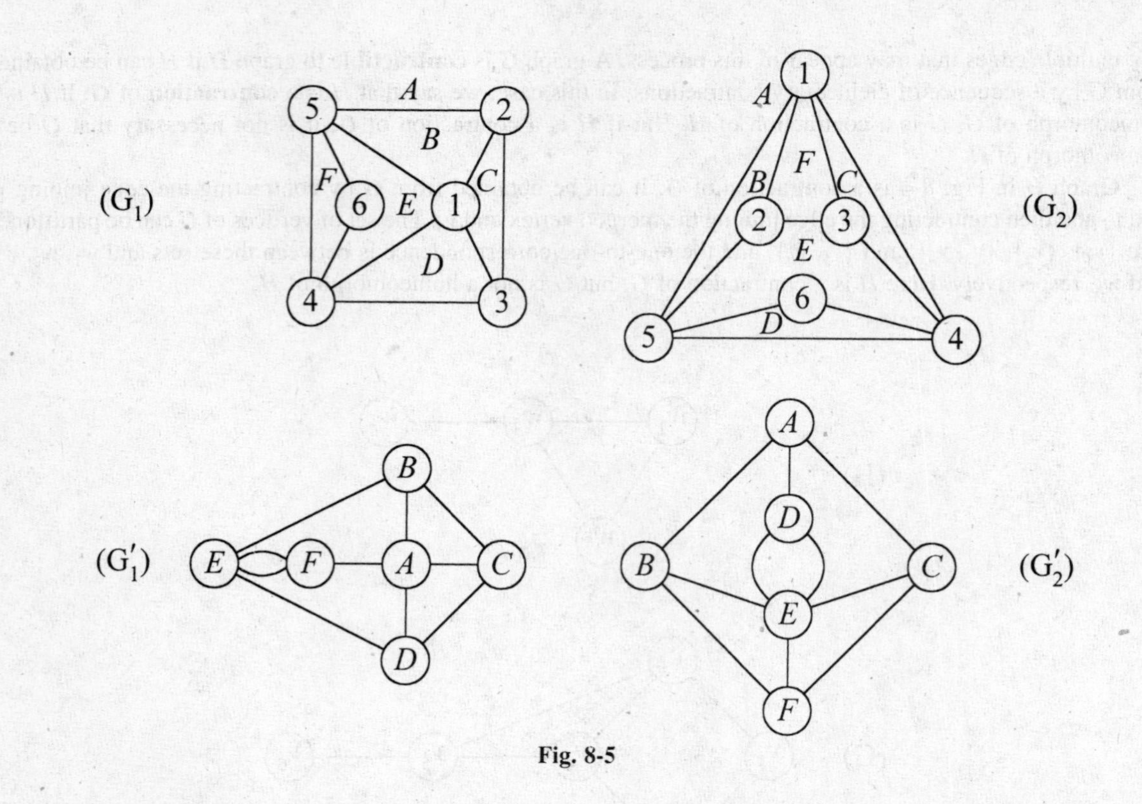

Fig. 8-5

This definition of a geometric dual makes sense only in the context of plane graphs and planar graphs. For a characterization of planar graphs using the concept of duality, it is desirable to have an abstract definition of duality for an arbitrary graph such that an abstract dual thus defined is the same as the geometric dual defined earlier. A graph $G^* = (V^*, E^*)$ is called an **abstract dual** of a graph $G = (V, E)$ if there is a bijection ϕ between E and E^* such that the set $C^* = \phi(C)$ of edges forms a cut set in G^* whenever the set C of edges forms a cycle in G. As in the case of geometric duals, it is possible for a graph to have two nonisomorphic abstract duals. In Fig. 8-6, both G_1^* and G_2^* are nonisomorphic abstract duals of G_1. It can be shown (see Solved Problem 8.67) that in the case of planar graphs, these two notions of duality are equivalent. It can also be shown

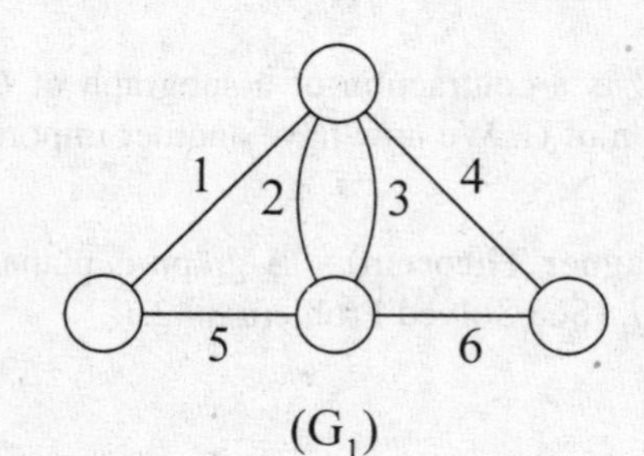

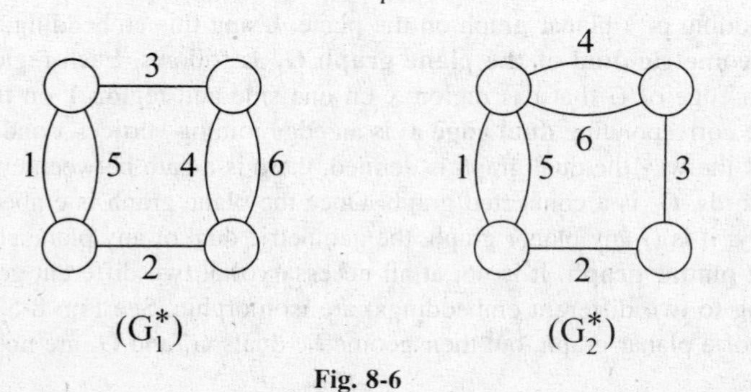

Fig. 8-6

(see Solved Problem 8.70) that if G^* is the abstract dual of G, G is the abstract dual of G^*. Thus two graphs are **abstract duals to each other** if there is a bijection between their sets of edges such that cycles in one graph correspond to cut sets in the other and vice versa.

Theorem 8.8 **(Whitney's Theorem).** A graph is planar if and only if it has an abstract dual. (See Solved Problem 8.75.)

Convex Polyhedra in Three Dimensions

We conclude this section with a brief look at convex polyhedra and their associated graphs. For each positive integer n greater than 2. there is a convex polygon in the plane with n sides. This notion of a convex polygon can be extended to higher dimensions. A **convex polyhedron** (in three dimensions) is a solid bounded by a finite number of surfaces (known as **faces**), each of which is a plane such that no point in the line joining any two points of the solid lies outside the solid. Corresponding to a convex polyhedron P that has a finite number of corners, sides, and faces, we can construct a simple connected plane graph $G(P)$ as follows. Any face of the solid can be considered as its base. Assume that the faces of the solid are made of rubber. Then we can hold the sides of the base and stretch them out to transform the three-dimensional solid into a two-dimensional flat sheet. Thus with every convex polyhedron P, there is an associated plane graph $G(P)$ called a **1-skeleton** of P in which each vertex corresponds to a corner of the solid and each edge corresponds to a side of the solid. The graph $G(P)$ is necessarily a connected graph. Furthermore, the exterior region corresponds to the base of the solid, and every interior region corresponds to a face of the solid. Each interior region of the plane graph can be represented as a convex polygon. The degree of each vertex is at least 3, and the degree of each region is also at least 3.

Figure 8-7(a) shows a convex polyhedron, and Fig. 8-7(b) shows its associated plane graph.

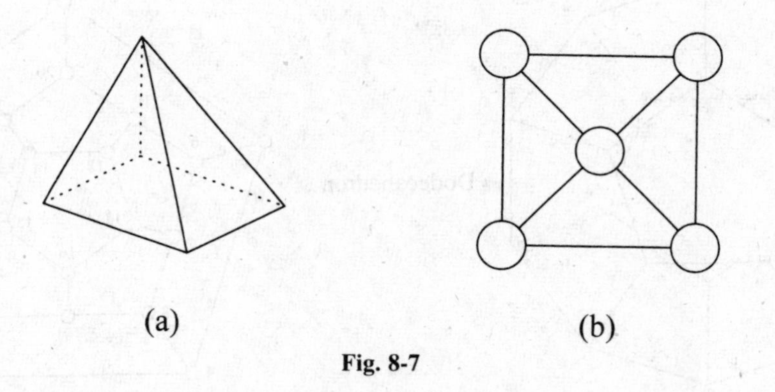

Fig. 8-7

If the convex polyhedron P has n corners, m sides, and f faces, the connected graph $G(P)$ has n vertices, m edges, and f regions. So by Theorem 8.1, we get the same relation $n - m + f = 2$. Thus this equation is also known as **Euler's polyhedron formula.**

Theorem 8.9 **(Steinitz's Fundamental Theorem on Convex Types).** A graph G is isomorphic to the graph $G(P)$ of a convex polyhedron P if and only if G is planar and 3-connected. (Hence, any planar 3-connected graph is known as a **polyhedral graph.**) [The necessity part of the proof is quite straightforward. The proof of the (intuitively obvious) sufficiency part, on the other hand, is rather complicated and is outside the scope of most books on graph theory. For an elegant treatment of this topic in a more general setting, the reader is referred to *Convex Polytopes* by B. Grunbaum.]

A **regular polyhedron** is a convex polyhedron, all of whose faces are congruent polygons and at each of whose vertices the same number of polygons meet. Even though the number of regular polygons is infinite, there are only five regular polyhedra (see Solved Problem 8.23), known as the **Platonic solids:** the regular tetrahedron (in which the four faces are congruent equilateral triangles), the cube (in which the six faces are congruent squares), the octahedron (in which the eight faces are congruent equilateral triangles), the dodeca-

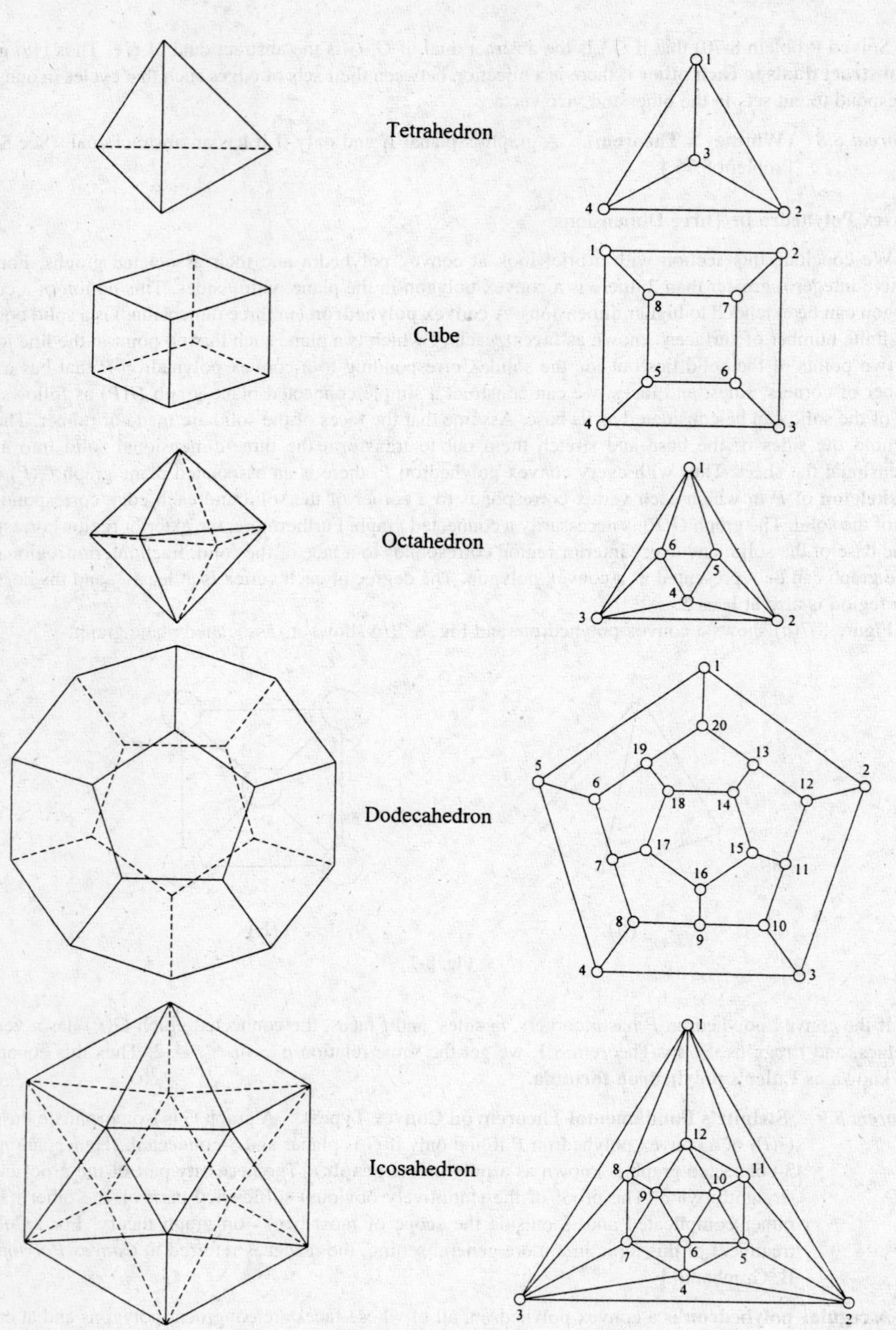

Fig. 8-8

hedron (in which the 12 faces are congruent regular pentagons), and the icosahedron (in which the 20 faces are congruent equilateral triangles). These five solids and their corresponding polyhedral graphs (the five **Platonic graphs**) are shown in Fig. 8-8.

The polyhedral graph $G(P)$ of each Platonic solid P has a unique embedding on the plane; therefore, it has a unique geometric dual. It can be easily verified that the (polyhedral) graph of a tetrahedron is its own dual, the graph of the cube and the graph of the octahedron are dual to each other, and the graph of the dodecahedron and the icosahedron are dual to each other. A regular graph is said to be **completely regular** if its dual also is regular. The polyhedral graph of any Platonic solid is completely regular.

8.2 HAMILTONIAN PLANAR GRAPHS

The 3-connected planar graphs corresponding to each of the five platonic solids are Hamiltonian: in each graph in Fig. 8-8, start from vertex 1 and return to vertex 1 after moving sequentially from one vertex to the next vertex. In general, however, an arbitrary 3-connected planar graph (known as a **3CP graph** in the lore, whereas a *cubic* 3-connected planar graph is a **C3CP graph**) need not be Hamiltonian; the **Herschel graph** shown in Fig. 8-9 is a counterexample. It is obviously a 3CP-graph, and since it has no odd cycles, it is a bipartite graph. But it cannot be a Hamiltonian graph since it has an odd number of vertices.

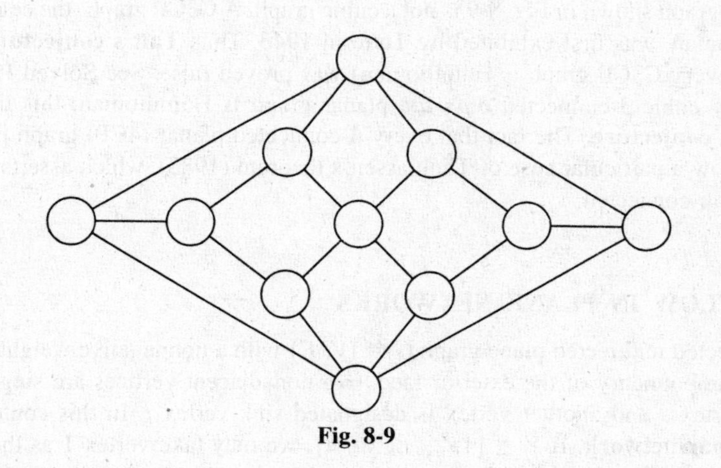

Fig. 8-9

In Chapter 3, several sufficient conditions for an arbitrary graph to be Hamiltonian were presented. If a graph is planar, we have a necessary condition (due to Grinberg and Kozyrev) to be satisfied by it if it is Hamiltonian. Needless to say, this theorem serves a very useful purpose when we would like to show that a given plane graph is not Hamiltonian. If r is the degree of a region in a plane graph, the **index** of that region is the integer $(r - 2)$. If the plane graph G has a Hamiltonian cycle C, C divides the plane into an interior part and an exterior part. The regions known as the **inner regions relative to** C are in the interior part, and the remaining regions, including the exterior region in the exterior part, are the **outer regions relative to** C.

Theorem 8.10 **(Grinberg–Kozyrev Theorem).** If C is any Hamiltonian cycle in a Hamiltonian plane graph of order n, the sum of the indices of the inner regions relative to C and the sum of the indices of the outer regions relative to C are both equal to $(n - 2)$. (See Solved Problem 8.76.)

Example 1. In the plane graph of order 11 shown in Fig. 8-10, each edge is either a thick line or a dashed line. The cycle sequentially passing through the 11 vertices starting from 1 and terminating in 1 is a Hamiltonian cycle; the 11 edges of this cycle are represented by thick lines. The inner regions are A, B, C, and D with degrees 4, 5, 4, and 4, respectively. So the sum of their indices is $2 + 3 + 2 + 2 = 9$. The outer regions are A', B', C', D', and the exterior region of the graph with degrees 3, 3, 3, 4, and 6, respectively. The sum of their indices is equal to $1 + 1 + 1 + 2 + 4 = 9$. Thus the theorem is easily verified.

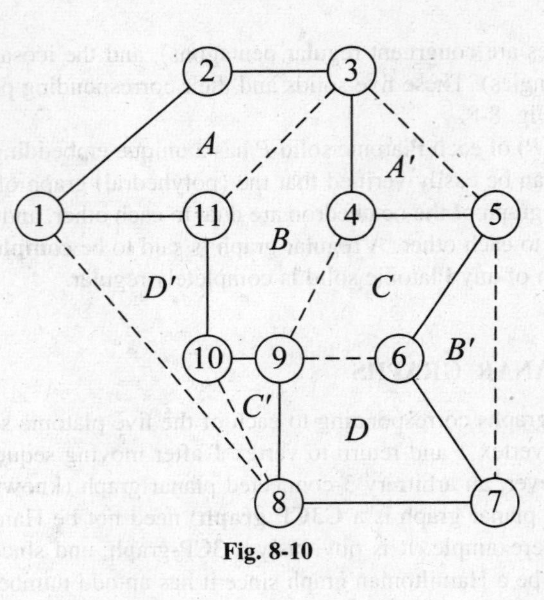

Fig. 8-10

The 3-connected graph shown in Fig. 8-9 is not a cubic graph. A C3CP graph (the celebrated **Tutte graph**), which is not Hamiltonian, was first exhibited by Tutte in 1946. Thus **Tait's conjecture** (which six decades earlier asserted that every C3CP graph is Hamiltonian) was proved false. See Solved Problem 8.80. It is not known whether every cubic 3-connected *bipartite* planar graph is Hamiltonian; this unresolved problem is known as **Barnette's conjecture.** The fact that every 4-connected planar (4CP) graph is Hamiltonian proved by Tutte in 1956 is now a particular case of Thomassen's theorem (1983), which asserts that every 4CP graph is actually Hamiltonian-connected.

8.3 MAXIMUM FLOW IN PLANE NETWORKS

Consider a connected undirected plane graph $G = (V, E)$ with a nonnegative weight defined on each edge as its capacity. On the boundary of the exterior face, two nonadjacent vertices are singled out: one vertex is designated source vertex s and another vertex is designated sink vertex t. In this context, the network G is known as an ***s*, *t* planar network.** If $V = \{1, 2, \ldots, n\}$, we may take vertex 1 as the source and vertex n as the sink in some cases. Then the optimization problem is the problem of finding a feasible s, t flow in the network such that the flow value is a maximum. Since vertices s and t are not adjacent, we construct an artificial edge e joining them that does not intersect any other edge so that the enlarged graph $G + e$ is also a plane graph. Then we construct the geometric (abstract) dual of the enlarged network $G + e$. The vertex in the dual that corresponds to the finite region in $G + e$ containing edge e is designated s^*, and the vertex that corresponds to the exterior face of $G + e$ is designated t^*. Once the dual is constructed, the edge joining s^* and t^* is deleted. The weight of dual edge f^* in the dual corresponding to edge f is, by definition, the weight of f. If there is more than one edge between two vertices, the one with minimum weight is chosen. The plane simple network thus defined is called the **dual network G^*.**

Recall that any s, t cut in G is a set of edges such that their deletion creates a subgraph in which there is no path between s and t. By our construction, the dual edges corresponding to the edges in an s, t cut in G constitute a path between s^* and t^* and vice versa. So the edges in a shortest path between s^* and t^* correspond to the edges in a minimum cut in G.

Example 2. In Fig. 8-11(a), the network G is an s, t network (where s is 1 and t is 6). An artificial edge joining s and t is constructed (as a dashed curve); Fig. 8-11(b) shows the dual network G^*. In each network, the capacities are indicated on their edges. For example, in G, the weight of the edge common to region A and region C is 8. In G^*, the weight of the dual edge joining vertices A and C is 8. Of the two edges between A and B with weights 3 and 6, the edge with weight 3 is chosen. A shortest path from s^* to t^* has edges with weights 5, 3, and 2. The corresponding minimum cut in G consists of edges {1, 2}, {1, 5}, and {1, 4}. Hence, the maximum flow value from the source to the sink (or the other way around) is 5 + 3 + 2 = 10.

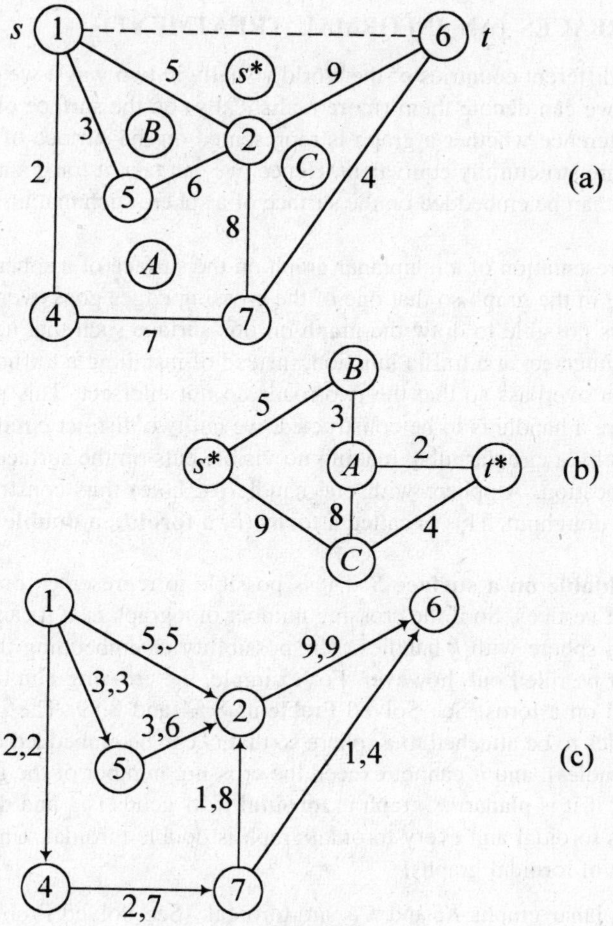

Fig. 8-11

The shortest distance between s^* and t^* in the dual network is the maximum flow value, as was seen in Example 2. Any shortest path between these two vertices in the dual network can be used to obtain a maximum flow in the given network; this is the content of the following assertion.

Theorem 8.11. Let G be an s, t network, and let G^* be its dual. Let $d(k^*)$ be the shortest distance between s^* and k^* in G^*. Define $x_{ij} = d(j^*) - d(i^*)$, where $\{i^*, j^*\}$ is the dual edge that corresponds to edge $\{i, j\}$ in G. Then the vector $[x_{ij}]$ is a maximum flow in G. (See Solved Problem 8.84.)

Example 3. In Fig. 8-11(*b*), the shortest distances are $d(s^*) = 0$, $d(A) = 7$, $d(B) = 5$, $d(C) = 9$, and $d(t^*) = 10$. The components of the maximum flow along the six edges are computed as follows:

$$x_{12} = d(B) - d(s^*) = 5 - 0 = 5 \qquad x_{14} = d(t^*) - d(A) = 10 - 8 = 2$$
$$x_{15} = d(A) - d(B) = 8 - 5 = 3 \qquad x_{25} = d(A) - d(B) = 8 - 5 = 3$$
$$x_{26} = d(C) - d(s^*) = 9 - 0 = 9 \qquad x_{27} = d(C) - d(A) = 9 - 8 = 1$$
$$x_{47} = d(t^*) - d(A) = 10 - 8 = 2, \qquad x_{76} = d(t^*) - d(C) = 10 - 9 = 1$$

Using this information, each edge is converted into an arc. The maximum flow is displayed in Fig. 8-11(c). Along each arc, are two numbers; the first is the flow along the arc, and the second is its capacity, which is actually the weight of the edge that has now become an arc.

8.4 GRAPHS ON SURFACES (AN INFORMAL TREATMENT)

We can represent the different countries of the world visually in two ways: we can denote them on a sheet of paper as a flat map, or we can denote them (more realistically) on the surface of a sphere as a global map. Likewise, it makes no difference whether a graph is represented on the surface of a globe or on a flat sheet: these two representations are structurally equivalent. Hence, we can take it for granted that a graph has a plane embedding if and only if it can be embedded on the surface of a sphere such that no two edges intersect, except possibly at vertices.

Now consider the representation of a nonplanar graph on the surface of a sphere. If we attach a **handle** to the sphere at each crossing in the graph so that one of the crossing edges goes over the handle while the other goes under the handle, it is possible to draw the graph on this surface such that no two edges intersect. (For example, when two roads intersect at a traffic junction, instead of installing a traffic light to control the traffic, we sometimes construct an overpass so that the two roads do not intersect. This results in a smooth flow of traffic.) At a location where a handle is to be constructed, we cut two distinct circular holes in the sphere and join the edges of the holes by a circular tube, making no visible cuts on the surface. Equivalently, we can dig a hole (a tunnel) at the location. A sphere with one handle (or hole) thus constructed can be continuously transformed to look like a doughnut. This is called a **torus** (or a **toroid**); a **double torus** is a sphere with two handles.

A **graph G is embeddable on a surface S** if it is possible to represent it on S such that no two edges intersect except possibly at vertices. So if the crossing number of a graph is k, it can be embedded on a surface S, which, in this case, is a sphere with k handles. The possibility of embedding the same graph on a sphere with fewer handles cannot be ruled out, however. For example, the crossing number of the Petersen graph is 2, but it can be embedded on a torus. See Solved Problems 8.54 and 8.89. The **genus of a graph G** is the minimum number of handles to be attached to a sphere so that G can be embedded on the surface S (consisting of the sphere and these handles), and it cannot exceed the crossing number of the graph. Of course, the genus of a graph is 0 if and only if it is planar. A graph is **toroidal** if its genus is 1 and **double toroidal** if its genus is 2. Every planar graph is toroidal and every toroidal graph is double toroidal. Unlike planar graphs, there is no known characterization of toroidal graphs.

Theorem 8.12. The nonplanar graphs K_5 and $K_{3,3}$ are toroidal. (See Solved Problem 8.86 for an actual embedding of these two graphs on a torus.)

For an embedding of a graph on a surface of positive genus, regions and boundaries are defined in the same way they are defined for an embedding on the plane. There is a vital difference between an embedding on a plane (or on a sphere) and on an arbitrary surface insofar as the nature of a region is concerned, however. A region R is called a **2-cell** if any simple closed curve (like a circle or ellipse) lying completely in R can be continuously shrunk into a single point in that region where at any stage, every point on the resulting curve continues to lie on R. (In other words, a region is a 2-cell if and only if it is homeomorphic to an open disk.) If a connected graph is embedded on a sphere, every region is obviously a 2-cell. This need not be the case if a graph is embedded on surface of positive genus, however. Consider an embedding of the complete graph with four vertices on a torus. It is easy to visualize a situation in which three of the four regions are 2-cells and one region is not. An embedding of a graph on a surface is a **2-cell embedding** (also known as a **cellular embedding**) if every region defined by the embedding is a 2-cell. Any embedding of a connected planar graph on the sphere is no doubt a 2-cell embedding. An embedding of a graph on a surface is a **minimal embedding** if the genus of the graph is equal to the genus of the surface. It turns out (see Solved Problem 8.91) that every minimal embedding of a connected graph is indeed a 2-cell embedding.

Theorem 8.13 **(Generalized Euler Formula).** If an embedding of a connected graph of order n and size m on a surface of genus g defines r regions and if each region thus defined by the embedding is a 2-cell, $n - m + r = 2(1 - g)$. (See Solved Problem 8.92.)

Solved Problems

PLANAR GRAPHS AND DUALITY

8.1 Prove Theorem 8.3 (Euler's formula for plane graphs): If a connected plane graph of order n and size m has f regions, $n - m + f = 2$.

Solution. The proof is by induction on m. If $m = 0$, $n = 1$ and $f = 1$. So the result holds when $m = 0$. Suppose the result is true for any connected graph with fewer than m edges, where $m \geq 1$. Let G be any connected plane graph with n vertices and m edges. If G is a tree, $n = m + 1$ and $f = 1$. So the result is true for any tree with m edges. If G is not a tree, it has a cycle C. Let e be an edge of C. The connected graph $G' = G - e$ has n vertices and $(m - 1)$ edges. By the induction hypothesis G', has f' faces, where $n - (m - 1) + f' = 2$. Hence, $f' = 1 - n + m$. This implies that G has f faces, where $f = 2 - n + m$. In other words, the connected planar graph with n vertices and m edges has f faces, where $n - m + f = 2$. Thus the result is true for m.

8.2 Show that if a plane graph of order n and size m has f regions and k components, $n - m + f = k + 1$.

Solution. Suppose the components are G_i with n_i vertices, m_i edges, and f_i faces (where $i = 1, 2, \ldots, k$). Then $n_i - m_i + f_i = 2$ for each i. The exterior region is the same for all components. If the exterior region is not considered, $n_i - m_i + f_i = 1$ for each i, and on summation, we get $n - m + f = k$. So with the inclusion of the common exterior region, we obtain the relation $n - m + f = k + 1$.

8.3 Verify Euler's formula for the plane graphs shown in Fig. 8-12.

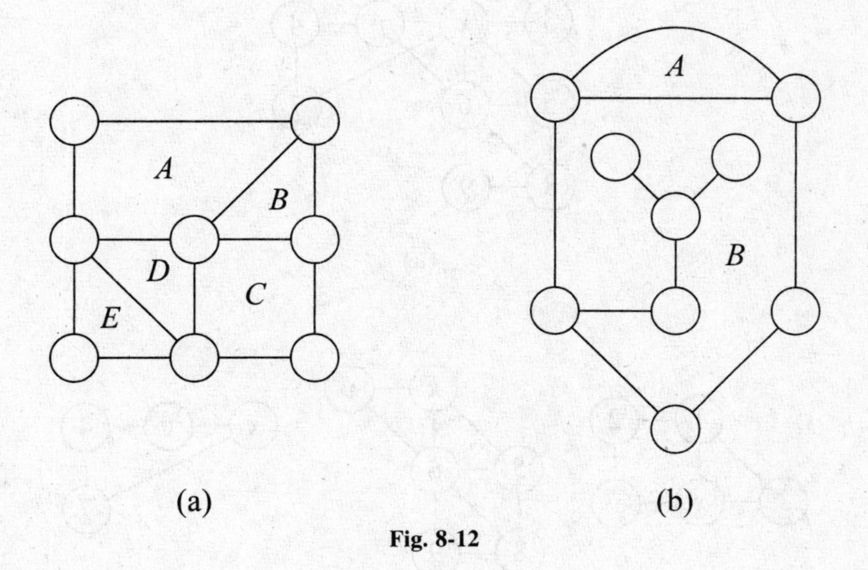

Fig. 8-12

Solution. (*a*) $n = 8$, $m = 12$, and $r = 6$; (*b*) $n = 9$, $m = 10$, and $r = 3$. In both cases, $n - m + r = 2$.

8.4 Find the degrees of the boundaries of the regions of the graphs shown in Fig. 8-12.

Solution. (*a*) The degrees of A, B, C, D, and E are 4, 3, 4, 3, and 3, respectively. The degree of the exterior region is 7. (*b*) The degree of A is 2, the degree of B is 13, and the degree of the exterior region is 5.

8.5 Show that a plane graph is 2-connected if and only if the boundary of each region is a cycle.

Solution. If the boundary of every region (including the exterior region) of a plane graph is a cycle, there will be at least two internally disjoint paths between every pair of vertices. So the graph is 2-connected. On the other hand, let G be a 2-connected plane graph. We have to establish that the boundary of every region of G is a cycle. Suppose there is a region of G whose boundary is not a cycle. Since G is 2-connected, it has at least one region bounded by a cycle. Let H be a maximal subgraph of G, the boundaries of whose regions are cycles. If there is a region in H whose boundary is a cycle and if two nonadjacent (in H) of vertices of this cycle are adjacent in G, H will have one more region whose boundary is a cycle, violating the maximality condition. We assume that this is not the case. $H = (W, F)$ is a proper subgraph of the 2-connected graph $G = (V, E)$. So whenever u and v are two vertices in W and w is a vertex in $V - W$, there exists a path between u and v that passes through w, as established in Solved Problem 6.32. This implies that there is a path joining two nonadjacent vertices in H such that every intermediate vertex in this path is in $V - W$. Any such path will partition some region of H into two regions, violating the maximality requirement. So the boundary of every region of a 2-connected plane graph is a cycle.

8.6 Let $G = (V, E)$ be a graph with vertex-connectivity number $\kappa(G) = 2$, and let x, y be two vertices such that $G - x - y$ is disconnected with components $H_i (i = 1, 2, \ldots, k)$, where $H_i = (V_i, E_i)$ and $W_i = V_i \cup \{x\} \cup \{y\}$. For each i, G_i is the graph induced by the set W_i if vertices x and y are adjacent in G. If these two vertices are not adjacent, G_i is the graph induced by W_i along with edge e joining x and y. The family $\{G_i : i = 1, 2, \ldots, k\}$ is called the **hammock decomposition** of G with respect to the separating set $\{x, y\}$. Obtain a hammock decomposition for the graph in Fig. 8-13(a) with respect to vertices x and y indicated in the diagram.

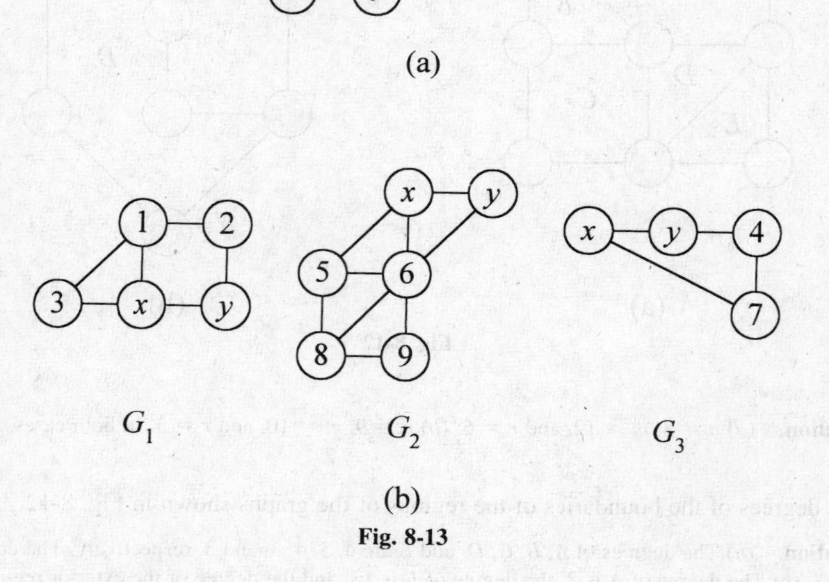

(a)

(b)

Fig. 8-13

Solution. The hammock decomposition consists of the three graphs shown in Fig. 8-13(b).

8.7 Show that a graph G with $\kappa(G) = 2$ is planar if and only if each graph in a hammock decomposition of G with respect to any separating set consisting of two vertices is planar.

Solution. Suppose G is planar and $\{G_i\}$ is a hammock decomposition with respect to a separating set consisting of x and y. If these two vertices are adjacent in G, each graph in the decomposition is a subgraph of G; therefore, it is planar. Let us examine the case when x and y are not adjacent. Observe that $\kappa(G_i) = 2$ for each i. So between vertices x and y are at least two paths in G_i for each i. Hence, graph G_i' has a path P_i between x and y, where G_i' is the subgraph of G induced by the set of vertices of G_i. Each G_i' is planar. Construct a new path Q_i in G_i' by introducing new vertices of degree 2 between x and y duplicating P_i such that $G_i' \cup Q_i$ is planar. Then G_i is homeomorphic to $G_i' \cup Q_i$. Thus each graph in the decomposition is planar. To prove the converse, assume that each graph in the decomposition $\{G_1, G_2, G_3, \ldots, G_k\}$ is planar. These graphs can be sequentially merged by edge merging to form a planar graph H; first merge G_1 and G_2 to form a planar graph. Then merge this new graph and G_3 to form another planar graph. Continue this process until all the k graphs are merged to obtain H, which is planar. Graphs G and H differ at most by an edge. So G is planar.

8.8 Show that any triangulation with at least four vertices is 3-connected.

Solution. Suppose the triangulation G is not 3-connected. This implies that $\kappa(G) = 2$, and there exist vertices u and v defining a Hammock decomposition $\{G_i : i = 1, 2, \ldots, k\}$ such that each G_i is planar and $\kappa(G_i) = 2$ for each i. Let $(k - 1)$ of these graphs be merged (by sequentially edge merging two at a time) to obtain a planar graph H. The remaining graph in the decomposition is G_k, which is also planar. Suppose the edge used in merging these two graphs joins vertices p and q. Let $C: p \to q \to v_1 \to v_2 \to \cdots \to v_m \to p$ be a cycle in H, and let $C': p \to q \to w_1 \to w_2 \to \cdots \to w_n \to p$ be a cycle in G_k. No vertex v_i is adjacent to vertex w_j. So we can have a planar embedding consisting of G and an edge joining nonadjacent vertices v_i and w_j. This violates that G is a maximal triangulated graph. So G is 3-connected.

8.9 If X is the set of blocks and Y is the set of cutvertices in a graph $G = (V, E)$, the **block–cut vertex graph (BC graph)** of G is the bipartite graph $H = (X, Y, F)$ in which there is an edge joining block B and cutvertex v if and only if v is a vertex in B. Construct the BC graph of graph G shown in Fig. 8-14(a).

Solution. Graph G shown in Figure 8-14(a) has six blocks as subgraphs induced by the following sets of vertices: $\{1, 2, 3\}$, $\{2, 4\}$, $\{2, 5\}$, $\{2, 6\}$, $\{6, 7, 8, 9\}$, and $\{8, 10, 11, 12, 13\}$. The corresponding BC graph is shown in Fig. 8-1(b).

8.10 Show that if a graph is connected, its BC graph is a tree.

Solution. Let H be the BC graph of graph G. If G is connected, H is also connected. Suppose C: $v_1, B_1, v_2, B_2, \ldots, v_i, B_i, v_{i+1}, \ldots, B_k, v_1$ is a cycle in H. In each block B_i is a path between v_i and v_{i+1}. The union of these paths form a cycle C' in G. Among the k blocks in C, there should be at least two blocks that have edges in common with C'. Now if a block contains two vertices x and y, it contains all the paths joining those two vertices. So these two blocks should contain C', implying that they have at least three vertices in common. This is a contradiction since two blocks can have at most one vertex in common. So H is acyclic.

8.11 A block B in a graph is called a **pendant block** if B contains exactly one cut vertex. Show that if a graph has a cut vertex, it has at least two pendant blocks.

Solution. If G has a cut vertex, its BC-graph is nontrivial and acyclic with at least two vertices of degree 1. Any vertex of degree 1 in the BC-graph is a pendant block. The graph shown in Fig. 8-14(a) has four pendant blocks, which are displayed as leaves of its BC-tree in Fig. 8-15.

8.12 Show that a graph G is planar if and only if each of its blocks is planar.

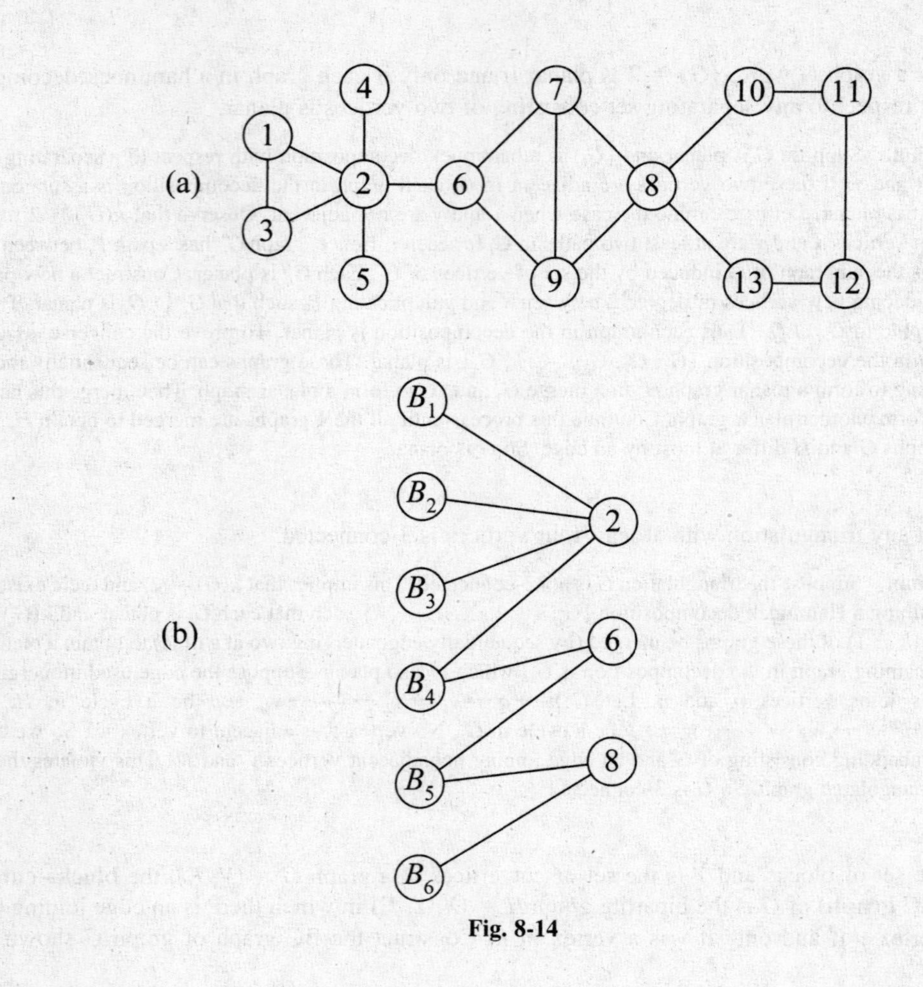

Fig. 8-14

Solution. We may assume without loss of generality that G is connected. Clearly, if a graph is planar, every block in it is planar. The converse is proved by an inductive argument on the number k of blocks in G. The result is true when $k = 1$. Assume that every graph G with fewer than k blocks (each being planar) is planar, where $k \geq 2$. Let G be a graph with k planar blocks, and let B be any pendant block (it exists, as proved in Problem 8.11) in G. Since it is a pendant block, there is a unique cutvertex v of G common to B. Let G' be the graph obtained from G after deleting all the vertices of B other than v. By the induction hypothesis, G' is planar. Since B is a pendant block, it can be adjoined to G', which is planar, by embedding both B and G' on the plane (where B is in the exterior region of G') with v as a connecting vertex. Thus G is planar. So the result is true for k as well.

8.13 Prove Theorem 8.4: The number of edges in a triangulation of order n (where $n \geq 3$) is $3n - 6$.

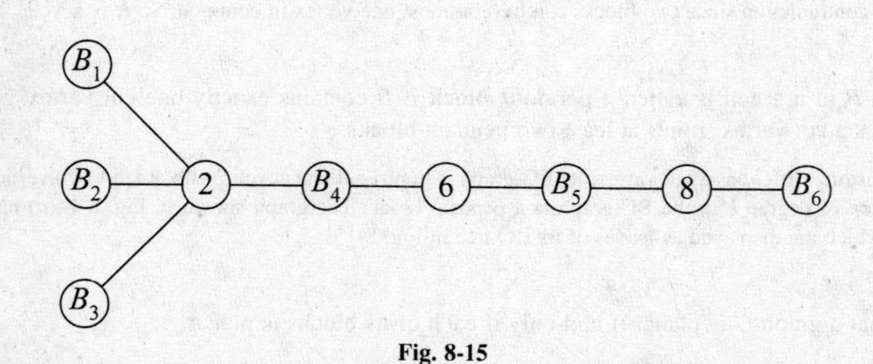

Fig. 8-15

Solution. Suppose there are m edges and f faces in a triangulation. The boundary of each face consists of three edges. If we add the edges of all the f faces, we get $3f$ edges. In this adding process, each edge is counted twice. So $3f = 2m$. On substituting $f = \frac{2}{3}m$ in Euler's formula $n - m + f = 2$, we get the relation $m = 3n - 6$.

8.14 Show that the number of edges in a simple planar graph of order n is at most $3n - 6$.

Solution. Let G be a planar embedding of the given planar graph, and suppose it has k components each with n_i vertices and m_i edges, where $i = 1, 2, \ldots, k$. Since the graph is simple, it follows from Problem 8.15 that $m_i \leq 3\,n_i - 6$ for each i. By adding over all their components, $m \leq 3n - 6k \leq 3n - 6$.

8.15 Prove Theorem 8.5: Both K_5 and $K_{3,3}$ are nonplanar.

Solution.

(*i*) If K_5 is a planar graph, it cannot have more than $(3)(5) - 5 = 9$ edges, according to the result established in Problem 8.14. But it has 10 edges, so it is not planar.

(*ii*) Since $K_{3,3}$ is bipartite, it cannot have an odd cycle. So the boundary of each face has at least four edges. If we add the edges of the boundaries of all the f faces, the sum is at least $4f$. But the number of edges is 9. So $4f \leq 18$, since each edge is counted twice in the adding process. But, by Euler's formula, f should be 5.

8.16 Show that in a simple planar graph G, there are at least four vertices with degrees at most 5. In particular, every convex polyhedron has at least four corners that are adjacent to three, four, or five corners.

Solution. We may assume that G is of order n and size m and is a plane graph. By joining nonadjacent vertices as and when needed, a maximal plane graph G' is constructed. Suppose the number of vertices of degree i in G' is n_i. Since the degree of each vertex in G' is at least 3, $3n_3 + 4n_4 + 5n_5 + \cdots = 2m = 2(3n - 6)$. Hence, $3(n_3 + n_4 + n_5) + 6(n_6 + n_7 + n_8 + \cdots) \leq 6n - 12$. But $(n_6 + n_7 + n_8 + \cdots) = n - (n_3 + n_4 + n_5)$. So $(n_3 + n_4 + n_5) \geq 4$, which implies that G' (and hence G) has at least four vertices of degree at most 5. In the case of the graph $G(P)$ of a convex polyhedron, the degree of each vertex is at least 3. So there are at least four vertices in $G(P)$ with degree 3, 4, or 5.

8.17 Exhibit a simple planar connected graph in which the degree of each vertex is at least 5.

Solution. In the plane graph corresponding to the icosahedron (see Fig. 8-8), the degree of each vertex is 5. Furthermore, it has 12 vertices.

8.18 Exhibit a simple planar connected graph in which the degree of each vertex is exactly 4.

Solution. In the plane graph of the octahedron (with eight vertices), the degree of each vertex is 4.

8.19 Show that if a simple graph G has at least 11 vertices, both G and its complement cannot be planar graphs.

Solution. Suppose G has n vertices and m edges and its complement has m' edges. Then $m + m' = n(n - 1)/2$. If both are planar, $n(n - 1)/2 \leq (6n - 12)$. This inequality is true if and only if $n < 11$.

8.20 Suppose n_i be the number of vertices of degree i in a triangulation G of order n. Establish the relation $3n_3 + 2n_4 + n_5 = n_7 + 2n_8 + 3n_9 + \cdots + (n - 6)n_k + 12$, where k is the maximum degree in G.

Solution. The degree of each vertex is at least 3. The sum of the degrees is twice the number of edges, which is equal to $6n - 12$. So $4n_3 + 4n_4 + \cdots + kn_k = 6n - 12$. But $n_3 + n_4 + \cdots + n_k = n$. Hence, $3n_3 + 2n_4 + n_5 = n_7 + 2n_8 + 3n_9 + \cdots + (n - 6)n_k + 12$.

8.21 Show that every convex polyhedron has at least one face whose boundary consists of three, four, or five edges. (This is similar to the fact established in Problem 8.16 that every convex polyhedron has a corner that is adjacent to three, four, or five corners.)

Solution. Assume that the convex polyhedron has n corners (vertices), m sides (edges), and f faces (regions). Let f_k be the number of faces of degree k. Suppose $f_k = 0$ when $k = 3, 4,$ or 5. Then $6f_6 + 7f_7 + \cdots = 2m$. Hence, $6(f_6 + f_7 + \cdots) \leq 2m$, which implies that $6f \leq 2m$.

Suppose there are n_i corners of degree i. Since each corner is adjacent to at least three corners, $n_i \geq 3$ for each i. The degree sum $3n_3 + 4n_4 + \cdots = 2m$ implies that $3n \leq 2m$. Now $n - m + f = 2$ implies that $m = n + f - 2 \leq \frac{2}{3}m + \frac{1}{3}m - 2 = m - 2$. This contradiction establishes that there is at least one face surrounded by at most five faces.

8.22 Show that the total number of corners of degree 3 and faces of degree 3 in convex polyhedron is at least eight.

Solution. Assume that the convex polyhedron has n corners (vertices), m sides (edges), and f faces (regions). Let f_k be the number of faces of degree k and n_k be the number of vertices of degree k. Then $(3n_3 + 4n_4 + \cdots) + (3f_3 + 4f_4 + \cdots) = 4m = 4n + 4f - 8$, implying that $n_3 + f_3 \geq 8$.

8.23 Show that there are only five regular polyhedra.

Solution. Suppose the regular convex polyhedron P under consideration has n corners, m sides and f faces. Then its associated simple planar graph G has n vertices, m edges and f regions. Since P is regular, the degree each vertex of G is equal to a fixed integer k (≥ 3) and the degree of each region of G is a fixed integer $r (r \geq 3)$. Then $kn = rf = 2m$. But $n - m + f = 2$. So $8 = 4n - 2m - 2m + 4f = 4n - kn - rf + 4f = n(4 - k) + f(4 - r)$. Hence either $(4 - k) \geq 0$ or $(4 - r) \geq 0$. In other words $3 \leq k \leq 4$ or $3 \leq r \leq 4$.

Case (i): Let $k = 3$. Then $rf = 3n = 2m$ and $n + f(4 - r) = 8$. Hence, $24 = 3n + 12f - 3rf = rf + 12f - 3rf = f(12 - 2r)$, implying that $f(6 - r) = 12$. Again, $8 = n + f(4 - r) = n + 4f - rf = n + 4f - 3n$, implying that $n = 2f - 4$. Three choices arise: *(a)* $r = 3, f = 4, n = 4$, and $m = 6$ (tetrahedron); *(b)* $r = 4, f = 6, n = 8$, and $m = 12$ (cube); and *(c)* $r = 5, f = 12, n = 20$, and $m = 30$ (dodecahedron).

Case (ii): Let $k = 4$. In this case, $rf = 4n = 2m$ and $(4 - r)f = 8$. This implies that $r = 3, f = 8, n = 6$, and $m = 12$ (octahedron).

Case (iii): Let $r = 3$. Then $3f = kn = 2m$ and $n(4 - k) + f = 8$. Hence, $24 = 12n - 3nk + 3f = 12n - 2kn$, implying that $n(6 - k) = 12$. Also, $8 = 4n - nk + f = 4n - 3f + f$, implying that $f = 2n - 4$. Three choices arise: *(a)* $k = 3, n = 4, f = 4$, and $m = 6$ (tetrahedron again); *(b)* $k = 4, n = 6, f = 8$, and $m = 12$ (octahedron again); and *(c)* $k = 5, n = 12, f = 20$, and $m = 30$ (icosahedron).

Case (iv): Let $r = 4$. In this case, $k = 3, n = 8, f = 6$, and $m = 12$ (cube again).

8.24 *(Kotzig's Theorem)* Show that every polyhedral graph has two adjacent vertices such that their degree sum is at most 13.

Solution. It is enough if this result is proved for a triangulation. Let p be the number of edges in the graph joining two vertices such that their degree sum is at most 12, and let q be the number of edges joining two vertices such that their degree sum is 13. We have to prove that $p + q$ is positive. The result is true if p is positive. So what remains to be proved is that q is positive when p is zero. As usual, let n_i be the number of vertices of degree i. Suppose the degree of vertex v is 3 and that e is the edge joining v and vertex w. Since $p = 0$, the degree of w is 10 or more than 10. Thus $3n_3 \leq q + \frac{1}{2}\Sigma_{k\geq 11} kn_k$. Similarly, $3n_3 + 4n_4 \leq \frac{1}{2}\Sigma_{k\geq 9} kn_k$ and $3n_3 + 4n_4 + 5n_5 \leq \frac{1}{2}\Sigma_{k\geq 8} kn_k$. Multiplying these three inequalities by 5, 3, and 2, respectively, and then adding gives $30n_3 + 20n_4 + 10n_5 \leq 5q + 8n_8 + 24n_9 + 25n_{10} + 5\Sigma_{k\geq 11} kn_k$. Using Euler's formula, we then obtain the inequality

$$120 + 10[\Sigma_{k\geq 7} (k - 6)n_k] \leq 5q + 8n_8 + 24n_9 + 25n_{10} + 5[\Sigma_{k\geq 11} kn_k]$$

It follows that $120 \leq 5q$, which implies that $q > 0$, as we wished to prove.

8.25 A graph G is said to be **minimally nonplanar** if every proper subgraph of G is planar. Show that a minimally nonplanar graph is a block.

Solution. If G is not connected, each component is planar, which implies that G itself is planar. So G is connected. Suppose G has cut vertex v. Then the deletion of v will give subgraphs that are planar, implying that G is planar. Hence, G is 2-connected. Since G is nonplanar, at least one of its blocks is nonplanar. If G has two blocks, both will be planar with at most one vertex in common, implying that G is planar. Hence, G is a block.

8.26 Suppose G is a nonplanar graph that does not have a K-subgraph such that the size of any other nonplanar graph that does not have a K-subgraph is more than the size of G. Show that G is 3-connected.

Solution. By our assumption, if there is a graph H such that the size of H is less than the size of G and if it has no K-subgraph, it has to be planar. So no edge can be deleted from G without violating its nonplanarity property. Hence, G is minimally nonplanar, therefore, it is a block, as established in Problem 8.25. Suppose G becomes disconnected by deleting two vertices, x and y. Let $G_1 = (V_1, E_1)$ and $G_2 = (V_2, E_2)$ be two subgraphs of G such that $G_1 \cup G_2 = G$ and $V_1 \cap V_2 = S$, where S is the separating set $\{x, y\}$. Let H_i be the graph obtained by joining vertices x and y to G_i for $i = 1, 2$ by edge e. Suppose both H_1 and H_2 are planar. Then each can be embedded on the plane such that e is a boundary of the exterior face. This implies that $H_1 \cup H_2 - e$ is a plane embedding of the nonplanar graph G. This contradiction implies that either H_1 or H_2 is nonplanar. The minimality assumption furthermore requires that one of these graphs, say H_1, is a nonplanar graph containing a K-subgraph. In that case, edge e is an edge of K; otherwise, K will be a subgraph of G, which is a contradiction. If we replace edge e by a simple path (in H_2) joining x and y, we get a homeomorph of K as a subgraph of G, which is a contradiction. So G does not have a 2-vertex separating set.

8.27 (*Thômassen's Theorem*) Show that if G is a 3-connected graph with at least five vertices, it has edge e such that $G.e$ is 3-connected.

Solution. Suppose $G.e$ is not 3-connected for every edge e of the graph. Let t be the vertex obtained by contracting edge e joining u and v. Since graph G is 3-connected, there are at least two internally disjoint paths between every pair of vertices in $G.e$. Hence, $G.e$ is 2-connected. Let x and y be two vertices in so that $G.e - x - y$ is disconnected. Suppose x, y, and t are distinct. Since G is 3-connected, $G - x - y$ is a connected graph. Notice that $G.e - x - y$ can be obtained from $G - x - y$ by contracting edge e, so it is a connected graph. This is a contradiction since $G.e - x - y$ is a disconnected graph. So x, y, and t cannot be distinct. In other words, whenever $G.e$ is constructed by contracting an edge into a single vertex, this contracted vertex is necessarily one of the two vertices in any separating set of $G.e$. Thus, corresponding to edge e joining u and v in G is vertex w (call it a mate of e) in G such that $G - \{u, v, w\}$ is disconnected. Let H_1 be a component of $G - \{u, v, w\}$ with the smallest number of vertices. By our hypothesis, these three vertices are adjacent to three distinct vertices in H_1. Suppose e_1 is the edge joining w and vertex p in H_1. As before, we use edge e_1 to contract G to obtain $G.e_1$. Let the mate of e_1 be q. Then $\{w, p, q\}$ is a minimum disconnecting set of G. The component H_2 of $G - \{w, p, q\}$ with the minimum number of vertices is a proper subgraph of H_1. By continuing this process, we ultimately get a component with a single vertex and edge f joining it to one of three vertices in the previous separating set. Then $G.f$ is 3-connected, which is a contradiction.

8.28 Show that if the contraction G.e of a graph G has a K-subgraph, G also has one.

Solution. Let a homeomorph H of K_5 or $K_{3,3}$ be a subgraph of $G.e$, where e is the edge joining u and v in G that are merged together to form the new vertex w in $G.e$. If w is not a vertex of H, H is a subgraph of G. If w is a vertex in H of degree 2, by expanding vertex w into the edge (in G) joining u and v, we get a Kuratowski graph H' (a homeomorph from H) as a subgraph of G. If w is a vertex in H of degree more than 2, either its degree in H is 3 (in which case H is a homeomorph of $K_{3,3}$) or its degree in H is 4 (in which case H is a homeomorph of K_5). Suppose H is a homeomorph of the complete bipartite graph with partite sets $X = \{1, 2, 3\}$ and $Y = \{4, 5, w\}$. If vertex w is expanded into edge e joining u and v in G, we get a homeomorph H' from H as a subgraph of G. This H' is a homeomorph from the complete bipartite graph (X, Y', E'), where $Y' = \{4, 5, u\}$ or $\{4, 5, v\}$. Finally, suppose H is a homeomorph of K_5. Consider the case when w is expanded into the edge joining u and v: two of the edges incident to w become edges incident to u, and the other two become edges incident to v. So there are two vertices in H (say u_1 and u_2) of degree 4 and two paths P_1 and P_2 (in G) between u and each of these two vertices such that these paths have no internal vertices of degree 4. Likewise, there are two vertices v_1 and v_2 in H with paths P_3 and P_4 joining v and each of these two vertices. Delete those edges that belong to H from these four paths. We then find that G has a subgraph H' that is a homeomorph of the complete bipartite graph with partite sets $\{u, v_1, v_2\}$ and $\{v, u_1, u_2\}$. In the only remaining case, when the vertex u has at most one edge incident to it out of the four edges incident to w, we see that G has a subgraph homeomorphic to K_5.

8.29 (*Tutte's Theorem and Its Proof by Thomassen*) Show that if a 3-connected graph G has no K-subgraph, it has a convex straight-line embedding on the plane.

Solution. The proof is by induction on the order n of G. The induction hypothesis is that any 3-connected graph of order less than n with no K-subgraph has a convex embedding. The order of any 3-connected graph is at least 4. The complete graph of order 4 is the only 3-connected graph with four vertices, and it has a convex straight-line embedding. So the theorem is true when $n = 4$. Let $n \geq 5$. Since G is 3-connected, it has edge e joining u and v such that $G.e$ is 3-connected, as shown in Problem 8.27. If $G.e$ has a K-subgraph, G should also have a K-subgraph, as shown in Problem 8.28, violating our hypothesis. So $G.e$ does not have a K-subgraph; hence, by the induction hypothesis, it has a convex embedding. The proof is complete if we show that G has a convex straight-line embedding. If w is the vertex obtained by merging u and v, graph $G' = G.e - w$ is 2-connected. So the boundary of every region (including the exterior region) is a cycle in G', as proved in Problem 8.5. Any cycle in G' is a cycle in G. If we delete the edges incident to w from $G.e$, vertex w will be in one of the regions of G'. Let C be the cycle in which lies the boundary of the region that contains w.

Case (i): There are three vertices p, q, and r in C that are adjacent to both u and v in G. See Fig. 8-16. Then these five vertices are the vertices of a subgraph of G, which is a homeomorph of K_5, which is a contradiction.

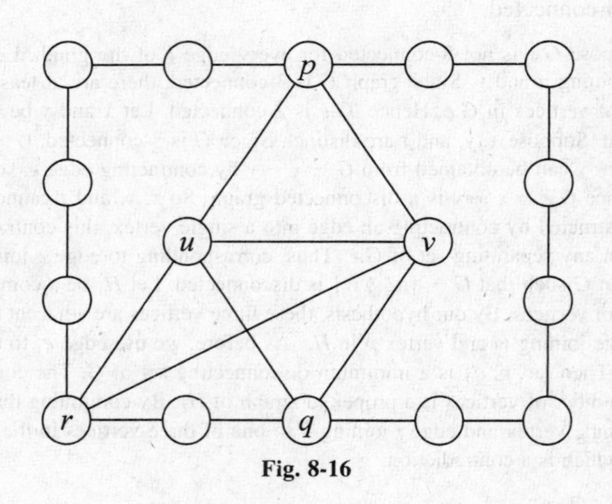

Fig. 8-16

Case (ii): There are four vertices p, q, r, and s that appear in that order on cycle C such that u is adjacent to p and r and v is adjacent to q and s. See Fig. 8-17. These six vertices are the vertices of a subgraph of G, which is a homeomorph of $K_{3,3}$, which is a contradiction.

Case (iii): This case is neither case (*i*) nor case (*ii*). So the number of vertices in C that are adjacent to both u and v is at most 2. We also have the situation in which the vertices adjacent to u and the vertices adjacent to v do not appear on the cycle as stipulated in case (*ii*). So if the vertices adjacent to u are $u_1, u_2, \ldots, u_k$ and the vertices adjacent to v are $v_1, v_2, \ldots, v_m$, these vertices are cyclically ordered on C as $u_1, u_2, \ldots, u_k$ first and then $v_1, v_2, \ldots, v_m$. (It is possible that $u_k = v_1$ or $v_m = u_1$, showing that the number of vertices adjacent to both u and v is at most 2.) See Fig. 8-18(*a*), with $k = 4$ and $m = 5$. This scenario as it is may give a convex straight-line

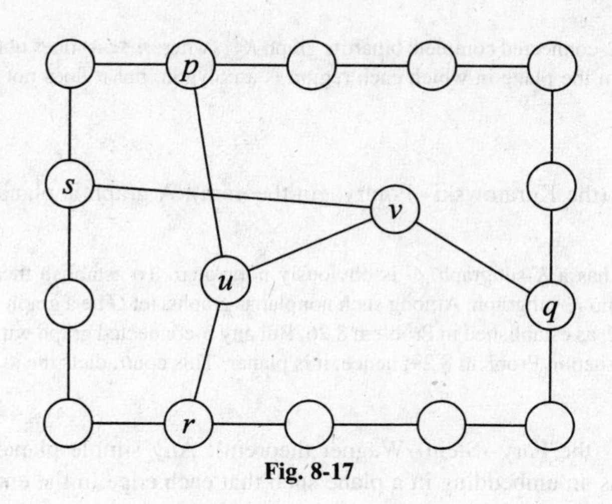

Fig. 8-17

embedding. Otherwise, we move vertex v to another region enclosed by a part of C that contains all the vertices adjacent to v, resulting a convex straight-line embedding for G. See Fig. 8-18(b). Thus, in this case, we have a convex embedding of G by placing v inside the appropriate region.

8.30 Show that 3-connectivity is necessary in Tutte's theorem.

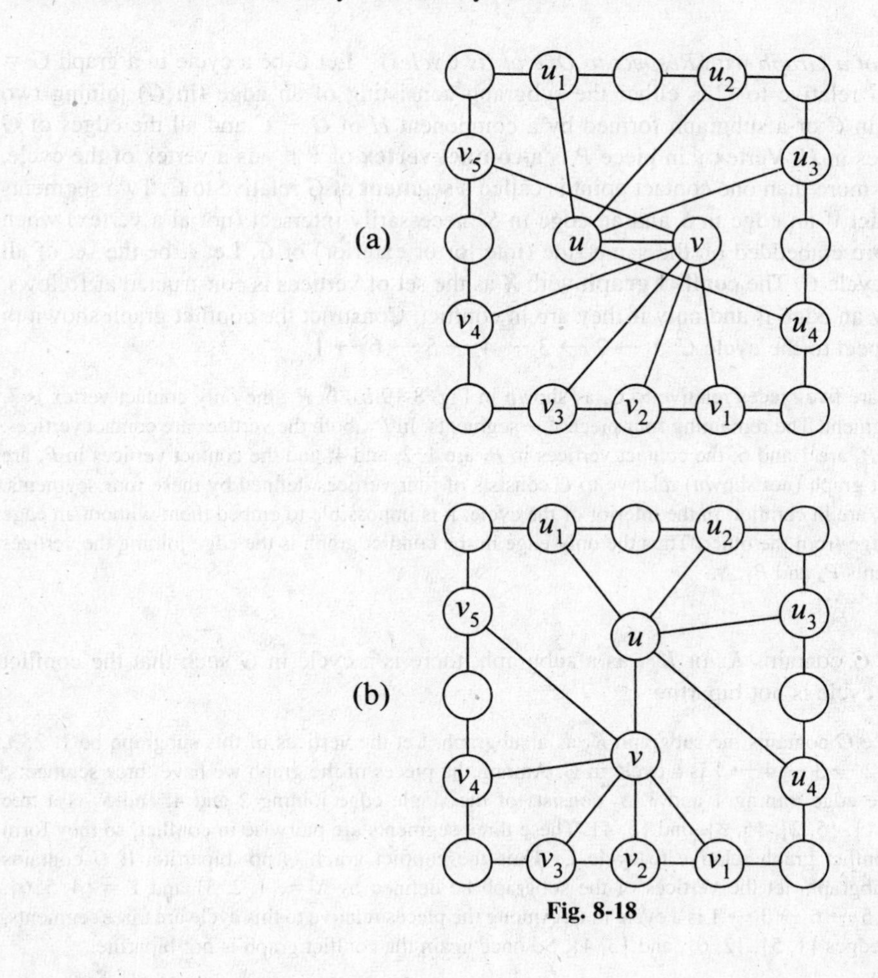

Fig. 8-18

Solution. The 2-connected complete bipartite graph $K_{2,n}$ (where $n \geq 4$) does not have a Kuratowski subgraph. It has an embedding in the plane in which each region is a polygon, but it does not have a convex embedding in the plane.

8.31 Prove Theorem 8.6 (the Kuratowski–Pontryagin theorem): A graph is planar if and only if it does not have a K-subgraph.

Solution. If G has a K-subgraph, G is obviously nonplanar. To establish the converse, suppose there is a nonplanar graph with no K-subgraph. Among such nonplanar graphs, let G be a graph with as few edges as possible. Then G is 3-connected, as established in Problem 8.26. But any 3-connected graph with no K-subgraph has a convex embedding, as established in Problem 8.29; hence, it is planar. This contradicts the assumption that G is nonplanar.

8.32 Prove Theorem 8.1 (the Fary–Stein–Wagner theorem): Any simple planar graph has a straight-line representation: it has an embedding in a plane such that each edge in the embedding is a straight line.

Solution. Every triangulation is a 3-connected graph, as shown in Problem 8.8. In other words, every triangulation is a 3-connected graph with no K-subgraph. So, by Tutte's theorem, it has a convex straight-line embedding, as established in Problem 8.27. But every simple planar graph is a spanning subgraph of a triangulation. So every simple planar graph has a straight-line embedding. The bipartite graph $K_{2,n}$ (where $n \geq 4$) is a simple planar graph that has a straight-line embedding but no convex embedding; it is a 2-connected graph that is not 3-connected.

8.33 (*The Conflict Graph of a Graph with Respect to One of Its Cycles*) Let C be a cycle in a graph $G = (V, E)$. A **piece** of G relative to C is either the subgraph consisting of an edge (in G) joining two nonadjacent vertices in C or a subgraph formed by a component H of $G - C$ and all the edges of G adjacent to the vertices in H. Vertex v in piece P is a **contact vertex** of P if v is a vertex of the cycle. Any piece containing more than one contact point is called a **segment** of G relative to C. Two segments S and S' are **in conflict** if an edge in S and an edge in S' necessarily intersect (not at a vertex) when these two segments are embedded on the same side (interior or exterior) of C. Let X be the set of all segments relative to cycle C. The **conflict graph** with X as the set of vertices is constructed as follows. Join two segments by an edge if and only if they are in conflict. Construct the conflict graph shown in Fig. 8-19(a) with respect to the cycle C: $1 \to 2 \to 3 \to 4 \to 5 \to 6 \to 1$.

Solution. There are five pieces relative to C, as shown in Fig. 8-19(b). In P_1, the only contact vertex is 3, so this piece is not a segment. The remaining four pieces are segments. In P_2, both the vertices are contact vertices. The contact vertices in P_3 are 1 and 6; the contact vertices in P_4 are 1, 2, and 4; and the contact vertices in P_5 are 3, 5, and 8. The conflict graph (not shown) relative to C consists of four vertices defined by these four segments. The segments P_4 and P_5 are in conflict; in the interior of the cycle, it is impossible to embed them without an edge from one crossing an edge from the other. Thus the only edge in the conflict graph is the edge joining the vertices corresponding to segments P_4 and P_5.

8.34 Show that if a graph G contains K_5 or $K_{3,3}$ as a subgraph, there is a cycle in G such that the conflict graph relative to that cycle is not bipartite.

Solution. Suppose G contains the subgraph K_5 as a subgraph. Let the vertices of this subgraph be 1, 2, 3, 4, and 5. Then C: $1 \to 2 \to 3 \to 4 \to 1$ is a cycle in G. Among the pieces of the graph we have three segments: S_1 consists of the single edge joining 1 and 3, S_2 consists of the single edge joining 2 and 4, and S_3 is a tree consisting of edges $\{5, 1\}$, $\{5, 2\}$, $\{5, 3\}$, and $\{5, 4\}$. These three segments are pairwise in conflict, so they form an odd cycle in the conflict graph relative to cycle C. Thus the conflict graph is not bipartite. If G contains $K_{3,3} = (X, Y, E)$ as a subgraph, let the vertices of the subgraph be defined by $X = \{1, 2, 3\}$ and $Y = \{4, 5, 6\}$. Then C: $1 \to 4 \to 2 \to 5 \to 6 \to 3 \to 1$ is a cycle in G. Among the pieces relative to this cycle are three segments, in conflict pairwise, as edges $\{1, 5\}$, $\{2, 6\}$, and $\{3, 4\}$. So once again the conflict graph is not bipartite.

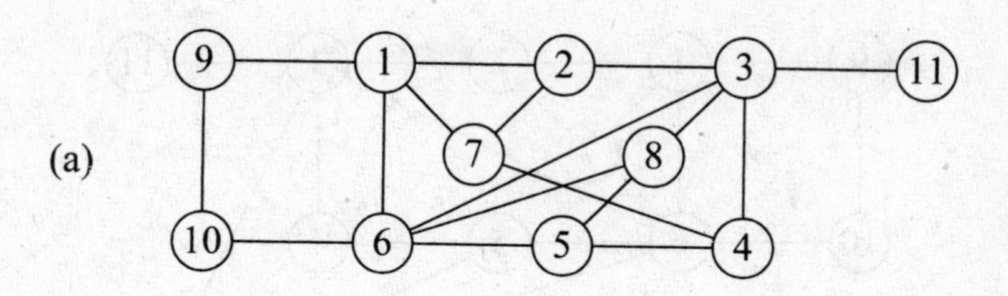

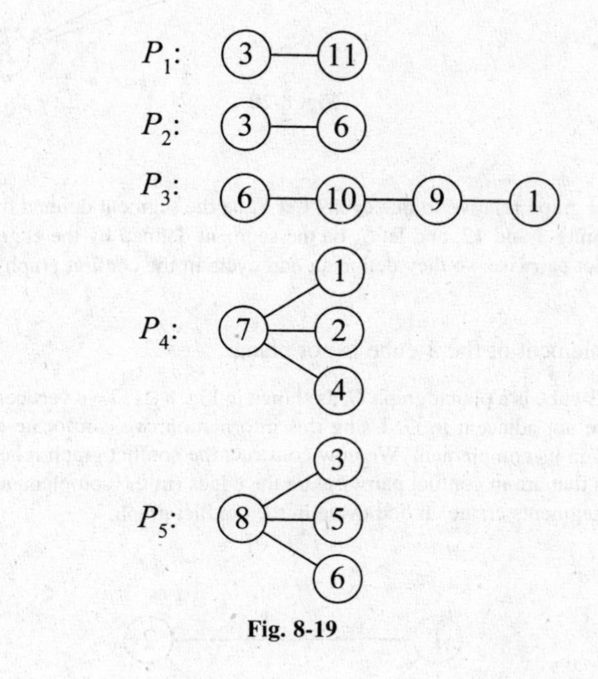

Fig. 8-19

8.35 (*Tutte's Conflict Graph Theorem*) Show that a graph is planar if and only if its conflict graph relative to every cycle in it is bipartite.

Solution. Suppose the graph is planar. Take any cycle C in it. First, C is drawn with no intersecting edges. The pieces that are not segments can be easily embedded so that the edges do not intersect. Since the graph is planar, the segments can be embedded in such a way that segments in conflict lie on either side of the cycle. So no two segments in the interior of C are adjacent vertices in the conflict graph. Similarly, no two segments in the exterior of C are adjacent. In other words, the conflict graph relative to any cycle in a planar graph is bipartite. To prove the converse, suppose G is a nonplanar graph. Then, by Kuratowski's theorem, it contains a subgraph homeomorphic to either K_5 or $K_{3,3}$. It follows from Problem 8.35 that there is a cycle in the graph such that the conflict graph relative to that cycle is not bipartite. In other words, if the conflict graph with respect to every cycle is bipartite, the graph is planar.

8.36 Prove that the conflict graph relative to any cycle in the graph shown in Fig. 8-19(*a*) is bipartite.

Solution. An embedding of the graph as a plane graph is shown in Fig. 8-20. So the given graph is planar. Hence, its conflict graph relative to any cycle is necessarily bipartite.

8.37 Show that the Heawood graph (see Fig. 7-25) is nonplanar.

Solution. This cubic graph has 14 vertices, as shown in Fig. 7-25(*a*). Consider the cycle C as shown passing consecutively through these 14 vertices. Each edge of the graph not belonging to the cycle is a segment defining

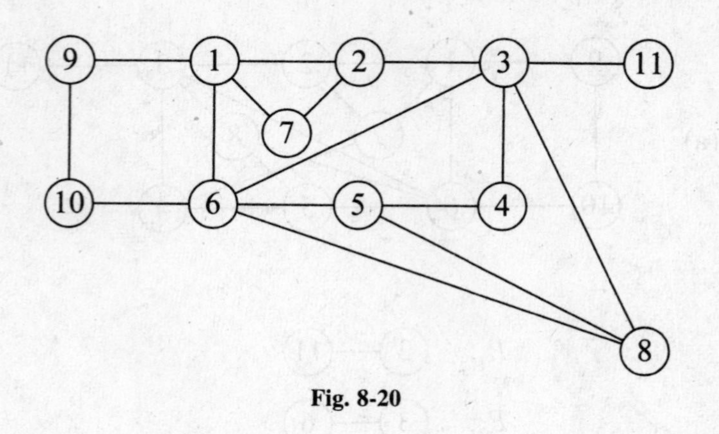

Fig. 8-20

a vertex in the conflict graph relative to this cycle. Let S_1 be the segment defined by the edge joining 1 and 10, let S_2 be the segment joining 3 and 12, and let S_3 be the segment defined by the edge joining 5 and 14. These three segments are in conflict pairwise, so they define an odd cycle in the conflict graph.

8.38 Show that the complement of the 3-cube is not planar.

Solution. The 3-cube is a planar graph G, as shown in Fig. 8-21. Two vertices in its complement are adjacent if and only if they are not adjacent in G. Using this information, we can locate a cycle C: $1 \to 3 \to 5 \to 4 \to 8 \to 2 \to 7 \to 6 \to 1$ in its complement. We now construct the conflict graph relative to this cycle in the complement. Three segments that are in conflict pairwise are the edges (in the complement) joining 1 and 8, 2 and 5, and 4 and 6. These three segments create an odd cycle in the conflict graph.

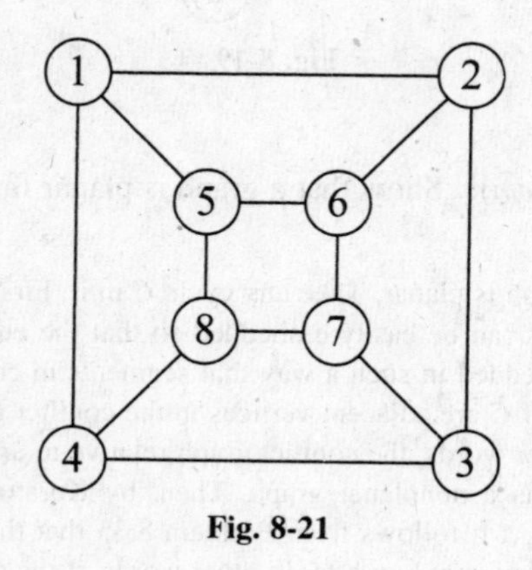

Fig. 8-21

8.39 Give an example of two homeomorphs of a graph such that each is not a homeomorph of the other.

Solution. In Fig. 8-22, both G' and G'' are homeomorphs of G, but G' is not a homeomorph of G'', and G'' is not a homeomorph of G'.

8.40 Show that if a graph G has $K_{3,3}$ as a subcontraction, G is nonplanar.

Solution. Suppose $G = (V, E)$ has a subgraph $H = (W, F)$ that is contractible to the complete bipartite graph in which the two partite sets are $X = \{u_i : i = 1, 2, 3\}$ and $Y = \{u_i' : i = 1, 2, 3\}$. So set W has a partition $\{W_1, W_2, W_3, W_1', W_2', W_3'\}$ such that u_i is matched with W_i and u_i' with W_i' for each i. Furthermore, since u_i and u_j' are adjacent, the subgraph G_i induced by W_i has vertex v_{ij}, and the subgraph G_i' induced by W_i' has vertex v_{ij}'

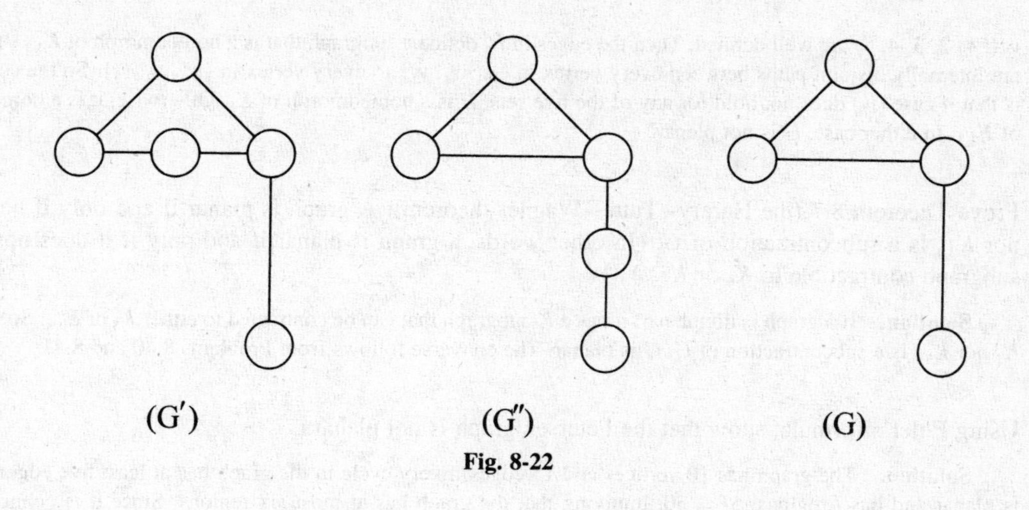

Fig. 8-22

such that these two vertices are adjacent. If the three vertices v_{i1}, v_{i2}, and v_{i3} are the same vertex, let each be labeled v_i. Otherwise, let v_i be a vertex in G_i that has internally disjoint paths connecting it to the distinct vertices among v_{i1}, v_{i2}, and v_{i3}. Thus each W_i has a vertex v_i with three internally disjoint paths from it to vertices in W'_1, W'_2, and W'_3. Similarly, each W'_i has vertex v'_i with three internally disjoint paths from it to vertices in W_1, W_2, and W_3. Graphs G_i and G_j are all connected graphs. Hence, G has a subgraph that is a homeomorph of $K_{3,3}$, so it is nonplanar.

8.41 Show that if a graph G has K_5 as a subcontraction, G is nonplanar.

Solution. Suppose $G = (V, E)$ has a subgraph $H = (W, F)$ that is contractible to a complete graph with a vertex set $\{u_i : 1 \le i \le 5\}$. Then $\{W_i : 1 \le i \le 5\}$ is a partition of W such that u_i is matched with W_i for each i. Let G_i be the subgraph induced by the set W_i. If there is a vertex v_i for each i such that there are four internally disjoint paths between v_i and v_j $(i \ne j)$, H is a homeomorph of K_5; hence, G is nonplanar. As in Problem 8.40, there exists vertex v_{ij} in G_i and vertex v_{ji} in G_j such that these two vertices are adjacent, where $i \ne j$, $1 \le i, j \le 5$. Let us fix $i = 1$. Then there are several mutually exclusive possibilities concerning the vertices v_{12}, v_{13}, v_{14} and v_{15}.

Case (i): These four vertices are the same, in which case let v_1 represent this common vertex; or three of the four vertices are the same, in which case also label this common vertex v_1. In both situations, there are four internally disjoint paths from v_1 to vertices in the other four sets.

Case (ii): Two of the four are the same (labeled v_1), and the other two are distinct. Also suppose that there are internally disjoint paths from v_1 to the other two vertices. In this case, there are also four internally disjoint paths from v_1 to vertices in the other four sets.

Case (iii): The four vertices are distinct. Also suppose there is vertex v_1 in W_1 such that there are four internally disjoint paths from v_1 to these four vertices so that there will be internally disjoint paths from v_1 to vertices in the other sets.

Case (iv): None of the above.

So if the scenario presented in case (*i*), (*ii*), or (*iii*) exists for each of five sets of the partition, we are done.

We now scrutinize case (*iv*). There is at least one graph, say G_1, among the five graphs in which case (*iv*) holds. This implies that there are two vertices w_1 and w'_1 in G_1 joined by path P_1 in which no intermediate vertex is v_{1j} such that each of them is connected to two of these four vertices by internally disjoint (probably trivial) paths. [Specifically, if $v_{12} = v_{13}$ and $v_{14} = v_{15}$ are two distinct vertices, we take $v_{12} = v_{13} = w_1$ and $v_{14} = v_{15} = w'_1$. If $v_{12} = v_{13} = w_1$ and if v_{14} and v_{15} are distinct, and since there are no internally disjoint paths from w_1 to these two vertices, w'_1 is the last common vertex in a path between w_1 and v_{14} and in a path between w_1 and v_{15}. If the four vertices are distinct and if there is no vertex with internally disjoint paths from that vertex to these four vertices as stipulated in case (*iii*), we can locate two vertices w_1 and w'_1 such that there is a path P_1 between these two vertices that does not pass through any of these four vertices. At the same time we can also locate two internally disjoint paths from w_1 to v_{12} and to v_{13} and two internally disjoint paths from w'_1 to v_{14} and to v_{15}.] Thus we have path P_1 joining w_1 and w'_1 and four internally disjoint paths (two from w_1 and two from w'_1) to the other four sets. Let E_1 be the set of edges of these paths. If case (*iv*) does not hold for any of the remaining four sets W_i, vertices

v_i (i = 2, 3, 4, 5) are well defined. Then the edges in E_1 define a subgraph that is a homeomorph of $K_{3,3}$ since there are internally disjoint paths between every vertex in $\{w_1, v_4, v_5\}$ to every vertex in $\{w_1', v_2, v_3\}$. So the conclusion is that if case (*iv*) does not hold for any of the five sets, H is a homeomorph of K_5. Otherwise, it is a homeomorph of $K_{3,3}$. In either case, G is not planar.

8.42 Prove Theorem 8.7 (the Harary–Tutte–Wagner theorem): A graph is planar if and only if neither K_5 nor $K_{3,3}$ is a subcontraction of G. (In other words, a graph is planar if and only if it does not have a subgraph contractible to K_5 or $K_{3,3}$.)

Solution. If a graph is nonplanar, it has a K-subgraph that can be contracted to either K_5 or $K_{3,3}$. So if neither K_5 nor $K_{3,3}$ is a subcontraction of G, G is planar. The converse follows from Problems 8.40 and 8.41.

8.43 Using Euler's formula, show that the Petersen graph is not planar.

Solution. The graph has 10 vertices and 15 edges. Every cycle in the graph has at least five edges. So if it is planar and has f regions, $5f \leq 30$, implying that the graph has at most six regions. Since it is connected and planar, it should have exactly seven regions.

8.44 Show that the Petersen graph is nonplanar by establishing that it has a K-subgraph.

Solution. Consider the subgraph of the Petersen graph obtained by deleting one edge. Fig. 8-23 shows a subgraph H of the Petersen graph after deleting the edge joining vertex 3 and vertex 4 as well as the edge joining vertex 7 and vertex 10. Vertices 3, 4, 7, and 10 become vertices of degree 2. It is easy to see that the subgraph H is a homeomorph of $K_{3,3} = (X, Y, E)$, where $X = \{1, 8, 9\}$ and $Y = \{2, 6, 5\}$. So the Petersen graph has a K-subgraph; hence, it is nonplanar.

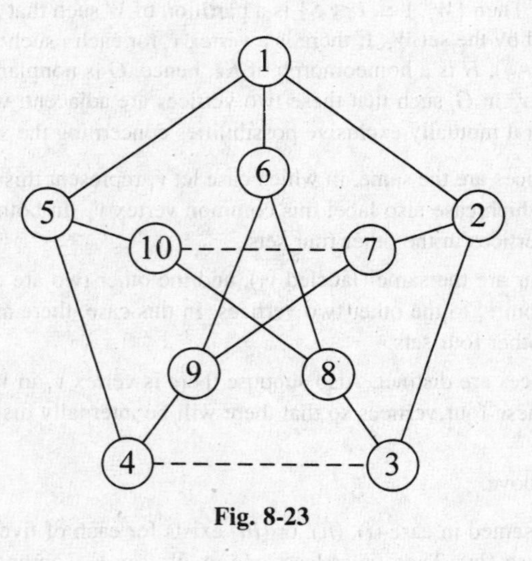

Fig. 8-23

8.45 Show that the Petersen graph is not planar by showing that the conflict graph relative to one of its cycles is nonbipartite.

Solution. In the graph shown in Fig. 8-23, consider the cycle (a longest possible cycle) $1 \to 2 \to 3 \to 4 \to 5 \to 10 \to 7 \to 9 \to 6 \to 1$. Edge e_1 joining 1 and 5 will intersect edge e_2 joining 2 and 7 if we embed both edges inside the cycle. So segment S_1 consisting of edge e_1 and segment S_2 consisting of edge e_2 are in conflict. These two segments are both in conflict with segment S_3, which consists of the edge joining 4 and 9. So these three segments form an odd cycle in the conflict graph. Thus the conflict graph relative to this cycle is not bipartite.

8.46 Show that the Petersen graph is nonplanar by showing that it is contractible to K_5.

Solution. Consider the representation of the Petersen graph with an outer pentagon and inner pentagon, as shown in Fig. 8-23. Contract this graph by merging a vertex in the outer pentagon with its mate in the inner pentagon. The resulting graph is K_5.

8.47 (*Theorem of H. Peyton Young*) Show that a 4-connected graph is nonplanar if and only if it has K_5 as a subcontraction.

Solution. Any graph is nonplanar if and only if it has either K_5 or $K_{3,3}$ as a subcontraction. It remains to be shown that if a 4-connected graph is nonplanar, it has K_5 as a subcontraction. Let G be a 4-connected graph with $K_{3,3} = (X, Y, E)$ as a subcontraction, where $X = \{A, B, C\}$ and $Y = \{D, E, F\}$. Let H be the subgraph consisting of nine internally disjoint paths joining the vertices in X and Y. These paths are called H-paths. By contracting the intermediate vertices, contract each of these H-paths into paths with two edges each. The contracted graph will have these six vertices (of degree 3) as before and nine intermediate vertices (of degree 2), and we also denote this graph by H. Since G is 4-connected, $G - Y$ is connected. Let P be a path between A and B in $G - Y$. Traversing P from A, let w be the last vertex on an H-path from A, and let x be the first vertex after w on an H-path from B or C, say from B. Let P_{wx} be the section of P between w and x, including terminal vertices w and x. No intermediate vertex in this subpath is a vertex of H. Among the three vertices in Y, there is a vertex (say F) such that neither w nor x is an intermediate vertex of any H-path emanating from F. Let Q be any path in $G - X$ from F to either D or E, say E. As we traverse Q from F, let y be the last vertex in this path belonging to any H-path emanating from F. Let z be the next vertex in this path that belongs to $P_{wx} \cup H$. (Here H is the contracted graph.)

If $z \in P_{wx}$, z is a vertex on an H-path from E. The four edges $\{A, w\}$, $\{B, x\}$, $\{E, z\}$, and $\{F, y\}$ are contracted into single vertices A, B, E, and F. At this stage, we have the contracted graph shown in Fig. 8-24(a). If we contract the edge joining C and D in this graph, we get the complete graph K_5.

If $z \in P_{wx}$, three cases need to be examined.

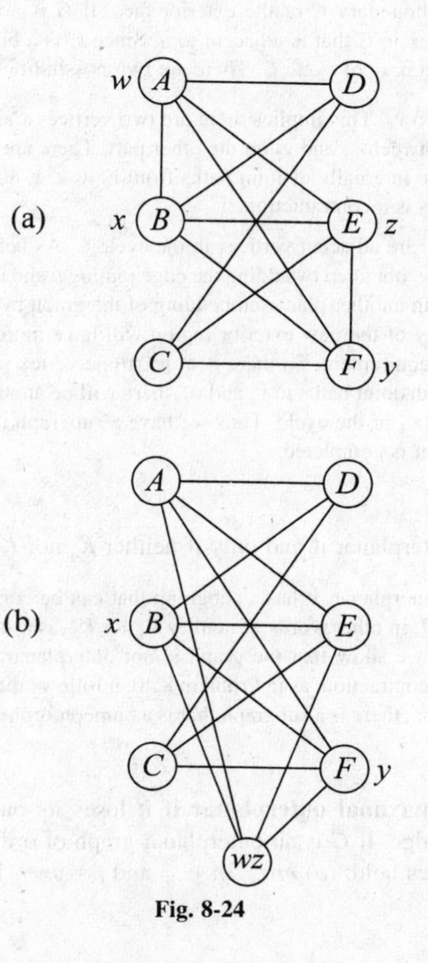

Fig. 8-24

***Case (i):* w ≠ A *and* z ≠ x.** If we contract edges $\{w, z\}$, $\{B, x\}$, and $\{F, y\}$, we get the graph shown in Fig. 8-24(b). If edges $\{A, D\}$ and $\{C, E\}$ are contracted in this graph, we get K_5.

***Case (ii)* w ≠ A *and* z = x.** Since z is a vertex in path Q that belongs to the graph $G - \{A, B, C\}$, vertices z and B are distinct. The common vertex $z = x$ is denoted as z. Contract edges $\{w, z\}$ and $\{F, y\}$. We get the graph shown in Fig. 8-24(b).

***Case (iii)* w = A *and* z = x.** While traversing the path Q from F to E, let v be the next vertex after z that is in H. This vertex v by definition is not in an F-path. Observe that z is a vertex in P between A (which coincides with w) and B (which coincides with x), and observe that it is also a vertex in Q between v and y. Let t be the vertex obtained by contracting all the intermediate vertices between A and B in P and all the intermediate vertices between v and y in Q so that t is adjacent to A, B, v, and y in the contracted graph. Then contract edges $\{E, v\}$ and $\{F, y\}$. We again get the graph shown in Fig. 8-24(b), which can be contracted to K_5.

Outerplanar Graphs

8.48 (*Chartrand–Harary Theorem*) A planar graph is an **outerplanar graph** if it has an embedding on the plane such that every vertex of the graph is a vertex belonging to the boundary of the same (usually exterior) region. Show that a graph is outerplanar if and only if it has no subgraph that is a homeomorph of K_4 or $K_{2,3}$. (These two complete graphs are the "forbidden graphs" for outerplanarity and play more or less the same role as K_5 and $K_{3,3}$ play in topics related to planarity.)

Solution. If a graph G has a subgraph that is a homeomorph of one of these two complete graphs, G is definitely not outerplanar. To establish the converse, assume that G has no subgraph that is a homeomorph of K_4 or $K_{2,3}$. Hence, it does not have a K-subgraph, so it is planar. Since a graph is planar if and only if every block in it is planar, we can assume that G is a block. Suppose G is embedded on the plane, with the maximum number of its vertices belonging to the boundary C of the exterior face. If G is not outerplanar, there is a vertex u in the interior of C. Let v_1 be a vertex in C that is adjacent to u. Since G is a block (with more than two vertices), there is a path P between u and vertex v_2 of cycle C. There are two possibilities to be considered.

Case (i): v_2 is not adjacent to v_1. This implies there are two vertices x and y in the cycle: x between v_1 and v_2 on one part of the cycle and y between v_1 and v_2 on the other part. There are three internally disjoint paths from v_1 to u, x, and y, and there are three internally disjoint paths from v_2 to u, x, and y. So the graph G has a subgraph that is a homeomorph of $K_{2,3}$. This is a contradiction.

Case (ii): Suppose v_1 and v_2 are adjacent vertices in the cycle C. As before, let path P be between u and v_2. Let P' be the path between v_1 and v_2 obtained by adding the edge joining u and v_1 to P. If the degree of every intermediate vertex in P' is 2, we can obtain another planar embedding of the graph by deleting P' and drawing it outside cycle C, in which case the boundary of the new exterior region will have more vertices than the number of vertices in C, violating the maximality requirement. So there is at least one vertex w in this path of degree 3. So from w, in addition to the two internally disjoint paths to v_1 and v_2, there will be another path (with no vertex in common with the other two paths) to vertex v_3 of the cycle. Thus we have a subgraph that is a homeomorph of K_4. This also is a contradiction. Thus the proof is completed.

8.49 Show that a graph G is outerplanar if and only if neither K_4 nor $K_{2,3}$ is a subcontraction of G.

Solution. If G is not outerplanar, it has a subgraph that can be contracted to one of these complete graphs, as established in Problem 8.57. In other words, if neither K_4 nor $K_{2,3}$ is a subcontraction, G is outerplanar. To prove the converse, it is enough if we show that the graph is not outerplanar whenever one of these two graphs is a subcontraction. If $K_{2,3}$ is a subcontraction, as in Problem 8.40, it follows that G has a subgraph that is a homeomorph of $K_{2,3}$. If K_4 is a subcontraction, there is a subgraph that is a homeomorph of K_4. In either case, G is not outerplanar.

8.50 An outerplanar graph is **maximal outerplanar** if it loses its outerplanarity if any two nonadjacent vertices are joined by an edge. If G is an outerplanar graph of order n and size m with f regions, show that the following properties hold: (a) $m = 2n - 3$ and $f = n - 1$, (b) there are at least three vertices of degree not exceeding 3, (*c*) there are at least two vertices of degree 2, and (*d*) the vertex-connectivity number $\kappa(G) = 2$.

Solution.

(*a*) The degree of the exterior region is n, and the degree of each interior region is 3. So $3(f - 1) + n = 2m$. Then, by applying Euler's formula, we find that $f = (n - 1)$ and $m = (2n - 3)$.

(*b*) Let n_i be the number of vertices of degree i. Obviously, $n_1 = 0$. Then $2n_2 + 3n_3 + 4n_4 + \cdots = 2n_2 + 2n_3 + n_3 + 4n_4 + \cdots = 2m = 4n - 6$. Hence, $2n_2 + 2n_3 + 4(n_4 + n_5 + \cdots) \leq 4n - 6$. But $n_4 + n_5 + \cdots = n - (n_2 + n_3)$. Thus $2n_2 + 2n_3 + 4(n - n_2 - n_3) \leq 4n - 6$, implying that $n_2 + n_3 \geq 3$.

(*c*) $n_2 = 0$ implies that $3n_3 + 4n_4 + \cdots = 4n - 6$. So $3n_3 + 4(n - n_3) \leq 4n - 6$, which implies that $n_3 \geq 6$, which is a contradiction. So there is at least one vertex of degree 2 that is obvious. Suppose $n_2 = 1$. Then $3n_3 + 4(n - n_3) \leq 4n - 4$, which implies that $n_3 \geq 4$ which is also a contradiction. Hence, $n_2 \geq 2$.

(*d*) Suppose all the vertices are situated along the corners of a polygon. Each vertex is adjacent to its two neighboring vertices. The deletion of a single vertex will not disconnect the graph. So $\kappa(G) > 1$. But $\kappa(G) < 3$ since there is a vertex of degree 2. Hence, $\kappa(G) = 2$.

8.51 Show that the three conditions listed in Problem 8.50 are necessary but not sufficient for a graph to be maximal outerplanar.

Solution. The graph in Fig. 8-25 satisfies parts (*a*) through (*d*). It is not outerplanar.

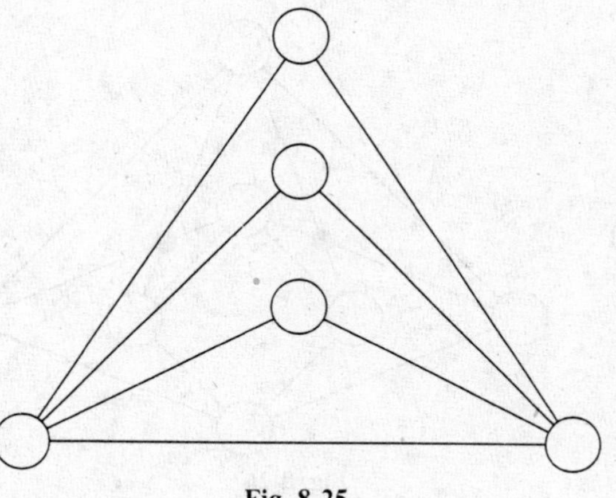

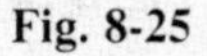

Fig. 8-25

Crossing Numbers and Thickness

8.52 Two edges in a graph create a **crossing** if there is an embedding of the graph in which the two edges intersect at a point that is not a vertex. If two edges meet at a crossing, a third edge should not pass through that crossing or it should not intersect either of these edges at a different point. The minimum number of crossings of the edges among all the embeddings of a graph G with this requirement is its **crossing number $\nu(G)$,** which is zero when G is planar. Find the crossing numbers of K_5, $K_{3,3}$, K_6, and $K_{3,4}$.

Solution. Both K_5 and $K_{3,3}$ are nonplanar, and each can be embedded with one crossing. So $\nu(K_5) = \nu(K_{3,3}) = 1$. Consider a drawing of K_6 on the plane with c crossings. We know that $c \geq 1$. At each crossing (the point of intersection), we introduce a new vertex, creating a planar graph. Each crossing will create two new edges. Thus we have $6 + c$ vertices and $15 + 2c$ edges. Since the graph is planar, $15 + 2c \leq 3(6 + c) - 6$, which implies that $c \geq 3$. But there is an embedding, as shown in Fig. 8-26(*a*) with three crossings. So its crossing number is 3. Any crossing in a bipartite graph (X, Y, E) involves two edges e and f, where e joins vertex p from X and q from Y. Similarly, f joins x from X and y from Y. If the crossing number is 1, the deletion of one of these vertices would have produced a planar graph, but that is not the case here. So the crossing number is more than 1. See Fig. 8-26(*b*), which shows an embedding of $K_{3,4}$ with two crossings. So the crossing number of $K_{3,4}$ is 2.

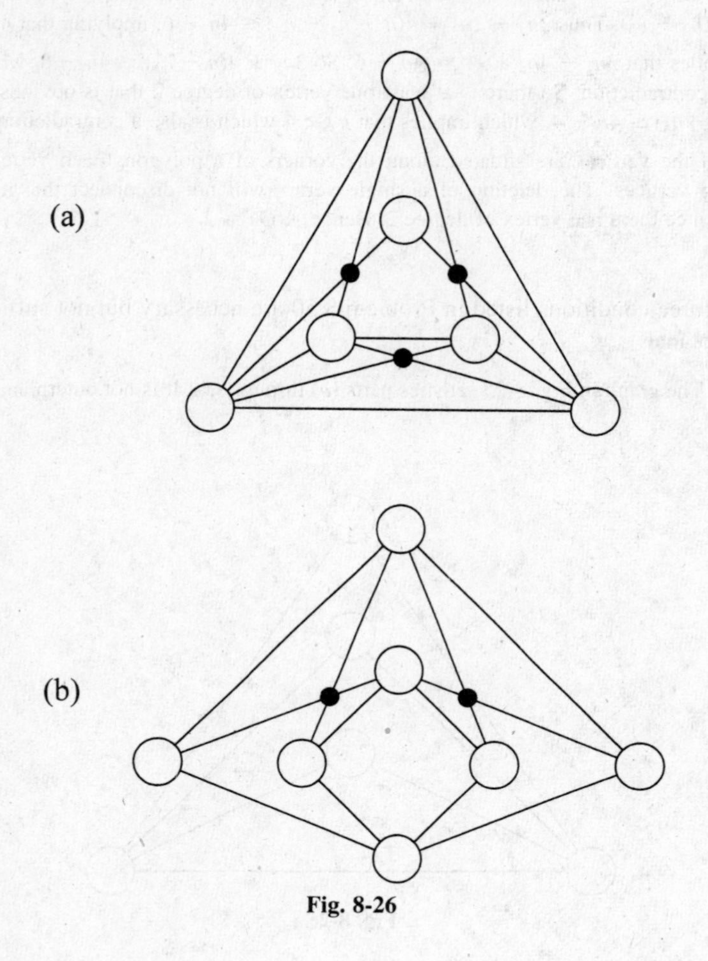

Fig. 8-26

8.53 A graph $G = (V, E)$ is ***k*-partite** if there is a partition of V into k subsets V_i such that every edge in E joins some vertex in V_i to some vertex in V_j, where $i \neq j$. If each V_i has r_i vertices and if there is an edge from each vertex in V_i (for every i) to every vertex in V_j for every j (where $i \neq j$), we have the complete k-partite graph $K(r_1, r_2, \ldots, r_k)$. Find the crossing number of $K(3, 2, 2)$.

Solution. It is easy to see that $K_{3,3}$, which can now be described as $K(3,3)$, is a subgraph of $G = K(3, 2, 2)$. So the crossing number c of G is at least 1. Suppose there is an embedding with just one crossing at the intersection of edges $\{p, u\}$ and $\{q, v\}$. At least one of these four vertices should belong to a partite set with two vertices. Suppose p is one such vertex. So $G - p$ is planar since there is only one crossing number. But $G - p$ is isomorphic to $K(1, 2, 3)$; hence, it is nonplanar. So the crossing number is at least 2. Figure 8-27 shows an embedding of the graph with two crossings. Thus the crossing number is 2.

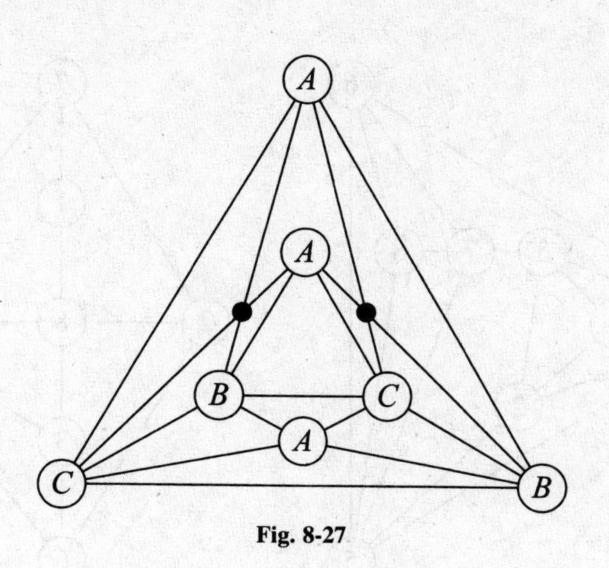

Fig. 8-27

8.54 Find the crossing number of the Petersen graph.

Solution. Since the Petersen graph is nonplanar, its crossing number is at least 1. It is not possible to embed it on the plane with one crossing since the deletion of any vertex leaves a homeomorph from $K_{3,3}$. An embedding with two crossings is shown in Fig. 8-28. So the crossing number is 2.

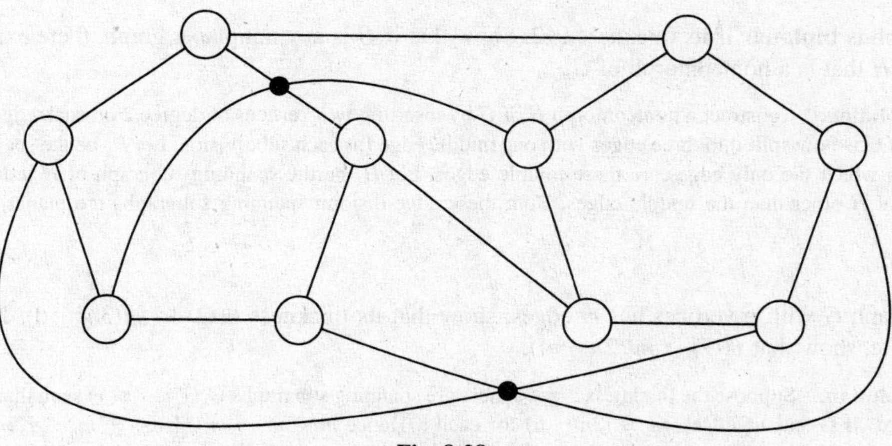

Fig. 8-28

8.55 The **thickness** $\theta(G)$ of a graph G is the minimum number of pairwise edge-disjoint spanning subgraphs in a decomposition of the graph. Find the thickness of the complete graph with n vertices, where $n \leq 8$.

Solution. The thickness of a graph is 1 if and only if the graph is planar. So the thickness of K_n is 1 if $n \leq 4$. Graph K_8 has a planar decomposition consisting of two spanning subgraphs, as shown in Fig. 8-29. So the thickness of K_n is 2 if $5 \leq n \leq 8$.

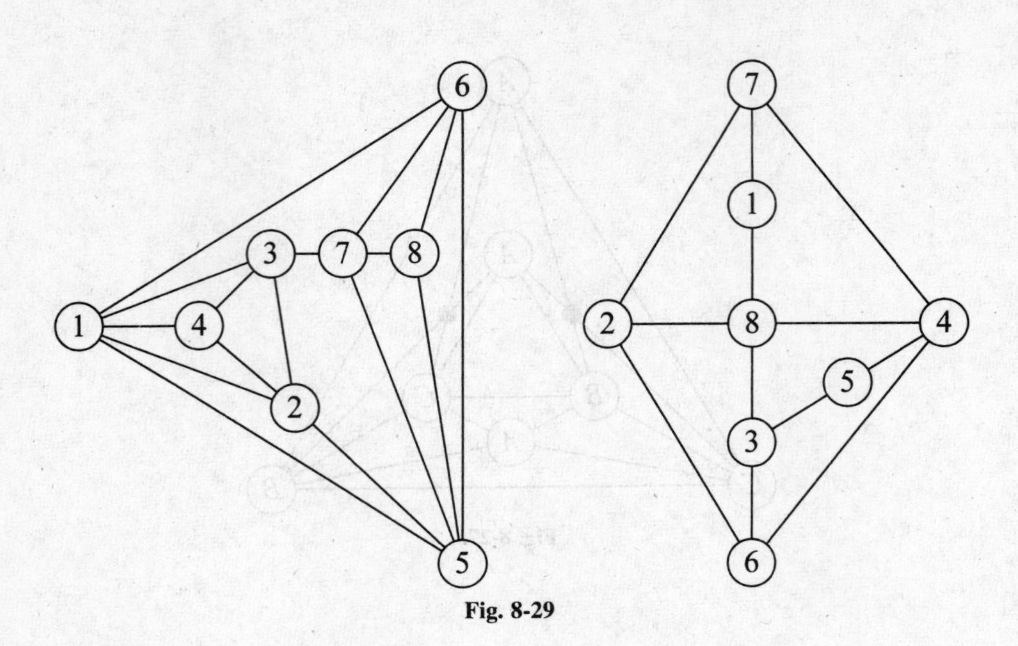

Fig. 8-29

8.56 Find the thickness of the Petersen graph.

Solution. The Petersen graph can be decomposed into two planar edge-disjoint spanning subgraphs: one planar spanning subgraph consisting of two disjoint cycles with five vertices each (the outer pentagon and the inner pentagon) and another planar spanning subgraph consisting of five copies of K_2 (the five "spokes"). So the thickness is 2.

8.57 A graph is **biplanar** if its thickness is 2. Show that if G is any nonplanar graph, there exists a biplanar graph H that is a homeomorph of G.

Solution. Construct a homeomorph H of G by inserting two vertices of degree 2 on each edge of G. So each edge of G is now split into three edges with one middle edge for each subdivision. Let H_1 be the spanning subgraph of H in which the only edges are these middle edges. Let H_2 be the spanning subgraph of H induced by all the edges of H other than the middle edges. Both these edge-disjoint spanning subgraphs are planar, and H is their union.

8.58 If a graph G with n vertices has m edges, show that its thickness $\theta(G) \geq m/(3n - 6)$. If the graph is bipartite, show that $\theta(G) \geq m/(2n - 4)$.

Solution. Suppose the thickness is r. So there are spanning subgraphs G_i $(1 \leq i \leq r)$ such that each subgraph is planar. If G_i has m_i edges, $m_i \leq (3n - 6)$ for each i. Hence $m = m_1 + m_2 + \cdots + m_r \leq r(3n - 6)$. So $r \geq m/(3n - 6)$. If the graph is bipartite, $m_i \leq (2n - 4)$ since the boundary of every region is at least 4.

8.59 Find a lower bound for the thickness of the complete graph of order n, where $n \geq 5$, using the result of Problem 8.58.

Solution. Here the size m is $n(n - 1)/2$. So from the result obtained in Problem 8.58, $\theta(K_n) \geq m/(3n - 6) = [n(n - 1)]/[6(n - 2)] = t$. Now $n(n - 1) = (n - 2)(n + 7) - 6(n - 2) + 2$. Hence, $t = (n + 7)/(6 - 1) + 1/[3(n - 2)]$. Let us write $(n + 7)/6 = q + r/6$, where $0 \leq r \leq 5$. Then $t = q + r/(6 - 1) + 1/[3(n - 2)]$. Now $n \geq 5$ implies that $1/[3(n - 2)] \leq \frac{1}{9} \leq \frac{1}{6}$ and $r \leq 5$ implies that $r/6 \leq \frac{5}{6}$. Hence, $r/6 + 1/[3(n - 2)] \leq 1$. Thus a lower bound for the thickness of the complete graph of order n is the smallest integer greater than or equal to t that is equal to the greatest integer less than or equal to $(n + 7)/6$. In particular, when $n \geq 5$, the thickness is at least 2, and when $n \geq 11$, the thickness is at least 3.

8.60 Find a lower bound for the thickness of the complete bipartite graph $K_{m,n}$ using the result of Problem 8.58.

Solution. The order is $(m + n)$, and the size is mn. Hence, a lower bound is $mn/2(m + n - 2)$.

Geometric Duals

8.61 Show that there is no plane graph with five regions such that there is an edge between every pair of regions.

Solution. If there is a plane graph G with five regions such that each pair shares a common edge, its geometric dual G' is nonplanar since it has K_5 as a subgraph. This is a contradiction.

8.62 Show that a planar graph is bipartite if and only if its dual is Eulerian.

Solution. Let G be an embedding of a planar bipartite graph on the plane. Since every cycle in G is an even cycle, the degree of each vertex in the dual graph (which is connected) is even. Hence the dual graph is Eulerian. Conversely, suppose the dual of a plane bipartite graph G is Eulerian. So every face in G is bounded by an even cycle. Suppose G is not bipartite. Then there is an odd cycle C in G that cannot be the boundary of a face. Suppose there are k faces in the interior of C. The edge sum of these k faces is even. In computing this edge sum, each edge in the interior of C is counted exactly twice and each edge in C is counted exactly once. So the number of edges belonging to C is even, which contradicts the assumption that C is an odd cycle. So G is bipartite.

8.63 Show that if G' is the geometric dual of a connected planar graph G, G is the geometric dual of G'.

Solution. Suppose G has m edges, r regions, and k components. Then it has $2 + k + m - r$ vertices. The dual graph G' is connected and has r vertices, m edges, and $2 + m - r$ regions. So connected graph $(G')'$ has $2 + m - r$ vertices. Thus both G and $(G')'$ are of the order if and only if G is connected. Each edge in the boundary of a region in G' intersects an edge of G joining two vertices of G, one of which has to be in the interior of the region. In other words, every region of G' "contains" at least one vertex of G. If there is a region in G' that contains more than one vertex of G, it implies that G is not connected. Thus the assumption that G is connected implies that each region X in G' contains exactly one vertex x from G. The number of edges in the boundary of X in G' is equal to the number of edges incident to vertex x since each boundary edge intersects an edge that is incident to x and vice versa. So if we construct the dual of the plane graph G' using the same procedure of constructing the dual of G from G, we recover the connected graph G intact as long as G is connected.

8:64 Show that if a planar graph is 3-edge-connected, its geometric dual is a simple graph.

Solution. If the geometric dual of a 3-edge-connected graph G is not simple, G will have a vertex of degree less than 3, and parallel edges in the dual will make a cut set of edges in G.

8.65 Show that a set of edges in a connected plane graph forms a spanning tree if and only if the set of duals of the remaining edges forms a spanning tree in a geometric dual of the graph.

Solution. Suppose the connected plane graph G has n vertices and m edges. Since any spanning tree in G will have $(n - 1)$ edges, there are $m' = (m - n + 1)$ remaining edges. The dual graph G' is a connected graph with $2 + m - n$ vertices. Hence, the set of m' dual edges forms a spanning tree in G'. The reverse implication holds since $G = (G')'$.

8.66 Show that a set of edges of a plane graph forms a cycle if and only if the set of dual edges forms a cut set in a geometric dual of the graph.

Solution. Suppose $G = (V, E)$ is a plane graph. If C is a cycle in the graph enclosing one or more regions of the graph, it contains in its interior a nonempty set S of vertices of the dual graph $G' = (V', E')$. The dual edges corresponding to the edges of cycle C are precisely those edges joining vertices in S and its complement $T = V' - S$, forming a cut set in the dual graph. The proof of the converse implication is similar.

Abstract Duals

8.67 Show that a geometric dual of any planar graph of G is the same as an abstract dual of G.

Solution. Let G be a planar embedding of a planar graph and let G' be a geometric dual of G. Then, as shown in Problem 8.66, there is a bijection from the set of edges in G to the set of edges in the dual graph G' such that a set of edges forms a cycle in C if and only if the corresponding set of dual edges forms a cut set in G'.

8.68 Show that the number of edges common to a cycle and a cut set in a graph is always even.

Solution. Let C be any cycle, and let $D = (X, Y)$ any cut set in G. If all the vertices of C are in X (or in Y), C and D have no edges in common. Suppose C has two vertices x and y, where x is in X and y is in Y. Then cycle C (which starts from x and ends in x) will necessarily have an even number of edges in common with the set of edges in D.

8.69 Let X be a set of edges in a graph $G = (V, E)$. Show that (*a*) if X has an even number of edges in common with every cut set of the graph, the edges in X constitute an edge-disjoint union of cycles, and (*b*) if X has an even number of edges in common with every cycle of the graph, the edges in X constitute an edge-disjoint union of cut sets.

Solution.

(*a*) Assume without loss of generality that G is connected. Let D be any cut set. By hypothesis, $|X \cap D)|$ is even. Let $T = (V, F)$ be any spanning tree in G, and let $X \cap (E - F) = \{e_1, e_2, \ldots, e_r\}$. Each e_i defines a unique cycle C_i (called a fundamental cycle) consisting of e_i and some edges of T. So, by Problem 8.68, $|C_i \cap D|$ is even for each i. Let C be the ring sum $C_1 \oplus C_2 \oplus \cdots \oplus C_r$ of these r cycles, and let $X' = X \oplus C$. [The **ring sum** of $G_1 = (V_1, E_1)$ and $G_2 = (V_2, E_2)$ is the graph $G_3 = G_1 \oplus G_2 = (V_3, E_3)$, where $V_3 = V_1 \cup V_2$ and $E_3 = (E_1 \cup E_2) - (E_1 \cap E_2)$. It can be easily verified that the ring sum operation is both commutative and associative.] Then $|X' \cap D|$ is even. If e is any edge in X', it has to be an edge belonging to the spanning tree. This edge e defines a unique cut set (called a fundamental cut set) in G, and the only edge common to this cut set and X' is edge e, contradicting that $|X' \cap D|$ is even. So set X' is empty, implying that set X and set C are the same. Hence, X is an edge-disjoint union of cycles.

(*b*) The proof is similar.

8.70 Show that if G^* is an abstract dual of G, G is an abstract dual of G^*. (Notice that G need not be connected here, in contrast to Problem 8.63.)

Solution. Let $D = (X, Y)$ be a cut set in G. Then $|D \cap C|$ is even for any cycle C in G, as shown in Problem 8.68. So C^* and D^* will also have an even number of edges in common. Since G^* is an abstract dual of G, C^* is a cut set in G^*. In other words, D^* has an even number of edges in common with every cut set in G^*, which implies that D^* is either a cycle or a disjoint union of cycles in G^*, as proved in Problem 8.69. Thus, using part (*a*) of Problem 8.69, we have shown that cut sets in G correspond to edge-disjoint unions of cycles in G^*. Similarly, using part (*b*) of Problem 8.69, it can be shown that cycles in G^* correspond to edge-disjoint unions of cut sets in G. So if D^* is not a single cycle, each cycle in the disjoint union will give a disjoint union of cut sets in G, which implies that D itself is an edge-disjoint union of cut sets instead of a single cut set. So D^* is a cycle. Thus cut sets in a graph correspond to cycles in the dual graph. So we have an isomorphism between G and $(G^*)^*$.

8.71 Give an example of a plane graph for which the geometric dual of the geometric dual and the abstract dual of the abstract dual are not the same.

Solution. The geometric dual G' and the abstract dual G^* are the same for any plane graph G. Also, $(G^*)^*$ is the same as G^*. Furthermore, if G is connected, both G and $(G')'$ are the same. Thus we are looking for a disconnected plane graph. If G is the plane graph that is the disjoint union of two cycles with three edges each, graph $(G')'$ has five vertices, whereas graph $(G^*)^*$ has six vertices.

8.72 Show that neither (*a*) $K_{3,3}$ nor (*b*) K_5 has an abstract dual.

Solution.

(*a*) Graph $K_{3,3}$ has no cut set consisting of two edges. So if it has an abstract dual G, it has no multiple edges. Graph $K_{3,3}$ has cycles of lengths 4 and 6. So G has no cut set consisting of fewer than 4 edges. This implies that the degree of each vertex in G is at least 4. Hence, it should have at least five vertices since G is simple. In that case, it should have at least 10 edges since the degree of each vertex is 4. But G can have only nine edges since $K_{3,3}$ has only nine edges. So $K_{3,3}$ has no abstract dual.

(*b*) The graph K_5 has no cut set consisting of two edges. So if it has an abstract dual G, it is a simple graph. It has cycles of length 3, 4, or 5. So the degree of each vertex in G is at least 3. Since the only cut sets of K_5 are sets of four edges, the cycles in G are all of size 4. So G is bipartite. If the order of G is less than 7, its size is at most 9. But it should have 10 edges since K_5 has 10 edges. So the order of G is at least 7. This implies that it should have at least (7)(3)/(2) edges. This is a contradiction again.

8.73 Show that if a graph G has an abstract dual, every subgraph of G has an abstract dual.

Solution. Let e be an edge of $G = (V, E)$ joining u and v, and let e^* be the dual edge in the abstract dual $G^* = (V^*, E^*)$ joining p and q. Let H be the subgraph of G obtained by deleting e, and let H' be the graph obtained from G^* by contracting e^*. Now every cycle C in H is a cycle in G; therefore, it corresponds to a cut set C^* in G^*. This cut set partitions V^* into two sets V_1^* and V_2^*. Since e^* is not an edge in C^*, both p and q are in one of these two sets. So C^* is a cut set in H'. Thus every cycle in H corresponds to a cut set in H'. Let D be a cut set in H'. Since e^* is not in D, set D is a cut set in G^* that should correspond to a cycle in H'. So every cut set in H' corresponds to a cycle in H. Since a subgraph H of G can be obtained from G by deletion of edges that do not belong to H, it follows that H has an abstract dual whenever G does.

8.74 Show that if G is a homeograph of H and if G has an abstract dual, H has an abstract dual.

Solution. Any two edges of G incident to a vertex of degree 2 correspond to two parallel edges in the abstract dual. If these two edges are replaced by a single edge, one of the two parallel edges will disappear. So it follows that if G has an abstract dual, H does also.

8.75 Prove Theorem 8.8 (Whitney's theorem): A graph is planar if and only if it has an abstract dual.

Solution. If a graph is planar, it has a geometric dual that is an abstract dual, as proved in Problem 8.67. To establish the reverse implication, we prove that a nonplanar graph G has no abstract dual. Suppose G has an abstract dual. Then any subgraph H of G has an abstract dual. If this H is a homeomorph of another graph H', H' also should have an abstract dual. Since G is nonplanar, there is a subgraph that is a homeomorph of either K_5 or $K_{3,3}$. Thus the existence of an abstract dual for a nonplanar graph would imply the existence of an abstract dual for K_5 or $K_{3,3}$. But there is no abstract dual for either of these two graphs, as shown in Problem 8.72. This contradiction proves that a nonplanar graph has no abstract dual.

HAMILTONIAN PLANAR GRAPHS

8.76 Prove Theorem 8.10 (the Grinberg–Kozyrev theorem): If C is any Hamiltonian cycle in a Hamiltonian plane graph of order n, the sum of the indices of the inner regions relative to C and the sum of the indices of the outer regions relative to C are both equal to $(n - 2)$.

Solution. Let the number of edges in the interior of the Hamiltonian cycle be q. Then the number of inner regions is $q + 1$. If the number of inner regions of degree k is r_k, the total number of inner regions is also equal to $(r_3 + r_4 + \cdots + r_n)$. Thus $q = (r_3 + r_4 + \cdots + r_n) - 1$. Now $(3r_3 + 4r_4 + \cdots + nr_n) = 2q + n$ by counting the edges in the boundary of each inner region. Hence, $(3r_3 + 4r_4 + \cdots + nr_n) = 2(r_3 + r_4 + \cdots + r_n) - 2 + n$. Hence, $(3 - 2)(r_3) + (4 - 2)r_4 + \cdots + (n - 2)(r_n) = n - 2$. The left-hand side is the sum of the indices of the inner regions. The proof is similar when we consider the outer regions.

8.77 Use the Grinberg–Kozyrev theorem to establish that the plane graph shown in Fig. 8-30 is not Hamiltonian.

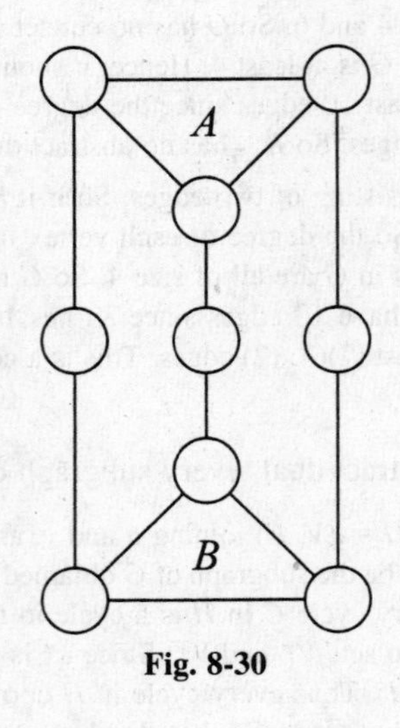

Fig. 8-30

Solution. There are five regions with indices 1, 1, 4, 4, and 4. Suppose there is a Hamiltonian cycle. There are two cases.

(*i*) Of two regions A and B, one is an inner region and the other is an outer region. Their indices are equal and cancel out each other in the counting process. There is no way that the indices of the three remaining regions cancel out.

(*ii*) Both regions A and B are either inner or outer. Suppose both are inner. Their indices add up to 2. There is no way of classifying the three remaining regions such that the sum of the indices of the inner regions is the same as the sum of the outer regions. So the graph is not Hamiltonian.

8.78 Show that the 3-connected cubic plane graph known as the **Grinberg–Kozyrev graph** shown in Fig. 8-31 is not Hamiltonian.

Solution. The degrees of the regions are 5, 8, or 9. So their indices are 3, 6, or 7. Suppose the graph is Hamiltonian. Let C be a Hamiltonian cycle. There is only one region with index 8, and it is an outer region with respect to C. Suppose the number of outer regions of indices 3 and 6 are x and y, respectively. Suppose the number of inner regions of indices 3 and 6 are a and b, respectively. Then we have the equation $3x + 6y + 7 = 3a + 6b$, which has no solution in nonnegative integers.

8.79 Show that in the Hamiltonian graph G shown in Fig. 8-32, any Hamiltonian cycle that contains edge e does not contain edge f.

Solution. Suppose there is a Hamiltonian cycle C in G that contains both the edges. Let G' be the graph obtained by subdividing these two edges by introducing one vertex (of degree two) in each of these edges. Since any Hamiltonian cycle in G that contains e and f is necessarily a Hamiltonian cycle in G', the graph G' is Hamiltonian. The graph G' has six regions with index 3 (out of which one is exterior) and one region with degree 2. Suppose there are x inner regions with index 3 and y inner regions with index 2. The sum of the indices for the inner regions is $3x + 2y$. The sum of the indices for the outer regions is $3 + 3(5 - x) + 2(1 - y)$. Since G' is

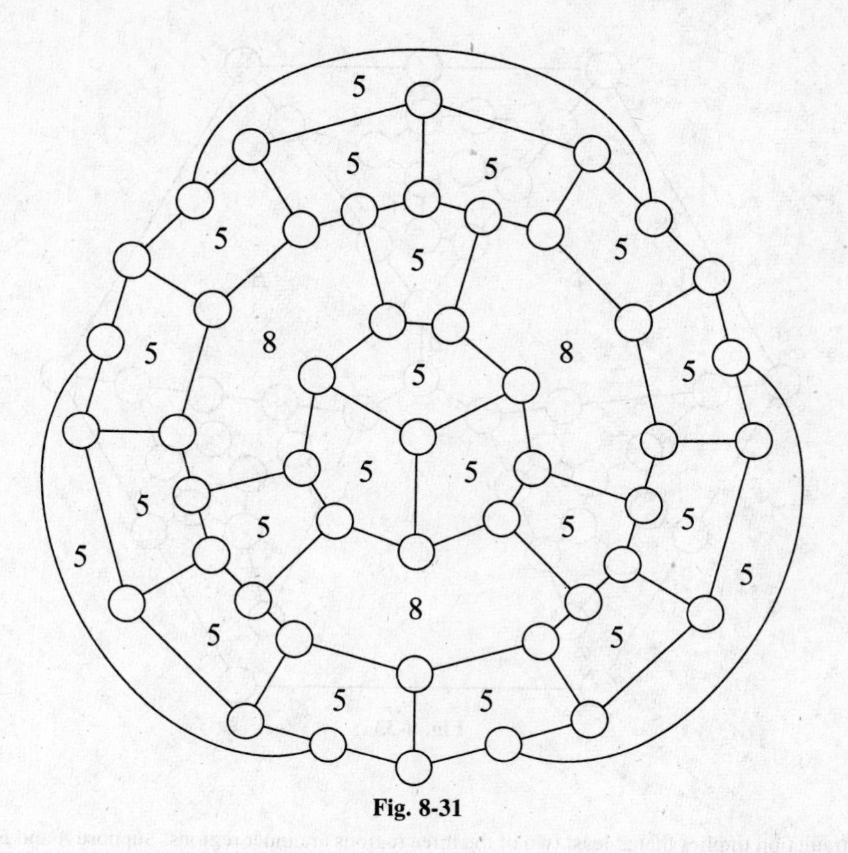

Fig. 8-31

Hamiltonian, these two sums are equal. So $3x + 2y = 10$, where $y = 0$ or 1. In either case, x is not an integer. So G' is not Hamiltonian, which is a contradiction.

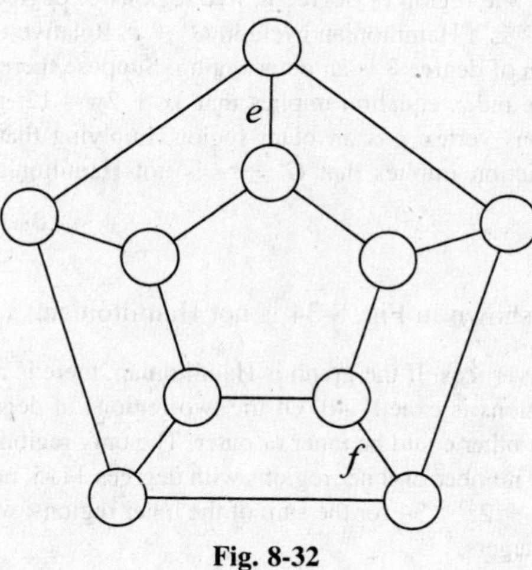

Fig. 8-32

8.80 Show that the Tutte graph shown in Fig. 8-33 is not Hamiltonian.

Solution. Suppose there is a Hamiltonian cycle D in the graph. Let two of the three regions marked A, B, and C (say regions A and B) be outer regions (relative to D) if possible. So edge a is not an edge of D. Since the exterior region is an outer region, both b and c are also not in D. But D should contain two of these three edges.

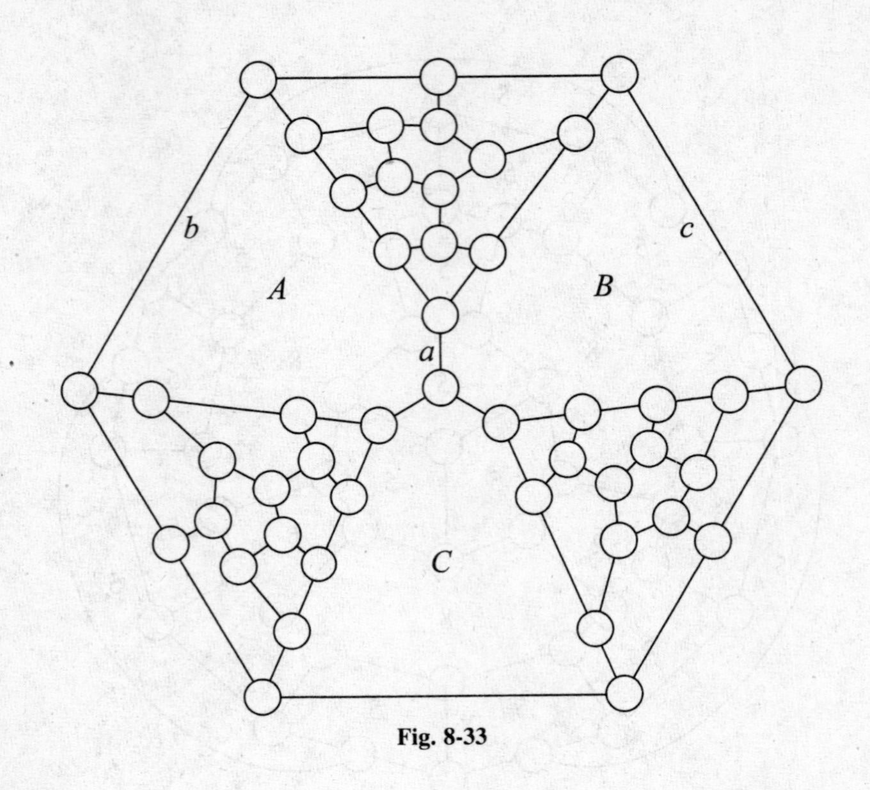

Fig. 8-33

This contradiction implies that at least two of the three regions are inner regions. Suppose A and B are inner regions. In this case also, a is not in the cycle D. Since the exterior region is an outer region, both b and c are edges of D. Let G' be the subgraph induced by the set of vertices in the "triangular section" of the graph whose end vertices are x, y, and z as marked in the figure. That edges b and c are edges of the Hamiltonian cycle D implies that there is a Hamiltonian path between x and y in G'. If we join x and y by a new edge e (not shown in the figure), $G' + e$ is a Hamiltonian graph with one region of degree 8, five regions of degree 5, two regions of degree 4, and one region of degree 3. Suppose C' is a Hamiltonian cycle in $G' + e$. Relative to this cycle, the region of degree 3 is an inner region, and the region of degree 8 is an outer region. Suppose there are x inner regions of degree 5 and y inner regions of degree 4. The index equation implies that $3x + 2y = 12$, giving the solution $x = 4$ and $y = 0$. Hence, the region that contains vertex z is an outer region, implying that the Hamiltonian cycle C' does not pass through z. This contradiction implies that $G' + e$ is not Hamiltonian. Hence, the Tutte graph G is not Hamiltonian.

8.81 Show that the plane graph shown in Fig. 8-34 is not Hamiltonian.

Solution. There are 42 vertices. If the graph is Hamiltonian, there is a Hamiltonian cycle such that the sum of the indices of the inner regions is exactly 40. Of the two regions of degree 11 in the graph, one has to be an outer region with index 9. The other could be inner or outer. The only region of degree 8 has to be an inner region, and its index is 6. Suppose the number of inner regions with degrees 11, 5, and 4 are x, y, and z, respectively. Then we have the equation $9x + 3y + 2z = 34$ for the sum of the inner regions, where x and z are binary variables. This equation has no solution in integers.

8.82 Show that is there is no Hamiltonian planar graph having regions of degrees 5 and 8 and one region of degree 7.

Solution. Suppose the graph is Hamiltonian. Then the order n is 8. Suppose there are m edges and r regions. Since the graph is Hamiltonian, it is connected. Therefore, by Euler's formula, $8 - m + r = 2$. If there are x

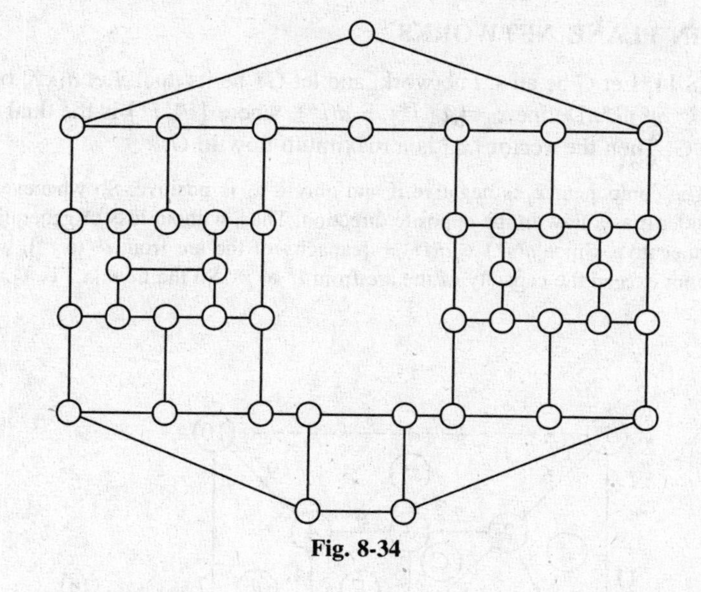

Fig. 8-34

regions of degrees 5 and y regions of degree 8, we have $5x + 8y + 7 = 2m$ and $x + y + 1 = r$. These three equations together will imply that $3x + 6y = 7$, which has no solution in nonnegative integers.

8.83 Give an example of a maximal planar graph that is not a Hamiltonian graph.

Solution. A maximal planar graph is a planar graph that becomes nonplanar whenever a pair of nonadjacent vertices is joined by a new edge. Graph G shown in Fig. 8-35 is a maximal planar graph. Suppose G is a Hamiltonian graph. If we delete the set W of five vertices labeled 1, 2, 3, 4, and 5, we get a disconnected graph $G - W$ with six components that each consist of a single vertex. Since G is Hamiltonian, the number of components of $G - W$ cannot exceed the cardinality of W. This contradiction implies that G is not Hamiltonian.

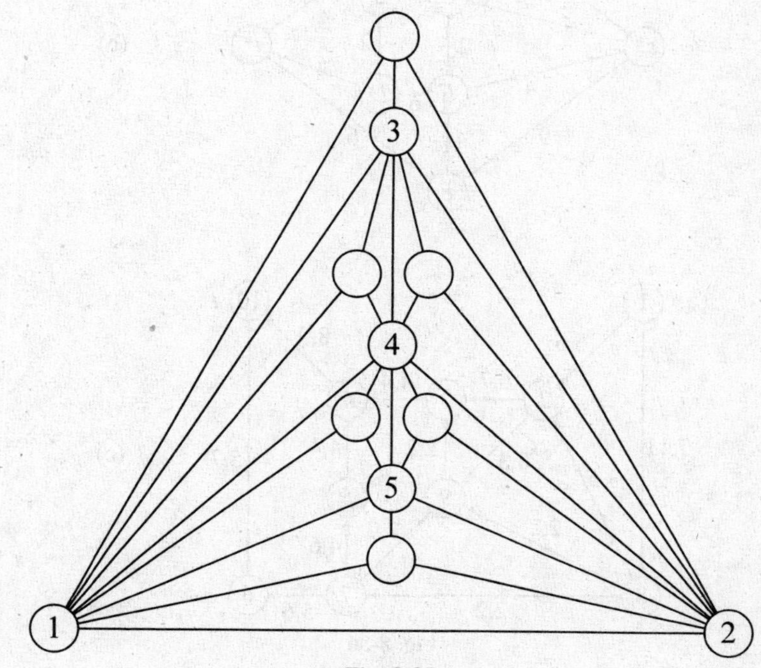

Fig. 8-35

MAXIMUM FLOW IN PLANE NETWORKS

8.84 Prove Theorem 8.11: Let G be an s, t network, and let G^* be its dual. Let $d(k^*)$ be the shortest distance between s^* and k^* in G^*. Define $x_{ij} = d(j^*) - d(i^*)$, where $\{i^*, j^*\}$ is the dual edge that corresponds to edge $\{i, j\}$ in G. Then the vector $[x_{ij}]$ is a maximum flow in G.

Solution. The component x_{ij} is negative if and only if x_{ji} is positive. So whenever the flow from i to j is negative, we consider it as a flow in the opposite direction. Thus, without loss of generality, we assume that each component is nonnegative. Since $d(j^*) \leq d(i^*)$ + [capacity of the arc from i^* to j^*], we can conclude that the component x_{ij} cannot exceed the capacity of the arc from i^* to j^*. So the flow $[x_{ij}]$ is a capacitated flow. Let k be

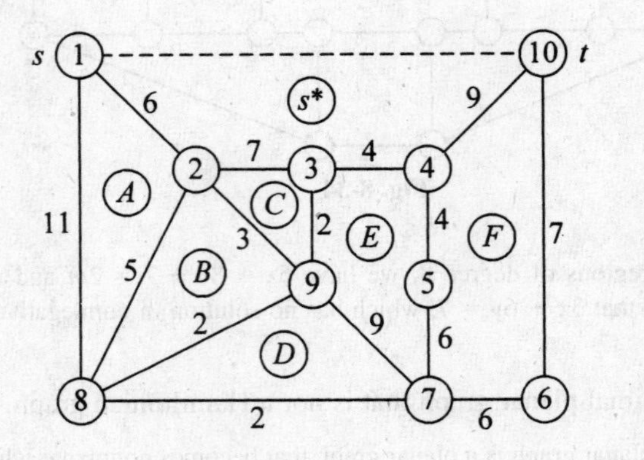

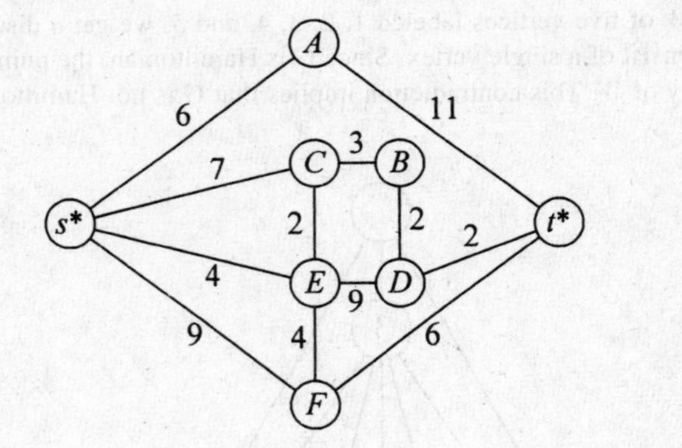

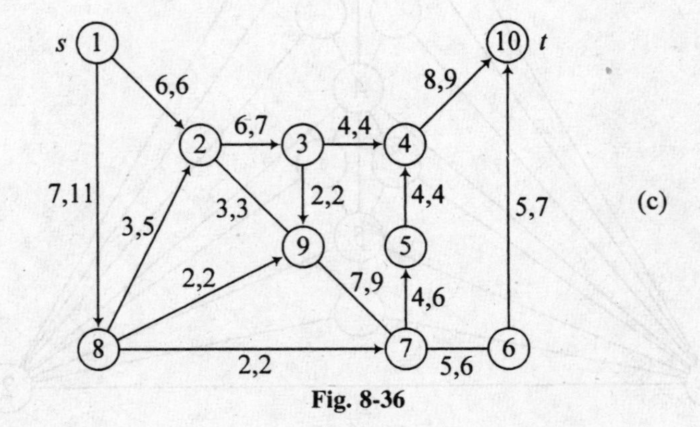

Fig. 8-36

any vertex in G other than the source or the sink. Consider the cut $\{\{k\}, V - k\}$ in G. The edges in this cut correspond to a cycle in G^*. The sum of the flow components along the edges of the cycle is zero. So the net flow at vertex k is zero. Suppose P^* is a shortest path between s^* and t^* in the dual network. Then $d(j^*) = d(i^*) +$ [capacity of the (i^*, j^*)]. So each edge in P^* is saturated. This implies that the flow is a maximum flow.

8.85 Obtain a maximum flow in the plane network shown in Fig. 8-36(a) with vertex 1 as the source and vertex 10 as the sink.

Solution. An artificial edge joining the source and the sink that encloses the finite region s^* is constructed. The exterior region is t^*. There are six other finite regions. The dual network G^* is shown in Fig. 8-36(b). The shortest distances from s^* to vertices A, B, C, D, E, F, and t^* are 6, 9, 6, 11, 4, 8, and 13, respectively. The maximum flow value is 13. The components of a maximum flow along the edges are computed using the formula established in Theorem 8.11. Using these components, each edge is converted into an arc. The undirected network now becomes a directed network, as shown in Fig. 8-36(c), displaying a maximum flow along the arcs with a flow value of 13.

GRAPHS ON SURFACES

8.86 Prove Theorem 8.12: The nonplanar graphs K_5 and $K_{3,3}$ are toroidal.

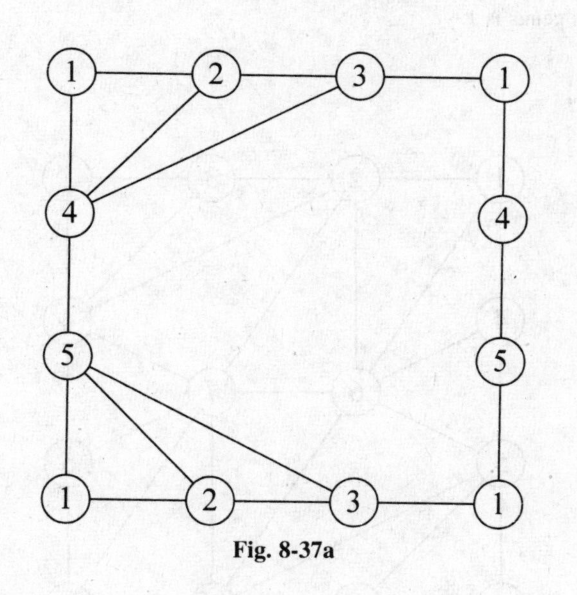

Fig. 8-37a

Solution. Any rectangle of length x and width y can be rolled into a circular tube of length x with two circular holes on either side. When these holes are glued together, we get a torus. This process is reversible. So embedding a graph of genus 1 on a torus can be easily visualized with the help of a rectangle. If vertex v is on a side of the rectangle, it is also represented symmetrically (as a duplicate) on the opposite side. If a vertex is inside the rectangle, it is not duplicated. Once the vertices are marked, the edges are identified. If the graph is toroidal, it can be embedded such that no two edges intersect. Figure 8-37(a) shows an embedding of K_5 on a torus, and Fig. 8-37(b) shows an embedding of $K_{3,3}$ on a torus.

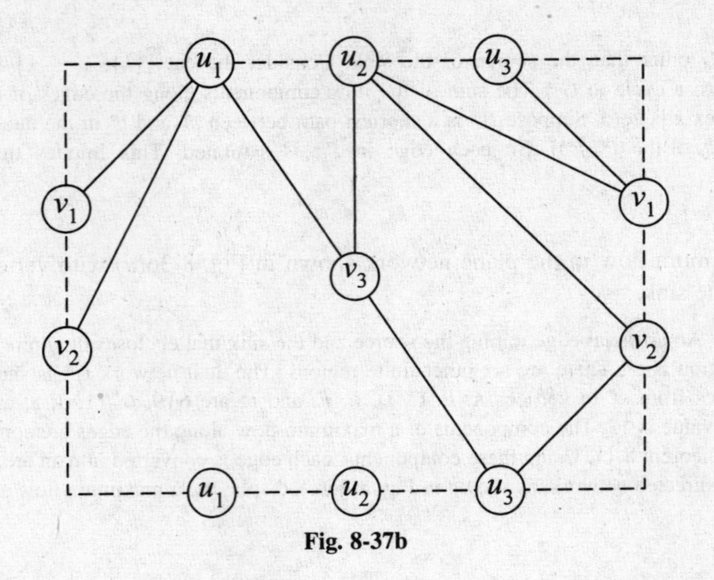

Fig. 8-37b

8.87 Find the genus of K_7.

Solution. Since K_7 is nonplanar, its genus is at least 1. Figure 8-38 shows an embedding of this complete graph on a torus. So its genus is 1.

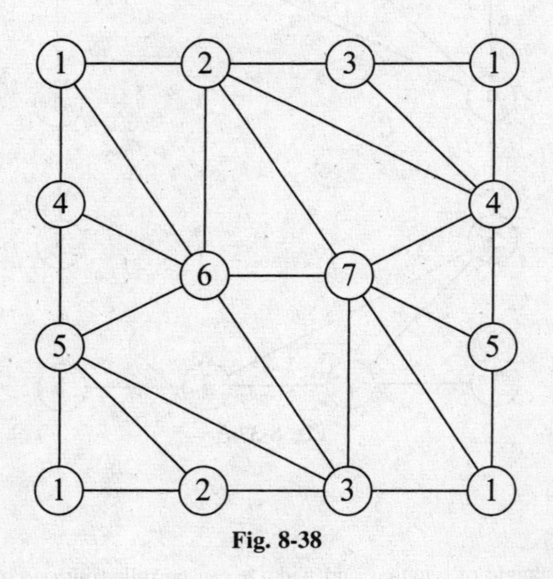

Fig. 8-38

8.88 Show that $K_{4,4}$ is a toroidal graph.

Solution. Graph $K_{4,4}$ can be embedded on a torus, as shown in Fig. 8-39. So it is a toroidal graph. Its genus is 1 since it is not planar.

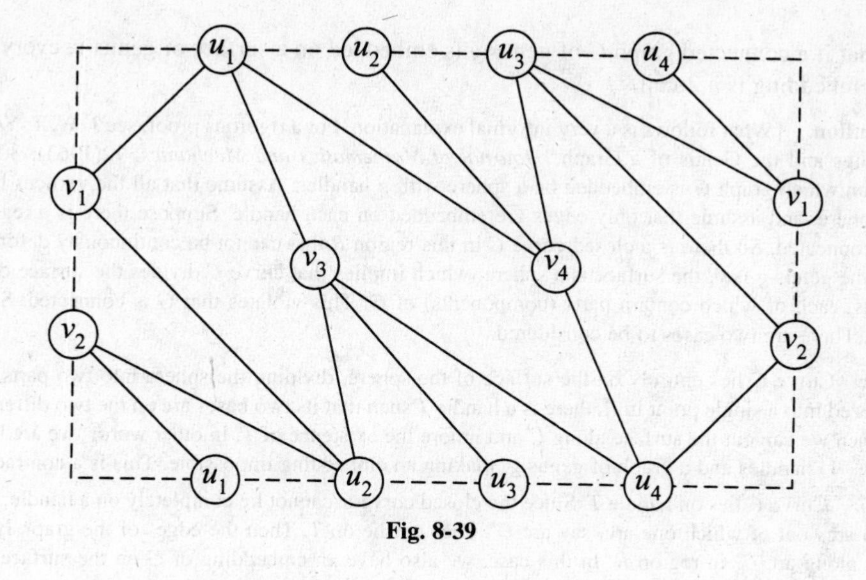

Fig. 8-39

8.89 Give an example of a graph for which the genus is less than the crossing number.

Solution. The Petersen graph comes to our rescue as usual. In Problem 8.54, it was shown that its crossing number is 2. But it is toroidal, as can be seen from Fig. 8-40. So its genus is only one. Since the crossing number of K_6 is 3 (see Problem 8.52), the crossing number of K_7 is at least 3. But its genus is only 1 since it can be embedded on a torus, as shown in Problem 8.88.

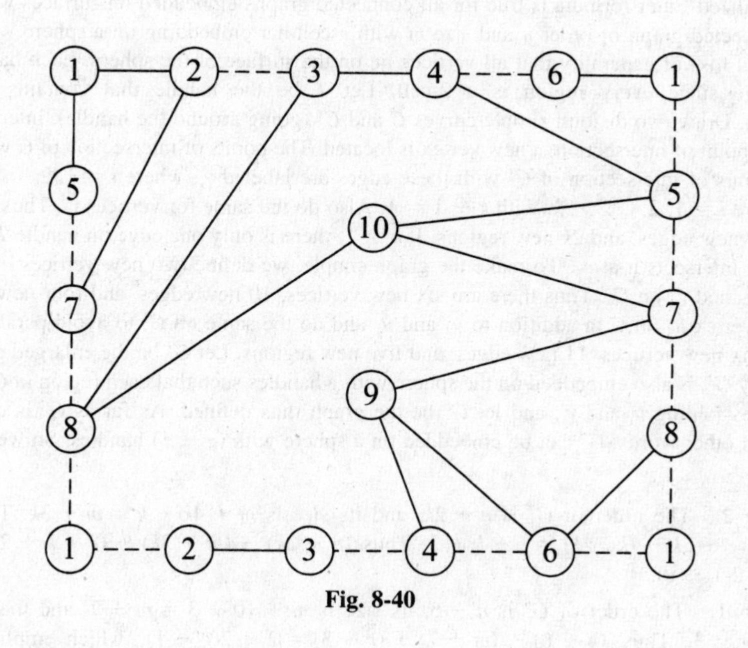

Fig. 8-40

8.90 Show that any graph can be embedded in a space of three dimensions (not on a surface) such that no two edges intersect except possibly at a vertex.

Solution. Let $V = \{1, 2, \ldots, n\}$ be the set of vertices in a graph. Each vertex i can be identified with a point (i, i^2, i^3) in 3-space. If i and j are adjacent in the graph, join the points (i, i^2, i^3) and (j, j^2, j^3) by a line segment. The resulting configuration is an embedding in 3-space.

8.91 Show that if a connected graph G of genus g is embedded on a surface of genus g, every region defined by the embedding is a 2-cell.

Solution. [What follows is a very informal explanation. For a rigorous proof, see J. W. T. Youngs, "Minimal Imbeddings and the Genus of a Graph," *Journal of Mathematics and Mechanics,* 12(1963): 303–315.] Let the surface on which graph G is embedded be a sphere with g handles. Assume that all the vertices lie on the surface of the sphere, and assume that only edges are embedded on each handle. Suppose there is a region R that is not simply connected. So there is a closed curve C in this region R that cannot be continuously deformed into a point in R. If the genus g is 0, the surface is a sphere, which implies that curve C divides the surface of the sphere into two parts, each of which contain parts (components) of G. This violates that G is connected. So the genus g is positive. There are two cases to be considered.

Case (i): Curve C lies entirely on the surface of the sphere, dividing the sphere into two parts. Since C cannot be squeezed into a single point in R, there is a handle T such that its two bases are on the two different parts defined by C. Then we can cut the surface along C and ignore the existence of T. In other words, we are left with a sphere with $(g - 1)$ handles and a graph of genus g, making an embedding impossible. This is a contradiction.

Case (ii): Curve C lies on handle T. Since the closed curve C cannot lie completely on a handle, it can be divided into two arcs out of which one arc, say arc C', does not lie on T. Then the edges of the graph lying on T can be redrawn along arc C' in region R. In this case, we also have an embedding of G on the surface without making use of handle T. This also is a contradiction, as in case (*i*).

8.92 Prove Theorem 8.13 (generalized Euler formula): If an embedding of a connected graph of order n and size m on a surface of genus g defines r regions and if each region thus defined by the embedding is a 2-cell, $n - m + r = 2(1 - g)$.

Solution. The proof is by induction on g. The result is true if $g = 0$, as established in Theorem 8.3. Assume that the generalized Euler formula is true for all connected graphs embedded on surfaces with genus $(g - 1)$. Let G be any connected graph of order n and size m with a cellular embedding on a sphere with g handles. We can assume without loss of generality that all vertices lie on the surface of the sphere. Each handle should contain at least one edge since every region is a 2-cell. Let T be the handle that contains edges labeled $e_i (i = 1, 2, \ldots, k)$. Draw two disjoint simple curves C and C' (going around the handle), intersecting each e_i exactly once. At each point of intersection, a new vertex is located. The points of intersection of C with k edges are labeled u_i, and the points of intersection of C' with these edges are labeled v_i, where $i = 1, 2, \ldots, k$. Join vertices u_i and u_{i+1}, where $i = 1, 2, \ldots, k$, with $k + 1 = 1$. Also do the same for vertices v_i. Thus if $k > 2$, there will be $2k$ vertices, $4k$ new edges, and $2k$ new regions. If $k = 1$, there is only one edge on handle T such that C intersects it at u_1 and C' intersects it at v_1. To make the graph simple, we define two new vertices u_2 and u_3 on C and two new vertices v_2 and v_3 on C'. Thus there are six new vertices, 10 new edges, and four new regions. If $k = 2$, we define a new vertex u_3 on C in addition to u_1 and u_2 and do the same on C' to avoid parallel edges. In this case, there will be six new vertices, 11 new edges, and five new regions. Let G' be the enlarged graph thus constructed.

Obviously G', is also embedded on the sphere with g handles such that each region in G' also is a 2-cell. Now delete all edges joining u_i and v_i, and let G'' be the graph thus defined. As far as G'' is concerned, handle T is superfluous. In other words, G'' can be embedded on a sphere with $(g - 1)$ handles. So we can use the induction hypothesis.

Case (i): $k > 2$. The order of G'' is $n + 2k$, and its size is $m + 4k - k = m + 3k$. The number of regions in G'' will be $r + 2k - (k - 2) = r + k + 2$. Thus $(n + 2k) - (m + 3k) + (r + k + 2) = 2 - 2(g - 1)$. So $n - m + r = 2(1 - g)$.

Case (ii): $k = 1$. The order of G'' is $n + 6$, its size is $m + 10 - 3 = m + 7$, and the number of regions is $r + 4 - 1 = r + 3$. Thus $(n + 6) - (m + 7) + (r + 3) = 2 - 2(g - 1)$, which implies that $n - m + r = 2(1 - g)$.

Case (iii): $k = 2$. The order of G'' is $n + 6$, its size is $m + 11 - 3 = m + 8$, and the number of regions is $r + 5 - 1 = r + 4$. Thus $(n + 6) - (m + 8) + (r + 4) = 2 - 2(g - 1)$, which also implies that $n - m + r = 2(1 - g)$.

8.93 Show that if a connected graph of order n, size m, and genus g is embedded on a surface of genus g and if the number of regions in the embedding is r, $n - m + r = 2(1 - g)$.

Solution. Since this is a minimal embedding, it is a cellular embedding, as proved in Problem 8.92. The result follows from what was established in Problem 8.92.

8.94 Find the number of 2-cells created if the Petersen graph is embedded on a torus.

Solution. The genus of the torous is 1, and the genus of the Petersen graph is also 1. So the embedding is minimal, and each region is a 2-cell. The graph is of order 10 and size 15. Using the generalized Euler's formula, there should be five regions on the surface.

8.95 If a simple connected graph G is embedded on a surface, find a lower bound for the genus of the graph. Examine the case when G is bipartite.

Solution. Let the genus of G (of order n and size m) be g. For embedding G on the surface, the genus of the surface should be at least g. If both have the same genus, it is a cellular embedding. Suppose there are r regions. Since the graph is simple, $3r \leq 2m$. Then, from the generalized Euler formula, we have the inequality $g \geq \frac{1}{6}(m - 3n + 6)$. If G is bipartite, there are no odd cycles; hence, $4r \leq 2m$. So the generalized Euler formula implies that $g \geq \frac{1}{4}(m - 2n + 4)$ in this case.

8.96 If a simple connected graph G is embedded on a surface, find an upper bound for the number of edges. Examine the case when the graph is bipartite.

Solution. The inequality obtained in Problem 8.95 also implies that $m \leq (3n - 2 + 2g)$. In the case of bipartite graphs, $m \leq (2n - 4 + 4g)$.

8.97 A simple connected graph of order n, size m, and genus g is a **maximal g-graph** if $m = (3n - 2 + 2g)$ if it is not bipartite and $m = (2n - 4 + 4g)$ if it is bipartite. Show that K_7 and $K_{4,4}$ are **maximal toroidal graphs** but K_6 is not.

Solution. In both cases, $g = 1$. For K_7, $n = 7$ and $m = 21$, and it is not bipartite. For $K_{4,4}$ $n = 8$ and $m = 16$, and it is bipartite. Both are maximal toroidal. In the case of K_6, which is toroidal, $m = 15$, which is less than 18. So it is not maximal toroidal.

8.98 Find a lower bound for the genus of the complete graphs (*a*) K_n and (*b*) $K_{m,n}$.

Solution.

(*a*) For K_n, $m = \frac{1}{2}(n)(n - 1)$. From Problem 8.95, the inequality is $g \geq \frac{1}{12}(n - 3)(n - 4)$.

(*b*) For $K_{m,n}$, the order is $(m + n)$, and the size is mn. This time, the inequality for the bipartite graph is $g \geq \frac{1}{4}(m - 2)(n - 2)$. [That the inequality is actually an equality (an achievement indeed) was established by Ringel and Youngs in 1968 for the nonbipartite case and by Ringel in 1965 for the other case.]

8.99 Find a lower bound for the genus of the k-cube.

Solution. The order is 2^k, and the size is $(k)(2^{k-1})$. The graph is bipartite. Using the same inequality as in Problem 8.95, $g \geq (k \cdot 2^{k-3} - 2^{k-1} + 1) = 1 + (2^{k-3})(k - 4)$. (In this case, that inequality is an equality was established by Beineke, Harary, and Ringel.)

Supplementary Problems

8.100 If the boundary of every face in a plane graph of order n has four edges, show that the size of the graph is $2n - 4$.

Ans. Suppose there are m edges and f faces in the graph under consideration. Then, as in Problem 8.99, $4f = 2m$. Using Euler's formula, $m = 2n - 4$.

8.101 If a plane graph of order n is 2-connected and if no face is a triangle, show that the size of the graph cannot exceed $2n - 4$.

Ans. By hypothesis, the boundary of each face has at least four edges. Thus if there are m edges and f faces, $4f \le 2m$, which implies that $m \le 2n - 4$.

8.102 If a 4-regular plane graph has eight faces, find the number of its vertices and edges.
Ans. If there are n vertices and m edges, $4n = 2m$. Using Euler's formula, $n = 6$ and $m = 12$.

8.103 If a 4-regular plane graph has 10 faces, find the number of its vertices and edges.
Ans. If there are n vertices and m edges, $4n = 2m$. Using Euler's formula, $n = 8$ and $m = 16$.

8.104 If the boundary of each region of a connected plane graph of order n and size m has r edges, show that $m(r - 2) = r(n - 2)$.
Ans. $rf = 2m$. Using Euler's formula, $r(2 + m - n) = 2m$. So $m(r - 2) = r(n - 2)$.

8.105 If a 3-regular connected plane graph has 12 regions, find the number of its vertices and edges.
Ans. If there are n vertices and m edges, $3n = 2m$. Using Euler's formula, $n = 20$ and $m = 30$.

8.106 If the girth (number of edges in a cycle with the minimum number of edges) of a connected of order n and size m is g, prove the inequality $m(g - 2) \le g(n - 2)$. [*Hint:* Use the inequality $gf \le 2m$ (f is the number of regions) and Euler's formula.]

8.107 Show that a graph with fewer than nine edges is planar. [*Hint:* Kuratowski says at least nine edges.]

8.108 Find the number of regions in a planar graph of order n if (a) it is a triangulation and (b) it is a maximal outerplanar graph.
Ans. Use Euler's formula. The number of edges is known.

8.109 Find the crossing number of $K_{3,4}$. *Ans.* 2

8.110 A plane graph is 2-connected if and only its geometric dual is 2-connected. [*Hint:* Use Solved Problem 8.5.]

8.111 Show that there is no Hamiltonian planar graph having regions of degrees 4 and 6 and one region of degree 9. [*Hint:* Proceed as in Problem 8.82.]

8.112 Show that the planar graph shown in Fig. 8-41 is not Hamiltonian. [*Hint:* Use the Grinberg–Kozyrev theorem.]

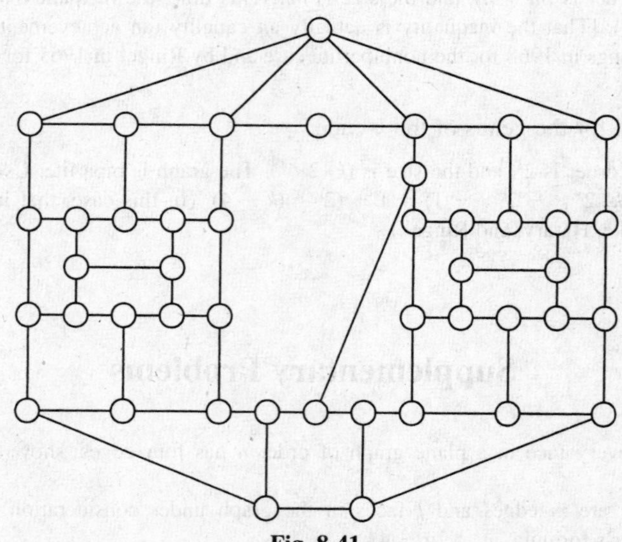

Fig. 8-41

8.113 Show that the planar graph shown in Fig. 8-42 is not Hamiltonian. [*Hint:* Use the Grinberg–Kozyrev theorem.]

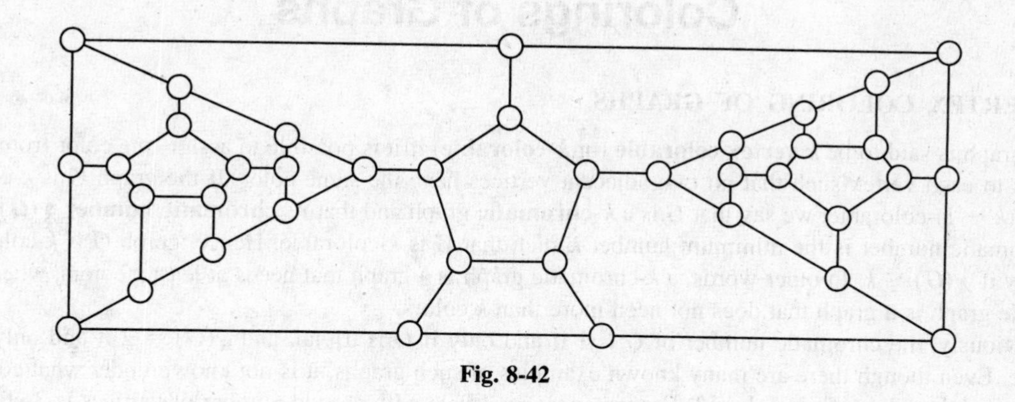

Fig. 8-42

8.114 Show that the planar graph in Fig. 8-43 is not Hamiltonian. [*Hint:* Use the Grinberg–Kozyrev theorem.]

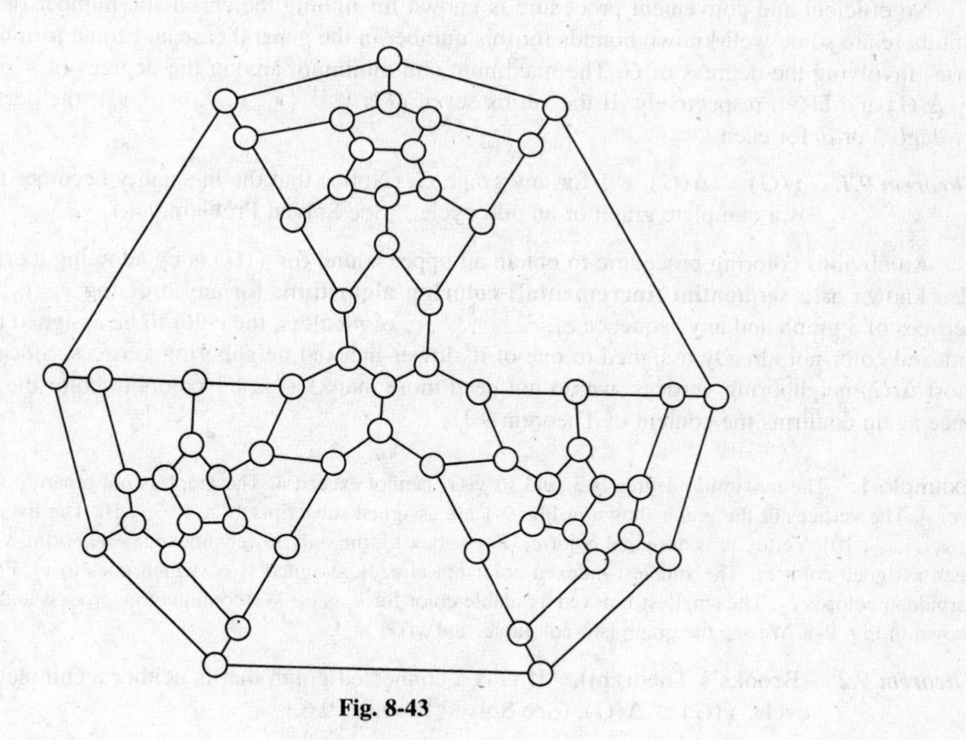

Fig. 8-43

Chapter 9

Colorings of Graphs

9.1 VERTEX COLORING OF GRAPHS

A graph is said to be ***k* vertex colorable** (or ***k*-colorable**) if it is possible to assign one color from a set of k colors to each vertex such that no two adjacent vertices have the same color. If the graph G is k-colorable but not $(k-1)$-colorable, we say that G is a ***k*-chromatic** graph and that its **chromatic number** $\boldsymbol{\chi(G)}$ is k. So the chromatic number is the minimum number k such that G is k-colorable. Hence, graph G is k-colorable if and only if $\boldsymbol{\chi(G)} \leq k$. In other words, a k-chromatic graph is a graph that needs at least k colors, whereas a k-colorable graph is a graph that does not need more than k colors.

Obviously, the chromatic number of G is 1 if and only if G is trivial, and $\chi(G) = 2$ if and only if G is bipartite. Even though there are many known examples of such graphs, it is not known under what conditions the chromatic number of a graph is 3. For instance, any cycle with an odd number of vertices is 3-chromatic. Likewise, it is easy to see that the chromatic number of a wheel of even order is 3. The complete graph K_n is n-chromatic, and if G has K_n as a subgraph, $\chi(G) \geq n$.

No efficient and convenient procedure is known for finding the chromatic number of an arbitrary graph. But there are some well-known bounds for this number in the general case and some formulas in a few special cases involving the degrees of G. The maximum and minimum among the degrees of a graph G are denoted by $\Delta(G)$ and $\delta(G)$, respectively. If the vertex set of G is $V = \{v_1, v_2, \ldots, v_n\}$, the degree of v_i is denoted by $\deg(v_i)$ or d_i for each i.

Theorem 9.1. $\chi(G) \leq \Delta(G) + 1$ for any graph G. (Notice that the inequality becomes an equality when G is a complete graph or an odd cycle.) (See Solved Problem 9.4.)

An obvious coloring procedure to obtain an upper bound for $\chi(G)$ is by adopting a greedy method that is also known as a **sequential (incremental) coloring algorithm:** for any ordering $v_1, v_2, \ldots, v_n$ of the n vertices of a graph and any sequence $c_1, c_2, \ldots, c_n$ of n colors, the color to be assigned to v_i is the smallest-indexed color not already assigned to one of its lower-indexed neighboring vertices. Since each vertex has at most $\Delta(G)$ neighboring vertices, we do not need more than $\Delta(G) + 1$ colors to color the vertices of G. This once again confirms the content of Theorem 9.1.

Example 1. The maximum degree is 3, and so $\chi(G)$ cannot exceed 4. The graph is not bipartite. Hence, $\chi(G)$ is either 3 or 4. The vertices of the graph shown in Fig. 9-1 are assigned subscripts $1, 2, \ldots, 10$. The list of 10 colors is c_i $(i = 1, 2, \ldots, 10)$. Vertex v_1 is assigned color c_1. For vertex v_2, the only lower-indexed neighboring vertex is v_1, which has been assigned color c_1. The smallest-indexed color not already assigned is c_2, which goes to v_2. For vertex v_3, the only forbidden color is c_2. The smallest-indexed available color for v_3 is c_1. We continue this process and color each vertex, as shown in Fig. 9-1. We see the graph is 3-colorable, and $\chi(G) = 3$.

Theorem 9.2 **(Brooks's Theorem).** If G is a connected graph that is neither a complete graph nor an odd cycle, $\chi(G) \leq \Delta(G)$. (See Solved Problem 9.6.)

Example 2. The Petersen graph is neither complete nor an odd cycle. Since it is not bipartite, it is not 2-colorable. Its maximum degree is 3. So, by Brooks's theorem, its chromatic number cannot exceed 3. Hence, the chromatic number is 3.

If we modify the sequential coloring method by ordering the set of vertices based on their degrees in nonincreasing order, we have a variation of the algorithm known as the **largest first sequential algorithm.**

Example 3. In the graph shown in Fig. 9-2, the set of seven vertices is an ordered set with degrees 4, 4, 3, 3, 3, 3, and 2. Since the graph is not bipartite, $\chi(G)$ is at least 3. By Brooks's theorem, $\chi(G)$ cannot exceed the maximum degree 4 since G is neither an odd cycle nor a complete graph. Thus $\chi(G)$ is either 3 or 4. The largest first sequential algorithm will not need more than five colors. Using this algorithm, the vertices get the colors

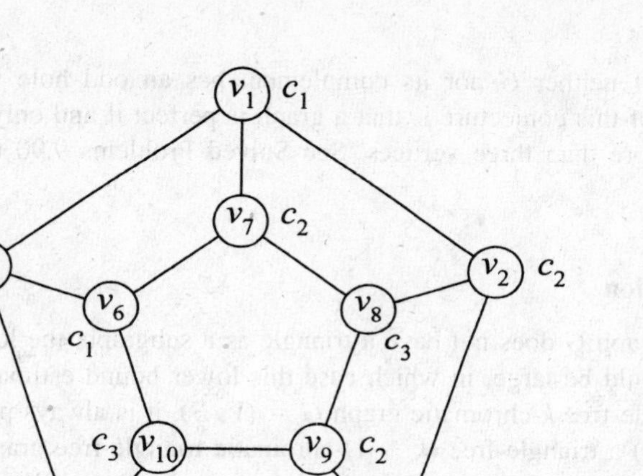

Fig. 9-1

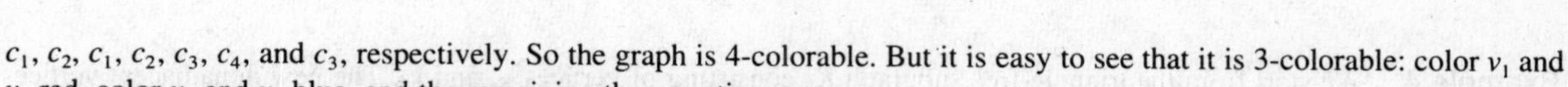
c_1, c_2, c_1, c_2, c_3, c_4, and c_3, respectively. So the graph is 4-colorable. But it is easy to see that it is 3-colorable: color v_1 and v_6 red, color v_2 and v_4 blue, and the remaining three vertices green.

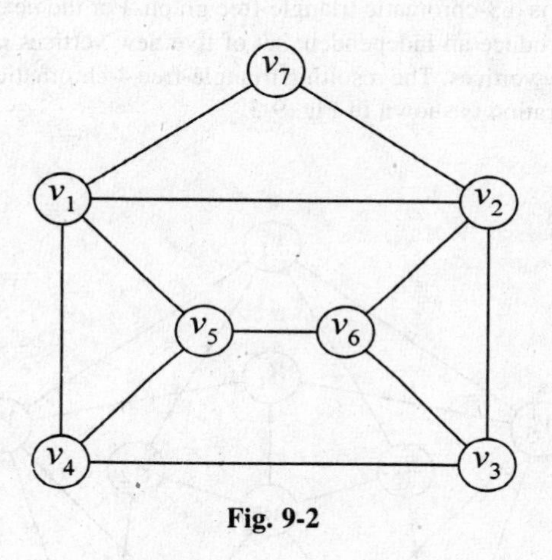

Fig. 9-2

Lower Bounds for the Chromatic Number

A set of A of vertices in a graph G of order n is an **independent set** in G if no two vertices in A are adjacent (in G), and the cardinality of a largest independent set in G is the **vertex-independence number** (or **independence** or **internal stability number),** denoted by $\alpha(G)$. See Problems 1.18 and 1.19. At the other extreme, a set Z of vertices in G is called a **clique** in G if there is an edge (in G) between every pair of vertices in Z, and the cardinality of a largest clique in G is the **clique number** of G, denoted by $\omega(G)$. Obviously, both $\omega(G)$ and $n/[\alpha(G)]$ are lower bounds for $\chi(G)$.

Perfect Graphs

A graph G is **perfect** if for every induced subgraph H of G, $\omega(H) = \chi(H)$. The **perfect graph theorem (PGT) of Lovasz** asserts that the complement of a perfect graph is perfect. See Solved Problem 9.100. An **odd hole** in a graph is a cycle that is an induced subgraph with k vertices, where k is odd. An **antihole** is the complement of an odd hole. The chromatic number of an odd hole with more than three vertices is 3, whereas its clique number is only 2. So any graph that has an odd hole with more than three vertices cannot be perfect. The **perfect graph conjecture** (also known as the **strong perfect graph** conjecture) asserts that a graph G is perfect if and only if neither G nor its complement has an odd hole with more than three vertices. An equivalent assertion of this conjecture is that a graph is perfect if and only if it does not have an odd hole or odd antihole with more than three vertices. See Solved Problems 9.90 through 9.101 for more on perfect graphs.

Mycielski Construction

If a connected graph G does not have a triangle as a subgraph, the lower bound $\omega(G)$ is only 2 and the chromatic number could be large, in which case this lower bound estimate is not very helpful. Specifically, starting from a triangle-free k-chromatic graph $G = (V, E)$, it is always possible to obtain (by the **Mycielski construction** method) a triangle-free $(k + 1)$-chromatic triangle-free graph for any choice of k. If $V = \{v_1, v_2, \ldots, v_n\}$, introduce two new sets $U = \{u_1, u_2, \ldots, u_n\}$ and $W = \{w\}$ so that the union of the three sets V, U, and W becomes the set of vertices for the new graph G' that contains G as a proper subgraph. Set U is an independent set in G'. Join w to each u_i. Also join each u_i to every vertex adjacent to v_i. It can be shown (see Solved Problem 9.33) that the graph G' thus constructed with $2n + 1$ vertices is a $(k + 1)$-chromatic graph that is triangle-free.

Example 4. We start from the triangle-free subgraph K_2 consisting of vertices v_1 and v_2. The new nonadjacent vertices are u_1 and u_2 along with w, which is adjacent to both of them. We join u_1 to v_2 and u_2 to v_1. At the end of the first iteration, we get a cycle with five vertices that is a 3-chromatic triangle-free graph. For the next iteration, we begin with a cycle with vertices v_i $(i = 1, 2, 3, 4, 5)$ and introduce an independent set of five new vertices u_i $(i = 1, 2, 3, 4, 5)$ and a new vertex w that is adjacent to each of these new vertices. The resulting triangle-free 4-chromatic graph (known as the **Grotsch graph**) obtained at the end of the second iteration is shown in Fig. 9-3.

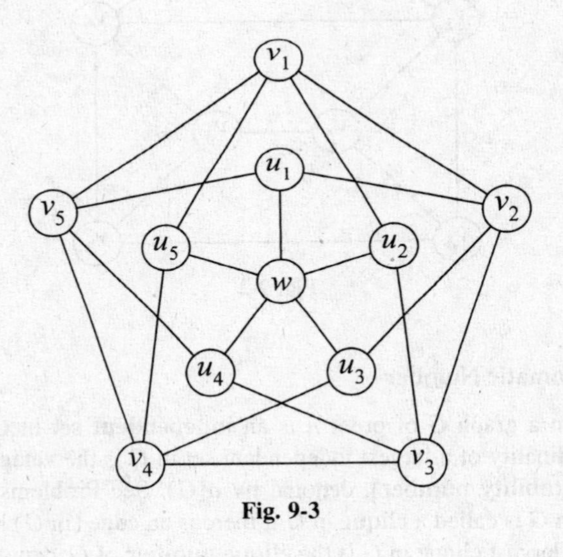

Fig. 9-3

Chromatic Polynomials (Chromials)

If the number of ways of coloring the vertices of a graph G (in such a way that no two adjacent vertices are assigned the same color) using at most x colors is denoted by $P(G, x)$, the smallest value of x such that $P(G, x)$ is not zero is the chromatic number of G since $P(G, x) = 0$ if and only if G is not x-colorable. For example, if the graph is K_2 and if there are x colors available, one of the vertices can be colored in x ways and the other can be colored in $(x - 1)$ ways. So $P(K_2, x) = x(x - 1)$. The smallest (positive integer) value of x such that the expression $x(x - 1)$ is not zero is 2, which is the chromatic number of K_2. More generally,

$P(K_n, x) = x(x-1)(x-2) \ldots (x-n+1) = x_{(n)}$. Suppose there are $f(r)$ ways of partitioning the vertex set of a graph into r independent sets. Then for each such partition, the number of ways of coloring the vertices of G is $x_{(r)}$; consequently, $P(G, x) = \Sigma f(r)\, x_{(r)}$. If the order of G is n, $f(r) = 0$ when $r > n$. Thus $P(G, x)$ is a polynomial in x (known as the **chromatic polynomial or chromial of** G) of degree n with *integer coefficients* in which the coefficient of the leading term x^n is 1 since there is only one way of partitioning the vertex set of n elements into n nonempty (independent) subsets. So it is a **monic polynomial.** Moreover, $f(0) = 0$; therefore, the constant term in the chromatic polynomial is 0. It can be shown that the absolute value of the coefficient of x^{n-1} is the size of the graph (see Solved Problem 9.39) and that the coefficients in the polynomial alternate (see Solved Problem 9.44) in sign. It is not known under what conditions a given polynomial will be the chromatic polynomial of a graph. See the Solved Problems for more on chromatic polynomials.

9.2 EDGE COLORING OF GRAPHS

A graph G with no loops is said to be ***k* edge colorable** if it is possible to assign to each edge one color from a set of k colors such that no two edges with a vertex in common get the same color. A k edge colorable graph is a ***k* edge chromatic graph** if it is not $(k-1)$ edge colorable and if its **chromatic index** $\chi'(G)$ is k. In other words, the chromatic index of G is the minimum number k such that it is k edge colorable. Obviously, the maximum degree $\Delta(G)$ of any graph G is necessarily a lower bound for its chromatic index, whereas by Brooks's theorem, $\Delta(G)$ is an upper bound of the chromatic number of any graph that is neither a complete graph nor an odd cycle. The surprising fact (see Theorem 9.4) is that $\Delta(G) + 1$, which is the chromatic number of G when G is complete or an odd cycle, is an upper bound for the chromatic index of any simple graph. Furthermore, the edges in an edge coloring of the graph that get the same color constitute a matching in that graph; therefore, the chromatic index of the graph is also equal to the minimum number of matchings into which the edge set of the graph can be partitioned. In particular, the edges of a regular graph G can be colored using $\Delta(G)$ colors if and only if G is 1-factorable. Notice also that the chromatic index of a simple graph G is the same as the chromatic number of its line graph $L(G)$ since two edges in G have a vertex in common if and only if the vertices corresponding to these edges are adjacent in $L(G)$. In particular, both the chromatic number and the chromatic index are the same for any cyclic graph.

Theorem 9.3. (*i*) The chromatic index of a bipartite multigraph G is $\Delta(G)$. In particular, the chromatic index of the complete bipartite graph $K_{m,n}$ is $\max\{m, n\}$. (*ii*) If G is the complete graph with n vertices, its chromatic index is $\Delta(G) = n - 1$ if n is even, and it is $\Delta(G) + 1 = n$ if n is odd. (See Solved Problems 9.52, 9.55, and 9.56.)

Theorem 9.4 **(Vizing's Theorem).** The chromatic index of a simple graph G is either $\Delta(G)$ or $\Delta(G) + 1$. (See Solved Problem 9.59.)

(In view of Vizing's theorem, the collection of all simple graphs can be grouped into two classes. A graph is in class 1 if its chromatic index is equal to its maximum degree. Otherwise it belongs to class 2. Even though the **classification problem** of determining which graphs belong to which class remains unsolved, P. Erdös and R. J. Wilson have proved that the more the number of vertices in a graph, the more its probability to claim membership in class 1.)

Example 5. In the graph shown in Fig. 9-4(*a*), the letters marked on the edges represent their colors in an optimal edge coloring, and the letters marked on the vertices represent their colors in an optimal vertex coloring. Since graph G is neither complete nor an odd cycle, its chromatic number, $\chi(G)$ by Brooks's theorem, cannot exceed the maximum degree, which is 3. Since the graph has an odd cycle, $\chi(G)$ is at least 3. The chromatic index $\chi'(G)$ is at least 3, and by Vizing's theorem, it cannot exceed 4. Any partitioning of the edge set will have at least four sets. The line graph is shown in Fig. 9-4(*b*), and its chromatic number is 4.

Theorem 9.5. If a cubic graph has a bridge, its chromatic index is 4. (See Solved Problem 9.62.)

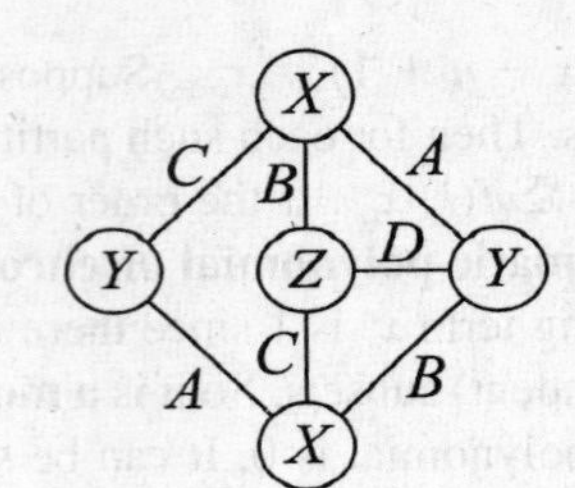

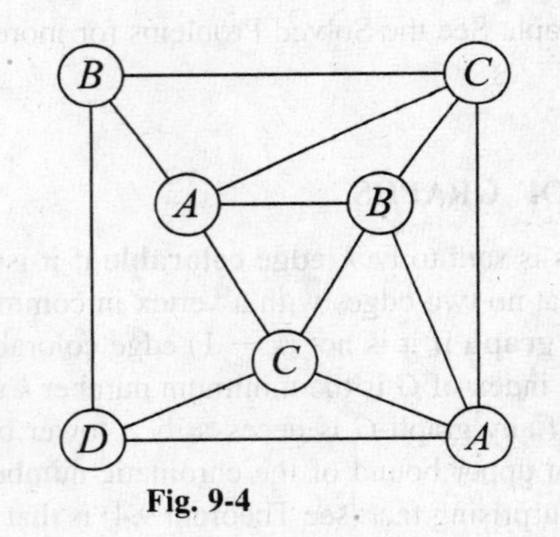

Fig. 9-4

9.3 COLORING OF PLANAR GRAPHS

In a political (or geographical) map showing various countries on a sheet of paper or on the surface of a globe, the boundary of a country is usually a closed simply connected curve, and two countries are neighbors if they share at least one boundary of nonzero length. The map is said to be colored if each country gets a color such that no two neighboring countries have the same color. Any map can be viewed as a connected plane graph in which each country corresponds to a region in the graph and where two countries are neighbors if and only if the corresponding regions in the graph have at least one edge in common. It is reasonable to assume that this connected graph has no loops and also no bridges; if there were a bridge after crossing it (which is actually a boundary), one would not have left the country. Thus in graph-theoretical terminology, a **map** is a connected bridgeless plane multigraph. The problem of determining the minimum number of colors needed to color a map is one of the most famous problems in graph theory. In the words of K. Appel, for almost 150 years, a "Holy Grail of graph theory has been a simple incisive proof" of the **four-color theorem (4CT)**, which asserts that no map needs more than four colors. In an attempt to prove the 4CT in 1890 (see Solved Problem 9.3), Heawood was able to show that no map needs more than five colors. Finally, in 1976, K. Appel and W. Hakken proved the 4CT, but theirs was not a "simple incisive" proof. At this point, nobody knows whether the grail exists. A presentation of the awe-inspiring proof (or even a sketch of the proof) of this famous theorem is beyond the scope of this book. A few important results related to the saga of the 4CT will be presented here, however.

If G is a map, its geometric dual G' is a connected plane graph in which the degree of each vertex is at least 2. Since G' has no loops, one can color its vertices such that no two adjacent vertices are of the same color. Any coloring of the regions of G defines a coloring of the vertices of G' if we assign each vertex of G' the same color that the corresponding region in G gets. In this coloring of the vertices, no two adjacent vertices have the same color since two vertices are adjacent if and only if the two corresponding regions are neighbors. So if the regions of G are colored using k colors, the vertices of G' can also be colored using k colors. The reverse implication is obvious. Thus an alternative statement of the 4CT is that the **chromatic number of a planar graph cannot exceed 4** since any planar graph is isomorphic to some plane graph. It is in this (dual) form the 4CT is usually stated. Notice that by the coloring of a *map,* we mean the coloring of its regions (including the unbounded exterior region), whereas by the coloring of a *graph* (unless otherwise explicitly stated), we mean the coloring of its vertices.

A planar graph is said to be **irreducible** if it is 5-chromatic and if the chromatic number of any graph with fewer vertices is less than 5. (The 4CT implies that there are no irreducible graphs.) A graph is **reducible** if it is not a subgraph of an irreducible graph. A **configuration** in a planar graph G is a subgraph of G consisting of a cycle C in G and the vertices, edges, and regions of G interior to C. A set of graphs is an **unavoidable set** if every planar graph contains at least one graph from the set as a configuration.

Suppose there exists a finite unavoidable set $X = \{H_1, H_2, \ldots, H_k\}$ of reducible configurations. Assume that there is a planar graph that needs at least five colors. Then there should be an irreducible planar graph G. Since G is planar, it should contain at least one graph, say H_i, from X as a subgraph. Since H_i is reducible, it cannot be a subgraph of the irreducible graph G.

So the existence of an unavoidable set of reducible configurations will imply that no map (planar graph) needs more than four colors. The starkest description of Appel and Hakken's proof is that they were able to exhibit an unavoidable set of about 2000 reducible configurations with the aid of a computer. (The size of the set has been reduced to 700 or so in the recent past.) It is not true that they first obtained a finite set of configurations and then proved that every configuration in that set is reducible. In fact, both concepts had their independent roles toward the construction of the final set of reducible configurations. For a lucid explanation of Appel and Hakken's truly impressive and monumental work, see their October 1977 *Scientific American* article, "The Solution of the Four-Color Map Problem."

More generally, when we consider a surface S_g of positive genus g, its chromatic number $\chi(S_g)$ is the maximum chromatic number among all graphs that can be embedded on S_g. It was proved (see Solved Problem 9.86) by G. Ringel and J. W. Youngs in 1968 that $\chi(S_g)$ is equal to the floor of the number $\frac{1}{2}\{7 + \sqrt{1 + 48g}\}$ for all $g > 0$. Known as the **Heawood Map Coloring Theorem,** this fact is true for all nonnegative g and is also a consequence of the four-color theorem.

Example 6. Figure 9-5(a) shows a map G depicting seven finite regions and one exterior region that can be labeled region 8 that is not labeled. Notice that between region 4 and region 5 are two frontiers. To investigate the

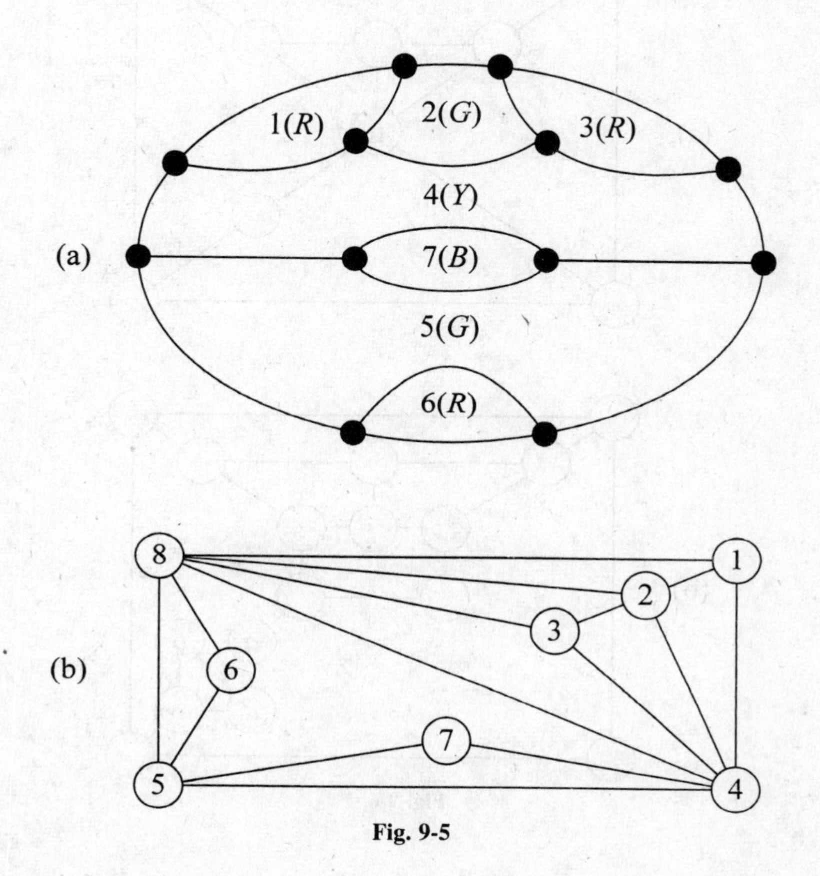

Fig. 9-5

vertex-colorability of the dual graph, we have to draw only one edge joining the vertices corresponding to these two regions. The connected graph G' shown in Fig. 9-5(b) is the dual graph thus modified. The subgraph induced by the set $\{2, 3, 4, 8\}$ of vertices in G' is K_4, and as such, at least four colors are needed to color the vertices of G'. So by the 4CT, the chromatic number of G' is 4. Consequently, four colors are needed to color all the regions, including the exterior region of the map. If the map represents an island, region 8 is the ocean, which can be colored blue, and the same color can be assigned to region 7, which could be a lake. Regions 1, 3, and 6 can be colored red. Regions 2 and 5 can be colored green. Finally, region 4 is colored yellow.

Cubic Maps and Edge-Colorings

A **cubic map** is a 3-regular bridgeless connected planar graph. Given any map G in which the minimum degree is at least 2, it is possible to construct a cubic map G' such that the k-colorability of one implies the k-colorability of the other. The procedure is as follows. If there is a vertex u in G whose degree r is more than 3, we introduce r new vertices "surrounding" u, construct a cycle passing through these r new vertices, and delete vertex u. Then join each vertex in the cycle to the vertex in G that was adjacent to u, making sure at the same time that the newly constructed graph continues to be planar. If v is a vertex of degree 2 joining two vertices, say p and q, insert a new vertex a on the edge joining p and u, insert a second new vertex c on the edge joining q and u, construct a third new vertex b and join it to a and c, construct a fourth vertex d and join it to a and c, delete u, and finally join b and d. Thus a vertex of degree 2 is effectively replaced by the planar graph K_4 minus one of its edges. When this process is completed (see Fig. 9-6) for each vertex of degree more

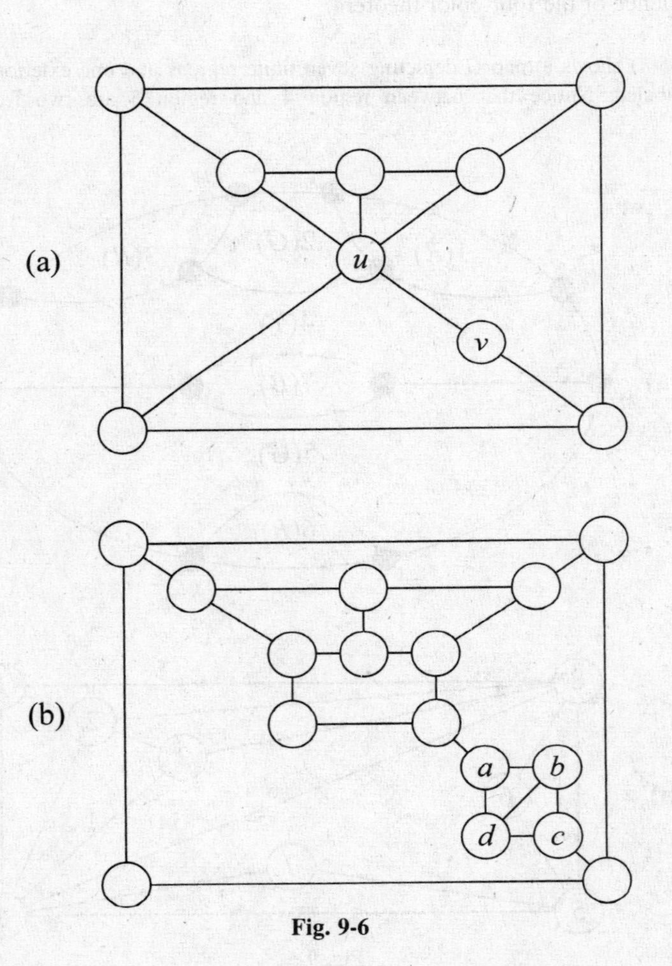

Fig. 9-6

than 3 and for each vertex of degree 2, we get a cubic bridgeless planar graph G'. This construction naturally leads to the following equivalence theorem.

Theorem 9.6. The four-color theorem is true if and only if every cubic map is 4-colorable.

Next we consider a cubic map G. Suppose its faces are 4-colored using colors A, B, C, and D. Then we can 3-color the boundaries of its regions using colors B, C, and D as follows. If there is a boundary between a region with color A and a region with color B, that boundary is assigned color B. This coloring assignment can be symbolically represented by the equation $A + B = B + A = B$. Similarly, we define $A + C = C$, $A + D = D$, $B + C = D$, $B + D = C$, and $C + D = B$. See the 4-coloring of the regions in the cubic graph shown in Fig. 9-7 and the 3-coloring of its edges. The color of the external region is B. On the other hand, suppose the edges of a cubic map are colored using the colors B, C, and D. To get a 4-coloring of its faces, we proceed as follows. Start from any region and label it A. When we move from that region to a neighboring region after crossing the edge colored, say B, we assign the color $A + B = B$ to that region. Thus, using the same addition, rule all the regions (including the external region) can be colored using these four colors. (This coloring scheme is taken from an article by M. Gardner's April 1976 *Scientific American* article under the column Mathematical Games). Thus we have the following characterization of the 4-colorability of planar graphs.

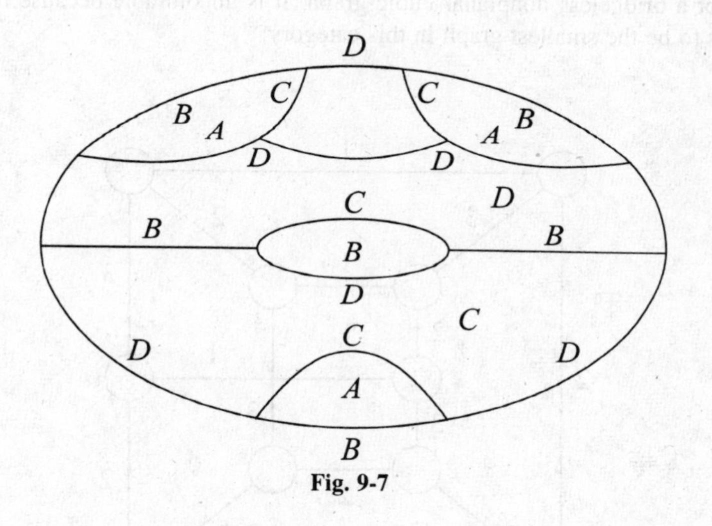

Fig. 9-7

Theorem 9.7 **(Tait's Theorem).** The following claims are equivalent:

(*i*) The chromatic number of a planar graph does not exceed 4.

(*ii*) The chromatic index of a cubic bridgeless planar graph (CBP graph) is 3.

(*iii*) Any CBP graph is 1-factorable.

(*iv*) Any CBP graph is a class 1 graph.

Any 3-coloring of the edges of a cubic graph is called a **Tait coloring,** and a cubic graph is said to be **uncolorable** if it does not have a Tait coloring. So by Vizing's theorem, the chromatic index of an uncolorable graph is 4. *Thus there are four kinds of cubic graphs:*

1. Cubic graphs with at least one bridge. Such graphs are uncolorable, as stated in Theorem 9.5. See Fig. 9-8(a).
2. Bridgeless planar cubic graphs. The four-color theorem implies that such graphs are colorable.

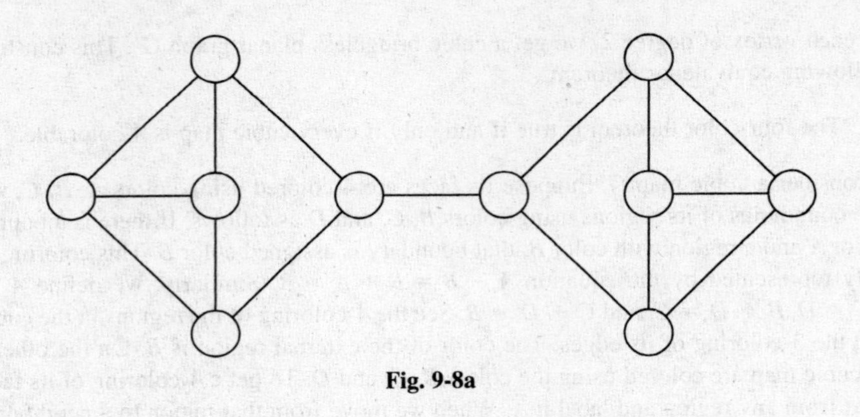

Fig. 9-8a

3. Bridgeless nonplanar cubic graphs that are colorable. See Fig. 9-8(*b*) in which the colors are marked 1, 2, and 3.
4. Bridgeless nonplanar cubic graphs that are uncolorable. The Petersen graph belongs to this category. It is no doubt a bridgeless nonplanar cubic graph. It is uncolorable because it is not 1-factorable. It also happens to be the smallest graph in this category.

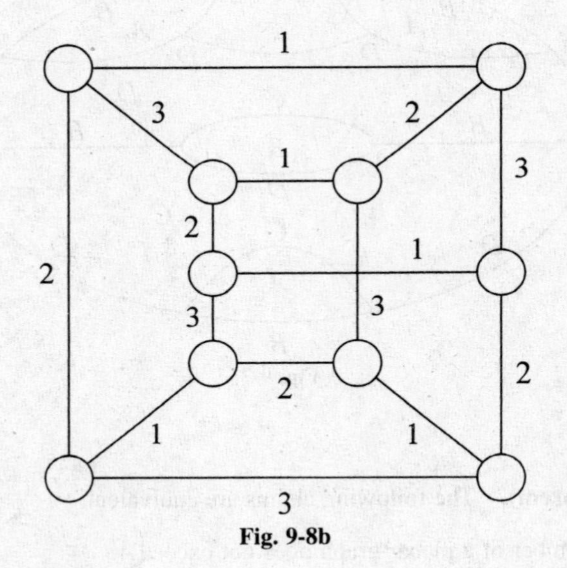

Fig. 9-8b

Finding such bridgeless cubic nonplanar uncolorable graphs is not easy. As R. Isaacs puts it, anyone who looks for them will be "vividly impressed with the maddening difficulty of finding" a single one. In the words of D. A. Holton and J. Sheehan, these graphs "do not appear to be thick on the ground." Such graphs were christened **snarks** by M. Gardner after Lewis Carroll's ballad entitled "The Hunting of the Snark." Given a cubic bridgeless graph G of chromatic index 4, one can obtain another such graph G' by a "trivial modification" (quoting B. Descartes) of G by replacing a vertex by a triangle or by inserting vertices on two of its edges having a vertex in common and joining them. Figure 9-9 shows two such trivial modifications of the Petersen graph. For a more exclusive definition of snarks (and the concepts leading to this definition) that precludes such

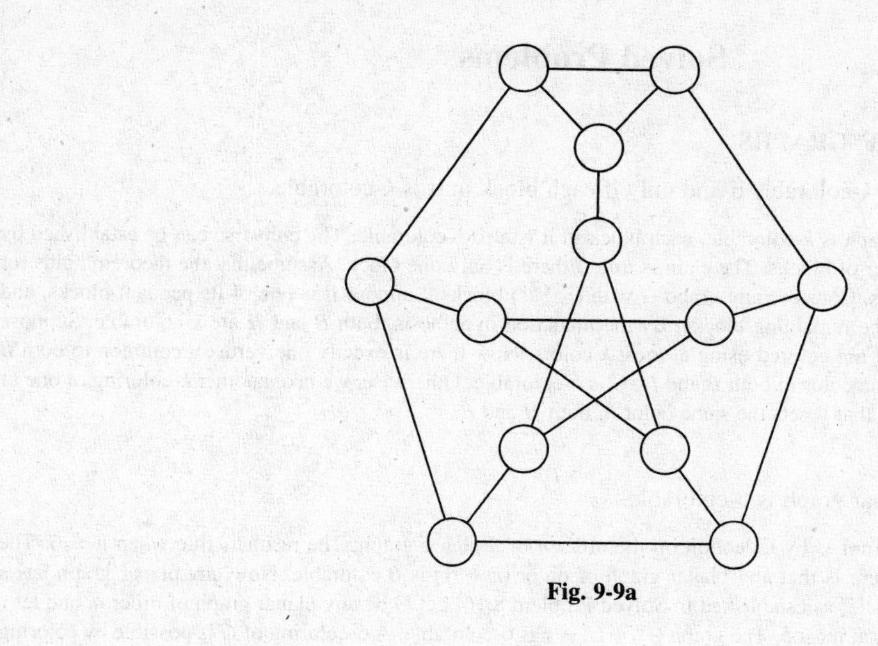

Fig. 9-9a

trivial generalizations, see solved Problems 9.61 through 9.70. We conclude this chapter with a statement of the celebrated **Tutte's conjecture:** Every snark has a subgraph that is homeomorphic (or contractible) to the Petersen graph. Observe that this conjecture implies the four-color theorem since the Petersen graph is non-planar; therefore, it is a genuine generalization of the theorem. Could an incisive proof of the verity of this conjecture be the Holy Grail of graph theory?

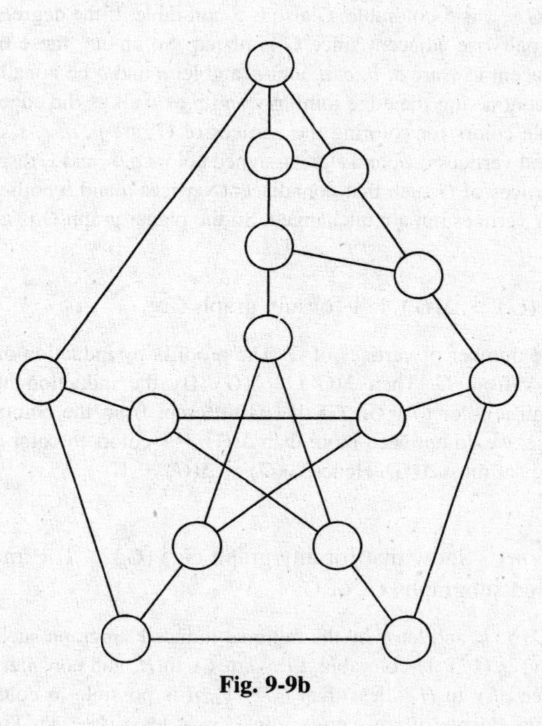

Fig. 9-9b

Solved Problems

VERTEX COLORING OF GRAPHS

9.1 Prove that a graph is k-colorable if and only if each block in it is k-colorable.

Solution. If a graph is k-colorable, each block in it is also k-colorable. The converse can be established by induction on the number of blocks. The clain is true if there is only one block. Assume that the theorem holds for any graph with r blocks. Consider any graph G with $(r + 1)$ blocks. Suppose B is one of its pendant blocks, and let H be the union of the remaining blocks. By the induction hypothesis, both B and H are k-colorable. Suppose the vertices of B and H are colored using at most k colors. Now there is exactly one vertex v common to both B and H. If v gets the same color in both B and H, G is k-colorable. Otherwise, we take another k-coloring of one of these components such that v gets the same color in both H and B.

9.2 Show that every planar graph is 6-colorable.

Solution. The proof is by induction on the order n of a planar graph. The result is true when $n \leq 6$. The induction hypothesis here is that any planar graph of order $(n - 1)$ is 6-colorable. Now any planar graph has a vertex of degree at most 5, as established in Solved Problem 8.16. Let G be any planar graph of order n, and let v be any vertex of degree at most 5. The graph $G' = G - v$ is 6-colorable. A 6-coloring of G is possible by coloring v with a color that is different from the colors of the (at most five) vertices adjacent to v. Hence, the chromatic number of a planar graph is at most 6.

9.3 Show that every planar graph is 5-colorable.

Solution. As in Problem 9.2, the proof is by induction on the order n of a planar graph. The result is true when $n \leq 5$. The induction hypothesis here is that any planar graph of order $(n - 1)$ or less is 5-colorable. Now any planar graph has a vertex of degree at most 5, as established in Solved Problem 8.16. Let G be any planar graph of order n, and let v be any vertex of degree at most 5. The graph $G' = G - v$ is 5-colorable. If the degree of v is less than 5, since $G - v$ is 5-colorable, G also is 5-colorable. If the degree of v is equal to 5, the five vertices adjacent to v cannot be pairwise adjacent since G is planar. So among these five, at least two are not adjacent. Suppose the vertices adjacent to v are a, b, c, d, and e, and let a and b be nonadjacent. Construct the planar graph G' of order $(n - 2)$ by contracting the edge joining v and a as well as the edge joining v and b. The graph G' is 5-colorable. The available colors for coloring the vertices of G' are p, q, r, s, and t. Suppose the merged vertex vab is assigned color p and vertices c, d, and e are assigned colors q, r, and s, respectively. Now unravel the merged vertex vab. Color the vertices of G such that nonadjacent vertices a and b both get color p and vertex v gets color t. The colors of the other vertices remain unchanged. So the planar graph G is also 5-colorable.

9.4 Prove Theorem 9.1: $\chi(G) \leq \Delta(G) + 1$ for any graph G.

Solution. Let n be number of vertices of G. The proof is by induction on n. Let G' be the graph obtained by deleting any vertex v from G. Then $\Delta(G') \leq \Delta(G)$. By the induction hypothesis, $\chi(G') \leq \Delta(G') + 1 \leq \Delta(G) + 1$. We can assign a color to v (in G) that is different from the colors (assigned in G') to the vertices adjacent to it. In that case, we do not need more than $\Delta(G) + 1$ colors to color the vertices of G since the number of vertices adjacent to v is at most $\Delta(G)$. Hence, $\chi(G) \leq \Delta(G) + 1$.

9.5 (*Szekeres–Wilf Theorem*) Show that for any graph G, $\chi(G) \leq 1 + \max \delta(G')$, where the maximum is taken over all induced subgraphs G' of G.

Solution. Let $\chi(G) = k$, and let H be the minimal induced subgraph such that $\chi(H) = k$. So for any vertex v in H, the graph $(H - v)$ is $(k - 1)$-colorable. Fix vertex v in H, and consider any $(k - 1)$ coloring of $(H - v)$. In that case, if the degree of v in H is less than $(k - 1)$, it is possible to color the vertices of H using at most $(k - 1)$ colors. Hence, the degree of any vertex v in H is at least $(k - 1)$. Thus $(k - 1) \leq \delta(H) \leq \max \delta(G')$.

[Observe that by replacing max $\delta(G')$ by $\Delta(G)$, we revert to the upper bound in Problem 9.4. Also, max $\delta(G')$ cannot be replaced by $\delta(G)$ for an arbitrary graph. See Problem 9.15.]

9.6 Prove Theorem 2 (Brooks's theorem): If G is a connected graph that is neither a complete graph nor an odd cycle, $\chi(G) \leq \Delta(G)$.

Solution. Let the order of G be n with $\chi(G) = k$ and $\Delta(G) = r$. If $r = 1$, G is complete. If $r = 2$, either the graph is bipartite, in which case $r = k$, or the graph is an odd cycle. So we assume that $r \geq 3$.

Case (i): The graph G is not r-regular. So there is a vertex v with degree less than r. Construct a spanning tree rooted at this vertex. Since there are n vertices, the root is labeled v_n, and the remaining vertices are labeled in descending order as and when a new vertex is added to the spanning tree, culminating at vertex v_1. Thus we have an ordered set of n vertices. Observe that each vertex v_i other than the root has a higher-indexed neighbor in the unique path (in the tree) from that vertex to the root. Therefore, the number of its lower-indexed neighbors is at most $(r - 1)$ since the maximum degree is only r. So if we use the greedy (sequential search) method to color the vertices of G, we need at most r colors; hence, $r \leq k$.

Case (ii): G is r-regular and w is a cut vertex of G. Suppose G' is one of the components of $G - w$. Let H be subgraph obtained by adjoining w and the edges joining w to the vertices in G'. Then the degree of w in H is less than r. As we saw in case (i), subgraph H is r-colorable. Then we can have a recoloring of the vertices in G so that G also is r-colorable. This argument holds for every cut vertex in G. Thus $r \leq k$.

Case (iii): G is r-regular and 2-connected. Since G is not complete, there are two nonadjacent vertices u, v, and a third vertex w that is adjacent to both u and v. If G is 3-connected, the graph $G - u - v$ is connected. If G is not 3-connected, there is a vertex w such that $(G - w)$ has a cut vertex. Each pendant block of $(G - w)$ has a vertex adjacent to w. We can then choose u and v from two different pendant blocks. Thus in any case, there are two nonadjacent vertices u and v and a vertex w adjacent to both of them such that $G - u - v$ is connected. Let us relabel the nonadjacent vertices u and v as v_1 and v_2, respectively, and their common neighbor as v_n. Now construct a spanning tree in the graph $G - v_1 - v_2$ with v_n as the root and terminating at vertex v_3 as in case (i). A coloring of the vertices using the greedy method can be now implemented. First, we have an ordering of the $(n - 3)$ vertices $\{v_3, v_4, \ldots, v_{n-1}\}$ defined by the tree. Each vertex in this set has at most $(r - 1)$ lower-indexed adjacent vertices. Since v_1 and v_2 are not adjacent, they can be assigned the same color. In addition to these two vertices, the root has $(r - 2)$ lower-indexed adjacent vertices in $\{v_3, v_4, \ldots, v_{n-1}\}$. So the greedy algorithm uses at most r colors to color G. Thus the proof is completed. (This proof is a modification by D. West of the proof given by Lovasz in 1975.)

9.7 (*Welsh–Powell Theorem*) Prove that if the vertex set $V = \{v_1, v_2, \ldots, v_n\}$ of a graph G has the nonincreasing degree sequence $d_1 \geq d_2 \geq \cdots \geq d_n$, $\chi(G) \leq \max_i \min\{d_i + 1, i\}$.

Solution. Apply the sequential search method to color the vertices with the given ordering of the vertices based on their degrees. When the ith vertex v_i is colored, the number of vertices already colored cannot exceed the minimum of $(i - 1)$ and its degree d_i for each i. So to color this vertex and the lower-indexed vertices, the number of colors needed does not exceed $1 + \min\{i - 1, d_i\}$. Hence, $\chi(G)$ cannot exceed the maximum among these n numbers. So $\chi(G) \leq 1 + \max_i \min\{d_i, i - 1\} \leq \max_i \min\{d_i + 1, i\}$.

9.8 Show that the number of colors needed to color the vertices of G using the largest first searching method does not exceed $1 + \Delta(G)$.

Solution. From Problem 9.7, $\chi(G) \leq 1 + \max_i \min\{d_i, i - 1\} \leq 1 + \Delta(G)$. So the number of colors needed if we use this algorithm cannot exceed $1 + \Delta(G)$.

9.9 Use the largest first algorithm to color the vertices of the graph shown in Fig. 9-10.

Solution. The set of vertices is ordered by degrees in nonincreasing order. Since there is an odd cycle, $\chi(G) \geq 3$. By Brooks's theorem, $\chi(G) \leq 4$. Using the algorithm, the vertices can be colored sequentially with colors 1, 2, 3, 1, 2, 3, and 4. So the graph is 4-colorable.

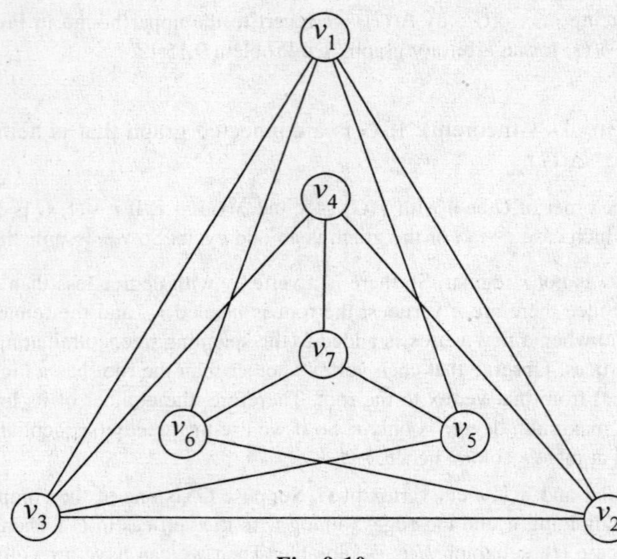

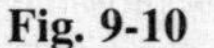

Fig. 9-10

Critical Graphs

9.10 A k-chromatic graph G is said to be a **critically k-chromatic** (or a **k-critical) graph** if $\chi(G - v) = k - 1$ for every vertex v of G. Show that a critically k-chromatic graph G is a block.

Solution. If G is not connected, the chromatic number of any component of G is less than k, which implies that the chromatic number of G also is less than k. So G is connected. If G has a cut vertex v giving a vertex partition $V_1, V_2, \ldots, V_r$ of the disconnected graph $G - v$, let G_i be the subgraph induced by $V_i \cup \{v\}$. Each G_i is $(k - 1)$-colorable. The only vertex common to any pair of these r subgraphs is v. G is also $(k - 1)$ colorable, which is a contradiction. So G has no cut vertex. Thus G is a block.

9.11 Give an example of a k-chromatic graph that is not critically k-chromatic.

Solution. Consider the graph consisting of a triangle and an edge joining one of its vertices to a fourth vertex. Its chromatic number is 3, and it is not critically 3-chromatic.

9.12 Characterize critically k-chromatic graphs when $k = 2$ and $k = 3$.

Solution. Obviously, a graph is critically 2-chromatic if and only if it is K_2, and it is critically 3-chromatic if and only if it is an odd cycle.

9.13 A k-chromatic graph G is said to be a **minimally k-chromatic graph** $\chi(G - e) = k - 1$ for every edge e of G. Give an example of (a) a minimally k-chromatic graph and (b) a critically k-chromatic graph that is not minimally k-chromatic.

Solution. Notice that every k-chromatic graph of minimum order is critically k-chromatic, whereas every k-chromatic graph of minimum size is minimally k-chromatic. Of course, every minimally k-chromatic graph without isolated vertices is a critically k-chromatic graph.

(a) Of course, K_2 and K_3 are minimally k-chromatic, where $k = 2$ and 3, respectively. For $k = 4$, the Grotzsch graph shown in Fig. 9-11(a) serves as an example.

(b) The Harary graph G shown in Fig. 9-11(b) is critically 4-chromatic but not minimally 4-chromatic since $\chi(G - e) = \chi(G) = 4$.

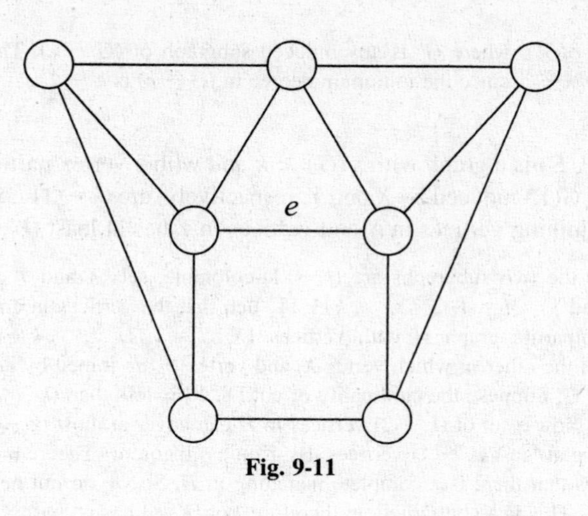

Fig. 9-11

9.14 Show that a k-chromatic graph contains a critically (minimally) k-chromatic graph.

Solution. Suppose the k-chromatic graph G is not critically k-chromatic. Then there exists a vertex v such that $G - v$ is k-chromatic. If $G - v$ is critically k-chromatic, we are done. Otherwise, there is a vertex w in $G - v$ such that $(G - v) - w$ is k-chromatic. Continue this process. We eventually arrive at a critically k-chromatic graph. The proof is similar for the case of minimally k-chromatic graphs.

9.15 Show that if G is critically k-chromatic, $\delta(G) \geq (k - 1)$. Give a counterexample to show that the converse does not hold.

Solution. Suppose $\delta(G) < (k - 1)$. Let v be a vertex in G such that the degree of v is equal to $d = \delta(G)$. The graph $G - v$ is $(k - 1)$-colorable, giving a vertex partition $\{V_1, V_2, \ldots, V_{k-1}\}$ of the graph G. By assumption, $d < (k - 1)$. So there is one set V_i of vertices in this partition such that no vertex in that set is adjacent to v. Thus it is possible to have another partition $\{V_1, V_2, \ldots, V_i \cup \{v\}, V_{k-1}\}$, which implies that G is $(k - 1)$-colorable. If e is an edge of K_4, $G = K_4 - e$ is 3-chromatic but not critically 3-chromatic. But the inequality holds in G.

9.16 Give an example of a k-chromatic graph G for which the inequality $\delta(G) \geq (k - 1)$ is not valid.

Solution. We are looking for k-chromatic graph (which is not critically k-chromatic) in which the inequality will not hold. Consider the example in Problem 9.11. The chromatic number is 3, but the inequality does not hold good because the minimum degree is 1.

9.17 Prove that a k-chromatic graph G has at least k vertices of degree at least $k - 1$.

Solution. Let H be a critically k-chromatic subgraph of G. The degree of every vertex in H is at least $k - 1$. Since the chromatic number of H is k, it should have at least k vertices.

9.18 Use the inequality obtained in Problem 9.15 to prove the Szekeres–Wilf theorem (Problem 9.5).

Solution. Let H be any induced critically k-chromatic subgraph of a k-chromatic graph G. Then $\chi(G) = \chi(H) \leq \delta(H) + 1 \leq 1 + \max \delta(G')$, where the maximum is taken over all the induced subgraphs G' of G.

9.19 Prove that a critically k-chromatic graph G with exactly one vertex whose degree exceeds $(k - 1)$ is minimally k-chromatic.

Solution. Since G is critically k-chromatic, the degree of each vertex is at least $(k - 1)$. In this case, the degree of each vertex is $(k - 1)$ except for one vertex. If e is any edge of G, $\delta(G - e) = k - 2$. Then

$\chi(G - e) \leq 1 + \max \delta(G')$, where G' is any induced subgraph of $(G - e)$. Thus $\chi(G - e) \leq 1 + (k - 2) = k - 1$. So $\chi(G - e) = k - 1$ since the minimum degree in $(G - e)$ is $k - 2$.

9.20 Show that if $G = (V, E)$ is a graph with $\chi(G) \geq k$ and with a vertex partition (X, Y) of V such that the subgraphs $G(X)$ and $G(Y)$ induced by X and Y, respectively, are $(k - 1)$-colorable, cut $[X, Y]$ consisting of those edges in E joining vertices in X and vertices in Y has at least $(k - 1)$ edges.

Solution. Since the two subgraphs are $(k - 1)$-colorable, sets X and Y can be partitioned into $\{X_i : i = 1, 2, \ldots, k - 1\}$ and $\{Y_i : i = 1, 2, \ldots, k - 1\}$ such that the vertices in any of these subsets get the same color. Consider the bipartite graph H with vertices $\{X_i : i = 1, 2, \ldots, k - 1\}$ on one side and $\{Y_i : i = 1, 2, \ldots, k - 1\}$ on the other in which vertex X_i and vertex Y_j are joined by an edge if and only if there is no edge between them in G. Suppose the cardinality of cut $[X, Y]$ is less than $(k - 1)$. Then H will have more than $(k - 1)(k - 2)$ edges. Now a set of $(k - 2)$ vertices in H can cover at most $(k - 2)(k - 1)$ edges. So any vertex cover of H should have at least $(k - 1)$ vertices. By Konig's theorem (Theorem 6.12), H has a matching of size $(k - 1)$, which implies that there is a complete matching in H. So we do not need more than $(k - 1)$ colors to color the vertices of G. This is a contradiction; therefore, cut $[X, Y]$ has at least $(k - 1)$ edges.

9.21 Show that a critically k-chromatic graph G is $(k - 1)$-edge connected. [Equivalently, any connected minimally k-chromatic graph is $(k - 1)$-edge connected.] Give an example to show that the converse is not true.

Solution. If $k = 2$, the graph is K_2, which is 1-edge connected. If $k = 3$, the graph is an odd cycle, which is 2-edge connected. Let $k \geq 4$. Suppose G is not $(k - 1)$ edge connected. Then there is a partition of the vertex set of G into two sets X and Y such that the cardinality of cut $[X, Y]$ is less than $(k - 1)$. So the subgraphs induced by X and Y are $(k - 1)$-colorable. Since the chromatic number of G is k, according to Problem 9.20, cut $[X, Y]$ should have at least $(k - 1)$ edges. This contradiction establishes that G is $(k - 1)$ edge connected. If e is an edge of K_4, $K_4 - e$ is a 2-edge connected graph that is not critically 3-chromatic.

9.22 Show that if G is a critically k-chromatic graph, there are no subgraphs G_1 and G_2 such that $G = G_1 \cup G_2$ and, at the same time, such that $G_1 \cap G_2$ is complete.

Solution. Suppose there are subgraphs G_1 and G_2 such that $G = G_1 \cup G_2$ and such that the intersection $H = G_1 \cap G_2$ is complete. The chromatic number of both G_1 and G_2 is at most $(k - 1)$. Consider a $(k - 1)$-coloring of G_1. In this coloring, no two vertices in H will have the same color. The same is also true in any $(k - 1)$-coloring of G_2. So the distinct colors that are assigned to the vertices in H (either in G_1 or in G_2) can be used for a coloring of these vertices in graph G. This implies that we do not need more than $(k - 1)$ colors for a coloring in G also, which is a contradiction.

9.23 Show that the graph induced by a separating set W of vertices of a critically k-chromatic graph is not a complete graph. In particular, if $|W| = 2$, the two vertices in W are not adjacent.

Solution. Let W be a separating set of vertices in a critical graph $G = (V, E)$ with $\chi(G) = k$, and let the vertex sets of $G - W$ be $V_1, V_2, \ldots, V_r$. Then the subgraph G_i induced by $(V_i \cup W)$, known as a **W-component** of G, being a proper subgraph of G for each i, is $(k - 1)$-colorable. Suppose each G_i has a $(k - 1)$-coloring in which no two vertices in W get the same color. If vertex w in W gets color c in a particular G_i, w can be assigned the same color c in each G_i and also in G. Then, when it comes to the coloring of the remaining vertices in set $(V - W)$, it is not necessary that two vertices in this set get different colors. This implies that G can be colored using at most $(k - 1)$ colors. This contradiction proves that there are two vertices in W that are not adjacent in G.

9.24 Give an example of a graph G with a separating set W consisting of two adjacent vertices.

Solution. Since the subgraph induced by W is complete, the graph G cannot be critically k-chromatic for any k. In Fig. 9-12(a), the separating set W consists of the two adjacent vertices u and v. There are three W-components for this graph, as shown in Fig. 9-12(b). Observe that G is 4-chromatic but not critically 4-chromatic.

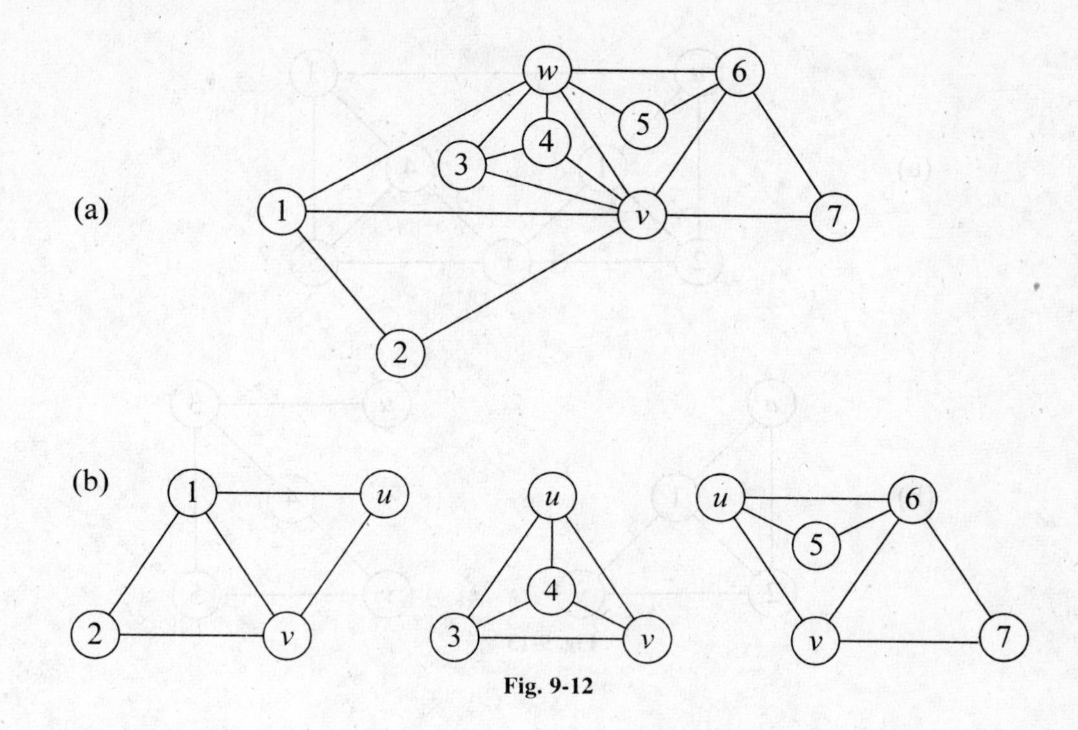

Fig. 9-12

9.25 (*Dirac's Theorem on Critical Graphs*) If a minimally k-chromatic graph G has a separating set W consisting of two vertices u and v, show that (a) there are exactly two W-components G denoted by G_1 and G_2 such that G is the union of these components, and (b) the graph $G_1 + e$ obtained by joining u and v by edge e and adding it to G_1 and the graph G_2' obtained from G_2 by joining u and v by an edge and then contracting it are both minimally k-chromatic.

Solution.

(a) u and v are not adjacent, as shown in Problem 9.24. Each W-component is $(k - 1)$-colorable. If there are $(k - 1)$-colorings of these components such that the two nonadjacent vertices u and v get the same color in each of these components, G also becomes $(k - 1)$-colorable. Thus there is a component, say G_1, such that both u and v get the same color under every $(k - 1)$-coloring, and there is a component, say G_2, such that these two vertices get different colors under any $(k - 1)$-coloring. In that case, the union of these components will need more than $(k - 1)$ colors. Since G is a critically k-chromatic graph, we then conclude that G is the union of these components and that there is no other component.

(b) In G_1, both vertices have the same color. If they are joined by edge e, they should have different colors. So $G_1 + e$ is k-chromatic. Let f be any edge of $G_1 + e$. We have to show that $G_1 + e - f$ is $(k - 1)$-colorable. This is obvious if $f = e$. In any $(k - 1)$-coloring of $G - f$, vertices u and v have different colors since component G_2 is a subgraph of G. The restriction of such a coloring to the vertices of G_1 is a $(k - 1)$-coloring of $G_1 + e - f$. So $G_1 + e$ is critically k-chromatic. For G_2', since any coloring assigns different colors to u and v, the graph obtained by merging them will need k colors. If f is any edge, $G - f$ is $(k - 1)$-colorable, and so G_2' is also.

9.26 Illustrate Dirac's theorem in the case of the graph shown in Fig. 9-13(a).

Solution. The graph shown in Fig. 9-13(a) is a minimally 4-chromatic graph. Set W of vertices consisting of u and v is a separating set defining the two components, as shown in Fig. 9-13(b). In the component where the degree of u is 2, any 3-coloring will assign the same color to both these vertices, whereas in the other component, any 3-coloring will have to assign different colors to these two vertices. Furthermore, if we join these two vertices in the first component, we get a minimally 4-chromatic graph. Likewise, if we merge these two vertices in the other component, we once again get a minimally 4-chromatic graph. Thus the theorem is verified.

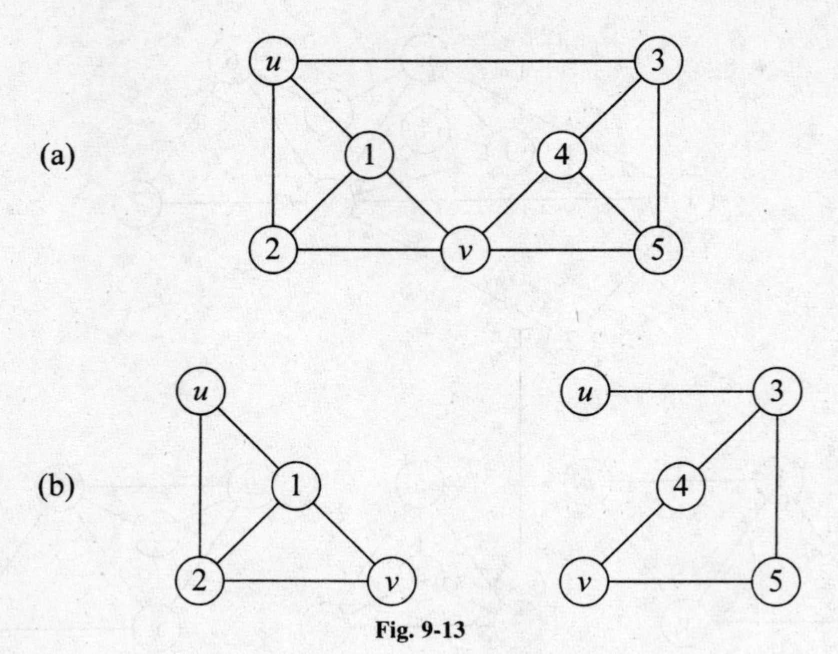

Fig. 9-13

9.27 Show that if a minimally k-chromatic graph has a separating set consisting of two vertices, the sum of their degrees is at least $(3k - 5)$.

Solution. Using the notation as in Problem 9.26, graph $G_1 + e$ is minimally k-chromatic. So (degree of u) + (degree of v) $\geq (k - 1) + (k - 1)$ in $(G_1 + e)$. This implies (degree of u) + (degree of v) $\geq (k - 2) + (k - 2) = (2k - 4)$ in G_1. Graph G_2' is minimally k-chromatic. So the degree of the vertex obtained by merging u and v is at least $(k - 1)$ in this graph. Now (degree of u) + (degree of v) $\geq (k - 1)$ in G_2. Thus (degree of u) + (degree of v) $\geq (2k - 4) + (k - 1) = (3k - 5)$ in G.

9.28 Use the inequality obtained in Problem 9.27 to prove that $\chi(G) \leq \Delta(G)$, where G is a connected graph that is neither a complete graph nor an odd cycle in the special case when G is not 3-connected. (This is a part of Brooks's theorem.)

Solution. We may assume without loss of generality that G is minimally k-chromatic. The hypothesis implies that $k \geq 4$ and G is 2-connected. Since it is not 3-connected, there is a separating set consisting of two vertices u and v. So $2\Delta(G) \geq$ degree of u + degree of v, which implies that $2\Delta(G) \geq (3k - 5) \geq (2k - 1)$ since $k \geq 4$. So $k \leq \Delta(G)$.

Uniquely Colorable Graphs

9.29 A k-chromatic graph G is **uniquely colorable** if any k-coloring of G induces the same partition of the vertex set of G. (*a*) List the k-chromatic graphs that are uniquely colorable when $k = 2$ and $k =$ the order of the graph. (*b*) Give an example of a graph that is not uniquely colorable. (*c*) Show that if the k-chromatic graph G is uniquely colorable, $\delta(G) \geq (k - 1)$.

Solution.

(*a*) The only uniquely colorable 2-chromatic graphs are the bipartite graphs. The complete graph of order n is the only uniquely n-colorable n-chromatic graph.

(*b*) The odd cycle of order 3 is a 3-chromatic graph that is not uniquely colorable.

(*c*) In any unique coloring, any vertex v should be adjacent to at least one vertex of every color different from that of v. Otherwise, it is possible to have a different vertex partition by changing the color of v.

9.30 Show that if a k-chromatic graph G is uniquely colorable, the subgraph induced by the union of any two sets in the vertex partition in a coloring is a connected graph.

Solution. Suppose V_1 and V_2 are two sets with colors c_1 and c_2, respectively, in a vertex partition of G, and let H be the subgraph induced by these two sets. If H is not connected, each component of H should have vertices of both colors c_1 and c_2. By interchanging these two colors, we have a different color assignment, contradicting the uniqueness assumption. The converse is not true, as seen from the 3-chromatic graph shown in Fig. 9-14.

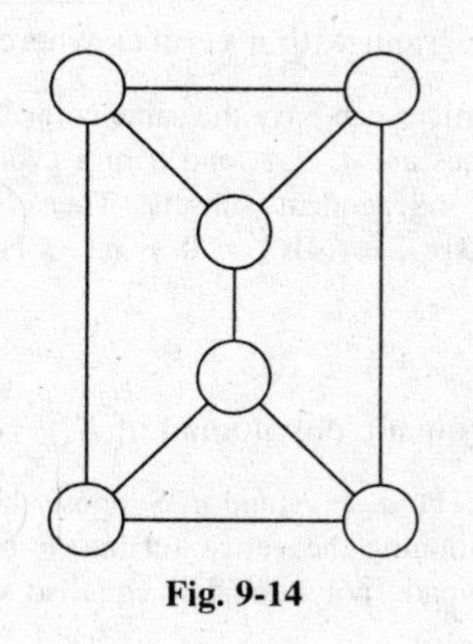

Fig. 9-14

9.31 Show that if a k-chromatic graph G is uniquely colorable, it is $(k - 1)$-connected.

Solution. This is true when G is the complete graph with k vertices. Consider the case when G is not complete. Suppose it is not $(k - 1)$-connected. So there is a set W of at most $(k - 2)$ vertices such that $G - W$ is disconnected. Since W has at most $(k - 2)$ vertices, there are at least two colors, say c_1 and c_2 (associated with sets V_1 and V_2 in a vertex partition of G), such that no vertex in W gets either of these two colors. At the same time, the subgraph H induced by the union of V_1 and V_2 is a connected subgraph contained in one of the components, say G_1, of the disconnected graph $G - W$. Now take any vertex v in a component of $G - W$ other than G_1 and assign it one of these two colors that no vertex in W has. So it is possible to have a new k-coloring of G, giving a different vertex partition. This is a contradiction.

9.32 Show that any 4-chromatic uniquely colorable planar graph G is a maximal planar graph.

Solution. Suppose the unique vertex partition corresponding to a 4-coloring defines the sets V_i ($i =$ 1, 2, 3, 4). Let G_{ij} be the subgraph (of order n_{ij} and size m_{ij}) induced by the union of V_i and V_j. There are six such connected subgraphs and six inequalities $m_{ij} \geq n_{ij} - 1$. By adding, we get $m \geq 3n - 6$, where m and n are the size and order of G. Since G is planar, $m \leq 3n - 6$. See Theorem 8.4. So equality that implies maximal planarity holds.

9.33 (*Zykov's Theorem*) Show that the graph obtained from a triangle-free k-chromatic graph by the Mycielski method is a triangle-free $(k + 1)$-chromatic graph.

Solution. Let $G = (V, E)$, where $V = \{v_i : i = 1, 2, \ldots, n\}$ be a triangle-free k-chromatic graph. Suppose $U = \{u_i : i = 1, 2, \ldots, n\}$ and $W = \{w\}$ are two new sets of vertices such that V, U, and W are pairwise disjoint. The new graph is G' with vertex set V', which is the union of V, U, and W. By definition, set U is an independent set in G', vertex w is adjacent to every vertex in U, and each u_i is adjacent to every vertex in G, which is adjacent to v_i. Obviously, the existence of a triangle in G' implies the existence of one in G. Thus G' is also triangle-free. Since the chromatic number of G is k and since both u_i and v_i can be assigned the same color, the same k colors that are used for a k-coloring of G can also be used for a k-coloring of $G' - w$. Once this is done, a new color can be assigned to w. Thus the chromatic number of G' cannot exceed $(k + 1)$. The crux of the problem then is to show that this is equal to $(k + 1)$. Suppose, to the contrary, that G' is k-colorable using the colors from the set $\{1, 2, \ldots, k\}$. Let the color of w be k. Then no vertex in U can have color k in any k-coloring of G'. Since the chromatic number of G is k, there should be at least one vertex in V with color k in any k-coloring of G using these

k colors. Once a k-coloring of G is accomplished, we can subsequently obtain a k-coloring of G' as follows. Let the color of w be k. If the color of v_i in G is k (in the k-coloring of G), we can recolor v_i (in G') using the color of u_i since these two vertices are not adjacent and since vertex v_j in V is adjacent to v_i in G if and only if v_j is adjacent to u_i. In that case, the restriction of this recoloring of G' to G uses at most $(k - 1)$ colors to color the vertices of G. This is a contradiction since the chromatic number of G is k.

Chromatic Polynomials (Chromials)

9.34 Find $P(G, x)$, where G is a cyclic graph with n vertices where $n = 3$ or $n = 4$.

Solution. If $n = 3$, no two vertices can have the same color. So the chromatic polynomial is $x(x - 1)(x - 2)$. If $n = 4$, suppose the four vertices are a, b, c, and d in a cyclic order. Let $f(r)$ be the number of ways of partitioning the vertex set into r independent subsets. Then $f(1) = 0$, $f(2) = 1$, $f(3) = 2$, and $f(4) = 1$. So $P(G, x) = f(1)x_{(0)} + f(2)x_{(2)} + f(3)x_{(3)} + f(4)x_{(4)} = 0 + x(x - 1) + 2x(x - 1)(x - 2) + x(x - 1)(x - 2)(x - 3) = x^4 - 4x^3 + 6x^2 - 3x$.

9.35 If e is an edge of K_4, find the chromatic polynomial of $K_4 - e$.

Solution. Let the four vertices be a, b, c, and d. Suppose the only nonadjacent vertices are b and d. Let $f(r)$ be the number of ways of partitioning the vertex set into r independent subsets. Then $f(1) = 0$, $f(2) = 0$, $f(3) = 1$, and $f(4) = 1$. So the chromatic polynomial is equal to $x(x - 1)(x - 2) + x(x - 1)(x - 2)(x - 3) = x(x - 1)(x - 2)^2$.

9.36 Show that if G is the union of two graphs G_1 and G_2 that share a single vertex, the chromatic polynomial of G is the product of the chromatic polynomials of G_1 and G_2 divided by x.

Solution. The number of ways of coloring the graph G_1 using x colors is $P(G_1, x)$. Once a coloring of this graph is done, the color that the common vertex gets remains the same for the coloring of the entire graph G. The number of ways of coloring the vertices of G that belong to G_2 (other than the common vertex) is $(1/x)P(G_2, x)$.

9.37 If G is the connected graph obtained by linking two triangles so that they share one vertex in common, find the chromatic polynomial of G.

Solution. The chromatic polynomial of each triangle is $x(x - 1)(x - 2)$. So, by Problem 9.36, $P(G, x) = (1/x)[(x)(x - 1)(x - 2)]^2 = x(x - 1)^2(x - 2)^2$.

9.38 Obtain the chromatic polynomial of the graph shown in Fig. 9-15.

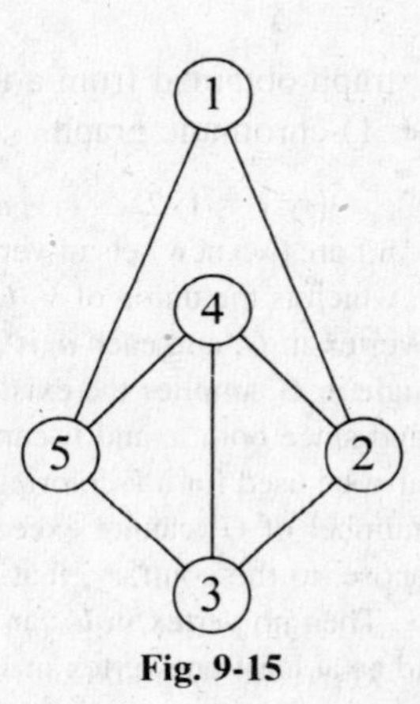

Fig. 9-15

Solution. Let $f(r)$ be the number of ways of partitioning vertex set $V = \{1, 2, 3, 4, 5\}$ into r independent subsets. There is at least one edge in the graph. So $f(1) = 0$. It is not possible to partition V into two independent subsets. So $f(2) = 0$. There are two ways of partitioning V into three independent subsets: $\{3\}, \{1, 4\}, \{2, 5\}$; and $\{4\}, \{1, 3\}, \{2, 5\}$. So $f(3) = 2$. There are three ways of partitioning V into four independent subsets: $\{1\}, \{3\}, \{4\}, \{2, 5\}$; $\{2\}, \{3\}, \{5\}, \{1, 4\}$; and $\{2\}, \{4\}, \{5\}, \{1, 3\}$. So $f(4) = 3$, and finally $f(5) = 1$. Thus $P(G, x)$ is $(2)(x)(x - 1)(x - 2) + (3)(x)(x - 1)(x - 2)(x - 3) + (x)(x - 1)(x - 2)(x - 3)(x - 4)$, which is equal to $x^5 - 7x^4 + 19x^3 - 23x^2 + 10x$.

9.39 Show that if G is a graph of order n and size m, the absolute value of the coefficient of x^{n-1} in the chromatic polynomial is equal to the size of the graph.

Solution. The coefficient of x^{n-1} in $f_{(n)}$ is $-\frac{1}{2}(n)(n - 1)$. So the coefficient of x^{n-1} in the chromatic polynomial is $f_{(n-1)} - \frac{1}{2}(n)(n - 1)$. Now $f_{(n-1)}$ is also equal to the number of nonadjacent pairs of vertices, so it is equal to $\frac{1}{2}(n)(n - 1) - m$. Thus the coefficient of x^{n-1} in the chromatic polynomial is $-m$.

9.40 Show that a graph of order n is a tree if and only if its chromatic polynomial is equal to $x(x - 1)^{n-1}$.

Solution. Let G be a tree with n vertices. We have to establish that the chromatic polynomial is $x(x - 1)^{n-1}$. The result is true if $n = 1$ and $n = 2$. Assume that this is true for all trees with $(n - 1)$ vertices. Let e be an edge joining vertex u to vertex v of degree 1 in a tree T of order $(n - 1)$. Then, by the induction hypothesis, the chromatic polynomial of $(T - v)$ is $x(x - 1)^{n-2}$. Vertex v can be assigned any color other than the color of u. So vertex v can be colored $(x - 1)$ ways. Thus the chromatic polynomial of T is $x(x - 1)^{n-2}(x - 1) = x(x - 1)^{n-1}$. On the other hand, let $x(x - 1)^{n-1}$ be the chromatic polynomial of a graph. Certainly the order of G is n. The coefficient of x^{n-1} is $-(n - 1)$, so the size is $(n - 1)$. If G is not connected, its chromatic polynomial is the product of the chromatic polynomials of its components; therefore, the coefficient of x in it has to be 0. But the coefficient of x in $x(x - 1)^{n-1}$ is not. So G is a connected graph with n vertices and $(n - 1)$ edges.

9.41 (*Reduction Theorem of Birkhoff and Lewis*) (*a*) If $G.e$ is the graph obtained from G by merging any two adjacent vertices u and v (joined by edge e) into a single vertex and by joining this new combined vertex to all those vertices to which either u or v were already adjacent, show that $P(G, x) = P(G - e, x) - P(G.e, x)$. (*b*) If $G.e$ is the graph obtained from G by merging any two nonadjacent vertices u and v into a single vertex and by joining this new combined vertex to all those vertices to which either u or v were already adjacent, show that $P(G, x) = P(G + e, x) + P(G.e, x)$, where $G + e$ is the graph obtained from G by joining u and v by new edge e.

Solution.

(*a*) The number of ways of coloring $G - e$ such that u and v do not get the same color is the same as the number of ways of coloring G. The number of ways of coloring $G - e$ such that u and v get the same color is the same as the number of ways of coloring $G.e$. Hence, $P(G - e, x) = P(G, x) + P(G.e, x)$.

(*b*) The proof is to (*a*).

9.42 Use the reduction theorem proved in Problem 9.41 to compute $P(G, x)$ of graph G in Problem 9.38.

Solution. The order of the graph is 5, and its size is 7. The complete graph with five vertices has 10 edges. We can delete the edges one at a time and apply the reduction rule (*a*) or add edges one at a time and apply the reduction rule (*b*). Let us use the second rule.

Iteration 1: Join vertices 2 and 5. Then contract the newly constructed edge. The resulting graphs are G_1 and G_2, as shown in Fig. 9-16.

Iteration 2: Join vertices 1 and 4 in G_1, and contract this edge. The resulting graphs are G_3 and G_4. Join vertices 1 and 4 in G_2, and contract the new edge. The resulting graphs are G_5 and G_6. Notice that G_4 is K_4 and G_6 is K_3.

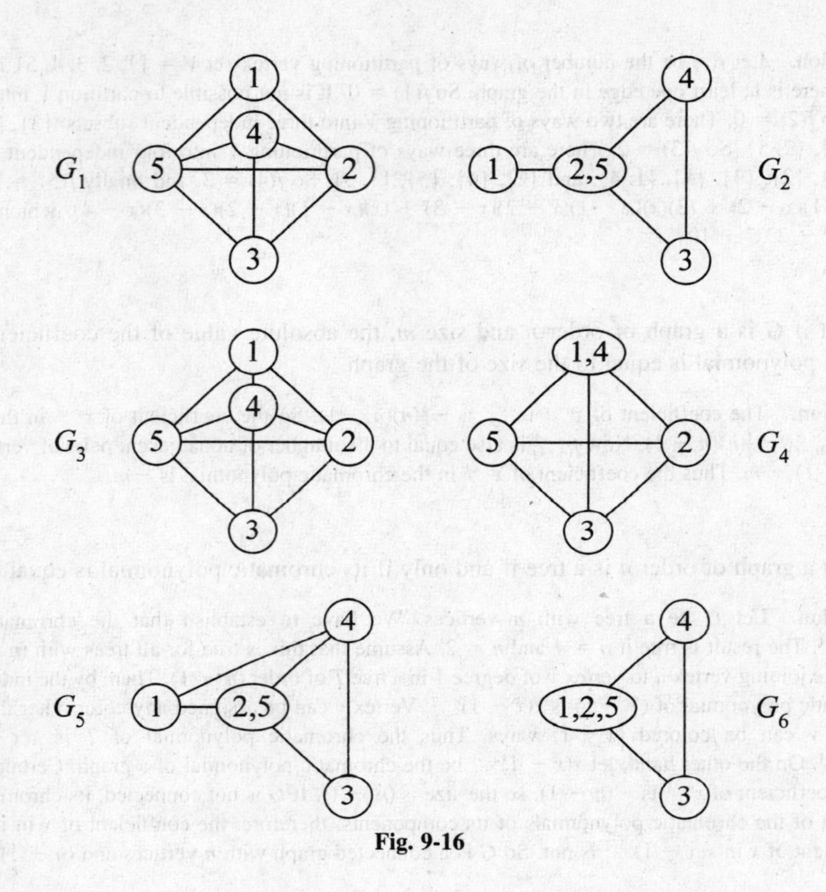

Fig. 9-16

Iteration 3: Join vertices 1 and 3 in G_3, and contract this edge. The resulting graphs are G_7 (which is actually K_5) and G_8 (which is K_4). Join vertices 1 and 3 in G_5, and contract this new edge. The resulting graphs are G_9 (which is K_4) and G_{10} (which is K_3). At this stage, all the graphs are complete graphs:

$$\begin{aligned} P(G, x) &= P(G_4, x) + P(G_6, x) + P(G_7, x) + P(G_8, x) + P(G_9, x) + P(G_{10}, x) \\ &= P(K_4, x) + P(K_3, x) + P(K_5, x) + P(K_4, x) + P(K_4, x) + P(K_3, x) \\ &= P(K_5, x) + 3\,P(K_4, x) + 2\,P(K_3, x) \\ &= x_{(5)} + 3x_{(4)} + 2x_{(3)} = x^5 - 7x^4 + 19x^3 - 23x^2 + 10x \end{aligned}$$

9.43 The chromatic polynomial of a cyclic graph of order n is $(x - 1)^n + (-1)^n(x - 1)$.

Solution. First, consider the case $n = 3$. The chromatic polynomial of the graph obtained by deleting an edge is $x(x - 1)^2$. The chromatic polynomial of the graph obtained by contracting an edge is $x(x - 1)$. So, by the reduction theorem, the chromatic polynomial of K_3 is $x(x - 1)^2 - x(x - 1) = (x - 1)^3 + (-1)^3(x - 1)$. Thus the formula is true when $n = 3$. Suppose this holds for all cyclic graphs of order $(n - 1)$. Let G be any cyclic graph of order n, and let e be one of its edges. Graph $G - e$ is a tree (that is, a path) with n vertices, so its chromatic polynomial is $x(x - 1)^{n-1}$. The graph obtained from G by contracting this edge is a cyclic graph of order $(n - 1)$, and its chromatic polynomial is $(x - 1)^{n-1} + (-1)^{n-1}(x - 1)$ by the induction hypothesis. So, by applying the reduction theorem, the chromatic polynomial of G is $[x(x - 1)^{n-1}] - [(x - 1)^{n-1} + (-1)^{n-1}(x - 1)] = (x - 1)^n + (-1)^n(x - 1)$. Thus the result is true for n as well.

9.44 Prove that the sum of the coefficients of the chromatic polynomial of a graph that has at least one edge is 0.

Solution. If a graph G has at least one edge, it is not possible to color its vertices with one color. In other words, $P(G, 1) = 0$. But $P(G, 1)$ is the sum of the coefficients of the chromatic polynomial of G.

9.45 Show that the coefficients of the chromatic polynomial alternate in sign.

Solution. The proof is by induction on the size m of graph G. If $m = 0$, the chromatic polynomial is x^n, and the result is true. Suppose the result is true for all graphs of order n and size less than m. Let G be any graph of order n and size m. Let e be any edge in G. Then, by the reduction theorem, $P(G, x) = P(G - e, x) - P(G.e, x)$. Now $(G - e)$ is a graph of order n with $(m - 1)$ edges, whereas $G.e$ is a graph of order $(n - 1)$ with $(m - 1)$ edges. So, by the induction hypothesis, the coefficients alternate in sign in their chromatic polynomials. Thus

$$P(G - e, x) = x^n - (m - 1)x^{n-1} + a_{n-2}x^{n-2} - a_{n-3}x^{n-3} + \cdots$$

and

$$P(G.e, x) = x^{n-1} - (m - 1)x^{n-2} + b_{n-3}x^{n-3} - b_{n-4}x^{n-4} + \cdots$$

where coefficients a_i and b_j are nonnegative.

On subtracting,

$$P(G, x) = x^n - mx^{n-1} + [a_{n-2} + (m - 1)]x^{n-2} - [a_{n-3} + b_{n-3}]x^{n-3} - \cdots$$

So the coefficients in $P(G, x)$ also alternate in sign.

9.46 Show that if $P(G, x) = x^n - a_{n-1}x^{n-1} + a_{n-2}x^{n-2} - a_{n-3}x^{n-3} + \cdots$ is the chromatic polynomial of a connected graph G, $1 < a_{n-1} < a_{n-2} < \cdots < a_r$, where r is the floor of $(n/2 + 1)$.

Solution. If G is a tree, $P(G, x) = x(x - 1)^{n-1}$, in which case the inequality is satisfied since the coefficients are the binomial numbers. So the theorem is true when $m = (n - 1)$. We now use induction on size m of the graph. Suppose the inequality condition is satisfied for any connected graph (with at least one cycle) with n vertices and m edges. Since G is not a tree, it has edge e belonging to a cycle; using this edge, we can construct two graphs $G - e$ and $G.e$, each with $(m - 1)$ edges, as in Problem 9.45:

$$P(G - e, x) = x^n - (m - 1)x^{n-1} + a_{n-2}x^{n-2} - a_{n-3}x^{n-3} + \cdots$$

and

$$P(G.e, x) = x^{n-1} - (m - 1)x^{n-2} + b_{n-3}x^{n-3} - b_{n-4}x^{n-4} + \cdots$$

where coefficients a_i and b_j are nonnegative and satisfy the inequality condition.

On subtracting,

$$P(G, x) = x^n - mx^{n-1} + [a_{n-2} + (m - 1)]x^{n-2} - [a_{n-3} + b_{n-3}]x^{n-3} - \cdots$$

It can be easily verified that the coefficients in $P(G, x)$ also satisfy the inequality condition, as stipulated in the problem. So the result holds for m as well.

9.47 Show that if $P(G, x) = x^n - a_{n-1}x^{n-1} + a_{n-2}x^{n-2} - a_{n-3}x^{n-3} + \cdots$ is the chromatic polynomial of a connected graph G, $a_i \geq \binom{n-1}{r-1}$ for each i.

Solution. The chromatic polynomial of any spanning tree T in the graph is $P(T, x)$, which is equal to $x(x - 1)^{n-1} = \sum_{r=1}^{r=n} (-1)^{n-r-1}\binom{n-1}{r-1}x^r$. If G' is the graph obtained by adding an edge of the graph to the spanning tree, the absolute value of the coefficient cannot decrease, as we saw in the reduction theorem. The entire graph G can be obtained from T by adding one edge at a time. This implies that $a_i \geq \binom{n-1}{r-1}$ for each i.

9.48 Show that the coefficient of x in the chromatic polynomial of a connected graph is not zero.

Solution. From Problem 9.47, this coefficient is at least 1.

9.49 Show that the smallest number k such that the coefficient of x^k is not zero in the chromatic polynomial of G is the number of components.

Solution. If G has k components, the chromatic polynomial of G is the product of k chromatic polynomials in each of which the coefficient of x is not zero.

9.50 Show that $x^5 - 7x^4 + 9x^3 - 3x^2$ cannot be the chromatic polynomial of a simple graph.

Solution. If it were the chromatic polynomial of a simple graph G, G should have two components, five vertices, and seven edges. No such graph exists.

9.51 The necessary conditions to be satisfied by the chromatic polynomial of a connected simple graph of order n and size m ($m > 0$) are (1) it should be a polynomial in x of degree n, (2) it is a monic polynomial, (3) the sum of the coefficients is zero, (4) the coefficients alternate in sign, (5) the constant term is zero, (6) the coefficient of x is not zero, (7) the coefficient of x^{n-1} is $-m$, and (8) the absolute values of the coefficients of $x^n, x^{n-1}, x^{n-2}, \ldots, x^r$ are strictly increasing, where r is the floor of $(n/2 + 1)$. Give an example of a polynomial that satisfies these eight conditions but is not the chromatic polynomial of a simple connected graph.

Solution. Consider the polynomial $x^5 - 11x^4 + 14x^3 - 6x^2 + 2x$. If this is the chromatic polynomial of a connected simple graph G, G should have five vertices and 11 edges. But the number of edges in a connected simple graph of order 5 is at least four and at most 10. So there is no graph for which this given polynomial is the chromatic polynomial.

EDGE COLORING OF GRAPHS

9.52 Show that if G is a bipartite multigraph, its chromatic index is $\Delta(G)$. In particular, show that the chromatic index of the complete bipartite graph $K_{m,n}$ is the maximum of $\{m, n\}$.

Solution. The proof is by induction on the number of edges in G. Let $\Delta(G) = r$. It is sufficient to prove that if every edge in $G - e$ (where e is an arbitrary edge in G) can be colored using colors from set S of r colors, the edges in G also can be colored using colors from this set. Suppose edge e joins vertices u and v. Let S_u be the set of colors in S that are not used to color the edges adjacent to u. Clearly, S_u and similarly S_v are nonempty subsets of S. If the intersection of these two subsets is nonempty, any color belonging to this intersection can be used to color edge e; therefore, the edges of G can be colored with r edges. Hence, $S_u \cap S_v$ is empty. Suppose $s \in S_u$ and $t \in S_v$. Let $H(s, t)$ be the connected subgraph of $G - e$ consisting of vertex u and all those vertices and edges in $G - e$ that can be reached from u by a path whose edges are either s-colored or t-colored. If v is a vertex in this component, edge e and any path joining u and v in H will form an odd cycle in G. So v is not a vertex in H. Now interchange the colors of the edges in H; every s-edge becomes a t-edge and vice versa. Then assign color s to edge e. This implies that the edges of G can be colored using r colors. Thus the chromatic index of a bipartite graph is its maximum degree. If the graph is $K_{m,n}$, the maximum degree is $\max\{m, n\}$.

9.53 If a graduate student in a department has taken k courses ($k \geq 0$) taught by a professor in that department, the professor has to give k oral examinations to the student at the end of the academic year. Each oral examination lasts exactly t hours. Find the minimum time required to complete all the departmental oral examinations if we know the number of courses each student has taken taught by each professor.

Solution. Construct a bipartite multigraph $G = (X, Y, E)$, where X is the set of graduate students and Y is the set of professors. Join vertex x in X and vertex y in Y by k edges if and only if x has taken k courses taught by y. The minimum number of time periods needed will be the chromatic index of G, which is the maximum degree $\Delta(G)$. So the minimum time taken is the product of t and $\Delta(G)$.

9.54 A **Latin square** of order n is an $n \times n$ matrix with entries from the set $\{1, 2, \ldots, n\}$ such that no entry appears twice in the same row and no entry appears twice in the same column. Show that a Latin square of order n can be constructed using an n edge coloring of the complete bipartite graph $K_{n,n}$.

Solution. Let $K_{n,n} = (X, Y, E)$, where $X = \{x_1, x_2, \ldots, x_n\}$ and $Y = \{y_1, y_2, \ldots, y_n\}$. The set of colors is $S = \{c_1, c_2, \ldots, c_n\}$. Assume that the edges have been colored using these n colors. If the color of the edge joining x_i and y_k is c_j, $a_{ij} = k$ in the $n \times n$ matrix $A = [a_{ij}]$. The matrix A is obviously a Latin square of order n.

9.55 Show that if G is the complete graph with $2n$ vertices, its chromatic index is $2n - 1$.

Solution. The number of vertices is even. Let $V = \{1, 2, 3, \ldots, 2n\}$ be the vertex set. The maximum degree is $2n - 1$. Consider set $S = \{c_i: i = 1, 2, 3, \ldots, (2n - 1)\}$ of colors. An actual coloring of the edges of the complete graph using each color from S exactly n times can be obtained by the following procedure.

Consider the following arrangement of the $2n$ numbers as a $(2n - 1) \times 2n$ matrix:

$$\begin{bmatrix} 1 & 2 & 3 & 4 & \ldots & 2n-3 & 2n-2 & 2n-1 & 2n \\ 2 & 3 & 4 & 5 & \ldots & 2n-2 & 2n-1 & 1 & 2n \\ 3 & 4 & 5 & 6 & \ldots & 2n-1 & 1 & 2 & 2n \\ \ldots & \ldots & \ldots & \ldots & \ldots & \ldots & \ldots & \ldots & \ldots \\ 2n-1 & 1 & 2 & 3 & \ldots & \ldots & 2n-3 & 2n-2 & 2n \end{bmatrix}$$

Take the elements in row 1 of the matrix and arrange them as pairs $(1, 2n)$, $(2, 2n - 1)$, $(3, 2n - 2)$, and so forth. There will be n such pairs, with each pair corresponding to a unique edge in the complete graph. Assign color c_1 to each edge in this collection.

Take the elements in row 2 of the matrix and arrange them as pairs $(2, 2n)$, $(3, 1)$, $(4, 2n - 1)$, and so forth. Assign color c_2 to the edge corresponding to each pair in this row. Continue this process until the elements in all the rows are paired and the corresponding edges are colored.

(Another proof is as follows: The complete graph with $2n$ vertices is a regular 1-factorable graph. So its chromatic index is its maximum degree, which is $2n - 1$.)

9.56 Show that if G is the complete graph with $2n - 1$ vertices, its chromatic index is $2n - 1$.

Solution. The number of vertices is odd. Color the edges of K_{2n} with $(2n - 1)$ colors, as in Problem 9.55. Delete one vertex. Then we have graph K_{2n-1} with a coloring of its edges using $(2n - 1)$ colors. Suppose it is possible to color the edges of this graph with $(2n - 2)$ colors. There are $(n - 1)(2n - 1)$ edges in this graph. When $(n - 1)(2n - 1)$ is divided by $(n - 1)(2n - 1)$, we get $(n - 1) + (n - 1)/(2n - 2)$, which is greater than $(n - 1)$. This implies that some color from a set of $(2n - 2)$ colors has to be assigned to at least n edges that will need $2n$ vertices. But the number of vertices is only $2n - 1$. So we need at least $(2n - 1)$ colors to color the edges of K_{2n-1}.

9.57 (*Chromatic Index and the Lucas Schoolgirls Problem*) There are $2n$ schoolgirls in a boardinghouse. Each morning they walk to school in groups of two, as doubles. Find the maximum number of consecutive morning walks they can undertake such that each girl forms a pair with every other girl exactly once during these walks.

Solution. Since there are $(2n - 1)$ ways of pairing a girl with the remaining girls, the number of consecutive morning walks is at most $(2n - 1)$ such that no two girls form a pair more than once during the walk. Construct a complete graph with $2n$ vertices. Since the chromatic index is $(2n - 1)$, the edges can be colored using $(2n - 1)$ colors. Each color represents a walk. So it is possible to have $(2n - 1)$ walks.

9.58 (*Chromatic Number and the Kirkman Schoolgirls Problem*) There are 15 girls in a boardingschool who walk to school in groups of three (as triples) all seven days a week. Is it possible to form triples such that no two girls walk together more than once?

Solution. The answer is yes. Any three girls can be chosen out of 15 girls in 455 ways. Each day, the set of 15 is partitioned into five sets of triples. Since there are seven days, we are interested in first obtaining a set S (if it exists) of 35 triples such that no two triples in this collection have more than one girl in common. One could construct a graph G with 455 vertices in which each vertice represents a triple. Join two vertices by an edge if and only if they have two elements in common. Any independent set of 35 vertices in this graph will be a candidate to be chosen as set S. Once an independent set S of 35 is located, construct a graph G' with S as a vertex set. Join two vertices in G' by an edge if and only if the two vertices have one element in common. The chromatic number of G' cannot be less than 7. If it is equal to 7, the problem is solved since the vertices having the same color form a partition of the 15 girls. Otherwise, choose another independent set of 35 vertices and proceed. (A discussion and analysis of the existence theorem in this context, which is in the realm of design theory, is beyond the scope of this book.) If the first 15 letters of the alphabet represent the girls, a solution of the problem is as follows:

Sunday: ABE, CNO, DFL, GHK, IJM

Monday: ACI, BHO, DKM, ELN, FGJ

Tuesday: AFK, BGL, CHM, DIN, EJO

Wednesday: ALM, BCF, DEH, GIO, JKN

Thursday: AHJ, BMN, CDG, EFI, KLO

Friday: AGN, BDJ, CEK, FMO, HIL

Saturday: ADO, BIK, CJI, EGM, FHN

9.59 Prove Theorem 9.4 (Vizing's theorem): The chromatic index of a simple graph G is either $\Delta(G)$ or $\Delta(G) + 1$.

Solution. Since the chromatic index $\chi'(G)$ of G cannot be less than $\Delta(G)$, it is enough if we show that $\chi'(G) \leq \Delta(G) + 1$ whenever G is simple. If this is not true, there is a simple graph G such that $\chi'(G) > \Delta(G) + 1$ and $\chi'(G - e) \leq \Delta(G - e) + 1 \leq \Delta(G) + 1$, where e is any edge of G. Let the maximum degree $\Delta(G) = r$, and let $H = G - e$, where e is some fixed edge of G joining vertices u and v. Suppose the edges of H are colored using colors from set S of $(r + 1)$ colors. Color s from S is said to be *missing at a vertex* if s is not assigned to any edge incident to that vertex. Since the degree of each vertex in H is at most r and the number of colors in S is $(r + 1)$, there is at least color from S missing at every vertex of H. If there is a color in S missing at both u and v, that color can be used to color edge e, which implies that the edges of G can also be colored using the colors from set S.

So if there is no color missing at both these vertices (joined by deleted edge e) under the existing coloring scheme in H, the crux of the problem is as follows. Show that the edges of H can be *recolored* using these $(r + 1)$ colors from S such that once the new coloring scheme is implemented, there will be a color in S missing at u and at some adjacent vertex w. In this case, the edge f joining v and w can be deleted to redefine the graph $H = G - f$. This shows once again that the edges of G can also be colored using at most $(r + 1)$ colors, thereby contradicting the assumption that more than $(r + 1)$ colors are needed to color the edges of G.

Let e_1 be the edge joining u and v_1 in the graph, and let $H = G - e_1$. Suppose s is a color that is missing at u and t_1 is a color that is missing at v_1. If there is no edge incident to u with color t_1, t_1 will be missing at both u and v_1. In that case, we are done. So we assume that there is edge e_2 (with color t_1) joining u and vertex v_2 and that t_2 is a color that is missing at v_2. Thus we inductively define a sequence $v_1, v_2, \ldots, v_i$ of vertices adjacent to u and a sequence of colors $t_1, t_2, \ldots, t_i$ such that color t_i is missing at v_i and such that edge e_i joining u and v_i has color t_{i-1}. There is at most one edge with color t_i joining u and vertex v. If $v \notin \{v_1, v_2, \ldots, v_i\}$, we label v as v_{i+1}, and a color missing at this vertex is labeled t_{i+1}. These sequences can have at most r terms. Suppose the sequences terminate with vertex v_k and color v_k. One reason for the termination at this stage could be that there is no new vertex v (adjacent to u) such that the edge joining this vertex and u has color t_k. Then we can recolor the edges so that e_i gets color t_i for $i = 1, 2, \ldots, k$. This implies that the edges of G can be colored using at most $(r + 1)$ colors.

The only *other* reason for the termination of the sequence is that for some $j < k$, the missing color t_k at v_k is the same as color t_{j-1} of edge e_{j-1} joining u and v_j. In this case, we recolor edges e_i with colors t_i, where $1 \leq i < j$. The edge joining u and v_j has lost its color. The missing color at v_j now becomes t_{j-1}, which is the same as t_k. The colors of the remaining edges in the sequence are unaffected. Thus every edge in G is colored using the colors from set S except edge e_{j-1} joining u and v_j in the sequence. Next we show that with some additional recoloring, we can locate a color that is missing at both u and v_j.

Consider the subgraph $G(s, t_k)$ consisting of all the edges (along with their incident vertices) of G that are colored either with s or with t_k. Each component of this subgraph is either a cycle or a path. Color s is missing at u, and color t_k is missing at both v_j and v_k. The degrees of three vertices in this subgraph are at most equal to 1. So these three vertices cannot belong to the same component of $G(s, t_k)$. There are two cases to be examined.

Case (i): Vertices u and v_j are in two different components. In the component that contains v_j, each s-edge is recolored as a t_k-edge and each t_k-edge is recolored as an s-edge so that s is a missing color at v_j. Since s is also missing at u, the uncolored edge e_{j-1} can be assigned color s.

Case (ii): The vertices u and v_k are in different components. Recolor the edge joining u and v_i with color t_i for each i, where $1 \le i < k$. This recoloring does not have any influence on the graph $G(s, t_k)$. At this stage, the only uncolored edge in G is the edge joining u and v_k. As before, interchange the colors of the edges in the component that contains v_k. As a result, color s becomes a missing color at v_k. Thus s becomes a missing color at the two vertices that join the only uncolored edge in G. This completes the proof in its entirety.

[This theorem has a generalization (also due to Vizing) as follows. If G is a multigraph, the chromatic index is at most equal to $\Delta(G) + m$, where m is the maximum of all edge multiplicities in G. Both Vizing's theorem and its generalization were independently proved by R. P. Gupta at about the same time.]

9.60 If G is an r-regular simple graph with an odd number of vertices, show that its chromatic index is $r + 1$. Is the converse true?

Solution. Since the maximum degree is r, the chromatic index is r or $r + 1$. The graph is not 1-factorable since it is of odd order. So the chromatic index is $r + 1$. The converse is not true. The chromatic index of the 3-regular Petersen graph is 4, but its order is not odd.

Uncolorable Cubic Graphs and Snarks

9.61 (*Blanusa's Theorem*) Let G be a 3-colorable cubic graph whose edges are colored using the colors $c_i (i = 1, 2, 3)$, and let F be a cut set in G. If the number of edges with color c_i in F are x_i, show that the three numbers x_1, x_2, and x_3 are all even or all odd.

Solution. Suppose set F disconnects G, partitioning the set of vertices of G into two sets X and Y of cardinalities x and y, respectively, such that each edge in F joins a vertex in X and a vertex in Y. Fix i. The x_i edges in F with color c_i join x_i vertices in X and x_i vertices in Y. Each of the remaining vertices in X is adjacent to another remaining vertex in X such that the color of the edge joining these two vertices is also c_i. In other words, $x - x_i$ is even for each i. So the three numbers x_1, x_2, and x_3 are all even or all odd.

9.62 Prove Theorem 9.5: If a cubic graph has a bridge, its chromatic index is 4.

Solution. Let G be a cubic graph with a bridge. So it has a cut set consisting of one edge. Suppose G is 3-colorable. The cut set has exactly one edge having one color, say color c_1, and no edges having the other two colors. So the number of edges in the cut set with color c_1 is odd, and the number of edges in the cut set with any one of the remaining two colors is even. This is a contradiction to the fact established in Problem 9.61. So its chromatic index is 4.

9.63 Let G' be the graph obtained from a cubic graph G by contracting a triangle (a cycle of three vertices) in it into a single vertex. Show that $\chi'(G) = 3$ if and only if $\chi'(G') = 3$.

Solution. Consider a cycle in G consisting of three vertices u_i that is condensed into a single vertex u in G'. Let each u_i be adjacent in G to vertex v_i, which is not in the cycle. If the edge joining u_i and v_i is denoted by e_i, the set $F = \{e_1, e_2, e_3\}$ of three edges forms a cut set in G. If the chromatic index of G is 3, the three edges in F are necessarily of different colors, as proved in Problem 9.61. Assign these three colors to the three edges joining u and vertices v_i. Thus G' is 3-edge colorable. The reverse implication is obvious.

9.64 Let e_1 and e_2 be two edges with no vertex in common in a cubic graph G. Insert two vertices x_1 and y_1 on e_1 and two vertices x_2 and y_2 on e_2. Join x_1 and x_2 by an edge. Join y_1 and y_2 by an edge. If the chromatic index of the new graph G' thus constructed is 4, show that the chromatic index of G also is 4.

Solution. Suppose it is possible to color the edges of G using three colors, red (R), blue (B), and green (G). Let e_1 be the edge joining u_1 and v_1, and let e_2 be the edge joining u_2 and v_2 such that the new edges are as shown in Fig. 9-17. There are two cases to be considered.

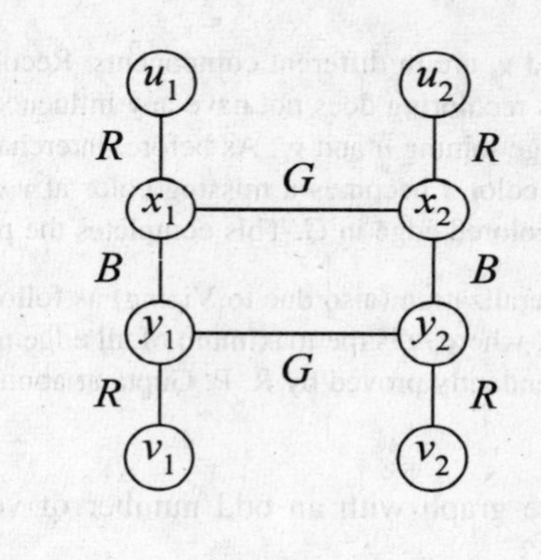

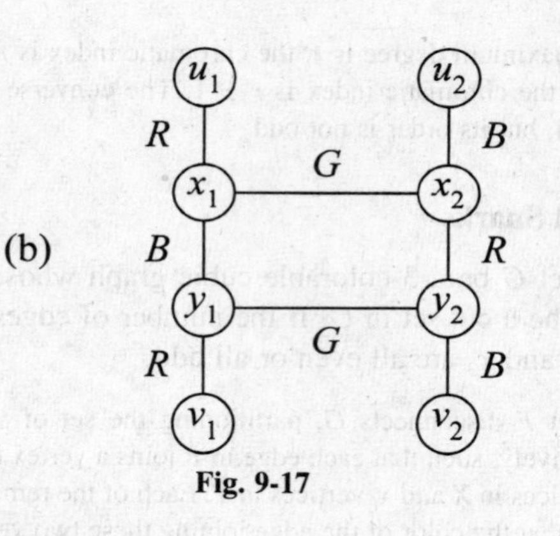

Fig. 9-17

(*i*) Both e_1 and e_2 have the same color, say R, in G. In this case, change the color scheme in G', as shown in Fig. 9-17(*a*).

(*ii*) Edges e_1 and e_2 have different colors; the first one is red and the other is blue. Now change the color scheme in G', as shown in Fig. 9-17(*b*).

In either case, the edges of G' can be colored using three colors, contradicting the hypothesis that the chromatic index of G' is 4.

9.65 Give a counterexample to show that the chromatic index of G' (in Problem 9.64) need not be 4 when the chromatic index of G is 4.

Solution. The converse is not true. In Fig. 9-18, we construct G' from the Petersen graph by inserting vertices on two nonadjacent edges. The chromatic index of G' is 3 (the three colors 1, 2, and 3, are indicated on the edges), but the chromatic index of the Petersen graph is 4.

9.66 Let G be a cubic graph with a cut set F consisting of three edges $e_i (i = 1, 2, 3)$ joining vertices u_i and v_i, where the six vertices are distinct such that $G - F$ has two subgraphs H_1 and H_2. Construct vertex u in H_1, and join it to the three end-vertices in H_1 of the cut set, creating a new cubic graph G_1. Likewise, create another cubic graph G_2 by constructing vertex v and joining it to the end-vertices in H_2 of the cut set. Show that the graph G is 3 edge colorable if and only if both G_1 and G_2 are 3 edge colorable.

Solution. Let G be 3-edge-colorable. By Problem 9.61, the three edges in the cut set are of three different colors. Assign the color of the edge joining u_i and v_i to the edge joining u and u_i as well as the edge joining v and v_i for each i. The coloring of the other edges remains unchanged. Thus we have a 3 edge coloring of the two new graphs. To prove the converse, assume both G_1 and G_2 are 3 edge colorable. Rearrange the coloring if necessary such that both the edge joining u and u_i and the edge joining v and v_i have the same color for each i. Then we have a 3-edge-coloring for G also.

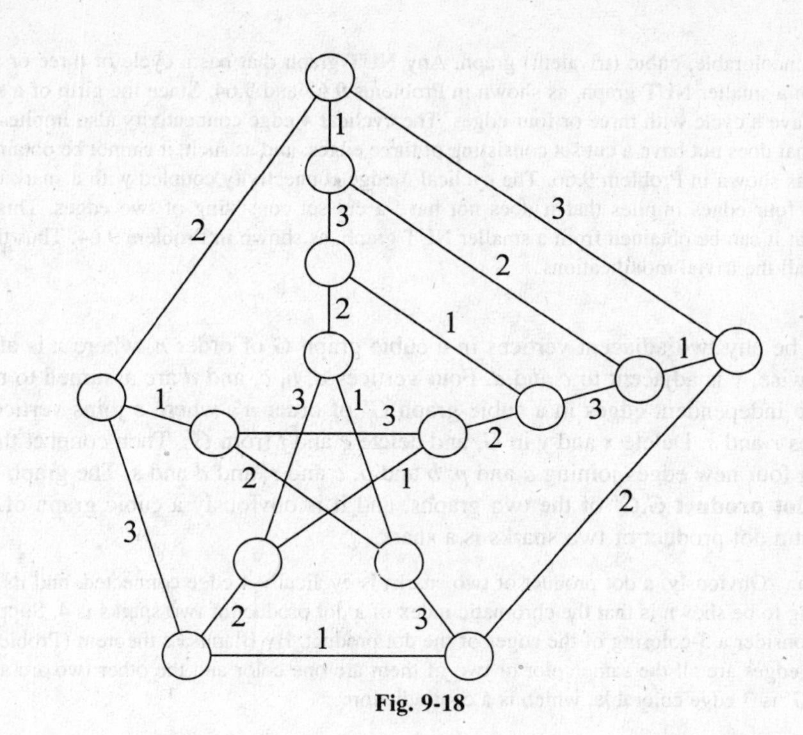

Fig. 9-18

9.67 Let G be a cubic bridgeless graph with a cut set F consisting of two edges, edge e_1 joining u_1 and v_1 and edge e_2 joining u_2 and v_2, so that $G - F$ has two components H_1 and H_2. Assume that u_1 and u_2 are in one component. Join u_1 and u_2 by an edge in H_1, creating a cubic (multi) graph G_1. Likewise, create G_2 from H_2 by joining v_1 and v_2. Show that the graph G is 3 edge colorable if and only if both G_1 and G_2 are 3 edge colorable.

Solution. Since the graph is bridgeless, the four vertices u_1, u_2, v_1, and v_2 are distinct. In any 3-coloring of G, two edges e_1 and e_2 are necessarily of the same color (say color c_1) as a consequence of the property established in Problem 9.61. Give the same color to the new edges. Thus both G_1 and G_2 are 3 edge colorable. On the other hand, if both these graphs are 3 edge colorable, we can always rearrange the colorings of their edges so that both the new edges have the same color, say c_2. Then we delete these two new edges and rebuild the deleted edges, which are assigned the same color c_1. Then we have a 3-coloring of the edges of G.

9.68 A cut set F in a graph G is called a **cyclic cutset** if $G - F$ has two components, each of which has a cycle. The **cyclic edge connectivity** $\boldsymbol{\lambda_c(G)}$ of G is the cardinality of the smallest cyclic cut set in G, and G is said to be **cyclically *k* edge connected** if $\lambda_c(G) \geq k$. Show that the Petersen graph is cyclically 4 edge connected.

Solution. Suppose the Petersen graph is not cyclically 4 edge connected. Since it is 3 edge connected, it is at least cyclically 3 edge connected. Let F be cyclic cutset consisting of three edges. Then $G - F$ has two components, each containing a cycle. Since the girth of G is 5, each component must have five vertices. Each component has three vertices of degree 2 and two vertices of degree 3, which implies that there is a cycle with fewer than five vertices in each component. This is a contradiction.

9.69 A **snark,** by definition, is an uncolorable, cyclically 4 edge connected cubic graph of girth at least 5. Show that this definition is more exclusive in the sense that a cubic bridgeless graph cannot be called a snark just because it is uncolorable. (The **girth conjecture** is the statement that every snark has a cycle consisting of five or six edges. This conjecture is now known to be false.)

Solution. First, the assumption that a snark is cyclically 4 edge connected implies that it has no bridges and by definition is not a multigraph. Following M. Gardner, let us use the notation NUT graph for a bridgeless (nontrivial), uncolorable, cubic (trivalent) graph. Any NUT graph that has a cycle of three or four edges can be obtained from a smaller NUT graph, as shown in Problems 9.63 and 9.64. Since the girth of a snark is more than 4, it cannot have a cycle with three or four edges. The cyclical 4 edge connectivity also implies that a snark is an NUT graph that does not have a cut set consisting of three edges, and as such, it cannot be obtained from a smaller NUT graph, as shown in Problem 9.66. The cyclical 4 edge connectivity coupled with a snark not having a cycle consisting of four edges implies that it does not have a cut set consisting of two edges. This gives rise to the possibility that it can be obtained from a smaller NUT graph, as shown in Problem 9.64. Thus the definition takes into account all the trivial modifications.

9.70 Let x and y be any two adjacent vertices in a cubic graph G of order n, where x is also adjacent to a and b. Likewise, y is adjacent to c and d. Four vertices a, b, c, and d are assumed to be distinct. Let e and f be two independent edges in a cubic graph G' of order n', where e joins vertices p and q and f joins vertices r and s. Delete x and y in G, and delete e and f from G'. Then connect the two graphs by constructing four new edges joining a and p, b and q, c and r, and d and s. The graph thus constructed is called a **dot product $G.G'$** of the two graphs, and it is obviously a cubic graph of order $n + n' - 2$. Show that a dot product of two snarks is a snark.

Solution. Obviously, a dot product of two snarks is cyclically 4 edge connected, and its girth is at least 5. The only thing to be shown is that the chromatic index of a dot product of two snarks is 4. Suppose the chromatic index is 3. Consider a 3-coloring of the edges of the dot product. By Blanusa's theorem (Problem 9.61), either all the four new edges are all the same color or two of them are one color and the other two are another color. This implies that G' is 3 edge colorable, which is a contradiction.

9.71 A **Blanusa snark** is a dot product of the Petersen graph with itself. Construct a Blanusa snark.

(a)

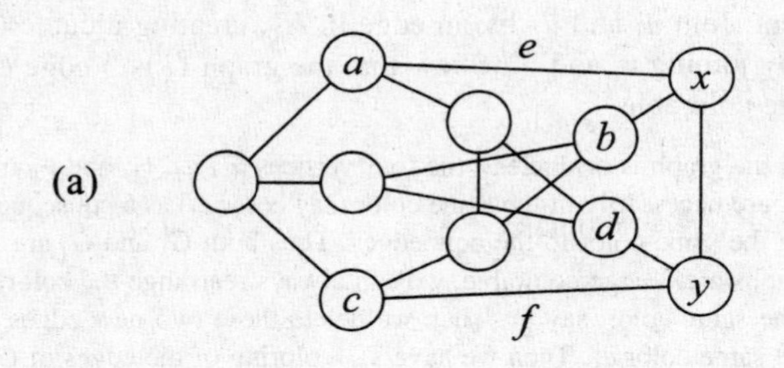

(b)

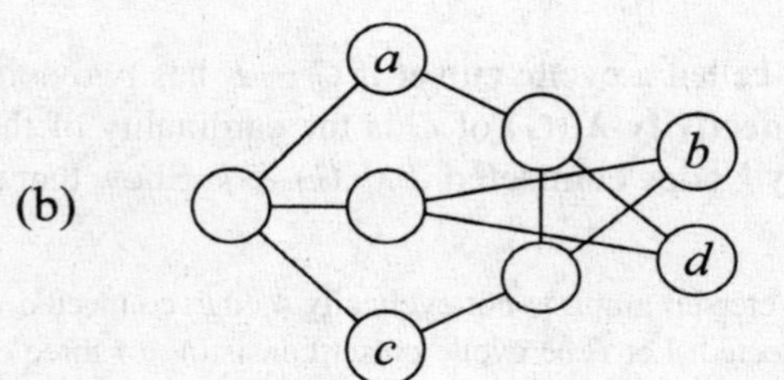

(c)

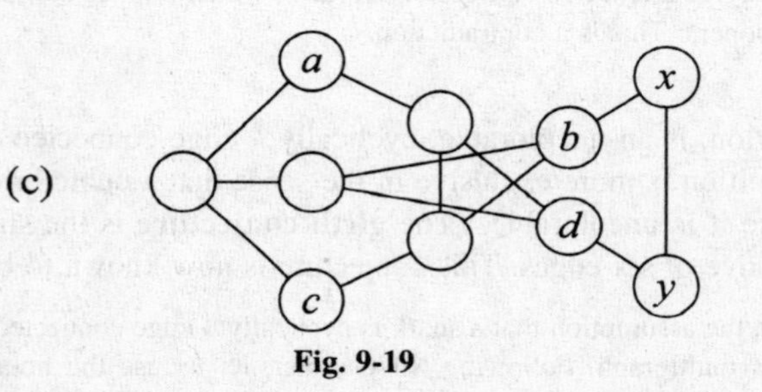

Fig. 9-19

Solution. The Petersen graph is shown in Fig. 9-19(a), with two adjacent vertices x and y and two independent edges e and f. Fig. 9-19(b) shows the graph obtained from G after deleting vertices x and y. Fig. 9-19(c) shows the graph obtained from G after deleting edges e and f.

A dot product of G with itself after making use of these deletions is shown in Fig. 9-20, in which the new edges are dashed lines. (It is possible to obtain another dot product by reconnecting the edges differently. Thus there are two Blanusa snarks.)

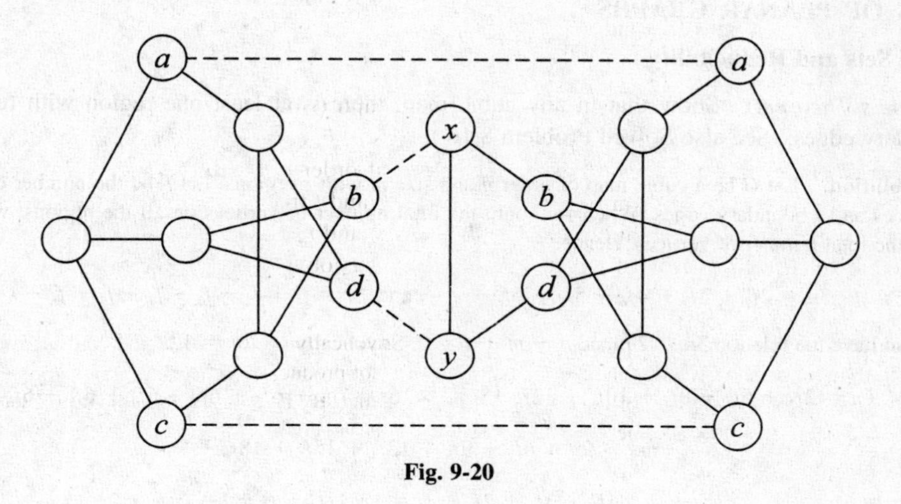

Fig. 9-20

9.72 Show that a snark is not a Hamiltonian graph.

Solution. Every snark has an even number of vertices. So if snark G is Hamiltonian, it has a Hamiltonian cycle with an even number of edges. The edges of this cycle can be colored using two colors. Once the edges of the cycle are colored, all the remaining edges can be colored using a third color. This implies that edges of snark G can be colored using three colors, which is a contradiction.

9.73 Give an example of a non-Hamiltonian cubic graph that is not a snark.

Solution. We are looking for a planar bridgeless cubic graph that is not Hamiltonian. Fig. 9-21 shows a graph of this kind.

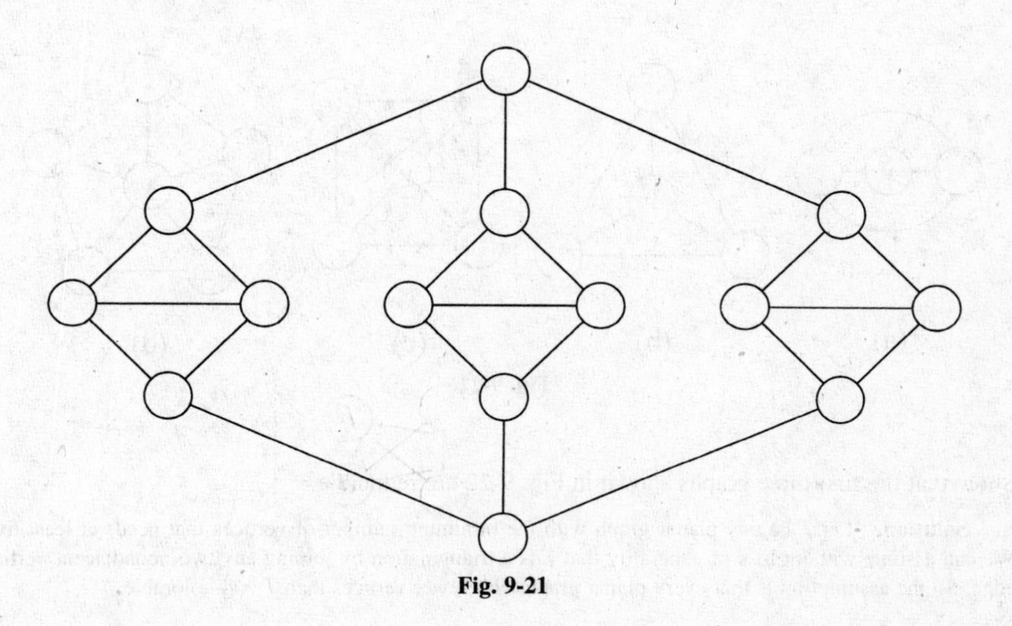

Fig. 9-21

9.74 Give an example of a "nonsnark," a cyclically 4-connected nonplanar cubic graph, the edges of which can be colored using three colors.

Solution. The graph shown in Figure 9-8(*b*) is a nonsnark.

COLORING OF PLANAR GRAPHS

Unavoidable Sets and Reducibility

9.75 (*Kempe's Theorem*) Show that in any cubic map, there is at least one region with fewer than six boundary edges. (See also Solved Problem 8.16.)

Solution. Let G be a cubic map of order n and size m with r regions. Let f_i be the number of regions with i vertices and i boundary edges. When we count the total number of vertices on all the regions, we obtain three times the total number of vertices. Hence,

$$3n = 2f_2 + 3f_3 + 4f_4 + 5f_5 + 6f_6 + \cdots \quad \text{and} \quad r = f_1 + f_2 + f_3 + f_4 + f_5 + f_6 + \cdots$$

We also have the relation $3n = 2m$ and $n - m + r = 2$. So $6n - 6m + 6r = 12$.

Now $\quad 6n = 2f_2 + 6f_3 + 8f_4 + 10f_5 + 12f_6 + \cdots, \qquad 6r = 6f_1 + 6f_2 + 6f_3 + 6f_4 + 6f_5 + 6f_6 + \cdots$

Again, $$6m = 9n = 6f_2 + 9f_3 + 12f_4 + 15f_5 + 18f_6 + \cdots$$

Thus $$4f_2 + 3f_3 + 2f_4 + f_5 = 12 + f_7 + f_8 + \cdots$$

Since the right-hand side of this equation is positive, the left-hand side also is necessarily positive. So there exists a region with at most five borders.

9.76 Show that there exists an unavoidable set of four configurations, and list them.

Solution. That there is an unavoidable set of four configurations is a consequence of Kempe's theorem or of the equivalent (dual) property that there is a vertex of degree at most 5 in any planar graph. Each configuration can be considered a graph (as in Fig. 9-22) with one vertex v in the interior surrounded by a cycle of two, three, four, or five vertices such that v is adjacent to each vertex in the cycle. Every triangulation should contain at least one of these graphs.

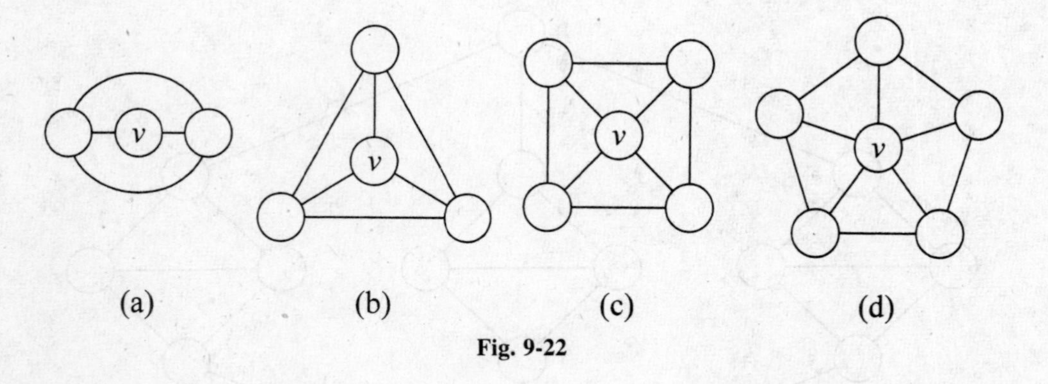

Fig. 9-22

9.77 Show that the first three graphs shown in Fig. 9-22 are reducible.

Solution. Let T be any planar graph with the minimum number of vertices that needs at least five colors. We can assume without loss of generality that T is a triangulation by joining any two nonadjacent vertices by an edge. So the assumption is that every planar graph with fewer vertices than T is 4-colorable.

(*i*) If T contains the configuration shown in Fig. 9-22(*a*) or (*b*), we could color $(T - v)$ with at most four colors and then assign a color to v that is not the same as any vertex adjacent to v in T. This implies that T also is 4-colorable, which is a contradiction. So both Fig. 9-22(*a*) and Fig. 9-22(*b*) are reducible.

(*ii*) If the four vertices a, b, c, and d together can be colored in fewer than four colors in any 4-coloring of $T - v$, T can also be 4-colored. If they have four different colors, we need a fifth color for vertex v if those four colors are not changed by a recoloring. So we have to recolor the vertices of $T - v$ such that a and c (or for that matter b and d) get the same color in the revised coloring. Suppose the colors of the four vertices, a, b, c, and d are A, B, C, and D, respectively. Let $H(A, C)$ be the subgraph of $T - v$ induced by the set of all vertices colored either A or B, and let $H(B, D)$ be the subgraph of $T - v$ induced by all vertices colored B or D. If both a and c belong to the same component of $H(A, C)$ and, at the same time, both b and d belong to the same component of $H(B, D)$, both these vertices have a vertex in common, which is not possible. So we may assume without loss of generality that a and c are not in the same component of the subgraph $H(A, C)$. In that case, we can interchange the colors in the component that contains a so that a get color c. Once this is done, we can assign color a to vertex v in T, which implies that T is 4-colorable.

[Any component of the subgraph induced by the vertices colored by two colors, as described in this problem, is called a **Kempe chain.** In a Kempe chain, one can always interchange the two colors of its vertices without changing the colors of the other vertices. This method of interchanging colors is called the **Kempe chain argument.** In 1879, Kempe erroneously claimed that the graph shown in Fig. 9-22(*d*) is also reducible, and he thereby presented an ostensible proof of the four-color theorem. Eleven years later, Heawood discovered a flaw in Kempe's proof of the reducibility of this last graph. The pentagon cannot be reduced. The four-color problem once again reverted to the status of a conjecture, and the quest began for a collection of reducible polygons in the place of a single pentagon.]

9.78 (*Wernicke's Unavoidable Set*) Show that the set consisting of the five graphs shown in Fig. 9-23 is an unavoidable set.

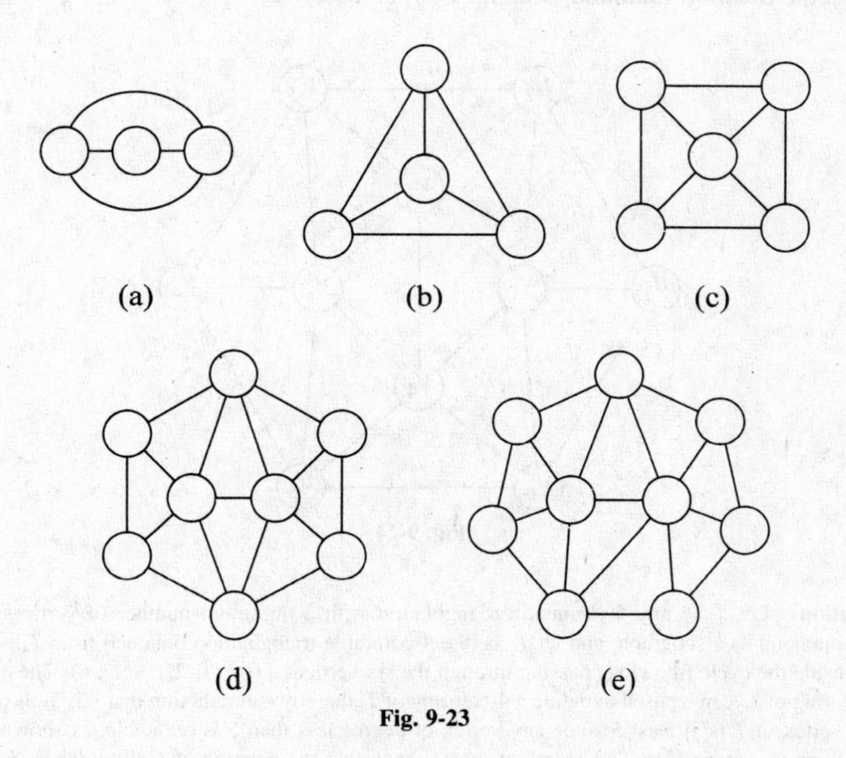

Fig. 9-23

Solution. Suppose T is a triangulation of order n and size m that does not contain any one of the five connected graphs shown in Fig. 9-23. So T has no vertices of degree 2, 3, or 4. If vertex v of degree 5 adjacent to another vertex w, the degree of w should be at least 7. Let n_i be the number of vertices of degree i. Then $n_5 + n_6 + \cdots = n$. Also

$$5n_5 + 6n_6 + 7n_7 + 8n_8 + \cdots = 2m$$

and since T is a triangulation, $2m = 6n - 12$. Thus

$$5n_5 + 6n_6 + 7n_7 + 8n_8 = 6(n_5 + n_6 + n_7 + n_8 + \cdots) - 12$$

Hence,

$$(6-5)n_5 + (6-6)n_6 + (6-7)n_7 + (6-8)n_8 + \cdots = 12 \quad \textbf{(Kempe's equation)}$$

Each vertex of degree i initially has a "charge" of $(6 - i)$. Vertices of degree 7 or more are called major vertices; others are minor vertices. Major vertices initially have negative charges. Only vertices of degree 5 have positive charge. Vertices of degree 6 have 0 charge. We redistribute the charges by taking the positive charge of one unit from every vertex of degree 5 and equally distributing it via the five edges incident to other vertices, making sure that the total charge remains the same positive number 12. This method of redistributing charges is known as **discharging.** Since the five vertices adjacent to a vertex of degree 5 are all major, at the end, the charge at each vertex of degree 5 is 0. Since a vertex of degree 6 is not adjacent to a vertex of degree 6, the charge at a vertex of degree 6 remains 0 during discharging. If v is any vertex of degree i more than 6, the updated charge at v is at most $[6 - i + (i/5)]$, which is negative if i is more than 7. So after the redistribution, the charges at vertices of degree 8 and more are still negative. Vertex w of degree 7 will end up with positive charge only if it is adjacent to at least six vertices of degree 5. This will imply that vertices of degree 5 become adjacent. So once the charges are redistributed, the total charge becomes negative, contradicting the requirement that the total charge is a positive number, namely 12. Thus there is no triangulation that does not contain at least one of the connected graphs from the set in Fig. 9-23 a subgraph.

9.79 Show that the **Birkhoff diamond** (see Fig. 9-24) is reducible.

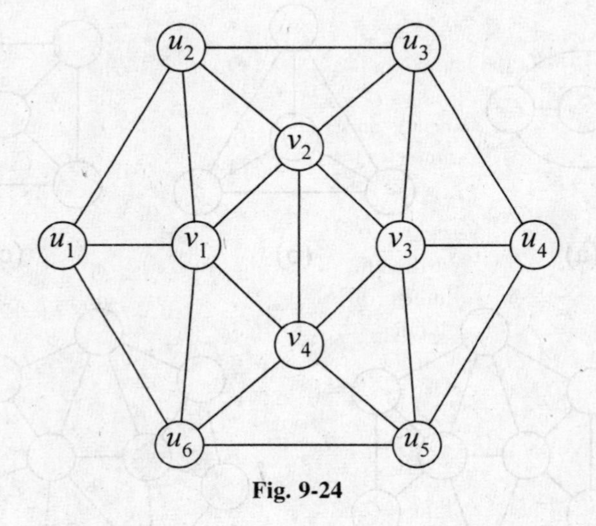

Fig. 9-24

Solution. Let T be any 5-chromatic triangulation with a minimum number of vertices that contains the Birkhoff diamond as a subgraph, and let T' be the 4-colorable triangulation obtained from T by deleting the four vertices inside the cycle (the **ring**) passing through the six vertices $u_i (i = 1, 2, \ldots, 6)$. The aim is to show that any 4-coloring of T' can be used to define a 4-coloring of T, thereby contradicting that it is 5-chromatic. The degree of every vertex in T is at least 5 since any vertex of degree less than 5 is reducible, as shown in Problem 9.77. Suppose there is a cycle C in T of length at most 4 such that the deletion of C disconnects T. This implies that there is a vertex in T of degree 3 or 4. So there cannot be a separating cycle of length at most 4 in T. If vertices u_3 and u_5 were adjacent, these two vertices and vertex v_3 together will form a separating set of length 3 whose deletion will cause a trivial component consisting of vertex u_4. Hence, u_3 and u_5 are not adjacent in T (and in T'); as such, there is no harm whatsoever in assuming (if necessary) that they both have the same color in any 4-coloring of T'. We can thus identify these two vertices in T' so that they both merge into a single vertex, say u. Then we join u and vertex u_1 by an edge. See Fig. 9-25. The resulting triangulation T'' is also 4-colorable.

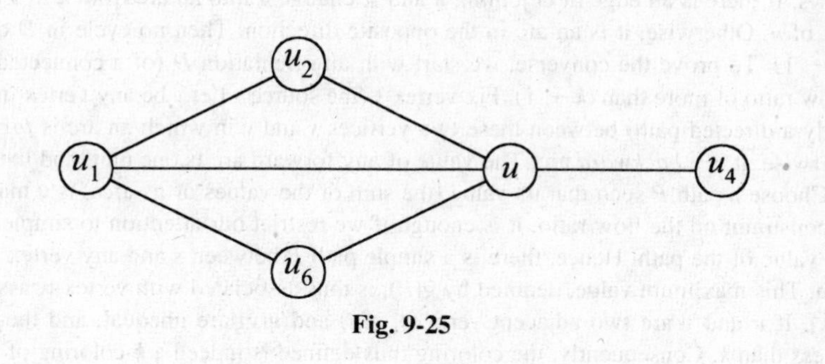

Fig. 9-25

Suppose the four available colors for coloring T'' are 1, 2, 3, and 4. There are exactly six (up to a permutation of these four colors) ways of coloring these five vertices $u_1, u_2, u_3 = u_5 = u, u_4, u_6$. They are 1, 2, 3, 1, 2; 1, 2, 3, 4, 2; 1, 2, 3, 4, 4; 1, 2, 3, 1, 4; 1, 2, 3, 2, 4; and 1, 2, 3, 2, 2. All these colorings except the last can be easily extended to a coloring of T. For example, the first coloring can be extended to the six vertices u_1, u_2, u_3, u_4, u_5, and u_6 by assigning colors 1, 2, 3, 1, 3, and 2, respectively. Then the four internal vertices v_1, v_2, v_3, and v_4 can be assigned colors 3, 1, 2, and 4, respectively. Thus we have a 4-coloring for T. But it is not possible to extend the last coloring—1, 2, 3, 2, 3, 2—of these six vertices to a 4-coloring of T without recoloring a vertex of the ring. At this stage, we invoke a Kempe chain argument. Suppose there is no path joining u_4 and u_2 (or joining u_4 and u_6) passing through vertices that are colored either color 2 or color 4. In that case, we change the color of u_4 from 2 to 4, defining the coloring scheme 1, 2, 3, 4, 3, 2 of the six vertices u_1, u_2, u_3, u_4, u_5, and u_6, which can be used to assign colors 3, 1, 2, and 4 to the four internal vertices v_1, v_2, v_3, and v_4. On the other hand, suppose there is a path P joining u_4 and u_6 in which the color of each vertex is either 2 or 4. If there is a path P' between u_1 and u_5 in which the color of every vertex is either 1 or 3, both P and P' will have a vertex in common. So if P exists, there is no P'; therefore, the color of u_5 can be changed from 3 to 1. Thus once again we have a coloring scheme 1, 2, 3, 2, 1, 2 for the six vertices u_1, u_2, u_3, u_4, u_5, and u_6 and a coloring scheme 4, 1, 4, 3 for the four internal vertices v_1, v_2, v_3, and v_4. The argument is analogous if there is a path Q joining u_4 and u_2 in which every vertex has either color 2 or color 4. This completes the proof.

Theorems of Gallai and Minty

9.80 (*Gallai's Theorem*) (*a*) Show that the chromatic number of a graph G cannot exceed $1 + t(D)$, where $t(D)$ is the number of arcs in a longest directed path in any acyclic orientation of G. (*b*) Show that for every graph G, there is an acyclic orientation D such that $\chi(G) = 1 + t(D)$. Hence, $\chi(G) = \min\{1 + t(D) : D$ varies through all the acyclic orientations of G.}

Solution.

(*a*) Observe that there is at least one acyclic orientation. If the order of G is n, assign labels $1, 2, \ldots, n$ to the vertices, and then draw an arc from i to j if and only if $i < j$ and there is an edge joining i and j in G. Let D be any acyclic orientation of a graph G. For every vertex v of G, let $t(v)$ be the number of arcs in a directed path that starts from v and with as many arcs as possible. Assign color $1 + t(v)$ to vertex v. Obviously, we do not need more than $1 + t(D)$ colors to color the vertices of G.

(*b*) Suppose the chromatic number of G is k. Color the vertices using colors $1, 2, \ldots, k$. If there is an edge joining vertices u and v, convert this edge into an arc from u to v if and only if the color of u is more than the color of v. The orientation D thus obtained is acyclic for which $t(D)$ is at most $(k - 1)$. So $1 + t(D) \leq k$. But $k \leq 1 + t(D)$.

9.81 (*Minty's Theorem*) The **flow ratio** of a cycle C in an orientation D of a graph G is the ratio p/q (where $p \geq q \geq 0$), p being the number of arcs in one direction and q being the number of arcs in the opposite direction in C. Show that the vertices of a G can be k-colored if and only if there is orientation D of the graph in which the flow ratio of no cycle exceeds $(k - 1)$.

Solution. Suppose the vertices of G are colored using colors $0, 1, 2, \ldots, k - 1$. Define an orientation D for G as follows. If there is an edge in G joining u and v, change it into an arc from u to v if the color of v is more than the color of u. Otherwise, it is an arc in the opposite direction. Then no cycle in D can have a flow ratio of more than $(k - 1)$. To prove the converse, we start with an orientation D (of a connected graph G) in which no cycle has a flow ratio of more than $(k - 1)$. Fix vertex s (the source). Let v be any vertex in D, and consider a path (not necessarily a directed path) between these two vertices s and v in which an arc is *forward* if it is directed to vertex v. Otherwise, it is a *backward* arc. The value of any forward arc is one unit, and the value of any backward arc is $1 - k$. Choose a path P such that its value (the sum of the values of its arcs) is a maximum. Because of the upper bound constraint on the flow ratio, it is enough if we restrict our attention to simple paths if the intent is to maximize the value of the path. Hence, there is a simple path P between s and any vertex v such that the value of P is maximum. This maximum value, denoted by $g(P)$, is thus associated with vertex v, assigned color $g(v)$, which is $g(P)(\text{mod } k)$. If u and v are two adjacent vertices, $g(u)$ and $g(v)$ are unequal, and the absolute value of their difference is less than k. Consequently, the coloring thus defined is indeed a k-coloring of the vertices of G.

Conjectures of Hajos and Hadwiger and Heawood's Theorem

9.82 **Hajos's conjecture** claims that every k-chromatic graph has a subgraph homeomorphic to the complete graph of order k. Show that the conjecture is true if $k \leq 4$.

Solution. The conjecture is obviously true if $k = 1$ or 2. If $k = 3$, the graph has an odd cycle that is a subdivision of K_3. For $k = 4$, the proof due to Dirac is along the following lines. Let G be a minimally 4-chromatic graph of order n, where n is at least 5. The proof is by induction on n. If there is a separating set consisting of two vertices u and v, the two vertices are not adjacent (see Problem 9.25), and, by the induction hypothesis, the graph $G_1 + e$, where e is the (new) edge joining these two nonadjacent vertices, contains a subdivision of K_4. If we now replace edge e by a path P in G_2 joining these two vertices, $G_1 \cup P$ contains a subdivision of K_4. Thus G also contains a subdivision of K_4. If G is 3-connected, it has a cycle C of length at least 4 since the degree of each vertex is at least 3. Let u and v be two nonadjacent vertices in this cycle. Then there are vertices p and q on C and a p, q path P in $G - \{u, v\}$. Likewise, there is a u, v path P' in $G - \{p, q\}$. If these two paths have no vertex in common, the union of these two paths and C constitutes a subgraph (of G) that is a subdivision of K_4. If the two paths have a vertex in common, let w be the first vertex of P that belongs to P'. Then we have a path P'' joining p and w as a subpath of P. In this case, the union of P', P'', and C forms a subgraph that is homeomorphic to K_4.

(The largest integer n such that a given graph contains a subgraph homeomorphic to K_n is called the **Hajos number** of G. Another way of stating Hajos conjecture is that the chromatic number of a graph cannot exceed its Hajos number.)

9.83 Show that if Hajos's conjecture is true for $k = 5$, the four-color theorem is true.

Solution. The hypothesis is that every 5-chromatic graph contains a subgraph that is homeomorphic to K_5. Suppose there is a planar graph G that is not 4-colorable. So the hypothesis implies that G has a subgraph that is homeomorphic to K_5, which is a contradiction since G is planar. Thus every planar graph is 4-colorable if Hajos's conjecture holds when $k = 5$.

9.84 Show that Hajos's conjecture is false if $k \geq 7$.

Solution. Let H be the multigraph obtained by replacing each edge of cycle C_5 by three edges. The degree of each vertex in H is 6. Let G be the graph obtained by deleting two nonadjacent vertices from the line graph of H. Since a maximum independent set in G (which is of order 13) has cardinality 2, its chromatic number is at least 7. Now it is easy to see that K_6 is a subgraph of G. If w is any vertex that is not a vertex of this K_6, the number of internally disjoint paths from w to any vertex in K_6 is less than 6. In other words, the Hajos number of G is 6. So G is a counterexample of Hajos's conjecture when $k = 6$. Let $h(G)$ be the Hajos number of any graph G. Then $h(G + v) = h(G) + 1$ and $\chi(G + v) = \chi(G) + 1$. So if G is a counterexample for k for Hajos's conjecture, $G + v$ is a counterexample for $k + 1$. Thus Hajos conjecture is false for all $k \geq 7$.

(The status of this conjecture is unsettled when $k = 5$ or 6. P. Erdös and S. Fajtlowicz, in a joint paper, have proved that this conjecture is false for almost all graphs.)

9.85 **Hadwiger's conjecture** claims that every k-chromatic graph has a subgraph contractible to the complete graph of order k. Show that (*a*) the conjecture is true if $k \leq 4$ and (*b*) (**Wagner's theorem**) the conjecture is true when $k = 5$ if and only if the four-color theorem is true.

Solution.

(*a*) First notice that Hajos's conjecture implies Hadwiger's conjecture. Thus Hadwiger's conjecture holds if $k \leq 4$.

(*b*) As in Problem 9.84, if Hadwiger's conjecture holds for $k = 5$, the four-color theorem is true. To prove the converse, assume that the four-color theorem is true. Let G be any 5-chromatic graph. The four-color theorem implies that G is nonplanar. So there is a subgraph that is contractible to K_5 or $K_{3,3}$. If it is contractible to K_5, we are done. Suppose G has no subgraph contractible to K_5. So by P. Young's theorem (see Solved Problem 8.47), G is not 4-connected. We may assume that G is minimal in the sense that every contraction of G is 4-colorable. There exists a set W of at most three vertices such that $G - W$ has components $H_1, H_2, \ldots, H_k (k \geq 2)$ with the property that each vertex in W is adjacent to some vertex in H_i for every i. Let G_i be the subgraph induced by W and the set of vertices of H_i. Suppose T is a maximal independent subset of W. Then each G_i can be 4-colored such that the vertices in T all get the same color (say color 1) and the vertices in $W - T$ get colors other than 1. Then these 4-colorings can all be combined to obtain a 4-coloring of G, contradicting the assumption that the chromatic number of G is 5.

(It has been proved by B. Bollobas, P. A. Catlin, and P. Erdös that Hadwiger's conjecture is true for almost all graphs.)

9.86 (*Heawood's Map-Coloring Theorem*) Prove that $\chi(S_g) = H(g)$, where $H(g)$ is the floor of $\frac{1}{2}\{7 + \sqrt{1 + 48g}\}$ for any surface of positive genus g.

Solution. Let G be any triangulation with n vertices and m edges on a surface of genus g. Let d' be the average degree. Then $(n)(d') = (2)(m) = (3)(r)$, where r is the number of regions. Applying Euler's formula, we find $d' = (12/n)(g - 1) + 6$, which implies that $(n - 1) \geq (12/n)(g - 1) + 6$. Thus $n^2 - 7n - 12(g - 1) \geq 0$. On solving this quadratic, we have the lower bound $n \geq H(g)$. If $n \leq H(g)$, obviously $\chi(G) \leq H(g)$.

Suppose $n > H(g)$. Then $d' < [12/H(g)](g - 1) + 6 = H(g) - 1$. Now $g = 0$ will imply that $d' < 3$. So the genus g is necessarily positive. Thus there is a vertex v in G of degree at most $H(g) - 2$. Identify v and any vertex w adjacent to it by an elementary contraction to obtain a graph G' of order $n - 1$. If $n - 1$ is equal to $H(g)$, G' is $H(g)$-colorable, which will induce an $H(g)$-coloring on G. Otherwise, $n - 1 > H(g)$. We continue this process. We ultimately end up with an $H(g)$-colorable graph, which implies that G itself is $H(g)$-colorable. Thus the chromatic number of any graph that can be embedded on a surface of positive genus cannot exceed $H(g)$. Hence, $\chi(S_g) \leq H(g)$, where $g > 0$.

Let us write $p = H(g)$. Then $p \leq \frac{1}{2}\{7 + \sqrt{1 + 48g}\}$. On simplifying this inequality we find that $g' = \frac{1}{12}(p - 3)(p - 4) \leq g$, where g' is the genus of K_p (see Solved Problem 8.98). Hence, $\chi(S_{g'}) \leq \chi(S_g)$. Since K_p is embeddable on $S_{g'}$, we also have the inequality $\chi(K_p) \leq \chi(S_{g'})$. But $\chi(K_p) = p = H(g)$. So $H(g) \leq \chi(S_{g'}) \leq \chi(S_g)$. Thus the reverse inequality is established because the genus of a complete graph of order n is $\frac{1}{12}(n - 3)(n - 4)$. This completes the proof.

Two Colorability and Three Colorability

9.87 Show that a plane graph G is 2-colorable if and only if the degree of each region in G is even.

Solution. If the plane graph G is 2-colorable, it is bipartite, hence, the degree of each region is even. On the other hand, suppose G is a plane graph in which the degree of each region is even. Let C be any cycle dividing the plane into two parts: a part consisting of regions inside the cycle and a part consisting of regions outside the cycle. Suppose the regions inside the cycle are $R_1, R_2, \ldots, R_k$ with degrees $d_1, d_2, \ldots, d_k$. The sum of these k degrees is even. Each edge in any of these regions that is not an edge of C is counted twice in this summation. So the number of edges in C is even. This implies that G is bipartite and hence 2-colorable.

9.88 Show that a map is 2-colorable if and only if it is Eulerian.

Solution. Let G be a map and let G' be its geometric dual. If G is 2-colorable, G' is 2-colorable, which implies that G is bipartite. Hence, G is Eulerian, as shown in Solved Problem 8.62. Conversely, if G is Eulerian, G' is bipartite and 2-colorable. Hence, G is also 2-colorable.

9.89 (*Krol's Theorem*) A plane graph is 3-colorable if and only if it is a subgraph of a triangulation in which the degree of each vertex is even.

Solution. Let G be any 3-colorable graph with four or more vertices. Once a certain 3-coloring using colors, 1, 2, and 3 is defined on G, the nontriangular regions of G can be partitioned into three classes and new vertices and edges can be constructed to obtain a triangulation G' as follows:

(*i*) Regions with an even number of edges in their boundaries in which the vertices are colored using two colors. Construct vertex v in the interior of any such region, and join that to every vertex on the boundary. The new vertex obviously gets the third color.

(*ii*) Regions with four edges in their boundaries such that the vertices in each region are colored using three colors. In each region, there will be two vertices, say u and v, whose colors are not the same. Construct a path that lies inside the region joining these two vertices, and insert two new vertices p and q on this path such that u and p are adjacent. Assign to p the color of v, and assign to q the color of u.

(*iii*) Regions with more than four edges in their boundaries in which the vertices are colored using all three colors. Introduce a new vertex v in the interior, and then join it to all vertices in the boundary that are colored 1 or 2. Then the region gets divided into subregions that are either triangles or with boundaries consisting of four edges. In the latter case, we proceed as in (*ii*). We ultimately have a 3-colorable triangulation G' that contains G as a subgraph. Let v be any vertex of G', and suppose its color is 1. So the vertices adjacent to v are all colored either 2 or 3, and these vertices constitute a cycle. So the degree of v is even. Thus the condition is necessary.

To prove the converse, assume that G is a subgraph of a triangulation T in which the degree of each vertex is even. Thus T is Eulerian. So each region can be colored either red or blue such that no two regions sharing an edge in common have the same color. Orient the edges of T such that each triangle becomes a directed triangle; specifically, the edges of a red triangle have a clockwise orientation, and the edges of a blue region have a counterclockwise orientation. Suppose C is any cycle in the graph enclosing a finite region R consisting of r red triangles and b blue triangles. If there are x edges in C in the clockwise direction, the total number of edges in the interior of R is $3r - x$. Likewise, if there are y edges in the counterclockwise direction in C, the total number of edges in the interior is also equal to $3b - y$. Thus $x - y = 3(r - b) \equiv 0 \pmod 3$. Let v be any vertex in T, and assign color 1 to it. Let w be any vertex in T, and let P be any path connecting these vertices. Because of the orientation, some arcs in this path will be directed toward w; these are the *forward* arcs in the path. The other arcs are known as the *backward* arcs. Suppose there are p forward arcs and q backward arcs in path P. Define $c(w, P) \equiv (1 + p - q) \pmod 3$. Let P' be another path between v and w with p' forward arcs and q' backward arcs. So $c(w, P') \equiv (1 + p' - q') \pmod 3$. If these two paths have no vertex in common, their union is a cycle with $p + q'$ arcs in the clockwise direction and $p' + q$ arcs in the other direction. So $(p + q') - (p' + q) \equiv 0 \pmod 3$, which implies that $(p - q) \equiv (p' - q') \pmod 3$. Hence, $c(w, P) = c(w, P')$. If paths P and P' have vertices in common, we can partition the union of these two paths into cycles and paths. This once again establishes that $C(w, P)$ is independent of the choice of P.

Thus each vertex w is assigned the unique color $C(w, P)$, which is either 0, 1, or 2, where P is any path between the fixed vertex v (with color 1) and w. It remains to be shown that no two adjacent vertices are of the same color. Let v be the same vertex as before with color 1, and let u and w be two adjacent vertices. Among all the v, u paths and v, w paths, let P be a path with as few arcs as possible, and assume without loss of generality that P is between v and w. Either there is an arc from u to w or from w to u. Let Q be the path consisting of P and the arc that is either from w to u or from u to w. In either case, color $c(u, Q)$ and color $c(w, P)$ cannot be the same.

(An easily verifiable sufficient condition for the 3-colorability of a plane graph is the result known as **Grunbaum's Theorem,** which states that a plane graph is 3-colorable if the number of triangles in G is at most three. The condition is not necessary; consider the wheel with a 4-cycle and a vertex in its interior adjacent to every other vertex.)

Perfect Graphs

9.90 Show that (*a*) a cyclic graph is perfect if and only if it has an even number of vertices, and (*b*) every bipartite graph is perfect.

Solution.

(*a*) Let C be a cyclic graph. If it is perfect, its chromatic number is 2. So its order is even. Conversely, let C be a cyclic graph of even order, and let H be any induced subgraph. If its clique number is 1, its chromatic number also is 1. If its clique number is 2, its chromatic number is also 2. So C is perfect.

(*b*) Let H be an induced subgraph of a bipartite graph G. If the clique number of H is 1, its chromatic number is also 1. If its clique number is 2, its chromatic number is also 2. So G is perfect.

9.91 The **clique covering number $\theta(G)$** (also known as the **partition** number) of a graph $G = (V, E)$ is the minimum number of pairwise disjoint cliques whose union is the set V. A graph G is **α-perfect** if for every induced subgraph H of G, $\theta(H)$ is equal to its internal stability number $\alpha(H)$. Show that a graph is perfect if and only if its complement is α-perfect.

Solution. By definition, the clique number of a graph is the internal stability number of its complement, and the chromatic number of a graph is equal to the clique covering number of its complement. The complement of the complement of G is G.

9.92 Show that the line graph of a bipartite graph G is perfect.

Solution. A clique in the line graph $L(G)$ corresponds to either a triangle in G or a set of edges having a vertex in common. Since G is bipartite, there are no triangles in G. Thus $\omega(L(G)) = \Delta(G)$. But $\Delta(G) = \chi'(G)$ (see Problem 9.52), and $\chi'(G) = \chi(L(G))$. So $\omega(L(G)) = \chi(L(G))$.

9.93 A directed graph is a **transitive digraph** if whenever there is an arc from vertex u to vertex v and an arc from v to vertex w, there is an arc from u to w. A graph G is a **transitively orientable graph** (also known as a **comparability graph**) if it is possible to orient its edges such that the resulting digraph is a transitive digraph. Show that a comparability graph is perfect.

Solution. Observe that if G is transitively orientable, any induced subgraph of G is also. Suppose P is a directed path in a transitive orientation of G with a maximum number of arcs. Let this path be from vertex v to vertex w consisting of k vertices, including u and v. The "length" of P is k. Then the k vertices in P constitute a clique in G, and no clique in G can have more than k vertices. So the clique number $\omega(G)$ is k. Next assign color $c(u)$ to vertex u, where $c(u)$ is the length of a directed path of maximum length that starts from u. Two adjacent vertices cannot have the same color; if there is an arc from vertex x to vertex y, $c(x) > c(y)$. Thus $\chi(G) \le k = \omega(G)$. But $\omega(G) \le \chi(G)$. So both numbers are equal to k.

9.94 Let W be a set of vertices in a connected graph $G = (V, E)$ such that the subgraph induced by W is complete and such that $G - W$ is a disconnected graph with components $G_i = (V_i, E_i)$, where $i = 1, 2, \ldots, r$. Let H_i be the subgraph induced by the union of V_i and W for each i. Show that if $\omega(H_i) = \chi(H_i)$ for each i, $\omega(G) = \chi(G)$.

Solution. Let $\omega(G) = k$. A clique in H_i is a clique in G. So $\omega(H_i) < \omega(G) = k$. If u and v are two vertices in two distinct components of $G - W$, there cannot be an edge joining them in G. So every clique in G is a clique in one of these induced subgraphs. In particular, there is an induced subgraph, say H_j, such that $\omega(H_j) = k$. Now $\chi(G) = \max\{\chi(H_i): i = 1, 2, \ldots, r\} = \max\{\omega(H_i)\} = \omega(H_j) = k = \omega(G)$.

9.95 Let W be a set of vertices in a chordal graph G such that W is a minimal separating set. Show that the graph induced by W is complete.

Solution. Suppose the subgraph H induced by W is not complete. So there are two vertices u and v in W that are not adjacent. Let $G_1 = (V_1, E_1)$ and $G_2 = (V_2, E_2)$ be two distinct components of $G - W$. If u is not adjacent to any vertex in G_1, set $W - \{u\}$ is a separating set violating the minimality requirement. So there are vertices in both V_1 and V_2 adjacent to u. The same argument goes for v as well. So there is a cycle C in G consisting of an edge joining vertex p in V_1 to u, an edge joining u to vertex q in V_2, a path joining q and vertex r in V_2, the edge joining r and v, then an edge joining v to vertex s in V_1, and finally a path joining s and p. That the graph is chordal implies that there is an edge joining u and v. This shows that H is complete.

9.96 Prove that a chordal graph is perfect.

Solution. Since an induced subgraph of a chordal graph is chordal, it is enough if we prove that $\omega(G) = \chi(G)$ for a chordal graph. The proof is by induction on the order of G. If G is complete, the result is true. Assume that G is not complete. So there exists a minimal separating set W such that the subgraph H induced by W is complete, as proved in Problem 9.95. Let the components of $G - W$ be $G_i = (V_i, E_i)$, and let H_i be the subgraph induced by $V_i \cup W$ for each i. By the induction hypothesis, each H_i is perfect. So $\omega(G) = \chi(G)$, as proved in Problem 9.94.

9.97 Show that a graph is perfect if and only if every induced subgraph H has an independent set W of vertices such that $\omega(H - W) < \omega(H)$.

Solution. Suppose G is perfect and H is any induced subgraph of G. Then $\omega(H) = \chi(H) = k$. Assume that the vertices of H are colored using the colors from the set $\{1, 2, \ldots, k\}$. Let W be set of vertices in H that have color 1. Then W is an independent set in H, and the vertices in $H - W$ can be colored using $k - 1$ colors. So $\chi(H - W) = k - 1$. But $\omega(H - W) \leq \chi(H - W)$. Hence, $\omega(H - W) \leq k - 1 < k = \omega(H)$. So the condition is necessary. To prove that the condition is sufficient, assume that every induced subgraph H of G contains an independent set W of vertices such that $\omega(H - W) < \omega(H)$. The proof is by induction on the order of H. By the induction hypothesis, the vertices of $H - W$ can be colored using $\omega(H) - 1$ colors since W is independent. Once this is done, the vertices in W can be assigned a new color. Thus the vertices of H can be colored using $\omega(H)$ colors. Hence, $\omega(H) = \chi(H)$ for every induced subgraph H of G. So G is perfect.

9.98 (*Lovasz Replacement Theorem*) Show that if the vertices of a perfect graph are "replaced" by perfect graphs, the resulting graph is perfect.

Solution. It is enough if we consider the case when we "replace" a single vertex v of a perfect graph H by another perfect graph G, where the two graphs have no vertices in common. To do this, construct a graph H' with vertex set $V(H') = V(G) \cup V(H) - \{v\}$. Vertices x and y are adjacent in H' if they satisfy one of the following conditions: (1) x and y are adjacent in G or in H; (2) x is in G, y is in H, and v and y are adjacent; and (3) x is in H, y is in G, and v and x are adjacent. Since G and H are perfect, G has an independent set A of vertices such that $\omega(G - A) < \omega(G)$, and the vertices of H can be colored using $\omega(H)$ colors. Under a specific $\omega(H)$-coloring of H, let B be the set of vertices that have the same color as v. Here B is the **color class** that contains v. Then $D = A \cup B - \{v\}$ is an independent set in H'. Let K be any maximum cardinality clique in H'. There are two cases:

(*i*) K does not intersect G. Then K is a complete subgraph of $H - v$. Now $\omega(H - v) \leq \omega(H) = \omega(H')$. So K is a maximum cardinality clique of size $\omega(H)$ in H, and it intersects every color class of every $\omega(H)$-coloring of $H - v$. In particular, K intersects the set $B - v$.

(*ii*) If K intersects G, since A is an independent set and K is a maximum clique, the sets A and K have a nonempty intersection.

So in either case, H' has an independent set D of vertices that intersects every maximum cardinality clique of H'. What is true of H' is true of any induced subgraph of H' in this regard. In view of the result established in Problem 9.97, to show that H' is perfect, it is enough to show that every induced subgraph of H' has an independent set of vertices that intersects every maximum cardinality clique in H''. That is exactly what has been done.

9.99 Show that if a graph G is perfect, it has a clique that intersects with every maximum cardinality independent set in G.

Solution. Suppose that for every clique, there is a maximum cardinality independent set such that the clique and the independent set have no vertices in common. Specifically, let B_i be the set of cliques in G, and let A_i be the corresponding maximum cardinality independent sets such that $B_i \cap A_i$ is empty for $i = 1, 2, \ldots, r$. For each vertex v in G, let $n(v)$ be the number of cliques that contain v. For each clique B_i, we define $n(B_i)$ to be the sum of all $n(v)$, where v runs through the vertices in the clique. Replace each vertex v in G by the perfect graph $K_{n(v)}$, thereby constructing the perfect graph G'. Obviously, every clique in G' is the union of complete graphs that replace the vertices of some clique in G. Consequently,

$$\omega(G') = \max\{n(B_i): 1 \le i \le r\} = \max\left\{\sum_{j=1}^{r} \left|B_i \cap A_j\right| : 1 \le i \le r\right\}$$

But A_i and B_i have no vertices in common, and the cardinality of $(A_i \cap B_j)$ is at most 1 if i and j are not equal. Hence $\omega(G') \le (r - 1)$.

Now the cardinality of each A_i is $\alpha(G)$. So the sum $|A_1| + |A_2| + \cdots + |A_r| = r\alpha(G)$. At the same time, consider the sequence S consisting of the vertices from $A_1, A_2, \ldots, A_r$ arranged one after the other. Each vertex v of G enters this sequence $n(v)$ times. So if n' is the number of vertices in G' we see that $n' = r\alpha(G)$. Now $n'/(\alpha(G')) \le \chi(G') = \omega(G')$. So $r\alpha(G) \le \omega(G')(\alpha(G'))$. But $\alpha(G) = \alpha(G')$. Hence, $r \le \omega(G')$. But $\omega(G') \le (r - 1)$. This contradiction establishes that there is a clique that intersects with every maximum cardinality independent set.

9.100 (*The Perfect Graph Theorem of Lovasz*) Show that the complement of a perfect graph is perfect.

Solution. Let G be a perfect graph, and let H be a nontrivial induced subgraph of the complementary graph G'. Then H' (the complement of H) is an induced subgraph of G. By Problem 9.99, H' has a clique B that intersects every maximum cardinality independent set of H'. Set B is a maximum cardinality independent set in H, and B meets every maximum cardinality clique. So H is perfect, as shown in Problem 9.97.

9.101 Show that a graph is perfect if and only if it is α-perfect.

Solution. This is an immediate consequence of the perfect graph theorem and the definition of α-perfect graphs.

Turan's Theorem

9.102 (*Erdös's Theorem*) If the graph $G = (V, E)$ does not contain K_{k+1} as a subgraph, show that there exists a k-chromatic graph $H = (V, F)$ such that $\deg_H(v) \ge \deg_G(v)$ for every v in V. (Graph G is said to be **degree-majored** by graph H.)

Solution. The proof is by induction on k. The result is true if $k = 1$ by letting $G = H$. In this case, the degree (in G and in H) of each vertex is 0. Suppose the theorem is true for all graphs that do not contain K_{k+1} as a subgraph. Consider a graph $G = (V, E)$ that does not contain K_{k+2} as a subgraph. Let v' be a fixed vertex of maximum degree in G, and let W be the set of vertices adjacent to v' in G. Subgraph $G' = (W, F)$ induced by W obviously does not contain K_{k+1} as a subgraph, so by the induction hypothesis, there exists a k-chromatic graph $H' = (W, F')$ such that $\deg_{H'}(v) \ge \deg_{G'}(v)$ for every v in W. Now construct a graph $H = (V, E'')$ as follows. (1) if both u and v are in W and if they are adjacent in H', they are adjacent in H. (2) In H, every vertex in W is adjacent to every vertex in $V - W$. Since H' is k-chromatic, H is $(k + 1)$-chromatic. If v is in W, $\deg_H(v) = |V| - |W| + \deg_{H'}(v) \ge |V| - |W| + \deg_{G'}(v) \ge \deg_G(v)$. Otherwise, $\deg_H(v) = |W| = \deg_G(v') \ge \deg_G(v)$.

Thus there exists a $(k + 1)$-chromatic graph $H = (V, E)$ such that $\deg_H(v) \ge \deg_G(v)$ for every v in V. The result holds for $k + 1$ also.

9.103 (*Turan Number and Turan Graph*) Find a nondecreasing sequence of k positive integers $a_i (i = 1, 2, \ldots, k)$ whose sum is n such that $\Sigma_{1 \le i \le j < k}\, a_i a_j$ is maximum.

Solution. What is true if $k = 2$ holds in the more general case? The desired quantity is a maximum when the numbers are "nearly equal." Thus $|a_i - a_j| \le 1$ for every i and j. Specifically, if k divides n, each a_i is equal to (n/k). Otherwise, let $n = qk + r$. In this case, $a_1 = a_2 = \cdots = a_{k-r} = q$, and the remaining numbers all equal

$q + 1$. For a fixed n and k, with $n \geq k \geq 2$, the number $\Sigma_{1 \leq i \leq k} a_i a_j$ with this choice of $a_i (i = 1, 2, \ldots, k)$ is called the **Turan number $t(n, k)$**. The complete k-partite graph of order n whose vertex set is partitioned into k sets V_i of cardinality $a_i (i = 1, 2, \ldots, k)$ is known as the **Turan graph $T(n, k)$.**

9.104 (*Turan's Theorem*) Show that if a graph $G = (V, E)$ of order n does not contain K_{k+1} as a subgraph. $|E| \leq t(n, k)$.

Solution. According to Problem 9.102, there exists a k-chromatic graph $H = (V, F)$ that degree-majorizes G. Since H is k-chromatic, set V can be partitioned into k nonempty sets V_i each of cardinality $a_i(i = 1, 2, \ldots, k)$ such that $a_1 + a_2 + \cdots + a_k = n$, where we assume that numbers a_i are in nondecreasing order without loss of generality. The maximum number of edges in the k-partite graph H is therefore $t(n, k)$. Hence, $|E| \leq |F| \leq t(n, k)$.

9.105 Find the maximum number of edges in (*a*) a 4-chromatic graph of order 20 and (*b*) a 6-chromatic graph of order 20.

Solution.

(*a*) Here $n = 20$ and $k = 4$. The graph does not contain K_5 as a subgraph. Each part in the 4-partite set contains five vertices. Any two of four can be chosen in six ways. Between any two parts are 25 edges. So the Turan number is 150. The number of edges is at most 150.

(*b*) The graph does not contain K_7 as subgraph. There are six partite sets consisting of 3, 3, 3, 3, 4, and 4 vertices, respectively. The Turan number is $(6)(9) + (8)(12) + 16 = 166$. The number of edges is at most 166.

Supplementary Problems

9.106 Find the chromatic number of a cubic nonbipartite graph.
Ans. If the graph is complete, the chromatic number is 4. Otherwise, it is 3.

9.107 Show that if G is a k-critical graph of order n and size m, $(k - 1)(n) \leq 2m$. [*Hint:* The degree of each vertex is at least $(k - 1)$, and the sum of the degrees is $2m$.]

9.108 Show that if a k-chromatic graph G is uniquely colorable, the subgraph induced by the union of any two or more subsets of a vertex partition defined by a k-coloring is $(k - 1)$-connected. [*Hint:* Such a subgraph is also uniquely colorable. Use Problem 9.31.]

9.109 Show that $\chi(G) \leq 1 + n - \alpha(G)$.
Ans. Assign the same color to each vertex in a maximum independent set. The number of uncolored (out of n vertices) vertices is $n - \alpha(G)$.

9.110 Find the chromatic polynomial of $K_{1,n}$. *Ans.* $x(x - 1)^n$

9.111 If G is the connected graph obtained by linking a triangle and a cyclic graph of order 4 so that they share one vertex in common, find the chromatic polynomial of G.
Ans. $(x - 1)(x - 2)(x^4 - 4x^3 + 6x^2 - 3x)$

9.112 If G is a connected graph of order n, prove that $P(G, x) \leq x(x - 1)^{n-1}$. [*Hint:* Any coloring of G is a coloring of any spanning tree in G.]

9.113 Show that Vizing's theorem need not be true if the graph under consideration is not simple. [*Hint:* Consider the multigraph G consisting of three vertices such that joining each pair of vertices are p edges, where $p > 1$.]

9.114 Show that the four-color theorem is true if and only if the following condition holds: Every planar graph has an orientation such that the flow ratio of any cycle in the orientation is at most 3. [*Hint:* Use Minty's theorem (Problem 9.81).]

9.115 Show that Hadwiger's conjecture is true for $k = 5$. [*Hint:* Use the four-color theorem and Problem 9.85.]

9.116 Show that any outerplanar graph is 3-colorable. [*Hint:* Use the definition of outerplanarity.]

9.117 Show that a triangulation is 3-colorable if and only if the degree of each vertex in it is even. [*Hint:* Use Problem 9.89.]

9.118 The complement of a comparability graph is known as an **incomparability graph.** Show that an incomparability graph is both perfect and α-perfect. [*Hint:* Use Problem 9.96.]

9.119 Show that the Petersen graph is not perfect. [*Hint:* It has an odd hole with five vertices.]

9.120 Show that an interval graph is perfect. [*Hint:* The complement of an interval graph G is a comparability graph (see Solved Problem 2.35.]

9.121 Find the size of the largest (*a*) 7-chromatic graph of order 21 and (*b*) 7-chromatic graph order 22.
Ans. (*a*) $49 + 49 + 49 = 147$; (*b*) $49 + 56 + 56 = 161$

Important Symbols

Symbol	Meaning	Page
$\alpha(G)$	Independence number (internal stability number)	13
$\alpha_1(G)$	Edge-independence number	14
$\beta(G)$	Vertex-covering number	13
$\beta_1(G)$	Edge-covering number	14
$\chi(G)$	Chromatic number	39, 244
$\chi'(G)$	Chromatic index	247
$\Delta(G)$	Maximum degree	4
$\delta(G)$	Minimum degree	4
$\delta(P)$	Flow capacity of the semipath P	133
$\kappa(G)$	Connectivity number	30
$\lambda(G)$	Edge-connectivity number	30
$\nu(G)$	Crossing number	225
$\theta(G)$	Thickness	227
$\theta'(G)$	Clique covering number (partition number)	281
$\sigma(G)$	Vertex domination number (external stability number)	13
$\omega(G)$	Clique number	245
(S, T)	Cut (source-sink cut)	131
$C(G)$	Center	31, 128
$c(G)$	Closure	80
$c(S, T)$	Capacity of the cut (S, T)	131
C_n	Cyclic graph	3
$d(G)$	Diameter	128
$e(v)$ or $e(i)$	Eccentricity of a vertex	31, 124
$f(G)$	Value of the feasible flow f	131
$f(S, T)$	Flow along the cut (S, T)	131
$\overline{G}$	Complement	9
G'	Geometric dual	201
G^*	Abstract dual	202
$G.e$	Contraction	52
$K_{m,n}$	Complete bipartite graph	1
K_n	Complete graph	1
$L(G)$	Line graph	44
$P(G, x)$	Chromatic polynomial	246
Q_k	Hypercube (k-cube)	19
$r(A)$	Rank of A in a matroid	104
$r(G)$	Radius	31, 128
$R(p, q)$	Ramsey number	19
$T(n, k)$	Turan graph	284
$t(n, k)$	Turan number	284

Select Bibliography

Ahuja, Ravindra K., Magnanti, Thomas L. and Orlin, James B. *Network Flows.* Prentice Hall, 1993.

Appel, K. and Haken, W. *Every Planar Map Is Four Colorable.* American Mathematical Society, 1984.

Balakrishnan, V. K. *Theory and Problems of Combinatorics.* McGraw-Hill, 1995.

Balakrishnan, V. K. *Network Optimization.* Chapman & Hall, 1995.

Behzad, M. and Chartrand, G. *Introduction to the Theory of Graphs.* Allyn and Bacon, 1971.

Berge, C. *Theory of Graphs and Its Applications.* John Wiley & Sons, 1958.

Berge, C. *Graphs and Hypergraphs.* North-Holland, 1973.

Bieneke, Lowell W. and Wilson, Robin J. (eds.). *Selected Topics in Graph Theory.* Academic Press, 1983.

Bondy, J. A. and Murty, U. S. R. *Graph Theory with Applications.* North-Holland, 1976.

Capobianco, M. and Molluzzo, J. C. *Examples and Counterexamples in Graph Theory.* North-Holland, 1978.

Chartrand, G. *Introduction to Graph Theory.* Dover Publications, 1977.

Chartrand, G. and Lesniak, L. *Graphs and Digraphs.* Wadsworth & Brooks/Cole, 1979.

Chartrand, G. and Oellerman, Ortrud R. *Applied and Algorithmic Graph Theory.* McGraw-Hill, 1993.

Clark, J. and Holton, D. A. *A First Look at Graph Theory.* World Scientific Publishing, 1991.

Foulds, L. R. *Combinatorial Optimization for Undergraduates.* Springer-Verlag, 1984.

Gibbons, A. *Algorithmic Graph Theory.* Cambridge University Press, 1985.

Gondran, M. and Minoux, M. *Graphs and Algorithms.* John Wiley & Sons, 1984.

Gould, Ronald. *Graph Theory.* The Benjamin-Cummings Publishing Company, 1988.

Grunbaum, B. *Convex Polytopes.* John Wiley & Sons, 1967.

Harary, F. *Graph Theory.* Addison-Wesley, 1969.

Hartsfield, N. and Ringel, G. *Pearls in Graph Theory.* Academic Press, 1990.

Holton, D. A. and Sheehan, J. *The Petersen Graph.* Cambridge University Press, 1993.

Lawler, Eugene L. *Combinatorial Optimization.* Rinehart & Winston, 1976.

Melnikov, O., Tyshkevich, R., Yemelichev, V. and Sarvanov, V. *Lectures of Graph Theory.* Wissenschafts-verlag, 1990.

Nemhauser, George L. and Wolsey, Laurence A. *Integer and Combinatorial Optimization.* John Wiley and Sons, 1988.

Ore, Oystein. *Graphs and Their Uses.* Mathematical Association of America, 1963.

Ore, Oystein. *The Four Color Problem.* Academic Press, 1967.

Papadimitriou, Christos H. and Steiglitz, Kenneth. *Combinatorial Optimization.* Prentice Hall, 1982.

Parthasarathy, K. R. *Basic Graph Theory.* Tata-McGraw-Hill, 1994.

Roberts, Fred R. *Graph Theory and Its Applications to Problems of Society.* Society for Industrial and Applied Mathematics, 1978.

Roberts, Fred R. *Applied Combinatorics.* Prentice-Hall, 1984.

Thulasiraman, K. and Swamy, M. N. S. *Graphs: Theory and Algorithms.* John Wiley & Sons, 1982.

Trudeau, Richard J. *Introduction to Graph Theory.* Dover Publications, 1993.

Tucker, A. *Applied Combinatorics.* John Wiley & Sons, 1994.

West, Douglas B. *Introduction to Graph Theory.* Prentice Hall, 1996.

Wilson, Robin J. *Introduction to Graph Theory.* Longman Scientific & Technical, 1990.

Yemelichev, V. A., Kovelev, M. and Kravtsov, M. K. *Polytopes, Graphs and Optimisation.* Cambridge University Press, 1984.

Zykov, Alexander A. *Fundamentals of Graph Theory.* BCS Associates, 1990.

Index